THE HELIOSPHERE IN THE LOCAL INTERSTELLAR MEDIUM

Space Science Series of ISSI

Volume 1

THE HELIOSPHERE IN THE LOCAL INTERSTELLAR MEDIUM

Proceedings of the First ISSI Workshop
6–10 November 1995, Bern, Switzerland

Edited by

R. VON STEIGER
International Space Science Institute, Bern, Switzerland

R. LALLEMENT
Service d'Aéronomie du CNRS, Verrières le Buisson, France

and

M. A. LEE
University of New Hampshire, Durham, NH, USA

Reprinted from Space Science Reviews, Vol. 78, Nos. 1-2, 1996

KLUWER ACADEMIC PUBLISHERS
DORDRECHT / BOSTON / LONDON

A.C.I.P. Catalogue record for this book is available from the Library of Congress

ISBN-13: 978-94-010-7296-0 e-ISBN-13: 978-94-009-1782-8
DOI: 10.1007/978-94-009-1782-8

Published by Kluwer Academic Publishers,
P.O. Box 17, 3300 AA Dordrecht, The Netherlands.

Kluwer Academic Publishers incorporates
the publishing programmes of
D. Reidel, Martinus Nijhoff, Dr W. Junk and MTP Press.

Sold and distributed in the U.S.A. and Canada
by Kluwer Academic Publishers,
101 Philip Drive, Norwell, MA 02061, U.S.A.

In all other countries, sold and distributed
by Kluwer Academic Publishers Group,
P.O. Box 322, 3300 AH Dordrecht, The Netherlands.

Printed on acid-free paper

TABLE OF CONTENTS

THE LOCAL INTERSTELLAR MEDIUM

INTERACTION BETWEEN HS AND LISM

First ISSI Workshop
The Heliosphere in the Local Interstellar Medium
Bern, November 6-10, 1995

Group Photograph

1. S. Wenger
2. J. Geiss
3. C. Gry
4. J. Linsky
5.A. Cummings
6. B. Bavassano
7. G. Nusser Jiang
8. V. Manno
9. H. Washimi
10. A. Galeev
11. R. Egger
12. I. Mann
13. A. Nishida
14. S. Grzedzielski
15. M. Baguhl
16. J. Lequeux
17. D. Breitschwerdt
18. N. Ness
19. R. v. Steiger
20. R. Lallement
21. D. Gurnett
22. J. Vallerga
23. J. Belcher
24. L. Fisk
25. M. Huber
26. E. Möbius
27. D. Rucinski
28. R. Marsden
29. Y. Whang
30. P. Frisch
31. E. Grün
32. V. Baranov
33. G. Zank
34. B. Hultqvist
35. M. Lee
36. T. Zurbuchen
37. R. Jokipii
38. M. Witte
39. D. Hovestadt
40. I. Axford
41. H. Fahr

not on picture:
J.-L. Bertaux
F. Bühler
G. Gloeckler

Foreword by the Series Editors

The present volume is the first in the "Space Sciences Series of ISSI", the main publication series of the International Space Science Institute in Bern, Switzerland. In this book series, which will appear as part of *Space Science Reviews* and as a hardcover volume, ISSI will publish the results of the study projects and workshops that form the main part of its scientific program.

ISSI's main task is to help the scientific community in carrying out interdisciplinary as well as multispacecraft / multiexperiment investigations of the very extensive and complex data sets already available or to become available in the coming years. To a large extent these data will be provided by the spacecraft programs coordinated by the Inter-Agency Consultative Group (IACG), made up of the major space agencies in the world. Data will also come from other satellites and from suborbital and ground-based measurements. Theorists and modellers will play important roles in the efforts to achieve interdisciplinary interpretations of these experimental data.

ISSIs study projects are organized around well defined scientific themes, selected after consultation with leading representatives of the scientific community and carried out by many scientists from all parts of the world. During the first several years, ISSI will concentrate on solar system sciences, where the need to bring scientists together for the synthesis and integration of knowledge and understanding of important scientific problems is considered greatest.

The theme of the present volume is a good example of the multidisciplinarity foreseen in the ISSI programme. About half of the participants were solar system scientists and the other half were astronomers, comparing in situ measurements of the interstellar gas and dust that had intruded into the heliosphere with remote observations of the local interstellar medium. Data from more than a dozen spacecraft were used in the deliberations of the workshop.

We want to express our appreciation and gratitude to the editors of this volume, Rudolf von Steiger, Rosine Lallement and Martin Lee, for their dedicated efforts in producing a well coordinated volume.

Finally, we express our hope that this first volume of the Space Sciences Series of ISSI, as well as those appearing in the future, will meet a need in the scientific community.

Bern, August 1996

Johannes Geiss and Bengt Hultqvist
Directors of ISSI

Foreword by the Volume Editors

When the newly created International Space Science Institute (ISSI) in Bern began operation in the summer of 1995, its first major activity was to organize a workshop on the topic of "The Heliosphere in the Local Interstellar Medium". This topic is a good illustration of present activities in space science for at least two reasons: first, many new results have been obtained very recently and there is now a phase of growing activity. Second, the development of the heliospheric interface and solar environment science makes use of a large number of spacecraft, including some not specifically designed for this field. As a matter of fact, besides Ulysses, the two Voyagers, Pioneer 10 and the HST, new results by the EUVE, ROSAT, and UARS have brought new insights, and we expect results from SOHO in the near future. Also, many "old" spacecraft results are still being analysed in this context, such as Prognoz 5-6, Pioneer-Venus, Pioneer 11, IMP, Helios, and others. The multi-spacecraft aspect – central to ISSI activities – is particularly fruitful and appealing. Moreover, this approach brings together two communities that otherwise work as separate entities, astronomers and heliospheric or solar system scientists. This workshop has demonstrated that there is now a real synergy between these two communities.

The workshop was convened by L. A. Fisk, J. Geiss, E. Grün, J. Lequeux, and E. Möbius. About 40 scientists were invited to ISSI for a week in November 1995. They presented some 25 invited talks and a few contributed talks, and subsequently discussed their views in nine topical working groups, each of which was concluded by a rapporteur presentation. The present volume is a collection of the invited, contributed, and rapporteur presentations given at this workshop (plus two papers by invitees who were unable to accept the invitation). Not all of the rapporteur presentations have resulted in a separate paper, as some of the working groups have chosen to integrate the material presented into the respective invited or contributed papers. All papers in the volume have been refereed. In editing this volume, we have attempted to make it more integrated than the average workshop proceedings by reading all the papers ourselves, making authors aware of undue duplications in the material presented and gaps in information between different articles, and pointing out cross references to other articles in this book. We hope that this has resulted in a volume providing a comprehensive, up-to-date, and self-contained overview of the topic, minimizing the need to consult external references unless much detail is required.

Since this was the first workshop at the new institute, it was concluded with the official inauguration ceremony of ISSI, at which Sir Herman Bondi gave the keynote address entitled "Science as an Adventure". We are very pleased that he agreed to the reproduction of his text in this volume.

We wish to express our sincere thanks to all those who have made this volume possible. First and foremost, we would like to thank the authors for writing original articles, for keeping to the various requests by the reviewers and by the editors, and for producing neat camera-ready papers. We would also like to thank the reviewers for their careful and critical reports, which have contributed considerably to the quality of this volume. Most of the communications for this volume have been done by the exchange of e-mail messages (well over 600 to and from Bern alone, plus a large number between other places), so we wish to thank the innumerable individuals who have made this possible by creating and maintaining the Internet. Finally, we express our sincere thanks to the ISSI directors, J. Geiss and B. Hultqvist, for their initiative and work that made this workshop possible, and to the ISSI staff, G. Bühler, V. Manno, G. Nusser Jiang, M. Preen, D. Taylor, and S. Wenger, for the local organization of the workshop and for their assistance in the preparation of this volume.

Bern, August 1996 R. von Steiger, R. Lallement, M. Lee

SCIENCE AS AN ADVENTURE

Keynote Address to the Inauguration of ISSI

H. BONDI
Churchill College, Cambridge, UK

Many of our fellow citizens have an image of science and of scientists we would find hard to recognise: They tend to think of science as something rigid, firm and soulless (and generally dull) created in an objective and often solitary manner by cool passionless persons. While they might be willing to see some nobility in 'pure' science, this reluctant generosity does usually not extend to its 'dirty' offshoot, technology. Absurd as all this picture looks to us, it is, I think, worth examining how these dangerous misconceptions arise and what might be done to improve the understanding of what we do.

These views are indeed dangerous in several respects. First there is the angry puzzlement that arises when no firm clear answer can be given to scientific queries that are of public interest (usually in the environmental or medical fields). When on some such issue different scientists hold differing views, journalists speak of "this extraordinary scientific controversy". When I tell them that controversy is normal in science and is indeed the lifeblood of scientific advance, they find it hard to believe me. The public tend to think that at least some of the scientists involved in such a controversy must be either venal or incompetent or both. These views arise partly from the conflict between the popular view that scientists are 'objective and dispassionate' and the normality of active arguments, yet people are unwilling to abandon their view. Moreover, the piece of science most people are familiar with is the Newtonian description of the solar system (Newton's clockwork, as I like to call it). The rigid predictability of this is taken as a model of what all science should be like. When this expectation is not fulfilled, there is disappointment.

A model, familiar to all, for many fields of science is weather forecasting, but this is not appreciated. The curse of rigidity is thought to apply to us and this view is reinforced by teaching mainly the examinable pieces of science where the right/wrong classification can readily be applied.

Secondly there is the denigration of technology arising from the widely held view that it is purely derivative and trails science. Thirdly and in some respects most importantly, there is the worry that it is on the basis of these widespread misconceptions that young people make their career choices of whether to become scientists or not. I feel sure that some who would have made excellent scientists were frightened off by the rigid image ("to every question there is just one right answer") so often conveyed at school, while some of those who become scientists guided by this image are disappointed to find their work full of uncertainties and question marks. We need the adventurous souls, but do little to attract them.

Space Science Reviews **78**: 1–5, 1996.

I sometimes comment that if we were a business with a prospectus as misleading as the one of science so often presented at school, we would be in trouble with the law.

How have these misconceptions arisen and what can we do to avoid generating them? I think there are a number of reasons. Foremost perhaps is the understandable desire to get the maximum quantity of science taught in the necessarily limited amount of time available (at school or even in undergraduate courses) with no attention paid to the need to convey something of its spirit. Coupled with this is the wish to confine instruction to the supposedly certain parts of science to avoid teaching something that later turns out to be incorrect. In fact it would be most educational to convey wonder and uncertainty. Neither the philosophy of science nor its history are considered to be parts of the normal syllabus. Yet it would be very beneficial to go through some of the very intelligent ideas of our predecessors that turned out to be wrong.

I totally accept that teaching hours are limited. If time is given to describing the evolution of scientific ideas and how they were shaped by technological developments, then clearly the total amount of science covered will be less than it is now. In my view this would be a price well worth paying for educational as well as for scientific reasons and would demonstrate the intensely human nature of science.

Perhaps an example will help. Children are taught that the Earth goes round the Sun, but rarely about tests of this hypothesis and probably never about the historical development which in fact is fascinating and, as will be seen, could readily be taught.

By the late seventeenth century the Copernican system was accepted by virtually all astronomers. The great prize and test would be measure a stellar parallax, the apparent change in the position of a near star due the Earth changing its position during its orbit about the Sun. The inaccuracies of the instrumentation of the time, coupled with the difficulty of choosing a sufficiently near star, led to numerous unsubstantiated claims until the first unambiguous parallax was at last established by F. W. Bessel in 1838. But already in 1725 James Bradley had discovered stellar aberration, the apparent change in the position of stars due to the change in the Earth's velocity during its orbit. This was the first clear test of the heliocentric system and should surely be widely taught at school (the effect on the inferred direction of a star due to the motion of the telescope is readily described). This would be far more helpful than the mere assertion that the Earth orbits the Sun. The aberration angle is many times greater than the parallax angle of the nearest stars, which accounts for it having been discovered much earlier. It is amusing to speculate that, had our civilisation developed on Jupiter, with its bigger orbit, but lower velocity, parallax would have been much larger and aberration rather less than here, so that presumably parallax would have been found first.

This story is a good example of scientific evolution, showing how it is driven by technological advances (the gradual improvement in the precision of astronomical measurements which eventually enabled Bessel to measure so small a parallax angle successfully), but also how a good scientist works: Bradley had originally not thought of the then unexpected phenomenon of stellar aberration, but very speedily worked it out to account for his otherwise inexplicable measurement of a stellar position shift at right angles to the parallax shift he was expecting to find.

In the philosophy of science I am a follower of Karl Popper. He sees the task of a scientist first to propose a theory that of course needs to be compatible with the empirical knowledge of the day, but that also must forecast what further, future experiments or observations will show. If such are performed and are incompatible with the theory, we say that it has been disproved. Liability to empirical disproof is the defining characteristic of science. If the tests turn out to be consonant with the forecasts of the theory, we must never regard this as a proof of it, since it remains scientific only if it continues to be liable to be disproved by further experiments. Thus all scientific insights should be viewed as provisional only. It is because the wholly unexpected can happen that science is such an adventure.

This analysis is appropriate because theories make general statements, whereas experiments and observations inevitably deal with the particular. This is also the reason why a theory can never be deduced from empirical knowledge. It necessarily requires a leap of the imagination to formulate one. Equally it is imagination that is needed to devise a novel experiment to test a theory. Thus imagination is essential in science, but do our fellow citizens appreciate this?

It is natural that a scientist will argue fiercely to defend a favoured theory, perhaps by criticising the accuracy or reliability of an incompatible experiment, which will be defended by its originator with equal passion. If relating this were part of ordinary teaching, perhaps the absurd popular picture of the cold, unimaginative, passionless scientist would gradually fade away. But there is another point on which we ourselves may be somewhat vulnerable. Communication is essential in science. We fully accept this. This need makes me think of a customer coming into a pet shop asking to buy a parrot of especially high intelligence. After some consideration he is sold a particular bird. Two weeks later this customer returns to the shop, absolutely furious because this reputedly so intelligent bird has not said a word in all this time. However the pet shop owner replies: This parrot is a thinker, not a talker. Indeed we do not regard any work as part of science until it has been widely communicated through being published in the accessible scientific literature. Yet do we consider the teaching of communication skills to have a legitimate claim on the time table of a science course? We all have the experience of a graduate student, highly competent in the relevant topic, yet finding it immensely difficult to convey the results in understandable form by the spoken or written word. Most of us eventually learn communication skills on the job, but do we give their early systematic acquisition the priority it deserves?

Nor do we often analyse the means of the conveying of information in depth. To me the printed word is of only modest effectiveness, though its permanence and wide distribution make it essential. The formal lecture is rather more efficient, but the less formal seminar or workshop (of which you have just had an excellent example) are far better for exchanging information. Yet to chat to a few colleagues with a glass in one's hand is superior to all other methods, for then one is willing to talk about one's doubts and failures as much as about one's successes.

I want to return now to the relation between science and technology, which is so often misunderstood. It is implicit in Popper's definition of science that to-morrow we can test our theories more searchingly and thoroughly than we can to-day. This means that the progress of science depends on the advance of the methods of empirical testing, i. e. of the available technology. Of course equally technology can get ahead by using novel insights of science. Thus it is a mutual relation in which neither science nor technology can be called primary, with the other secondary. Perhaps I can illustrate my thinking with an example of this interaction:

Physics made enormous strides during the last quarter of the nineteenth century with the discovery of the electron, of ions, of radio-activity, etc. Why were such discoveries made then and not earlier or later? Most of such work requires the use of evacuated vessels. Their availability depends on the efficiency and reliability of the pumps needed to extract the air. It so happened that the machining of brass pistons and cylinders improved considerably in the 1850s and 1860s. Though this was an essential pre-condition, it was not sufficient. Any vacuum system inevitably develops leaks that have to be plugged. For non-moving parts, sealing wax is an old established efficient means, but it is rather rigid and thus cannot be used in the links between the vibrating pump and the experimental vessel. A reasonably suitable material became available at the time, namely plasticine. (One could therefore say that much of physics is ultimately based on plasticine). The availability of such a vacuum was a major technological step that allowed much scientific work to be done.

In due course the use of such reliably evacuated vessels permitted Roentgen to make his great scientific discovery of X-rays a hundred years ago. Their importance for medicine was soon appreciated (though it took much longer before their dangers were understood). Accordingly, a new technology of X-ray machines came into being that in due course made them affordable, reliable, precise and safe. Some fifty years after Roentgen these machines were used to study the structure of organic materials and thus the new science of molecular biology come into being. This in turn gave birth, in time, to a wholly new technology, bio-technology. This is a clear example in which each advance of science or technology leads to an advance in the other. Neither can claim primacy.

The international nature of science is so strong and pervasive, because science is well tailored to our universal human characteristics, above all to our fallibility. Similarly it suits our sociability and our need to communicate. We value imagination and ingenuity highly, but the supreme yardstick of empirical test is recognised by all.

I would like to conclude with a personal story which illustrates some of these features. Many years ago my late colleague R. A. Lyttleton and I investigated the consequences that would arise if the electric charges of the electron and the proton were not exactly equal and opposite. (At the time this was only known to one part in 10^{13}). We showed that there would be very interesting astronomical and cosmological consequences if the discrepancy were as small as one part in 10^{19}. This paper irritated many. In their desire to prove us wrong, several very ingenious experiments were devised which showed that the maximum permissible discrepancy was less than one part in 10^{22}, far too small for the effects we had calculated. So within a very few years we had been disproved. However, I am proud of this paper and in now way ashamed. Thanks to the work which it provoked, an important constant of nature is known to much higher accuracy than before.

Following Popper, we know that empirical disproof is the seminal event to science. One can be right only for a limited time, but to be original and stimulating is the essential contribution a scientist can make to the unending adventure that is science.

It is a particular pleasure for me to be involved in the Inauguration of ISSI, the International Space Science Institute. This Institute is splendidly conceived to be international (as science is) right from the start, to be devoted to studying questions of our astronomical neighbourhood with the technology of instruments in space, with the highest quality expertise to interpret the results. I am sure it will flourish and greatly add to our knowledge.

Address for correspondence: Sir H. Bondi, Churchill College, Cambridge CB3 0DS, United Kingdom.

THE HELIOSPHERE

THE HELIOSPHERE

W.I. AXFORD
Max-Planck-Institut für Aeronomie, D-37189 Katlenburg-Lindau, Germany

Abstract. The history and current status of heliospheric physics are outlined very briefly. It is emphasized that the existing Voyager Interstellar Mission should go a long way to answering many of the outstanding questions during the next one or two decades, especially those concerning the nature and structure of the solar wind termination shock.

1. Introduction

The concept of the heliosphere was first developed by Davis about 40 years ago as part of his pioneering consideration of the solar modulation of galactic cosmic rays (Davis, 1955; Axford, 1972, 1990). During the past 25 years there have been significant advances in our understanding of the subject, in particular as a result of observations made by spacecraft missions such as Pioneers 10/11, Voyagers 1/2 and Ulysses. There have been a number of major conferences which have marked the progress which has been made since 1972, notably the regular "Solar Wind" conferences and COSPAR Meetings, including the first COSPAR Colloquium in 1989 (Grzedzielski and Page, 1990). The Proceedings of these meetings contain quite complete summaries of the continued improvement of our knowledge of the heliosphere that has taken place during this time.

The distant heliosphere is currently being investigated by the spacecraft Voyagers 1/2, which are at 50-65 AU from the Sun and which are expected to reach about 130 AU around the year 2020 AD before they fail as a result of loss of RTG power. These spacecraft are suitably instrumented for making in situ investigations of the state of the interplanetary and local interstellar media (although one can always think of improvements) and for this reason it has been recommended that they should comprise NASA's future Voyager Interstellar Mission (Axford et al., 1995).

2. General Properties of the Heliosphere

The size and shape of the heliosphere is determined by stress balance and the need for the supersonic solar wind in the inner solar system to accommodate to the presence of the flowing local interstellar medium (LISM) (see Figure 1). In general this accommodation should be made possible as the result of a shock wave which forms at a point where the local ram pressure of the wind is comparable to the external LISM pressure. However, the contributions of relatively long range components of

Space Science Reviews **78**: 9–14, 1996.

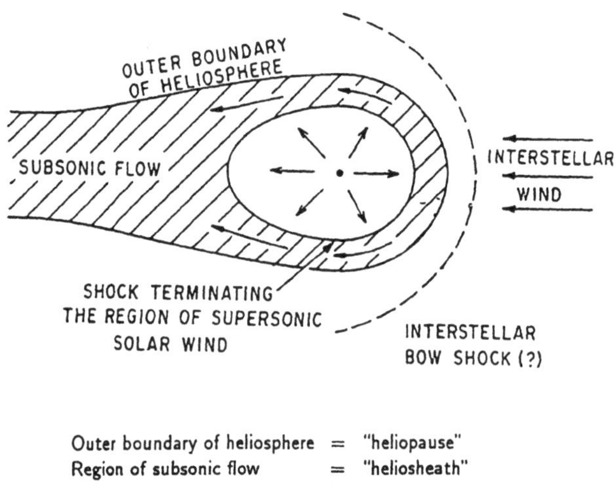

Figure 1. The configuration of the heliosphere shown schematically.

the stress balance can blur the shock to a considerable extent so that the concept of a "shock" must be broadened. In particular cosmic rays, including the anomalous component, and the neutral interstellar gas can in principle produce a smoothed shock structure quite different to the usual collisionless shock expected in a dilute plasma. Nevertheless, the overall result is the same as far as the need for a transition from "supersonic" to "subsonic" flow is concerned, if these terms are taken to refer to the appropriate mode of wave propagation. In any case, it is expected that the minimum distance to the "shock" should be about 100 AU ($\pm$ 20%) from the Sun on the basis of estimates of the total pressure of the LISM. Fortunately the direction of motion of the two Voyager spacecraft is such that they should encounter the shock at approximately this minimum distance.

In general we expect the minimum distance to the outer boundary of the region containing solar wind plasma, the "heliopause", to be about 30% or so larger than the minimum distance to the shock terminating the supersonic solar wind. On the downstream side of the heliosphere, however, i.e. in the "tail", solar plasma must be present to very large distances, although it will be cooled by charge exchange of the proton component with interstellar neutral hydrogen atoms and perhaps disrupted by reconnection with the local interstellar magnetic field. Beyond the boundary of the heliosphere (the "heliopause") there may be a further bow shock upstream in the LISM if the Sun is travelling supersonically with respect to the medium. It is unclear at present whether or not this is the case, as the magnetic field strength is unknown; however, there is evidence for the existence of a sheath of partly neutral plasma around other nearby stars (Linsky, 1996).

3. The Local Interstellar Medium

In order to make any progress in a quantitative sense it is necessary to know the essential parameters of the LISM. In fact, we know quite a lot about the neutral hydrogen and helium from observations of backscattered solar HI 1216 and HeI 584 lines, from the pattern of pick up ions of interstellar origin in the solar wind and from direct measurements by the GAS experiment on Ulysses. The temperature and velocity vector of the neutral gas are well established: the velocity vector lies close to the ecliptic plane, more or less in the direction of the projection of the solar apex on this plane, and the speed is about 25 km/sec (i.e. somewhat more than the speed of proper motion of the Sun relative to local stars). The temperature of the gas is about 7000 K and the densities of hydrogen and helium about 0.015 and 0.07/cm^3, respectively (see Witte et al., 1996; Geiss and Witte, 1996).

We know nothing directly of the magnitude and direction of the magnetic field in the LISM nor of the number density of electrons. However, we can be reasonably sure that the velocity and temperature of the ionized component is the same as that of the neutrals, since the collision and equilibration times are reasonably short. On the other hand we cannot consider the ionization state of the plasma as being in equilibrium in any sense simply because the recombination times at these temperatures and assuming an electron density $< 0.1/\text{cm}^3$ are large ($> 10^6$ years, equivalent to > 25 parsecs at 25 km/sec). The frequency cut-off of the radio emissions observed from Voyager 1/2 provides a guide to the electron density but this may not be definitive in view of our lack of understanding of the generation process. However, since it might be reasonably assumed that the H:He ratio in the LISM is of the order of 12, the difference between the H and He neutral densities suggests a lower limit to the electron density of about 0.1/cm^3, which allows sufficient ram pressure from the plasma alone to place the solar wind termination shock close to 100 AU (see Figure 2). [Note that this electron density is considerably larger than the average value obtained for the region within 100 pc of the Sun from pulsar dispersion measurements.]

Pulsar measurements have provided us with a reasonable idea of the strength and direction of the general interstellar magnetic field in the galactic arm in which the Sun is situated, namely about 2 microgauss and perpendicular to the direction to the galactic centre respectively (Ruzmaikin et al., 1988). However, the Sun is situated near the edge of what appears to be an old supernova bubble and not far from a "wall" of neutral gas, perhaps accounting for the low density and temperature of the LISM (Holzer, 1989). It is not possible to even guess what the magnetic field might be like in such a bubble, although some information might be obtained from considerations of the anisotropy of high energy cosmic rays, which are largely unaffected by the presence of the heliosphere (Axford et al., 1990).

The question of whether or not there is a shock wave in the interstellar medium beyond the heliopause depends on whether or not the interstellar medium is moving supersonically with respect to the Sun. As far as the plasma component

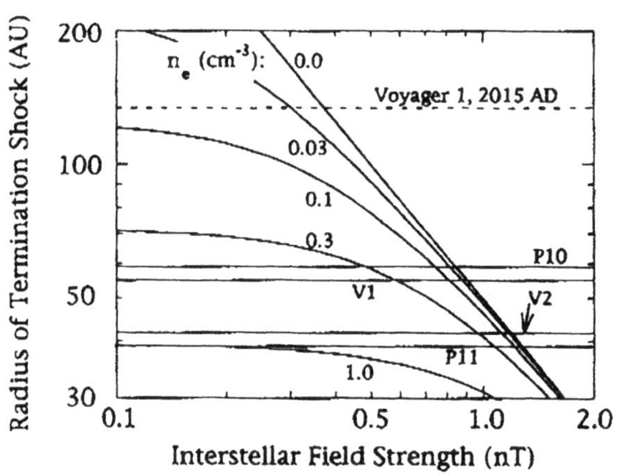

Figure 2. The minimum distance to the termination shock for various LISM plamsa densities and magnetic field strengths but omitting the possible effects of the neutral component. The current positions of the Voyager and Pioneer spcaecraft are shown approximately.

alone is concerned, the sound speed is about 14 km/sec and thus the flow may be supersonic. Coupling to the neutral component can only reduce this but if the magnetic field strength is as high as 4 microgauss the effective speed of sound is large enough to render the flow subsonic so that a shock will not occur. Regardless of whether there is a shock or not, coupling to the neutrals will partly "screen" the inner heliosphere from LISM neutral hydgrogen causing a change in our estimate for the neutral density in the LISM and thus reducing the limit on the value of the proton density and increasing the Alfvén speed. The situation is quite confusing!

4. Cosmic Rays

From the beginning, the study of the heliosphere has been closely connected with the study of cosmic rays. As far as galactic cosmic rays with rigidities above about 1 GV are concerned, there is modulation associated with the solar wind and interplanetary magnetic field but, although interesting in itself, these cosmic rays probably do not have any major effects on the heliosphere. This is because the total variation in their pressure is at most comparable with the ram pressure of the solar wind at the termination shock and much of the variation occurs within the region of supersonic flow. The effects are at most likely to be comparable to those of interstellar pick up ions on the solar wind, which are perhaps at the 5-10% level in terms of reduction of the momentum flux.

It is possible that galactic cosmic rays of lower rigidity than 1 GV in the LISM have a pressure which could influence the position of the termination shock, although it seems unlikely. The pressure of cosmic rays at their source tends to maximize in the 1-10 GV range and the lower energy particles suffer rather severe energy losses during propagation so that they would not be expected to have a significant pressure. However, if we are indeed located in a supernova bubble in which several supernovae have exploded in the last 10^6 years or so it might well be argued that the low energy component has locally an abnormally high pressure. On the other hand this would produce strong heating of the LISM and its observed characteristics might be difficult to explain.

The most interesting aspect of cosmic rays in the present context is the anomalous component, which is evidently the result of the acceleration of interstellar pick up ions since their composition corresponds to that expected for the neutral component of the LISM (Fisk, 1996). The most likely process for the acceleration of these particles involves the termination shock (Jokipii, 1996) but the possible contribution of CIR shocks and the interaction of these shocks with the termination shock should not be disregarded. It may be the case that the pressure of the anomalous cosmic rays at the termination shock is sufficiently large that the shock is a broad "cosmic ray" shock rather than the usual thin shock to be expected elsewhere in the interplanetary plasma.

5. Future Developments

As will be evident from the Proceedings of this Conference, there is a high level of activity connected with the above problems both from the side of theorists (notably simulations of the large-scale flow) and those analysing data from the Voyager, Pioneer and Ulysses spacecraft in particular. The likely encounter of both Voyagers with the termination shock during the next 5/10 years is probably the greatest stimulus for this work at the present time and there is no question that the in situ detection of the shock will cause widespread interest.

In my view it is essential that NASA's Voyager Interstellar Mission should be supported until the spacecraft are no longer serviceable, since the cost of repeating the mission, perhaps with improved experiments, is likely to be so great that it would never happen within any reasonable time. The Voyager spacecraft and experiments were designed and built more than 20 years ago and there are undoubtedly several things which, with hindsight and with the benefit of subsequent improvements of technique, could be done better. Nevertheless, we can be well satisfied with the existing missions, since they are capable of providing us with the most essential observations, notably of the position and nature of the termination shock. It would be of great significance if we could examine the structure of a cosmic ray shock in situ, not only for heliospheric physics but for the development of our ideas about the acceleration of cosmic rays generally. It would also be an achievement to be able to enter into interstellar space, free of the influence of the solar wind, for the first time, although one should keep in mind that measurements made in the

LISM are not "fundamental" in the sense that the state of the interstellar medium is highly variable and what prevails in our vicinity may not have any remarkable implications for future studies.

References

Axford, W.I.: 1972, 'The Heliosphere', in Solar Wind, NASA SP-308, 609.

Axford, W.I.: 1990, 'Introductory Lecture - The Heliosphere', in Cospar Colloquiua Series, Vol. 1, Eds. S. Grzedzielski and D.E. Page, Pergamon Press, London, 7-15.

Axford, W.I., M.K. Dougherty and J.F. McKenzie: 1990, 'Hydromagnetic waves in the local interstellar medium and the high energy cosmic ray anisotropy', in Proc. 21st ICRC, Adelaide **3**, 311.

Axford, W.I., M.A. Lee, J.H. Adams, C.J. Fishman, S.A. Fuselier, D.T. Hall, P. Király and G. Squibb: 1995, 'Exploring our Connection to the Galaxy: Report of the Heliospheric Missions Science Review Panel', NASA, Washington DC.

Davis, L.E., Jr.: 1955, 'Interplanetary magnetic fields and cosmic rays', *Phys. Rev.* **100**, 1440.

Fisk, L. A.: 1996, 'Implications of a weak termination shock', *Space Sci. Rev.,* this issue.

Geiss, J. and M. Witte: 1996, 'Properties of the interstellar gas inside the heliosphere', *Space Sci. Rev.,* this issue.

Grzedzielski, S. and D.E. Page: 1990, Physics of the Outer Heliosphere, COSPAR Colloquiua Series, Vol. 1, Pergamon Press, London.

Holzer, T.E.: 1989, 'Interaction between the solar wind and the interstellar medium', *Annual Reviews of Astron. Astrophys.* **27**, 199.

Jokipii, J. R. and J. Giacalone: 1996, 'The acceleration of pickup ions', *Space Sci. Rev.,* this issue.

Linsky, J. L.: 1996, 'GHRS observations of the LISM', *Space Sci. Rev.,* this issue.

Ruzmaikin, A.A., A.M. Shukurov and D.D. Sokoloff: 1988, 'Magnetic Fields of Galaxies', Astrophysics and Space Science Library, Kluwer, Dordrecht.

Witte, M., M. Banaszkiewicz and H. Rosenbauer: 1996, 'Recent results on the parameters of the interstellar Helium from the Ulysses/GAS experiment', *Space Sci. Rev.,* this issue.

THE HELIOSPHERIC MAGNETIC FIELD

ANDRÉ BALOGH
The Blackett Laboratory, Imperial College, London SW7 2BZ, U.K

Abstract. The heliospheric magnetic field (HMF) is an important component of the heliospheric medium. It has been the subject of extensive studies for the past thirty five years. There is a very large observational data base, mostly from the vantage point of the ecliptic plane, but now also from the solar polar regions, from the Ulysses mission. This review aims to present its most important large scale characteristics. A key to understand the HMF is to understand the source functions of the solar wind and magnetic fields close to the sun. The development of new modelling techniques for determining the extent and geometry of the open magnetic field regions in the corona, the sources of the solar wind and the HMF has provided a new insight into the variability of the source functions. These are now reasonably well understood for the state of the corona near solar minimum. The HMF at low-to-medium heliolatitudes is dominated, near solar minimum, by the Corotating Interaction Regions (CIRs) which arise from the interaction of alternating slow and fast solar wind streams, and which, in turn, interact in the outer heliosphere to form the large scale Merged Interaction Regions. The radial component of the HMF is independent of heliolatitude; the average direction is well organised by the Parker geometry, but with wide distributions around the mean, due, at high latitudes, to the presence of large amplitude, Alfvénic fluctuations. The HMF at solar maximum is less well understood, due in part to the complexity of the solar source functions, and partly to the lack of three dimensional observations which Ulysses is planned to remedy at the next solar maximum. It is suggested that the in-ecliptic conditions in the HMF, largely determined by the dynamics of transients (Coronal Mass Ejections) may also be found at high latitudes, due to the wide latitude distribution of the CMEs.

1. Introduction

The magnetic field in the heliosphere originates in the sun and its corona and is carried out into space by the highly conducting solar wind. This simple consequence of the existence of the solar wind was proposed by Parker (1958) who also calculated the geometry of the magnetic field lines in the heliosphere, assuming a radial magnetic field at the sun. The result, based on a uniform solar wind flow, is the well-known Parker spiral in the solar equatorial plane; out of the equatorial plane, the model yields field lines which spiral on the surface of cones aligned with the rotation axis of the sun and with a half angle equal to the co-latitude of the origin of the field lines on the sun. This geometrical framework provided by Parker's equations of the HMF has proved to be a fundamentally useful one to organise a description of the large scale structure of the HMF.

The solar wind, however, is not emitted uniformly from the sun; nor is the magnetic field distributed uniformly either on the photosphere or in the corona. It is instructive to carry out a brief examination to show how these non-uniformities affect the simple geometrical derivation of the HMF first used by Parker. We can define, conceptually, source functions for both the solar wind, $V_{SW}(r_0,\theta,\varphi)$, and the magnetic field $B(r'_0,\theta,\varphi)$ on the surface of spheres (not rotating with the sun) of heliocentric radius r_0

Space Science Reviews **78**: 15–28, 1996.

and r'_0, respectively, with θ and φ being the heliographic latitude and longitude. For the solar wind, we can take, again conceptually, the radius r_0 to be close to the sun, but at a distance where most of the acceleration of the solar wind has already taken place. For the magnetic field, r'_0 is normally taken to be about 2 to 3 solar radii, based, as discussed below, on potential magnetic field models of the corona.

Both functions are, in fact, strongly dependent on time on many scales. Along a radius vector from the sun (i.e. for a given θ and φ), the time dependence is introduced not only by the solar rotation, bur also, on a range of both longer and shorter timescales, by the evolution of the source regions of the solar wind, and on short timescales by dynamic phenomena such as Coronal Mass Ejections (CMEs).

Setting aside, in the first instance, the temporal and spatial dependencies implied by the function $B(r'_0,\theta,\varphi)$, the time dependencies in $V_{SW}(r_0,\theta,\varphi)$ introduce a range of dynamic effects in the heliosphere which, through the evolving interactions of the non-uniform flows along a streamline of the solar wind, introduce a corresponding structuring of the HMF. In these interactions, the magnetic field plays only a peripheral role on the large scale (because of the dominance of the flow energy of the solar wind), but it is an essential ingredient of the small scale processes which provide the physical basis for the MHD treatment of interacting solar wind streams.

Two timescales of major importance for both the solar wind and the HMF are the solar rotation period and the solar cycle. Around solar minimum, the corona is in a relatively stable state, with coronal holes, the regions of origin of the fast solar wind, covering both polar regions. The coronal streamer belt, a region of mostly closed magnetic field lines, is restricted to within about 15° to 30° of the solar equatorial plane. Slow solar wind streams are associated with the streamer belt. The relative stability of the coronal configuration, which persists over several solar rotations, introduces a periodicity in the function $V_{SW}(r_0,\theta,\varphi)$ through its dependence on φ. For an interplanetary observer within about 15° to 25° of the solar equatorial plane, V_{SW} at the source surface switches from high speeds (above 700 km/s) to low speeds (below 500 km/s), corresponding to solar longitudes with coronal holes and the streamer belt, respectively. This periodicity generates a quasi-steady pattern of fast solar wind streams interacting with the preceding slow solar wind streams which result in the formation of the Corotating Interaction Regions (CIRs), as discussed below.

The source functions, simplified but appropriate to solar minimum conditions and in agreement to the first approximation with the Ulysses observations (Phillips et al., 1995, Balogh et al., 1995b), are illustrated in Fig. 1. In particular, the existence of a sharp boundary between fast and slow solar wind streams has been shown through the analysis of freezing-in temperatures and abundance ratios of the solar wind ions (Geiss et al., 1995). For interplanetary observers within the ecliptic, the source functions in the lower panels of Fig. 1 generate the alternating fast and slow solar wind streams and magnetic polarities with the period of the solar rotation.

The regions of origin of the solar wind undergo a significant change between solar minimum and solar maximum. At solar maximum, there are no dominant large scale coronal holes and therefore the details of the source function are far more difficult to discern than at solar minimum; in terms of the function V_{SW} there are no clear periodicities at the solar rotation period. Although stream-stream interactions may still occur, persistent, corotating patterns no longer form at solar maximum.

The commonly used definition of the source function $B(r'_0, \theta, \varphi)$, conceptually defined above, is based on the extrapolation of measured photospheric magnetic fields to a source surface of two or so solar radii. The extrapolation is generally based on the assumption of either a source-free (potential field) corona, or, additionally, with localised model current systems, to match the observed latitudinal spreading of the solar wind (and therefore of the field on the source surface for each solar rotation (in Solar Geophysical Data, see Hoeksema, 1991, and references therein). Such models in fact derive a quantitative definition of the function $B(r'_0, \theta, \varphi)$ in a self-consistent manner, but without explicit reference to a corresponding function $V_{SW}(r_0, \theta, \varphi)$.

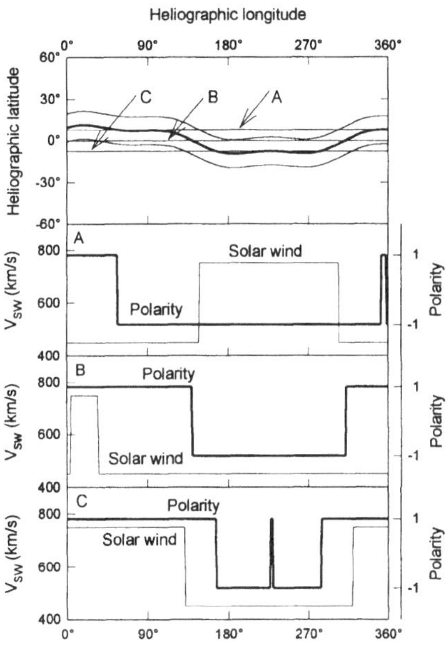

Figure 1. A schematic representation of the source functions of the solar wind and the heliospheric magnetic field at solar minimum. The top panel shows the magnetic neutral line (heavy line) on the source surface, determined by the potential model of the field applied in the corona. The slow solar wind (at 450 km/s) originates in a latitude band of $\pm 10^{\circ}$ (indicated by the thin lines) following the magnetic neutral line. Outside this band, the velocity of the solar wind is assumed constant at 750 km/s. The magnetic field has a constant magnitude over the whole surface, but has opposite polarities on the two sides of the neutral line. Samples of the source functions are shown in the three lower panels, at, respectively 7.25°, 0° and -7.25° latitude. The large scale topology of the HMF is the result of the dynamic evolution of the interacting solar wind streams, superimposed on the transport of the magnetic flux alternately from the two hemispheres.

An empirical relationship has been proposed between the areal expansion factor of open magnetic field flux tubes in the corona, more specifically in coronal holes and their boundaries, and the solar wind speed (Wang and Sheeley, 1990). These authors have found an anticorrelation between the expansion factor and the solar wind plasma speed: the smaller the expansion factor, the greater is the speed of the solar wind which is emitted from the corona. Combining magnetic field models in the corona which yield the geometry of coronal holes and the distribution of expansion factors, Wang and Sheeley (1994, 1996) have in effect provided the basis for the determination of both magnetic field and solar wind source functions.

Potential field models have been successful in predicting the polarity of the magnetic field measured in space, e.g. at 1 AU, but have been unsuccessful in predicting its magnitude. Results from the solar polar passes of Ulysses have shown that, contrary to expectations, the radial component of the magnetic field is essentially independent of heliolatitude inside the fast solar wind streams from the polar coronal holes (Smith and Balogh, 1995). A subsequent revision of the model calculations (Wang and Sheeley, 1995) have shown that the Ulysses result is consistent with polar field strengths of about 10 G. Furthermore, sheet currents need to be introduced in the models to ensure that the radial component of the field remains constant in the corona, i.e. that the source function $B(r'_0,\theta,\varphi)$ is simply a constant, with sign changes only across current sheets. This is equivalent to what has been called a "split monopole" type configuration, with uniform field distributions, but with a polarity reversal across a neutral line (in effect a current sheet). This is the basis for the source function shown in Fig. 1.

Reverting to the terminology of the potential field models, on the source surface defined for the magnetic field, the two polarities of the field are separated by a neutral line (as above). The shape of the neutral line is a function of the phase of the solar cycle. Near solar minimum, the neutral line remains close to the solar equatorial plane, as illustrated in Fig. 1. This is the consequence of the dominance of a close-to-axial dipole term in the source surface magnetic field. A tilted dipole corresponds to a sinusoidal neutral line on projections similar to Fig. 1. As solar activity builds up, the dipole becomes more inclined with respect to the rotation axis, and higher order, quadrupole and octupole terms also increase in magnitude. The inclination of the neutral line increases to high values, 70o or more at solar maximum. Finally, shortly after the peak in solar activity indicators, the polarities of the two hemispheres reverse and the dipole term becomes again dominant (Hoeksema, 1995).

Given a magnetic field which is purely radial at the source surface, a condition normally imposed on it by the potential field models, it is expected that the HMF would follow the Parker geometry at high latitudes where the solar wind has a relatively uniform speed and therefore interaction regions do not form. If the magnetic field retains any azimuthal component (due to the random motions of field lines in the photosphere and corona) as it is carried away from the sun by the solar wind, then due to the slower decay of the transverse component which varies as the inverse of the heliocentric distance, a very large distortion can potentially be introduced in the geometry of the magnetic field at large distances at high heliographic latitudes (Jokipii and Kota, 1989).

2. The HMF near Solar Minimum

The largest scale structure in the heliosphere is the Heliospheric Current Sheet (HCS), the extension of the magnetic neutral line, in effect the sun's magnetic equator, into interplanetary space. Interplanetary probes within the latitude range of the north-south excursion of the neutral line cross the HCS normally twice per solar rotation, as implied by Fig. 1. However, if there are higher order terms which generate ripples on the neutral line, hence in the HCS, there can be four or even six crossings of the HCS in each solar rotation. Away from the HCS crossings, the polarity of the magnetic field ("towards" or "away" from the sun, approximately along the Parker spiral) corresponds to that of the solar hemisphere which is connected to the spacecraft. The resulting magnetic sector structure, first discovered by Ness and Wilcox (1964), has been extensively studied over the past thirty years (e.g. Behannon et al., 1989, Smith, 1989, Hoeksema, 1995). These studies have demonstrated the close, although not perfect correspondence between the simple geometric extension of the neutral line into the heliosphere. Small, but significant differences have been noted and attributed to the dynamic deformation of the HCS by the stream-stream interaction process (e.g. Suess and Hildner, 1985, Behannon et al., 1989). The sector structure was observed to disappear at 16° heliographic latitude close to solar minimum by Pioneer 11 which spent several solar rotations in a unipolar magnetic field regime which corresponded to the polarity of the sun's hemisphere at that epoch (Smith et al., 1978).

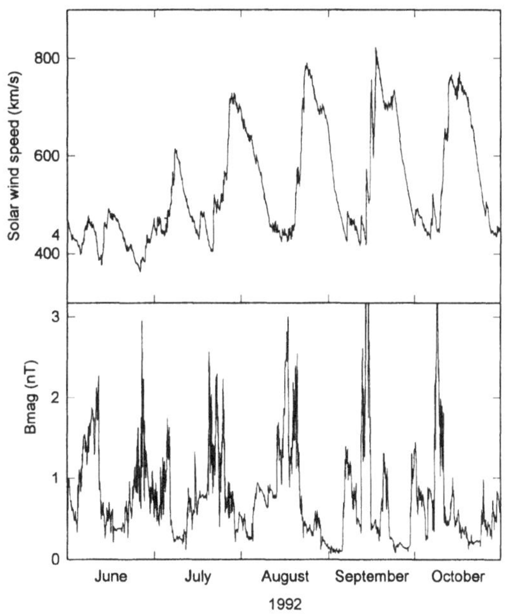

Figure 2. The development of Corotating Interaction Regions seen by Ulysses between 15° and 20° heliolatitude. The upper panel shows the solar wind velocity, with the high speed stream originating in the developing southern polar coronal hole. The lower panel shows the periodic compression of the magnetic field leading the peaks in solar wind velocity. (Solar wind data used courtesy of the Ulysses plasma team.)

The evolution of the sector structure and its disappearance at 30° south heliographic latitude during the 1990-1993 epoch following the solar maximum was observed by the Ulysses spacecraft (Smith et al., 1993). During the in-ecliptic and early out-of-ecliptic parts of the Ulysses orbit, from 1 to 5.4 AU, the sector structure was stable, with two sectors per solar rotation. The sector structure underwent a major evolutionary step in mid-1992, following a major CME event (Balogh et al., 1993). The period following this event was accompanied by the appearance of a high speed stream at Ulysses, originating in the developing southern coronal hole (Bame et al., 1993). The westwards drift of the sector structure after mid-1992, until its disappearance in May 1993, was characteristic not so much of the climb of Ulysses to high latitudes as of the development of the stable and recurring pattern of the solar wind source function on the sun which characterised the observations of Ulysses over the following 9 months. The early part of the alternance of high and low speed solar wind streams, and the development of the magnetic compression regions is shown in Fig. 2.

Shock waves form at the leading and trailing edges of CIRs, normally at heliocentric distances beyond 1 AU (Smith and Wolfe, 1979). These shocks appear to corotate, as do the CIRs themselves, in the stable interplanetary regime near solar minimum. Reverse shocks are invariably caught up by forward shocks of the following CIRs (see Whang, 1991, and references therein); this process initiates the eventual merging of the CIRs at heliocentric distances beyond 5 or 10 AU. Out of the solar equatorial plane, the tilted dipole configuration of the sun at solar minimum introduces a tilting of the CIRs themselves and, as a consequence, the propagation direction of shock waves. As was modelled by Pizzo (1991), Pizzo and Gosling (1994), and observed by Ulysses (Gosling et al., 1993, Balogh et al., 1995a, and more extensively described by Gonzales-Esparza et al., 1996), forward shocks are propagating towards lower latitudes, while reverse shocks propagate to higher latitudes. As Ulysses travelled to higher latitudes, the forward shocks associated with CIRs disappeared earlier than did the reverse shocks.

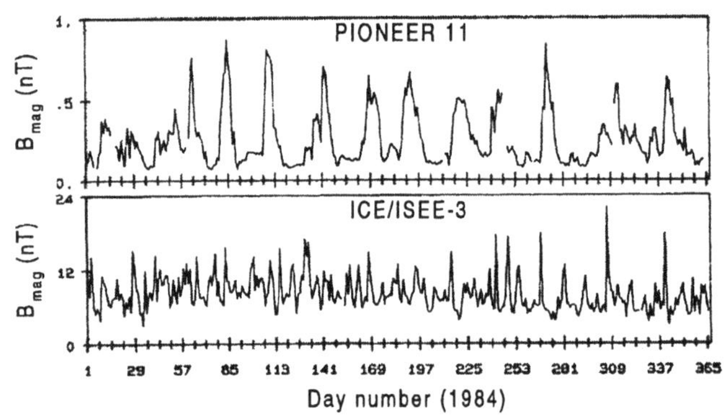

Figure 3. Comparison of the magnetic field magnitude observed in 1984, close to solar minimum, by ICE (ISEE-3) at 1 AU and Pioneer 11 at 16 AU, showing the coalescence of CIRs into Merged Interaction Regions. (From Smith, 1989)

The merging of CIRs is the main feature of the low latitude, outer heliosphere near solar minimum. In addition to the interaction of forward and reverse shocks associated with CIRs which follow each other, propagating outwards along the (approximately) radial direction, the CIRs themselves appear to merge and coalesce into wider structures (see, e.g. Burlaga, 1984, Smith, 1989, Belcher et al., 1993). For the HMF, this results in a wider compression region, with enhanced field magnitude. The effect of merging is illustrated in Fig. 3, in which a whole year of magnetic field data from 1984, close to solar minimum, is shown from ISEE-3 at 1 AU and from Pioneer 11 at 16 AU. It seems that the compression process continues with radial distance, and that solar wind plasma pressure continues to build up inside the Merged Interaction Regions.

While simple, one-dimensional simulations have qualitatively indicated the nature of the CIR merging process, and observations from both Pioneer and Voyager support the concept of the Merged Interaction Regions, the consequences of the process in three dimensions have yet to be examined in detail. Given the three-dimensional structure of the CIRs as modelled by, e.g. Pizzo, 1991, and confirmed by the Ulysses observations, the consequences for the structure of the large scale HMF away from the solar equatorial plane will need to be examined before a three dimensional model of the HMF at solar minimum, valid in the outer heliosphere, can be established.

The direction of the HMF near the solar equatorial plane follows in general the geometry of the Parker spiral configuration, but has a wide spread about the average (Thomas and Smith, 1980, Burlaga et al., 1982, Forsyth et al., 1996a). The distributions of hourly averaged magnetic field azimuth angles observed by Ulysses in the ecliptic and over the southern and northern polar regions of the sun are shown in Fig. 5. In these plots, the magnetic field vectors have been rotated into coordinate system which is aligned with the expected Parker spiral direction at the location of Ulysses, using the measured values of the solar wind (courtesy of the Ulysses plasma team). In this coordinate system, the magnetic field aligned with the Parker direction will have an azimuth angle of 0° or 180°, depending on the positive or negative polarities of the source field near the sun. The high latitude plots show single-peaked distributions, corresponding to the inward and outward polarities in the northern and southern solar hemispheres, respectively. The widths of the distributions at high latitudes are significantly larger than in the ecliptic. This is caused by the presence, in the high latitude observations, of high amplitude Alfvénic fluctuations (Forsyth et al., 1995, 1996a) which are discussed further below. However, the statement in the Introduction concerning the usefulness of the Parker geometry of the HMF for organising the observations is clearly justified (at least in the first approximation) by these distributions.

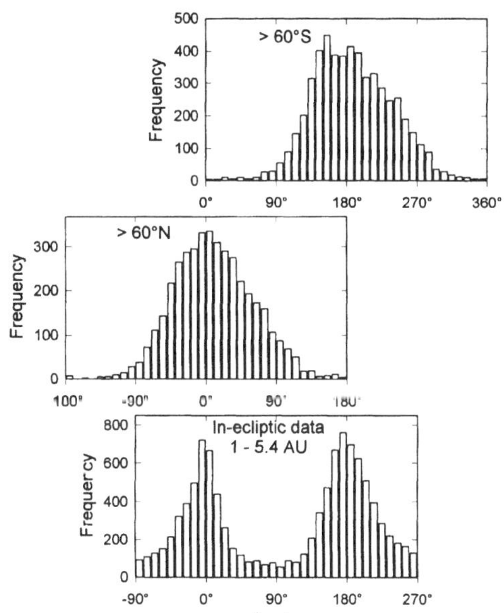

Figure 4. Distribution of the azimuthal direction of the HMF measured by Ulysses. The lower panel shows data from the ecliptic, where both northern and southern magnetic polarities were observed. The two upper panels show data from the northern and southern polar regions, where, in each case, only one polarity was present. Note the greater width of the distributions at the high latitudes; this is caused by the presence of large, long-period transverse fluctuations in the HMF. (From Forsyth et al., 1995, 1996a.)

The radial dependence of the magnetic field strength remains controversial. In a series of papers (Winterhalter et al., 1990 and references therein) the magnetic field strength has been reported to decrease faster than implied by the Parker model of the field. This result was based on a comparison of data at 1 AU and measurements by Pioneer 11 out to 21 AU, and led to the conclusion that there was a large scale flux deficit near the ecliptic, of order 1% / AU. The analysis of Voyager 1 and 2 observations, on the other hand, (Burlaga and Ness, 1993 and references therein) have shown no measurable deviation from the expected decrease as a function of heliocentric distance based on Parker's model. There is little prospect that the existence of a flux deficit will be resolved without further measurements over similar heliocentric distance ranges.

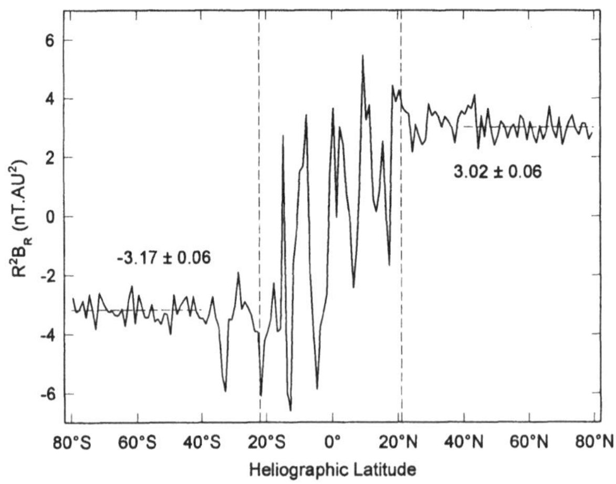

Figure 5. The radial component of the HMF, normalised to 1 AU, as a function of heliolatitude. This component does not depend on latitude in the uniform, fast polar flows, but reverses sign between the two solar hemispheres. The equatorial band extends to about 20^{0} on either side of the solar equatorial plane. (From Forsyth et al., 1996b.)

The study of the radial and latitudinal dependence of the long-term averaged radial component of the magnetic field has shown, as already discussed in the Introduction that it is essentially independent of heliolatitude (Smith and Balogh, 1995, Forsyth et al., 1996b), and varies (as expected) as r^{-2} where r is the heliocentric radial distance. This is shown in Fig. 5, in which the radial component of the magnetic field made during the fast latitude scan by Ulysses in 1994-95, normalised to 1 AU, is plotted against heliolatitude (Forsyth, 1996b).

A detailed discussion of magnetic fluctuations is beyond the scope of this review. However, large amplitude fluctuations, present over a large region of the heliosphere, do affect the large scale properties of the HMF. The high latitude HMF is strongly affected by the presence of large amplitude transverse, Alfvénic fluctuations (Smith et al., 1995). Such fluctuations had been first observed in high speed solar wind streams in the ecliptic (Belcher and Davis, 1971, Bavassano et al., 1982, Denskat and Neubauer, 1982). In the uniformly high speed solar wind flows observed by Ulysses over the polar regions, large amplitude waves are always present and, at low frequencies, appear to be largely unevolved even out to 4 AU radial distance (Horbury et al., 1995). The periods of the waves are very long, 10 hours or more; such waves have wavelengths of a significant fraction of an AU. Given that the Alfvén speed is only a small fraction of the solar wind speed, these waves are effectively convected in the solar wind. Given their large amplitudes and ubiquity in the polar regions, these waves significantly affect those properties of the HMF which are relevant to the scattering and transport of cosmic rays. The variances of the transverse magnetic components in these waves vary, as a function of heliocentric distance, as r^{-n}, where n = 3 for shorter periods, of order of an hour, to n = 2.5 or so for periods of 10 hours (Balogh et al., 1996). At the longest periods, of a day or more, the exponent approaches n = 2, as calculated by Jokipii and Kota (1989) for the convected structures which would be due to the residual transverse component of the magnetic field near the sun and which would significantly distort the high latitude HMF at large heliocentric distances.

3. The HMF at Solar Maximum

The state and topology of the solar corona present a much more complex picture near solar maximum; the source functions $V_{SW}(r_0,\theta,\varphi)$ and $B(r'_0,\theta,\varphi)$ of the solar wind and magnetic field, respectively, are in fact more difficult to define. There are no large scale coronal holes, as much of the sun all the way to the poles is covered by closed loop structures. The solar wind comes from isolated, small coronal holes; the large expansion factor generated by the necessarily large divergence of the magnetic field in such holes implies that no large scale fast wind streams are formed (Wang and Sheeley, 1990). The near-equatorial orientation of the magnetic dipole term in the potential field models of the corona, together with the importance of the higher order terms (Hoeksema, 1991) yields a heliomagnetic equator (i.e. the neutral line in the potential field model) which reaches to high latitudes, and has a complex topology.

In interplanetary space, crossings of the heliospheric current sheet and the sector structure can still be discerned (e.g. Behannon et al., 1989), but the agreement between the neutral line pattern in the corona and the sector structure is tenuous. The CIRs characteristic of the near-equatorial heliosphere at solar minimum cannot form in the absence of stable, large scale high speed solar wind streams.

The dominant cause of dynamics in the heliosphere at solar maximum is likely to be the large number of CMEs which are emitted at all heliolatitudes and at a rate which is about ten times greater than during the years of solar minimum (e.g. Hundhausen, 1993). Based on solar observations, the rate is about one CME about every ten days at solar minimum, and it increases up to about one per day at solar maximum. CMEs have a large apparent latitudinal width in the corona, with a median of about 40°. They have a very wide speed distribution, with the fastest ones up to 1000 km/s. In the presence of this dominant factor in coronal dynamics at solar maximum, it is expected that their interplanetary manifestations will effectively fill much of the inner heliosphere.

The conclusion for the HMF is that transient interplanetary disturbances, driven by CMEs, are shaping its large scale structure near solar maximum. Two properties are therefore important: first, the transient and random nature of the solar events, and second, their prevalence at all latitudes. These properties are superimposed on a generally slow and variable solar wind source at all latitudes, and a background magnetic field which originates from a far from dipole-like sun, with a complex neutral line separating the dominant polarities.

These points imply that the three-dimensional heliosphere, and therefore the HMF, are relatively insensitive to heliolatitude at solar maximum. The source functions for the solar wind and the HMF have no dominant, low order terms, and little stability against the coronal activity manifested by the frequency of the CMEs. It is likely that, in the first approximation, the high latitude heliosphere (which will be explored by Ulysses in 2000-2002) will bear some resemblance to the observations made in the vicinity of the ecliptic at previous solar maxima.

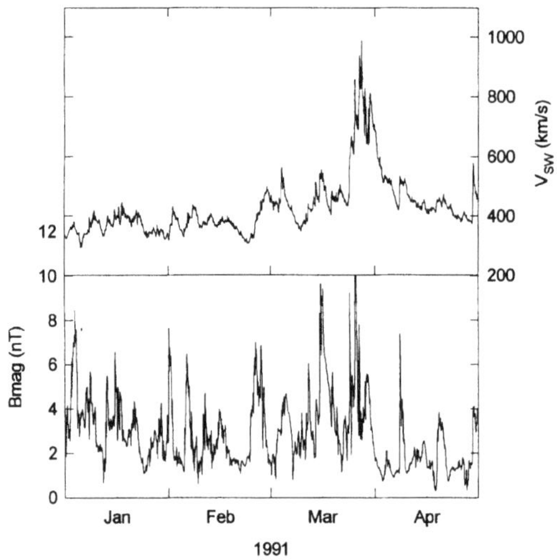

Figure 6. Solar wind and magnetic field observations by Ulysses during the first four months of 1991, close to solar maximum, in the ecliptic, covering the radial distance range from 1.5 to 3 AU. The magnetic field magnitude (lower panel) shows the random nature of the transient compression regions, due to enhanced solar activity. The solar wind is, in general slow, but fluctuating, until the major set of CMEs in the second half of March. (Solar wind data used courtesy of the Ulysses plasma team.)

If this assumption is valid, then observations made, for instance, by Ulysses in the early phase of its mission are representative of the HMF in the heliosphere as a whole, out to some relatively large (20 or more AU) heliocentric distance. In the first quarter of 1991, just past the peak of the last solar maximum, Ulysses observed a series of solar transient events, culminating in a group of large CMEs in the second half of March. Analysis of the solar wind and magnetic observations (Phillips et al., 1992) concluded that between 24 March and 4 April there were at least two but possibly three CMEs which apparently merged their signatures at 2.5 AU. Fig. 7 illustrates the solar wind velocity and magnetic field magnitude during the first four months of 1991, showing the generally slow and variable speed of the solar wind, the major events at the end of March, the magnetic compression regions ahead of the CMEs and the aperiodicity of the observations. This lack of a recognisably recurrent signature at the period of solar rotation is one of the main characteristics of the structure of the HMF near solar maximum; this should be contrasted with the observations in Fig. 2.

The Ulysses observations in March 1991 point to a phenomenon in the HMF which is qualitatively similar at both solar maximum and solar minimum. As discussed above for the CIRs observed at large heliocentric distances, CMEs can also merge and even overtake each other. Fig. 7 illustrates this phenomenon (Smith, 1990); much of the structure observed at 1 AU (presumably most of them due to individual CMEs in 1981, a year close to solar maximum) has disappeared by 12 AU where Pioneer 11 observed large scale Merged Interaction Regions, the result of a coalescence of CMEs. It is also to be noted that the disturbances created by the CMEs appear to survive, at least in this merged form, well beyond 1 AU. In fact, the large events in March 1991 (as observed at Ulysses) could also be observed at Voyager 2, at 38 AU (Belcher et al., 1993). This identification at Voyager 2 seems secure, based on considerations of transit times to the outer heliosphere, but the degree of merging is even more apparent than in the equivalent case for Pioneer 11. While these observations were made near the ecliptic plane, one can assume that at least some degree of merging also may occur with CMEs at high latitudes.

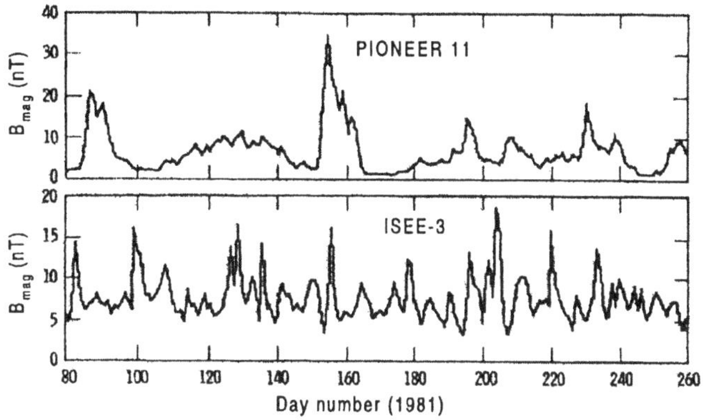

Figure 7. Magnetic field magnitude observations in 1981, close to solar maximum, from ISEE-3 at 1 AU and Pioneer 11 at 12 AU, showing the coalescence of the solar transients at large radial distances. This figure is similar to Fig. 4, except that it shows merging CMEs, rather than merging CIRs. The dynamics of the two merging processes are similar, but the spatial distribution of CMEs at solar maximum may present a similar picture of merging at high heliolatitudes, as well as near the ecliptic. (From Smith et al., 1990.)

The interplanetary extent of CMEs remains a key question for the large scale structure of the HMF at solar maximum. While their average latitudinal width is of order 40°, close to the sun, their interplanetary extent may be significantly greater than this. Earlier studies, using shock wave and energetic particle observations have concluded that their extent was as much as 120° in helio-longitude, although it is possible that highly energetic CMEs capable of generating shock waves and associated particle acceleration may have had a much larger than average extent already close to the sun.

Any differential propagation characteristics in latitude would introduce a distortion in the shape of the CMEs as a function of heliocentric distance. As it appears that CMEs are mostly emitted from above, or in the vicinity of the streamer belt (Hundhausen, 1993), the high latitude CMEs at solar maximum would be affected by the same distorting effects, indeed causing some of these, as the high latitude current sheets which are expected at that time. However, the solar wind regime is likely to be significantly different from the large scale uniform flows near solar minimum, therefore the distortions may not only be caused by largely geometric effects due to the diminished winding of the HMF, but may also depend more on the initial structure of the CMEs.

4. Conclusions

The HMF, studied extensively over the past thirty-five years, still remains a topic of active scientific interest. This is partly because of its significance as an important component of "laboratory astrophysics" in which direct observations in the solar system can provide a basis for the study of collisionless plasma phenomena in much wider astrophysical contexts. But an increasingly important motivation for the study of the HMF comes from the need to understand solar-terrestrial processes, in their widest sense.

Extensive observations have been made in the past decades in the ecliptic over heliocentric distances covered by interplanetary probes from 0.3 (Helios 1 and 2) to several tens of AU (Pioneer 10 and 11, Voyager 1 and 2), with an important normalisation point at 1 AU (e.g. IMP-8, ISEE-1, 2 and 3, and numerous others). Indeed, there are observations covering all phases of the solar cycle, but largely restricted to the proximity of the ecliptic plane and, hence, to within the $\pm$ 7.25° band of the solar equatorial plane. The exploratory journey of Ulysses has changed this restricted perspective into a three-dimensional one, providing not only the first and only observational base for the study of the high heliolatitude HMF, but also the motivation to re-visit previous studies and thus to re-evaluate the existing in-ecliptic data base.

The Ulysses observations of the high latitude HMF have been made at a time of low solar activity and have provided many of the key results which were needed for completing a good understanding of the heliosphere at solar minimum. In particular, the observation that the radial component of the magnetic field is independent of heliolatitude has helped to revise the magnetic modelling of the corona, and has therefore contributed to a better understanding of the source functions of the heliospheric medium. The tilted geometry of the CIRs, with the non-radially propagating shock fronts is an important example of the fundamentally three-dimensional nature of structure and dynamics in the heliosphere. Given all the in-ecliptic observations, together with the Ulysses results, we are now able to evaluate the goodness-of-fit of the Parker model of the HMF at solar minimum.

The HMF at solar maximum is less well understood. The complexities of the solar wind and coronal magnetic fields, together with the fact that heliospheric dynamics is driven by transient events, make the three-dimensional heliosphere at high solar activity difficult to model. The forthcoming Ulysses observations at the next solar maximum will be needed to provide the observational basis for the models. At present, the best guess is that the high latitude HMF will be affected in a similar manner to the low-latitude regions at the time of solar maximum. However, it is not known how the dynamics of interacting transient events depends on heliolatitude.

Acknowledgements

I wish to acknowledge the contributions of R.J. Forsyth, E.J. Smith and G. Erdos to this review, through the provision of material help and many useful discussions. I also wish to thank the Ulysses plasma team (PIs: S.J. Bame, J.L. Phillips and D.J. McComas) for their permission to use their data explicitly and implicitly in the preparation of this review.

References

Altschuler, M.D., and Newkirk, G.: 1969, *Solar Phys.* **9**, 131.
Balogh, A., Erdos, G., Forsyth, R.J., and Smith, E.J.: 1983, *Geophys. Res. Lett.* **20**, 2331.

Balogh, A., Gonzalez-Esparza, J.A., Forsyth, R.J., Burton, M.E., Goldstein, B.E., Smith, E.J., and Bame, S.J.: 1995a, in *The High Latitude Heliosphere*, ed. By R.G. Marsden, Kluwer Academic Publishers, Dordrecht, 171.
Balogh, A., Smith, E.J., Tsurutani, B.T., Southwood, D.J., Forsyth, R.J., and Horbury, T.S.: 1995b, *Science* **268**, 1007.
Balogh, A., Forsyth, R.J., Horbury, T.S., and Smith, E.J.: 1996, *Solar Wind Eight*, in press.
Bame, S.J., Goldstein, B.E., Gosling, J.T., Harvey, J.W., McComas, D.J., Neugebauer, M., and Phillips, J.L.: 1993, *Geophys. Res. Lett.* **20**, 2323.
Bavassano, B.M., Dobrowolny, M., Mariani, and Ness, N.F.: 1982, *J. Geophys. Res.* **87**, 3617.
Behannon, K.W., Burlaga, L.F., Hoeksema, J.T., and Klein, L.W.: 1989, *J. Geophys. Res.*, **94**, 1245.
Belcher, J.W., and Davis, L.: 1971, *J.Geophys. Res.* **76**, 3534.
Belcher, J.W., Lazarus, A.J., McNutt, R.L., Jr., and Gordon, G.S., Jr.: 1993, *J. Geophys. Res.* **98**, 15,177.
Burlaga, L.F.: 1984, *Space Sci. Rev.* **39**, 255.
Burlaga, L.F., Lepping, R.P., Behannon, K.W., Klein, L.W., and Neubauer, F.M.: 1982, *J. Geophys. Res.* **87**, 4345.
Burlaga, L.F., and Ness, N.F.: 1993, *J. Geophys. Res.* **98**, 3539.
Denskat, K.U., and Neubauer, F.M.: 1982, *J. Geophys. Res.* **87**, 2215.
Forsyth, R.J., Balogh, A., Smith, E.J., Murphy, N., and McComas, D.J.: 1995, *Geophys. Res. Lett.* **23**, 3321.
Forsyth, R.J., Balogh, A., Smith, E.J., Erdos, G., McComas, D.J.: 1996a, *J. Geophys. Res.* **101**, 395.
Forsyth, R.J., Balogh, A., Horbury, T.S., Erdos, G., Smith, E.J., and Burton, M.A.: 1996b, *Astron. Astrophys.*, in press.
Geiss, J., G. Gloeckler, G., von Steiger, R., Balsiger, H., Fisk, L.A., Galvin, A.B., Ipavich, F.M., Livi, S., McKenzie, J.F., Ogilvie, K.W., and Wilken, B.: 1995, *Science* **268**, 1033.
Gonzalez-Esparza, J.A., Balogh, A., Forsyth, R.J., Smith, E.J., and Phillips, J.L.: 1996, *J. Geophys. Res.* in press.
Gosling, J.T., Bame, S.J., McComas, D.J., Phillips, J.L., Balogh, A., and Strong, K.T.: 1995, in *The High Latitude Heliosphere*, ed. by R.G. Marsden, Kluwer Academic Publishers, Dordrecht, 133.
Hoeksema, J.T.: 1991, *Adv. Space Res.* **11**, (1)15.
Hoeksema, J.T.: 1995, in *The High Latitude Heliosphere*, ed. By R.G. Marsden, Kluwer Academic Publishers, Dordrecht, 137.
Horbury, T.S., Balogh, A., Forsyth, R.J., and Smith, E.J.: 1995, *Geophys. Res. Lett.*, **22**, 3401.
Hundhausen, A.J.: 1973, *J. Geophys. Res.* **78**, 1528.
Hundhausen, A.J.: 1993, *J. Geophys. Res.* **98**, 13,177.
Jokipii, J.R., and Kota, J.: 1989, *Geophys. Res. Lett.* **16**, 1.
Ness, N.F., and Wilcox, J.M.: 1964, *Phys. Rev. Lett.* **13**, 461.
Parker, E.N.: 1958, *Astrophys. J.* **128**, 664.
Phillips, J.L., Bame, S.M., Gosling, J.T., McComas, D.J., Goldstein, B.E., Smith, E.J., Balogh, A., Forsyth, R.J.: 1993, *Geophys. Res. Lett.* **19**, 1239.
Phillips, J.L., Bame, S.J., Feldman, W.C., Goldstein, B.E., Gosling, J.T., Hammond, C.M., McComas, D.J., Neugebauer, M., Scime, E.E., and Suess, S.T.: 1995, *Science* **268**, 1030.
Pizzo, V.J.: 1982, *J. Geophys. Res.* **87**, 4374.
Pizzo, V.J.: 1991, *J. Geophys. Res.* **96**, 5405.
Pizzo, V.J., and Gosling, J.T.: 1994, *Geophys. Res. Lett.* **21**, 2063.
Schatten, K.H., Wilcox, J.M., and Ness, N.F.: 1969, *Solar Phys.* **6**, 442.
Smith, E.J.: 1989, *Adv. Space Res.* **9**, (4)59.

Smith, E.J.: 1990, in *Physics of the Outer Heliosphere*, ed. By S. Grzedzielski and D.E. Page, Pergamon Press, Oxford, 253.
Smith, E.J., and Wolfe, J.H.: 1979, *Space Sci. Rev.* **23**, 217.
Smith, E.J., Tsurutani, B.T., and Rosenberg, R.L.: 1978, *J. Geophys. Res.* **83**, 717.
Smith, E.J., Neugebauer, M., Balogh, A., Bame, S.J., Erdos, G., Forsyth, R.J., Goldstein, B.E., Phillips, J.L., and Tsurutani, B.T.: 1993, *Geophys. Res. Lett.* **20**, 2327.
Smith, E.J., and Balogh, A.: 1995, *Geophys. Res. Lett.* **22**, 3317.
Smith, E.J., Balogh, A., Neugebauer, M., and McComas, D.J.: 1995, *Geophys. Res. Lett.* **22**, 3381.
Suess, S.T., and Hildner, E.: 1985, *J. Geophys. Res.* **82**, 2405.
Thomas, B.T., and Smith, E.J.: 1980, *J. Geophys. Res.* **85**, 6861.
Wang, Y.-M., and Sheeley, N.R., Jr.: 1990, *Astrophys. J.* **355**, 726.
Wang, Y.-M., and Sheeley, N.R., Jr.: 1994, *J. Geophys. Res.* **99**, 6597.
Wang, Y.-M., and Sheeley, N.R., Jr.: 1995, *Astrophys. J.* **447**, L143.
Wang, Y.-M., Hawley,S.H., and Sheeley, N.R., Jr.: 1996, *Science* **271**, 464.
Whang, Y.C.: 1991, Space Sci. Rev. 57, 339.
Winterhalter, D., Smith, E.J., and Slavin, J.A.: 1990, J. Geophys. Res. 95, 1.

THE SOLAR WIND: A TURBULENT MAGNETOHYDRODYNAMIC MEDIUM

B. BAVASSANO
Istituto di Fisica dello Spazio Interplanetario, Consiglio Nazionale delle Ricerche, 00044 Frascati, Italy

Abstract. MHD turbulence properties in the solar wind are briefly reviewed. The evolution of fluctuations of alfvénic type in near-ecliptic regions of the inner heliosphere is described. The role of interplanetary sources and the influence of interactions with structures convected by the solar wind are discussed. Turbulence features at high latitudes and in the outermost regions of the heliosphere are finally highlighted.

1. Introduction

Fluctuations in the magnetohydrodynamic (MHD) regime are a relevant component of the solar wind variations. They generally have a random character, hence the name 'turbulence', and their amplitude is large enough to suggest the presence of nonlinear effects. In the following sections some aspects of the MHD turbulence in the solar wind will be discussed. For left out topics reference can be made to the recent review by Tu and Marsch (1995).

2. Turbulence in the Inner Heliosphere

A large fraction of the solar wind turbulence in the inner heliosphere (i.e., inside 1 AU) has an alfvénic character, namely a high correlation between fluctuations in solar wind ($\mathbf{V}$) and Alfvén ($\mathbf{V}_A$) velocities. In this section we will focus on these fluctuations, although also other turbulent populations (e.g., two-dimensional turbulence and compressive turbulence) may be relevant to solar wind physics.

The great majority of the interplanetary alfvénic fluctuations propagates, in the solar wind frame, in outward directions (with respect to the Sun). It is well accepted that these outgoing modes are mainly of solar origin, generated inside the Alfvén critical point. As regards the much less abundant inward travelling fluctuations, their generation can only occur outside the critical point, otherwise the wind flow would not be able to carry them out into the interplanetary space and they would fall back on the Sun.

Inward modes, in spite of their minority character, are expected to be a fundamental ingredient in the alfvénic turbulence evolution, due to the nonlinear interactions that they are able to develop with the outward population. Elsässer variables $\mathbf{z}^{\pm} = \mathbf{V} \pm \mathbf{V}_A$ will be used in the following to indicate the two different modes,

Space Science Reviews **78**: 29–32, 1996.

with the sign of $\mathbf{V}_A$ taken in such way that $\mathbf{z}^+$ ($\mathbf{z}^-$) fluctuations always correspond to outgoing (ingoing) modes, independently on the magnetic field polarity.

Properties of $\mathbf{z}^+$ and $\mathbf{z}^-$ fluctuations ($\delta\mathbf{z}^+$ and $\delta\mathbf{z}^-$, respectively) have been extensively studied in near-ecliptic regions of the inner heliosphere (e.g., see Tu and Marsch, 1995). The dominant $\delta\mathbf{z}^+$ population exhibits higher power levels in fast than in slow wind streams. For both speed regimes a strong decline of $\mathbf{z}^+$ fluctuation amplitudes with increasing heliocentric distance is observed. This radial decrease is accompanied by a change in the spectral shape, with near-the-Sun flat spectra progressively steepening toward a Kolmogorov-like ($-5/3$) spectrum.

The noticeably smaller $\delta\mathbf{z}^-$ power levels are, in sharp contrast, much less variable, especially at intermediate and large wavelengths (approximately above $2 \cdot 10^6$ km). The $\delta\mathbf{z}^-$ spectra appear almost independent on both wind speed and radial distance and exhibit spectral indices very close to $-5/3$. Thus, the alfvénic turbulence evolution in the inner heliosphere is substantially driven by a decline of $\mathbf{z}^+$ modes.

Nonlinear MHD processes developed at velocity-shear layers have been proposed to account for the observed $\delta\mathbf{z}^+$ and $\delta\mathbf{z}^-$ behaviors (Roberts *et al.*, 1991). Numerical simulations indicate that, although shear-driven turbulence injects the same power in $\delta\mathbf{z}^+$ and $\delta\mathbf{z}^-$, the different initial conditions lead to a different evolution for the spectra of the two populations. It is also seen that these processes proceed rapidly if the velocity-shear layer is in proximity of a current sheet, such that the background magnetic field has only a small component along the shear. In these conditions a well developed and very slowly evolving $\delta\mathbf{z}^-$ spectrum could already exist around 0.5 AU.

However, there are turbulence features that may not be accounted for by shear-driven processes. In particular, observations clearly indicate that the large-scale $\mathbf{z}^-$ fluctuations are strongly reminiscent of convected structures and discontinuities (Marsch and Tu, 1994; Bruno *et al.*, 1996). Then, in an alternative point of view, the turbulence evolution might be driven by interactions that alfvénic modes can develop with structures embedded in the wind flow. Numerical simulations (e.g., Schmidt, 1995) show that this kind of approach is able to qualitatively reproduce many properties of the turbulence evolution.

A final remark is about the above mentioned two-dimensional (2D) turbulence. The 2D turbulence (Matthaeus *et al.*, 1990) is made up of fluctuations with wave-vectors transverse to both the background and the fluctuating magnetic fields. The relative weight of 2D contributions, as compared to those of alfvénic type, increases at larger scales. It is expected that the 2D turbulence becomes increasingly important in the outer heliosphere.

3. Polar Turbulence

The first in-situ observations of the high-latitude solar wind have been done very recently by the Ulysses spacecraft. The fast flows emerging from the large polar coronal hole provide an unique environment to study turbulence properties in the absence of typical aspects of the near-ecliptic solar wind as, for instance, strong shear flows and large compression regions.

It has been found (e.g., Goldstein *et al.*, 1995) that polar alfvénic turbulence is less evolved (or, with higher and flatter spectra for $\delta\mathbf{z}^+$) than in near-ecliptic regions at the same heliocentric distance. This appears mainly due to the absence of large velocity gradients. The spectral shape is not too different from that observed in near-ecliptic fast flows close to the Sun (0.3 AU) by the Helios spacecraft. Once Ulysses is inside the polar hole the turbulence properties appear to be determined by the heliocentric distance rather than by the heliolatitude.

4. Turbulence in the Outer Heliosphere

Beyond 1 AU the evolution of the $\delta\mathbf{z}^+$ spectrum towards a Kolmogorov-like $-5/3$ slope continues (Bavassano and Smith, 1986), with power levels becoming closer and closer to those of the $\delta\mathbf{z}^-$ spectrum (Roberts *et al.*, 1987). For increasing distance the alfvénic fluctuations lose their dominant character and solar wind structures become increasingly important. Out of $\sim$ 15 AU an f^{-2} spectrum is observed at low latitudes, indicating that sudden steps (or discontinuities) have assumed a relevant role in the near-ecliptic solar wind. However, at latitudes above 20° an $f^{-5/3}$ spectrum is still found (Burlaga *et al.*, 1987).

At distances greater than $\sim$ 30 AU the solar wind energy budget is largely affected by contributions from interstellar pickup ions. The deposition of energy into the solar wind is accompanied by a generation of MHD waves. Effects of this kind have been extensively observed in proximity of comets (e.g., Lee, 1989), where power spectra appear strongly excited at and above the cyclotron frequency of cometary pickup ions.

In conclusion, in the outermost regions of the heliosphere the MHD turbulence should be characterized by an f^{-2} spectrum at the low frequencies, where abrupt solar wind changes are dominant, and by a less steep spectrum at the high frequencies, where freshly-generated ions induce enhanced spectral levels and a subsequent turbulent cascade.

Acknowledgements

I wish to thank J. Geiss and V. Manno for their kind invitation to the workshop on *The Heliosphere in the Local Interstellar Medium*. Workshop attendance has been financially supported by the International Space Science Institute (ISSI) of Bern (Switzerland).

References

Bavassano, B. and Smith, E. J.: 1986, 'Radial variation of interplanetary alfvénic fluctuations: Pioneer 10 and 11 observations between 1 and 5 AU', *J. Geophys. Res.* **91**, 1706.

Bruno, R., Bavassano, B., and Pietropaolo, E.: 1996, 'On the nature of Alfvén inward modes in the solar wind', in D. J. McComas, J. L. Phillips, N. Murphy, and D. Winterhalter (eds.), *Proceedings of the Solar Wind 8 Conference*, Jet Propulsion Laboratory, Pasadena (CA), in press.

Burlaga, L. F., Ness, N. F., and McDonald, F. B.: 1987, 'Large-scale fluctuations between 13 AU and 25 AU and their effects on cosmic rays', *J. Geophys. Res.* **92**, 13647.

Goldstein, B. E., Smith, E. J., Balogh, A., Horbury, T. S., Goldstein, M. L., and Roberts, D. A.: 1995, 'Properties of magnetohydrodynamic turbulence in the solar wind as observed by Ulysses at high heliographic latitudes', *Geophys. Res. Lett.* **22**, 3393.

Lee, M. A.: 1989, 'Ultra-low frequency waves at comets', in B. T. Tsurutani and H. Oya (eds.), *Plasma Waves and Instabilities at Comets and in the Magnetospheres*, AGU Geophys. Monogr. Ser., Vol. 53, p. 13.

Marsch, E. and Tu, C.-Y.: 1994, 'Non-Gaussian probability distributions of solar wind fluctuations', *Ann. Geophys.* **12**, 1127.

Matthaeus, W. H., Goldstein, M. L., and Roberts, D. A.: 1990, 'Evidence for the presence of quasi-two-dimensional nearly incompressible fluctuations in the solar wind', *J. Geophys. Res.* **95**, 20673.

Roberts, D. A., Ghosh, S., Goldstein, M. L., and Matthaeus, W. H.: 1991, 'Magnetohydrodynamic simulation of the radial evolution and stream structure of solar wind turbulence', *Phys. Rev. Lett.* **67**, 3741.

Roberts, D. A., Goldstein, M. L., Klein, L. W., and Matthaeus, W. H.: 1987, 'Origin and evolution of fluctuations in the solar wind: Helios observations and Helios-Voyager comparisons', *J. Geophys. Res.* **92**, 12023.

Schmidt, J. M.: 1995, 'Spatial transport and spectral transfer of solar wind turbulence composed of Alfvén waves and convective structures II: Numerical results', *Ann. Geophys.* **13**, 475.

Tu, C.-Y. and Marsch, E.: 1995, 'MHD structures, waves and turbulence in the solar wind: Observations and theories', *Space Sci. Rev.* **73**, 1.

Address for correspondence: B. Bavassano, IFSI/CNR, via G. Galilei, CP 27, 00044 Frascati, Italy. (e-mail: bavassano@roma1.infn.it)

VOYAGER OBSERVATIONS OF THE MAGNETIC FIELD, INTERSTELLAR PICKUP IONS AND SOLAR WIND IN THE DISTANT HELIOSPHERE

L.F. BURLAGA
Laboratory for Extraterrestrial Physics, NASA-Goddard Space Flight Center, Greenbelt, MD 20771

N.F. NESS
Bartol Research Institute, The University of Delaware, Newark, DE 19716

J.W. BELCHER, A.J. LAZARUS and J.D. RICHARDSON
Department of Physics and Center for Space Research, Massachusetts Institute of Technology, Cambridge, MA 02139

July 10, 1996

Abstract. Voyagers 1 and 2 are now observing the latitudinal structure of the heliospheric magnetic field in the distant heliosphere (the region between $\simeq$ 30 AU and the termination shock). Voyager 2 is observing the influence of the interstellar medium on the solar wind. The pressure of the interstellar pickup protons, measured by their contribution to pressure balanced structures, is greater than or equal to the magnetic pressure and much greater than the thermal pressures of the solar wind protons and electrons in the distant heliosphere. The solar wind speed is observed to decrease and the proton temperature increase with increasing distance from the sun. This may result from the production of pickup ions by the charge exchange process with the interstellar neutrals. The introduction of the pickup ions into the dynamics of the magnetized solar wind plasma appears to be an important new process which must be considered in future theoretical studies of the termination shock and boundary with the local interstellar medium.

1. Introduction

This paper provides a summary of some recent results obtained by the plasma and magnetic field experiments on Voyager 2 (V2) and the magnetic field experiment on Voyagers 1 (V1) and V2. These results were presented at the First ISSI Workshop in Bern, Suisse, on "The Heliosphere in the Local Interstellar Medium" in November 1995, and many have or will be published elsewhere in the scientific literature. Therefore, this summary of the results is a brief overview, and we direct the reader to the papers referenced below for detailed discussions of the results and for references to the relevant literature.

Space Science Reviews **78**: 33–42, 1996.

2. Large-scale Magnetic Field Polarity

During 1994, V2 was at a heliocentric distance $\langle R \rangle = 43.4$ AU and latitude $\langle \delta \rangle = 11.9°$ S, while V1 was at a heliocentric distance $\langle R \rangle = 56.3$ AU and latitude $\langle \delta \rangle = 32.5°$ N. The spacecraft positions are illustrated in a cross-sectional view in Figure 1 from Burlaga et al. (1996a).

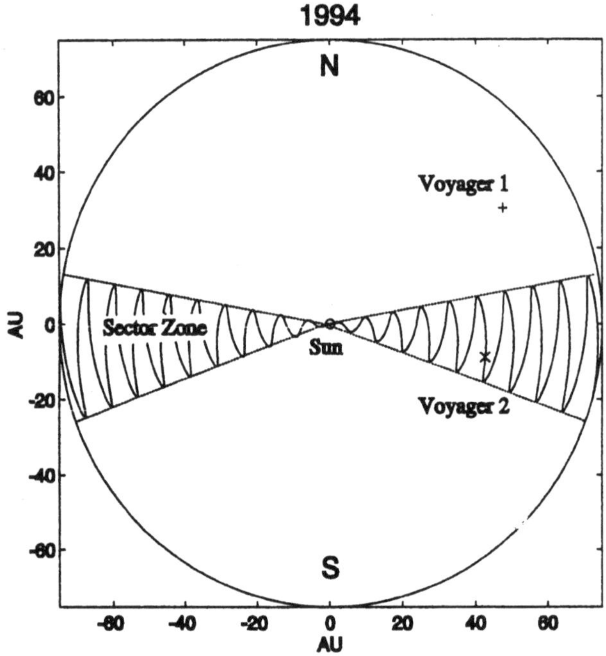

Figure 1. Heliocentric positions of Voyager 1 and Voyager 2 in 1994 and the latitudinal extent of the heliospheric current sheet (illustrated schematically by the wavy curve) which is used to define the "sector zone". (Burlaga et al., 1996a)

Computations based on the solar magnetic field observations obtained at the Stanford Solar Observatory from July, 1993 to June, 1994 show that the footpoints of the heliospheric current sheet (HCS) on a source surface at 3.5 solar radii had a most northerly latitude of $\approx (10° \pm 5°)$ N and a most southerly latitude of $\approx (20° \pm 5°)$ S. These limits define the upper and lower limits, respectively of the "sector zone", delineated by the dashed radial lines in Figure 1.

The HCS is shown schematically by the oscillatory curve in Figure 1 and defines the "sector zone". Thus, a spacecraft located in the sector zone should observe both "toward" and "away" magnetic polarities. But a spacecraft located above the sector zone, such as V1 in 1994, should observe a single ("toward") polarity if the maximum and minimum positions of the HCS do not change appreciably with distance from the sun.

The distributions of hourly averages of the azimuthal angles observed by V1 and V2 during 1994 are shown in Figure 2, from Burlaga et al. (1996a). A single-peaked distribution corresponding to magnetic fields directed toward the sun was observed by V1. A double-peaked distribution corresponding to both "toward" and "away" magnetic polarities was observed by V2. These distributions are consistent with the picture shown in Figure 1 and conclusions drawn therefrom.

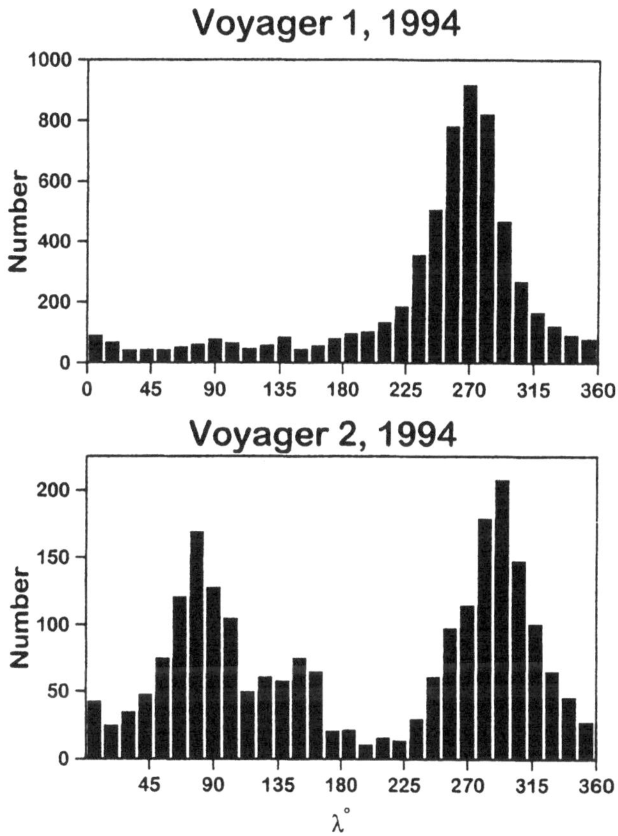

Figure 2. Distributions of azimuthal magnetic field direction observed by Voyager 2 and Voyager 1 during 1994. (Burlaga et al., 1996a)

3. Interstellar Pickup Protons

Interstellar pickup protons cannot be measured directly by the instruments on the spacecraft Voyager 1,2 and Pioneer 10,11. "Pressure-balanced structures" are defined as features with a characteristic "apparent" length of

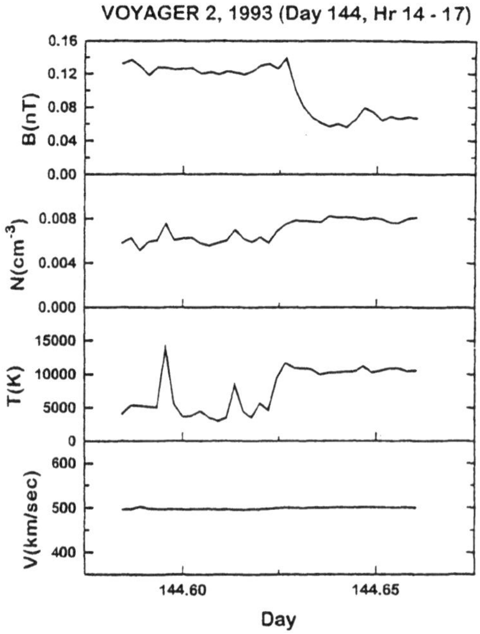

Figure 3. Example of a pressure balanced structure, observed by Voyager 2 in 1993 at 40.1 AU, showing the variations of the magnetic field intensity, B, solar wind proton number density, N, solar wind proton temperature, T, and solar wind speed, V.

the order of a few hundredths of an AU across which the magnetic field strength, solar wind density and solar wind temperature can change, but the total pressure of all the components (including that of pickup protons) is constant (Burlaga, 1995). By "apparent" is meant the inferred length scale associated with the convected solar wind structure.

Pressure balanced structures are observed at all distances from the sun explored by spacecraft. The presence of interstellar protons can be detected and their pressure and density can be calculated when their pressure is significant, using the method described by Burlaga et al. (1994, 1996b).

An example of a pressure balanced structure is shown in Figure 3, based on observations from Voyager 2 during 1993 at 40.1 AU. This event appears to be a tangential discontinuity across which the magnetic field strength decreases, the solar wind proton density increases, the solar wind proton temperature increases, and the solar wind speed is constant.

Figure 4 shows that the increase in the solar wind proton pressure cannot balance the decrease in magnetic pressure. (The contributions of the solar wind electrons and alpha particles are negligible.) However, one can obtain pressure balance by assuming that the interstellar pickup protons are present with a density proportional to the solar wind density (as expected from

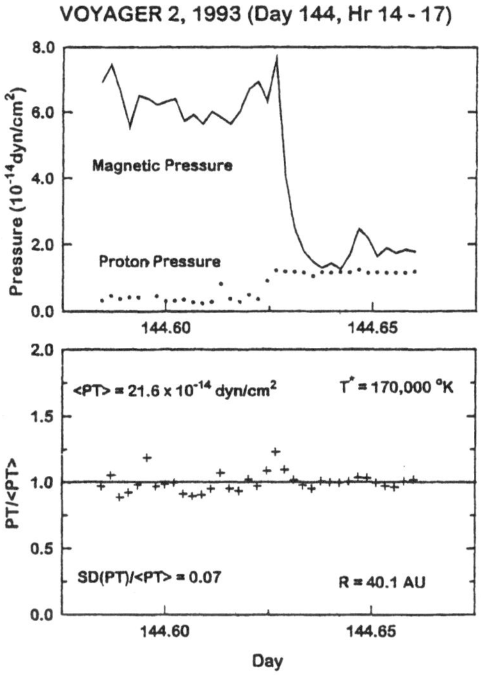

Figure 4. Magnetic pressure, thermal proton pressure and total pressure across the pressure balanced structure shown in Figure 3.

charge exchange) and an effective temperature determined by thermalization of the bulk flow of 170,000 °K (Burlaga et al., 1996b).

From an analysis of the pressure balanced structures observed at R = 39 to 41 AU during 1993 and at R = 43 AU during 1994, Burlaga et al., (1996b) determined the following:

1. The ratio of the pressure of interstellar pickup protons to the magnetic pressure was 1.7 ± 0.2 at 39 - 41 AU and 1.7 ± 0.5 at 43 AU.
2. The ratio of the pressure of interstellar pickup protons to the thermal solar wind proton pressure was 10 ± 2 at 39 - 41 AU and 8 ± 2 at 43 AU.

Clearly, the interstellar pickup protons provide a major contribution to the pressure of the solar wind near 40 AU. Hence, the interstellar pickup protons should play a major role in the dynamics of the "distant heliosphere" (the region between 30 AU and the termination shock). A similar conclusion was arrived at by Burlaga et al. (1994) based on an analysis of pressure balanced structures at 35 AU in a global merged interaction region.

The pressure of the interstellar pickup protons determined from an analysis of the V2 data are plotted as squares in Figure 5 which also includes

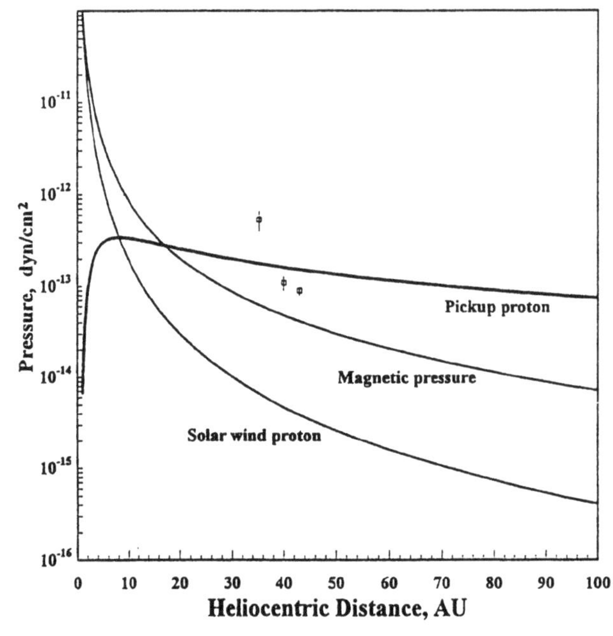

Figure 5. Model curves for the solar wind proton pressure, the magnetic pressure, and the pickup proton pressure together with observations of the pickup proton pressure derived from pressure balanced structures.

the magnetic pressure and solar wind thermal pressure based on a fit of the observations to 30 AU, extrapolated to 100 AU, and the interstellar pickup proton pressure (Whang et al. 1996).

In agreement with that model, the pressure of the pickup protons exceeds the magnetic field pressure and greatly exceeds the solar wind ion and electron thermal pressures in the distant heliosphere. Pickup ions have been studied directly in the inner heliosphere ($< 5.3AU$) by Ulysses (see Gloeckler et al., 1995, Gloeckler, 1996). Extrapolation of derived parameters to the outer heliosphere compare favorably with Voyager results.

4. Solar Wind Speed

One day averages of speed as measured by V2 from launch in 1977 to May 28, 1996 are shown in Figure 6. At the later time, V2 was at 48.9 AU from the sun and −15.8° S of the heliographic equator. A comparison of these speeds with those measured at 1 AU by IMP 8 suggests that there may be some slowing of the solar wind in the outer heliosphere, possibly due

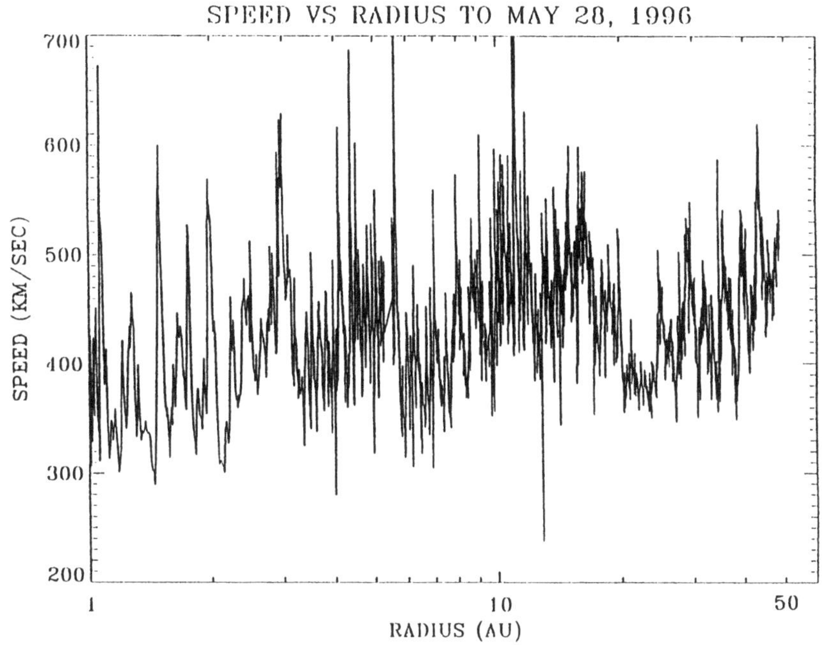

Figure 6. One day averages of speed as measured by Voyager 2 from launch to May 28, 1996.

to mass loading associated with the interstellar pick-up ions (Richardson et al., 1995a). The decrease in the solar wind speed is $\approx$ 30 km/s in the outer heliosphere, which corresponds to a pickup ion density $\approx$ 6% of the solar wind density implying an interstellar ion density of about 0.05 (Isenberg, 1987). Also evident are time variations in the speed starting in 1986 (at 20 AU) with an amplitude of $\approx$ 100 km/s and a period of $\approx$ 1.3 years. Similar variations have been identified in PVO, IMP 8, and Pioneer 10 and 11 data, suggestive of global variations in the solar wind speed occurring at the Sun.

5. Solar Wind Temperature

One day averages of solar wind proton temperature as measured by V2 from launch in 1977 to May 28, 1996 are shown in Figure 7. The fall-off in the solar wind proton temperature is much slower than adiabatic, $r^{-4/3}$, indicating some form of heating. The non-adiabatic temperature variations in the outer heliosphere could be associated with the heating by shocks, stream-stream interactions, turbulence and/or the pick-up of interstellar ions (Richardson et al., 1995b). There is no evidence for any merging of the distributions of the pickup ions and the solar wind protons.

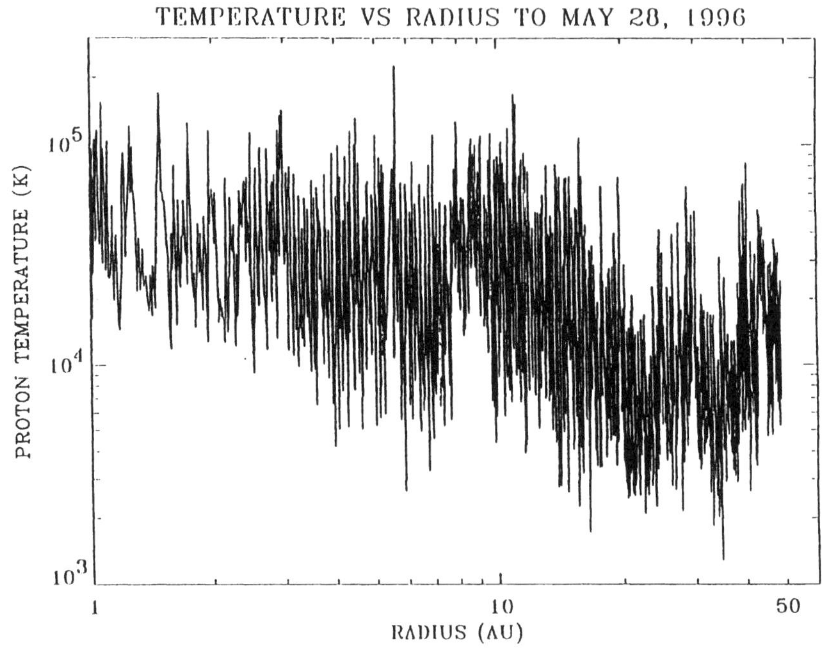

Figure 7. One day averages of proton temperature as measured by Voyager 2 from launch to May 28, 1996.

6. Solar Wind Ram Pressure

The ram pressure of the solar wind as measured by Voyager 2 from launch to May 28, 1996 is shown in Figure 8. The fall-off of ram pressure with distance from the sun is very close to being simply an inverse radius squared dependence, as expected. The line marked "Interstellar Pressure" is drawn at 2.24×10^{-12} dynes/cm^2. This pressure corresponds to the magnetic pressure associated with an interstellar magnetic field magnitude of 5 micro gauss, using the maximum enhancement factor possible due to draping (Belcher et al., 1993).

7. Summary

The Voyager spacecraft have been and continue to provide detailed in-situ information on the three-dimensional structure of the solar wind. Radial extrapolations of the position of the heliospheric current sheet continue to explain the polarity distribution of the heliospheric magnetic field in the distant heliosphere.

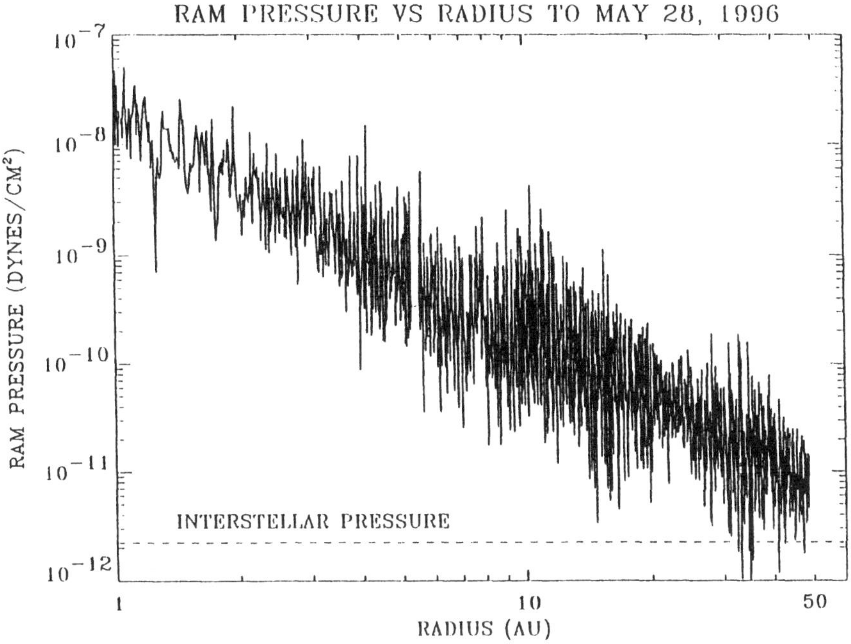

Figure 8. The ram pressure of the solar wind as measured by Voyager 2 from launch to May 28,1996.

The effects of the interstellar medium on the solar wind are now being observed by Voyager 2. The pressure of the interstellar pickup protons is greater than or equal to magnetic and thermal pressures of the solar wind beyond 30 AU. Thus, the interstellar medium can influence the dynamics of the solar wind sampled by Voyager 2. The interstellar pickup protons might also contribute to a small deceleration and a significant heating of the solar wind protons with increasing distance from the sun.

Acknowledgements

N.F. Ness acknowledges NASA support by Jet Propulsion Laboratory contract No. 959167. L.F. Burlaga and N.F. Ness also appreciate discussions of these results with many colleagues including but not limited to P.A. Isenberg, M.A. Lee and Y.C. Whang. J.W. Belcher, A.J. Lazarus and J.D. Richardson acknowledge NASA support by Jet Propulsion Laboratory contract No. 959203 and support by the NASA grant NAGW-1550.

References

Belcher, John W., Lazarus, Alan J., McNutt, Ralph L., Jr. and Gordon, George S., Jr.: 1993, 'Solar wind conditions in the outer heliosphere and the distance to the termination shock', *J. Geophys. Res.* **98**, pp. 15,177–15,183.

Burlaga, L.F., *Interplanetary Magnetohydrodynamics* Oxford University Press, New York.

Burlaga, L.F., and Ness, N.F.: 1996a, 'Magnetic fields in the distant heliosphere approaching solar minimum: Voyager 1 & 2 observations during 1994', *J. Geophys. Res.* **101**, pp. 13,473–13,481.

Burlaga, L.F., Ness, N.F., Belcher, J.W., Szabo, A., Isenberg, P.A., and Lee, M.A.: 1994, 'Pickup protons and pressure-balanced structures: Voyager 2 observations in merged interaction regions near 35 AU', *J. Geophys. Res.* **99**, pp. 21,511–21,524.

Burlaga, L.F., Ness, N.F., Belcher, J.W., Whang, Y.C.: 1996b, 'Pickup protons and pressure balanced structures from 39 to 43 AU', *J. Geophys. Res.* **101**, pp. 15,523–15,532.

Gloeckler, G., et al.: 1995, 'Acceleration of interstellar pickup ions in the disturbed solar wind obsrved on Ulysses', *J. Geophys. Res.* **99**, pp. 17,623–17,636.

Gloeckler, G.: 1996, 'The abundance of atomic ^{1}H, ^{4}He and ^{3}He in the local interstellar cloud from pickup ions observations with SWICS on Ulysses', *Space Sci. Rev.*, this issue.

Isenberg, P.A.: 1986, 'Evolution of interstellar pickup ions in the solar wind', *J. Geophys. Res.* **91**, pp. 9,965–9,972.

Richardson, J.D., Paularena, K.I., Belcher, J.W., and Lazarus, A.J.: 1994, 'Solar wind oscillations with a 1.3 year period', *Geophys. Res. Lett.* **21**, pp. 1559–1560.

Richardson, J.D., Paularena, K.I., Lazarus, A.J., and Belcher, J.W.: 1995a, 'Evidence for a solar wind slowdown in the outer heliosphere?' *Geophys. Res. Lett.* **22**, pp. 1169–1172.

Richardson, J.D., Belcher, J.W., Lazarus, A.J., Paularena, K.I., Gazis, P.R. and A. Barnes: 1995b, 'Plasmas in the outer heliosphere', *Solar Wind 8 Conference Proceedings*, Dana Point, California, June 1995.

Whang, Y.C., Burlaga, L.F., and Ness, N.F.: 1996, 'Pickup protons in the heliosphere', *Proceedings of the First ISSI Workshop on the Heliosphere in the Local Interstellar Medium, Space Science Rev.*, this issue.

Address for correspondence: Bartol Research Institute, University of Delaware, Newark, DE 19716

ORIGIN OF C^+ IONS IN THE HELIOSPHERE

J. GEISS
International Space Science Institute, Hallerstrasse 6, 3012 Bern, Switzerland
G. GLOECKLER
Department of Physics and Astronomy, University of Maryland, College Park MD 20742, USA
R. VON STEIGER*
Physikalisches Institut, University of Bern, Sidlerstrasse 5, 3012 Bern, Switzerland

Abstract. C^+ pickup ions were investigated with the SWICS instrument along the trajectory of Ulysses, covering a broad range of solar latitude and distance. Whereas nearly all the observed H^+, He^+, N^+, O^+ and Ne^+ pickup ions are created from the interstellar gas penetrating deep into the heliosphere, C^+ comes primarily from an "inner source" which is located at a solar distance below a few AU and extends over all heliospheric latitudes investigated up to now. We present evidence that the C^+ originates from carbon compounds evaporating from interstellar grains. This inner source also produces some O^+ and N^+ with estimated relative abundances of $C^+/O^+ \sim 1$ and $N^+/O^+ \sim 0.2$. However, the total amount of O^+ and N^+ produced by this inner source is only of the order of 10^{-3} as compared to the total production of O^+ and N^+ from the interstellar gas in the heliosphere, respectively. Thus the inner source does not significantly contribute to oxygen or nitrogen in the anomalous cosmic rays (ACR) but its contribution to ACR-carbon may not be negligible.

1. Introduction

Interplanetary space is filled with two types of ions which, under undisturbed conditions, can be readily distinguished by their velocity distribution functions: (1) Dominant are the regular solar wind ions observed in a characteristically narrow Mach angle that corresponds to Mach numbers of typically 10 to 20. These ions are of solar origin. (2) Next in abundance are the pickup ions which have broad "suprathermal" distribution functions that in the undisturbed solar wind show a sharp upper velocity limit at twice the solar wind speed (Möbius *et al.*, 1985; Gloeckler *et al.*, 1993). In free interplanetary space at solar distances of a few to several AU, their number densities are of the order of 10^{-3} relative to the number densities of the solar wind ions. All pickup ions identified up to now are of non-solar origin.

For presenting and discussing pickup ion results, it has become customary to relate $\mathbf{v}$, the velocity of an incoming ion to $\mathbf{v}_{SW}$, the solar wind velocity measured at that time. We thus introduce

$$\mathbf{W} = \frac{\mathbf{v}}{v_{SW}}, \tag{1}$$

the relative speed in the reference frame of the spacecraft, and

$$\mathbf{w} = \frac{\mathbf{v} - \mathbf{v}_{SW}}{v_{SW}}, \tag{2}$$

the relative velocity in the reference frame of the solar wind.

For identifying the sources of individual pickup ion species, three criteria are available:

* Now at International Space Science Institute

Space Science Reviews **78**: 43–52, 1996.

(1) *The velocity distribution:* Pickup ions are created from slowly moving neutrals ($W \ll 1$ or $\mathbf{w} \cong -\mathbf{v}_{SW}/v_{SW}$) by some combination of solar UV ionisation and charge exchange with solar wind ions. Prominent among the neutrals in interplanetary space are interstellar gas atoms, cometary constituents or molecules and atoms liberated from grains by evaporation or sputtering. The newly formed ions are immediately picked up by the electromagnetic field of the solar wind, and subsequent pitch-angle scattering would eventually give them a spherically symmetric, shell-like distribution in **w**. When the average direction of the magnetic field **B** is nearly perpendicular to $\mathbf{v}_{SW}$, i.e. for $r > 1$ AU and low heliocentric latitude, approximately spherically symmetric distributions may indeed be achieved (Gloeckler *et al.*, 1993; Gloeckler, 1996). On the other hand, at high latitudes where the **B** field is more radially oriented, the **w** distribution remains biased in the sunward direction, i.e. ions in the neighbourhood of $W = 2$ remain underrepresented (Gloeckler *et al.*, 1995; Gloeckler, 1996). In either case, the pickup ions are adiabatically cooled in the expanding solar wind. Thus when moving away from their source region, the w distribution of these ions is shrinking. For r^{-2} expansion and an adiabatic coefficient of 5/3, $\langle w^2 \rangle \sim r^{-4/3}$ is obtained.

(2) *The charge states:* Solar wind ions of heavy elements have high charge states resulting from the coronal temperature of $\sim 10^6$ K. On the other hand, the solar EUV spectrum and the cross sections for ionisation or charge exchange with solar wind particles are such that most of the pickup ions are produced with one charge only. In interplanetary space they are removed quickly from the regions of high EUV or solar wind flux, and therefore the overwhelming majority of pickup ions produced from neutral atoms or molecules outside magnetospheres remains singly charged. The only interstellar pickup ion with two or more charges identified sofar is He^{2+}, with a low relative abundance of $He^{2+}/He^{+} \sim 2 \times 10^{-4}$ (Gloeckler and Geiss, 1994; Gloeckler, 1996). Inside magnetospheres, where lifetimes are long and electron energies are high, ions produced from neutral atmospheres may carry two or more charges (Young *et al.*, 1977; Gloeckler *et al.*, 1985; McNutt *et al.*, 1981; Geiss *et al.*, 1992), and therefore, in the wake of planets, multiply charged ions of planetary origin may be found in the solar wind.

(3) *The spatial distribution in the heliosphere:* Ions produced from the interstellar gas have a very characteristic distribution in space, which is governed by the direction of the solar apex (cf. Rucinski and Bzowski, 1996), and the solar distance of maximum abundance is roughly proportional to the ionisation rate. On the other hand, the density of ions of local origin (e.g. comets, Jupiter) decrease with distance from the source (cf. Neugebauer *et al.*, 1987; Geiss *et al.*, 1994b; Ogilvie *et al.*, 1995).

Using the three criteria given above, the interstellar origin of several pickup ions has been established: H^+ (Gloeckler *et al.*, 1993), He^+ (Möbius *et al.*, 1985; Gloeckler *et al.*, 1993); He^{2+} (Gloeckler and Geiss, 1994; Gloeckler, 1996), N^+, O^+ and Ne^+ (Geiss *et al.*, 1994a). All these ions are produced from the neutral interstellar gas flowing through the heliosphere. Recently, Geiss *et al.* (1995) have presented evidence for interstellar grains as a source of pickup ions, in particular the C^+ ions that were found at all solar latitudes and all solar distances visited by Ulysses. In this paper we present further evidence for pickup ions produced from grains in interplanetary space and discuss the origin of these grains.

2. Experimental Results

The trajectory of Ulysses combined with the high mass/charge resolution and the extremely low background of the time-of-flight mass spectrometer SWICS (Gloeckler *et al.*, 1992) on board this ESA/NASA spacecraft is providing a unique global and three dimensional

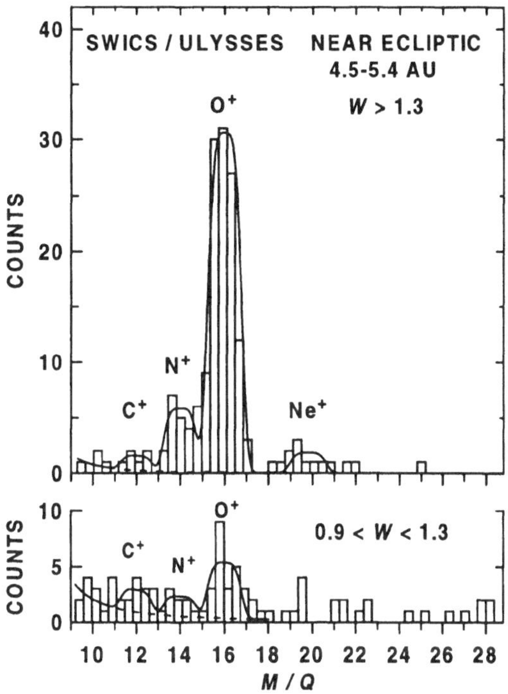

Figure 1. Mass/charge spectra in the slow solar wind near the ecliptic plane at 4.5 – 5.4 AU (139 days of data). The upper panel is for $W > 1.3$ and all azimuth directions, the lower panel is for $0.9 < W < 1.3$. Note that the upper velocity range covers more phase space than the lower one. In order to reduce residual counts from solar wind ions, one eighth of the spin period was omitted in the lower panel. No other background correction was made. Solid lines are fits to the peaks with widths proportional to M/Q. Background estimates are indicated by dashed lines. There is much less C^+ than O^+ at this solar distance, especially in the upper velocity range.

picture of densities and velocity distribution functions of the various pickup ion species. Figures 1, 4, and 6 show mass/charge (M/Q) spectra giving an overview of the occurrence of pickup ions in the M/Q range $\sim$ 9 to 30. During the time interval covered in Figure 1 Ulysses was in the slow solar wind. Since SWICS has an upper E/Q limit of 60 keV/e, corresponding to speed limits of 990 km/s and 860 km/s for C^+ and O^+, respectively, the entire velocity range of C^+ up to the limit $W = 2$ is covered in the slow solar wind most of the time, and for O^+ the velocity coverage is sufficient to allow a good extrapolation to $W = 2$ (cf. Geiss *et al.*, 1994a). O^+ can be readily recognized in both panels of Figure 1. The high abundance of O^+ in the velocity range $W > 1.3$ is in agreement with the copious production expected from ionisation of interstellar O in the vicinity of 5 AU (Geiss *et al.*, 1994a). C^+ is much less abundant at this solar distance. For $W > 1.3$ (upper panel) it cannot be clearly identified, and at lower velocity (lower panel) it can just be distinguished from the background. Geiss *et al.* (1995) have estimated total pickup ion number densities as a function of solar distance in the slow solar wind near the ecliptic plane (cf. Figure 2). From these densities we have calculated source strengths $q(r)$ for C^+ and O^+, using

$$q(r) = \frac{1}{r^2}\frac{\partial(r^2\phi)}{\partial r}, \qquad (3)$$

where $\phi(r)$ (= number density times solar wind speed) is the bulk flux of the pickup ion species. The result for $q(r)$ is shown in Figure 3. Although differentiating data points with relatively large statistical errors leads to even larger uncertainties, several conclusions may be drawn from Figure 3: (a) the source locations of C^+ and O^+ are very different, (b) within the uncertainties, all C^+ ions come from an inner source which has its maximum inside 3 AU, (c) there is also an inner source of O^+ with a strength similar to that of C^+, (d) for $r > 4$ AU, O^+ is dominantly interstellar, and (e) the interstellar O^+/He^+ ratio

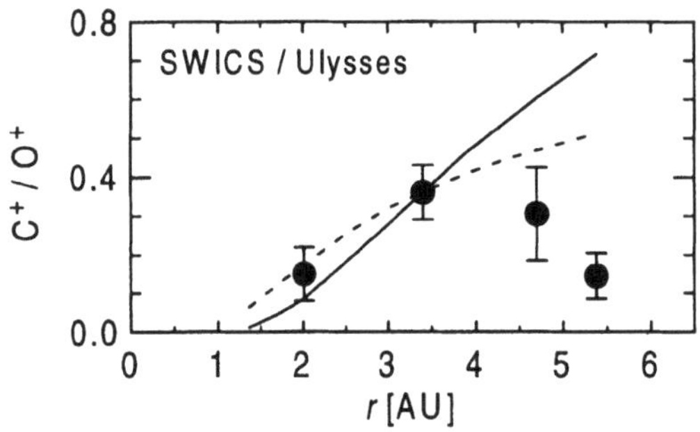

Figure 2. Ratio of the number densities of C^+ and O^+ measured near the ecliptic plane (solid circles), after Geiss *et al.* (1995). The theoretical curves show the radial dependency that would result if interstellar atomic C and O were the source of these ions. These curves were calculated for the Ulysses trajectory, using the following ionisation rates (in units of 10^{-7} s^{-1} at 1 AU): 6 for O^+ and for C^+ (dashed line), and 6 for O^+ and 12 for C^+ (solid line). The decrease of the C^+/O^+ ratio above $\sim$ 4 AU is incompatible with an interstellar gas source for C^+ and implies an inner source for this species.

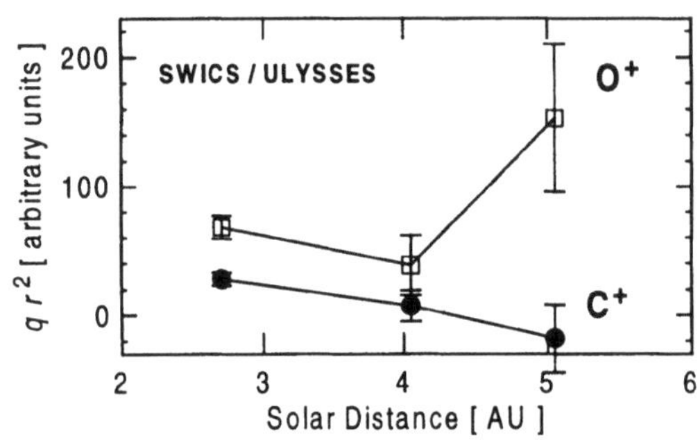

Figure 3. Location of the sources for O^+ and C^+. Given is qr^2 which was calculated from Eq. (3). Whereas $q(r)$ is compatible with the expected production of O^+ from interstellar atomic oxygen, C^+ comes from an inner source which has its maximum at $r < 3$ AU.

derived by Geiss *et al.* (1994a) is unaffected by this inner source because it was essentially based on $W > 1.3$ data taken at > 4.5 AU.

Figure 4 shows the pickup ion M/Q spectra in the high speed streams coming out of the southern and northern coronal holes, giving very similar relative abundances. The typical solar wind speed in these streams was $\sim$ 750 km/s, which corresponds to W limits of 1.33 and 1.15 for C^+ and O^+, respectively. Thus in these high speed streams, coverage of the pickup velocity spectrum is limited, and the comparison between two major W ranges that we show for the slow solar wind (Figure 1) cannot be made. Still, the observations of pickup C^+ and O^+, so clearly identifiable in Figure 4, are very important for characterising the source of C^+. This ion species is found at all the solar latitudes visited by Ulysses, showing that it is produced from an inner non-local source.

3. Discussion

The evidence presented in Figures 1 to 4 demonstrates that the observed C^+ ions are not derived from the interstellar gas, but from a source located inside of 3 AU. The persistence of C^+ over four years of observation and its occurrence at all latitudes is not compatible with one or a small number of local sources such as comets. The abundant occurrence of C^+ far away from the ecliptic plane speaks also against the asteroidal belt as the main source. Instead, the distribution in space and time of the C^+ ions points either to a relatively

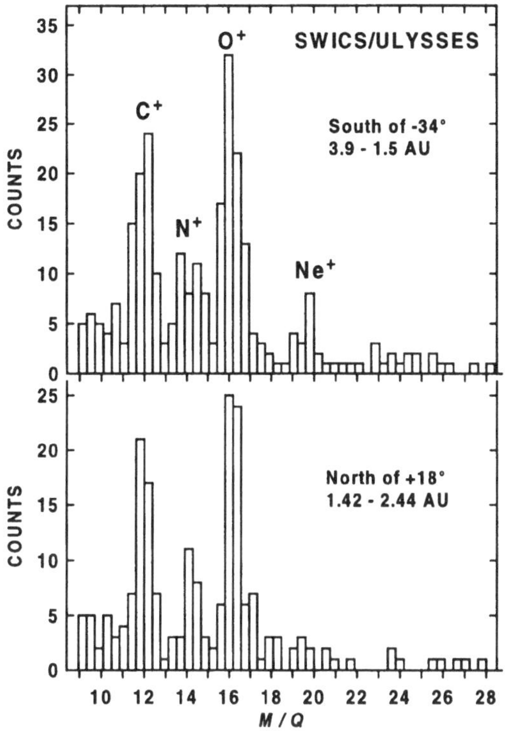

Figure 4. Mass/charge spectra of pickup ions with $W > 0.8$ in the high speed solar wind streams ($v_{SW} > 600$ km/s) originating in the polar coronal holes of the sun. Upper panel: southern high speed stream, 3.9 – 1.5 AU south of -34° (days 93.344 to 94.017). Lower panel: northern high speed stream, 1.42–2.44 AU, north of 18° (days 95.088.5–95.273.6). In order to reduce solar wind related background (e.g. accidental coincidences), ions with $W < 0.8$ were excluded. For the same reason, ions detected during the 12.5% of the spin period when the SWICS aperture included the solar direction were omitted. No further background corrections were made. The M/Q spectra are very similar, an indication that O^+, N^+ and C^+ are derived from extended sources.

large number of small comets ("mini-comets", cf. Brandt *et al.*, 1996) or to dust grains as the primary source. With their dust detector on Ulysses, Grün *et al.* (1993) and Baguhl *et al.* (1995, 1996) have measured the fluxes and arrival directions of dust grains as a function of solar latitude and distance, and they concluded that at solar distances > 3 AU and/or away from the ecliptic plane most of the grains with masses $> 2.5.10^{-14}$ g are of interstellar origin, for which they estimate a flux of 3×10^{-17} g m^{-2}s^{-1}. Carbon compounds evaporating from these grains seem to provide a good explanation for our C^+ observations. A few AU is a reasonable distance for loss of carbon compounds from which C^+ is then produced by dissociation/ionisation.

The calculation of partial and total densities from the observed counts involves uncertainties that are not small. They cancel, however, to a large extent, if abundance ratios are considered and if conclusions are drawn from radial trends. Thus we consider the radial trends presented in Figures 2 and 3 as convincing evidence for an inner gas source from which C^+ is produced from carbon compounds or atoms by solar UV and solar wind charge exchange.

In an attempt to distinguish between a finite number of local sources such as mini-comets and a truly distributed source such as free grains, we have studied the distribution of time intervals between individual pickup ion registrations. The result, shown in Figure 5, is that the events, both for all pickup ions with $M/Q > 10$ and for C^+ are compatible with an exponential distribution, i.e. the counts are Poisson distributed. Thus, within the statistical uncertainties we have no indication of a contribution of C^+ ions from point sources, or else we should observe deviations from the exponential distribution. Even if statistically superior data would be available, the case for mini-comets is difficult to falsify,

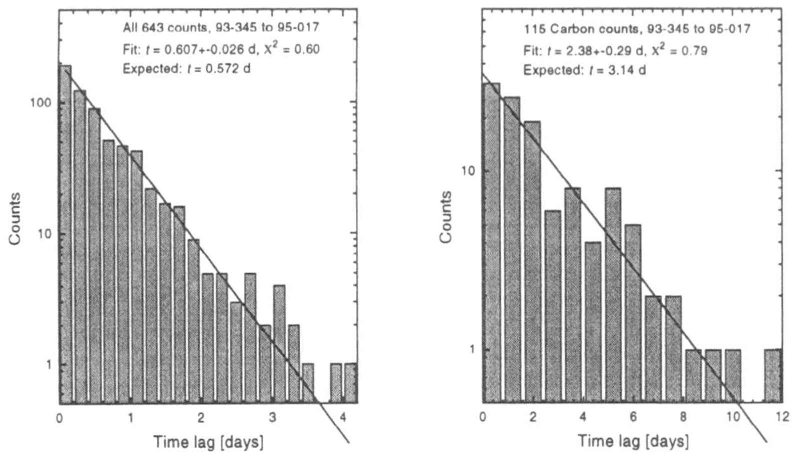

Figure 5. Distribution of time intervals between two adjacent pickup ions detected by SWICS (All ions with $M/Q > 10$ on the left, C^+ ions on the right). For both ion populations, the exponentials are shown that would result for Poisson distributed events, assuming constant a priori probabilities. Within the statistical uncertainties the distributions of events are in agreement with the Poisson distribution. Thus no evidence for one or a few local sources was found.

because of its largely hypothetical nature. We therefore conclude that the best and most consistent explanation of our results is that the C^+ ions observed outside the ecliptic plane are produced from interstellar grains.

For discussing the relative strengths of heliospheric ion sources, Geiss *et al.* (1995) introduced a solar ionisation cross section, for both the interstellar gas and grain sources. If grains are the source, evaporation of a volatile component, followed by dissociation and ionisation give an effective solar distance of ion formation, R_{eff}, which defines a geometric cross section

$$\sigma_{\text{ion}} = \pi R_{\text{eff}}^2 (1 + 2R_f/R_{\text{eff}}). \tag{4}$$

The term $(1+2R_f/R_{\text{eff}})$ allows for gravitational focussing. R_f, a characteristic "focussing distance" is given by $R_f = G(1-\mu)M_{\text{Sun}}/v_0^2$ (v_0 is the grain speed in the distant heliosphere). The factor $1-\mu$ allows for a reduction of focussing due to radiation pressure which is important for small grains. With $v_0 = 25$ km/s and neglecting radiation pressure, we have $R_f = 1.44$ AU.

From Figures 3 and 4, we can estimate only an upper limit for the solar distance of maximum C^+ production. Adopting $R_{\text{eff}} \cong 2.0$ gives $\sigma_{\text{ion}} \cong 30$ AU^2 for the inner C^+ source we have identified here.

Assuming $R_{\text{eff}} = 3$ AU and disregarding gravitational focussing, Geiss *et al.* (1995) estimated the C^+ flux at R_{eff} and obtained $\phi_{C^+}(R_{\text{eff}}) \sim 1 \times 10^{-18}$ g m^{-2}s^{-1}. Using their procedure we get the same value for $\phi_{C^+}(R_{\text{eff}})$, using $R_{\text{eff}} = 2$ and including gravitational focussing. A C^+ flux of 1×10^{-18} g m^{-2}s^{-1} corresponds to a few percent of the interstellar grain flux of 3×10^{-17} g m^{-2}s^{-1} (Grün *et al.*, 1993; Baguhl *et al.*, 1995, 1996). This is a reasonable percentage if a significant fraction of the carbon in the grains is lost at a few AU.

Figures 2, 3, and 4, indicate that inner sources also supply some O^+ and a small amount of N^+. In fact, comparison of our data with the predictions for the production of O^+ and N^+ from the interstellar gas confirms that we have an excess of these ions for

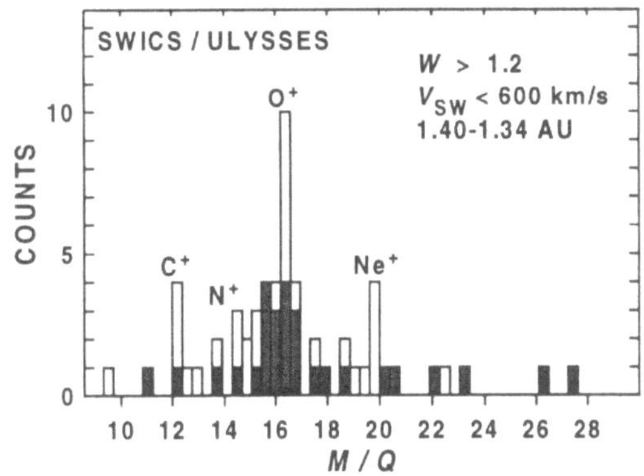

Figure 6. M/Q spectra of ions with $W > 1.2$ in the slow solar wind ($v_{SW} < 600$ km/s) during Ulysses' crossing of the ecliptic plane in the spring of 1995. The data were taken during 49 days when Ulysses travelled from $-22°$ to $+18°$ solar latitude (1.34 – 1.4 AU). The shaded areas represent ions observed near the solar equator, ($-8°$ to $+8°$ solar latitude). The O^+/C^+ ratio is increased in this equatorial region relative to higher latitudes.

solar distances below $\sim$ 3 AU. We can only give here a rough estimate of the relative abundances of C^+, N^+ and O^+ supplied by inner sources, because at high latitude, in the fast streams, SWICS covers only a limited part of the velocity distribution, and near the ecliptic plane in the slow solar wind, where the coverage is more complete, we have a somewhat higher background in the E/Q ranges where these pickup ions occur. Assuming that the radial source distributions for C^+ N^+ and O^+ are similar, we estimate $C^+/O^+ \cong 1$ and $N^+/O^+ \cong 0.2$ at high latitude. Near the ecliptic plane the O^+/C^+ ratio appears to be higher (cf. Figure 3). This is confirmed by the data of Figure 6 which gives the pickup ion counts obtained during Ulysses' crossing through the region of low heliographic latitude at 1.34 AU in the spring of 1995. In particular, the region from $-8°$ to $+8°$ gives a low C^+/O^+ ratio, an indication that near the ecliptic plane and inside $\sim$ 3 AU, grains of solar system origin might contribute to the production of O^+. (assuming that such grains produce a higher C^+/O^+ ratio than interstellar grains). In fact, Grün (1996) did find an increase in the number density of grains by a factor of $\sim$ 3 during Ulysses' ecliptic crossing in 1995 which may be due to grains of solar system origin.

4. Conclusions

The solar ionisation cross section for the interstellar gas source may be written as

$$\sigma_{\text{ion}} = \int \beta\, n\, r^2\, dr\, d\Omega / f_0. \tag{5}$$

$\beta(r) = \beta_e(R_e/r)^2$ is the local ionisation rate, β_e is this rate at 1 AU (R_e), n is the local number density of the neutral species, and $f_0 = n_0 v_0$ is the neutral flux at large distance. Assuming a spherical shape of the termination shock, Geiss *et al.* (1995) have evaluated this integral and obtained for the ion production cross section in the supersonic part of the heliosphere

$$\sigma_i = 4\pi R_{\text{ion}}\{R_{\text{sh}} - \kappa R_{\text{ion}}\}, \tag{6}$$

where R_{sh} is the distance to the termination shock, $R_{\text{ion}} = \beta_e R_e^2/v_0$ is a characteristic ionisation distance, and κ is a slowly varying function of $R_{\text{sh}}/R_{\text{ion}}$ that is obtained numerically. Equation (6) is valid for $R_{\text{sh}} > R_{\text{ion}} > R_f$ and thus applicable for C, N and O.

The cross section for the production of interstellar ion species, σ_{ion} (Eq. 6) increases linearly with R_{sh}, which explains the dominance of the interstellar gas as the source of

Table I

Estimates of total production, in tons per second, of C and O ions from major sources in the heliosphere (1 ton = 1000 kg).

	C [tons/s]	O [tons/s]
Solar wind ($C^{4+,5+,6+}$, $O^{6+,7+,8+}$)	1×10^4	2×10^4
Interstellar gas (from O^+ pickup ions)[a]		2×10^3
Interstellar grains, inner source (C^+, O^+)[b]	2	2
Interstellar gas (from ACR)[c]	10	2×10^3
Halley, outbound at 0.9 AU (C^+, O^+)[d]	2	15
Io-Torus, fast atoms (O^+)[e]		< 0.3
Mars, total O^+ release[f]		10^{-3}

References for data and theoretical work on which these estimates are based:
[a] Geiss *et al.* (1994a)
[b] this work
[c] Cummings and Stone (1990, 1995, 1996)
[d] Krankowsky *et al.* (1986), Eberhardt *et al.* (1987), Jessberger and Kissel (1991), Geiss (1987), Altwegg *et al.* (1993)
[e] Cheng (1986), Luhmann (1994), Geiss *et al.* (1994b), Ogilvie *et al.* (1995)
[f] Lundin *et al.* (1990)

heliospheric ions and of the anomalous cosmic rays (Fisk *et al.*, 1974; Cummings and Stone, 1990). Assuming $R_{sh} = 80$ AU (Cummings *et al.*, 1993; Gurnett, 1995) and $R_{ion} = 3.6$ AU (corresponding to $\beta_e = 6 \times 10^{-7}$ s^{-1}) numerical integration yields $\kappa = 5.8$, leading to $\sigma_{ion} = 2700$ AU^2 for the production of O^+ from interstellar atomic O, which is nearly two orders of magnitude larger than the cross section (Eq. 4) we have estimated for the inner source.

The cross sections given in Eqs. (4) and (6) allow us to compare the strengths of the interstellar gas source with grain sources and local sources. Estimates for some relevant sources are given in Table I. The production of C^+ and O^+ by Halley's comet was calculated assuming 100% of C and O in the gas and 80% of C and 50% of O in the grains were eventually transformed into C^+ and O^+ by evaporation, dissociation and ionisation. The ion production from the interstellar gas derived from ACR measurements were normalized to the O^+ production derived from the pickup ion measurements. With the exception of the O^+ loss from Mars, we have not included here ions leaking out of magnetospheres of planets. Their occurrence in space is essentially limited to the wake region of the planet, which can be avoided when studying ions from extended sources.

Table I indicates that the interstellar gas is the main source of O^+ ions in the heliosphere. The same ought to be true for H^+, He^+, N^+ and Ne^+, because these elements are strongly depleted in asteroids, comets and small planets. In the case of H^+ ions the separation from solar wind protons is obtained from the velocity distribution (Gloeckler *et al.*, 1993; Gloeckler, 1996).

Thus the measurements of the chemical composition, charge states and spatial distribution of pickup ions confirm that the source of hydrogen, helium, nitrogen, oxygen and neon in the anomalous cosmic rays is the interstellar gas (Fisk *et al.*, 1974). The situation with carbon is more complex. Atomic carbon (CI) is observed in the diffuse interstellar medium (Jenkins, 1987). Thus the gas in the local interstellar cloud (LIC) is probably an important source for ACR carbon. However, the abundance of C in the ACR is very low (Cummings and Stone, 1990, 1995, 1996), so that other sources, such as interstellar grains, may be significant (cf. the estimates for C sources from the ACR and from the inner source in Table I). Our measurements imply that a measurable fraction of the carbon in the local interstellar cloud is contained in grains. Thus the carbon in the LIC occurs in three forms: in grains, in the neutral gas, and in ionized form, and none of these forms seems to be negligible. This makes a determination of the degree of ionisation of the carbon in the gas phase of the LIC difficult.

The inner source of pickup ions located at a solar distance of a few AU could significantly contribute to the particles accelerated in corotating interaction regions (Geiss *et al.*, 1995). The C/O ratio of these particles is higher than the solar ratio (Gloeckler *et al.*, 1979; Fränz *et al.*, 1995). Since pickup ions are much more efficiently accelerated by shocks than solar wins ions (Gloeckler *et al.*, 1994) the relatively high abundance of pickup C^+ produced by the inner source could increase the C/O ratio of the CIR particles that are accelerated between 1 and a few AU.

Acknowledgements

The SWICS instrument was developed by a collaboration of the universities of Maryland, Bern and Braunschweig, and the Max-Planck-Institut für Aeronomie. The authors are indebted to Eberhard Grün of the MPK Heidelberg for discussions and to Christine Gloeckler of the University of Michigan for her assistance in the data reduction. This work was supported by grants of the Swiss National Science Foundation and the National Aeronautics and Space Administration of the United States/Jet Propulsion Laboratory contract No. 955460.

References

Altwegg, K., Balsiger, H., Geiss, J., Goldstein, R., Ip, W.-H., Meier, A., Neugebauer, M., Rosenbauer, H., and Shelley, E.: 1993, *Astron. Astrophys.* **279**, 260–266.
Baguhl, M., Grün, E., Hamilton, D. P., Linkert, G., Riemann, R., Staubach, P., and Zook, H. A.: 1995, *Space Sci. Rev.* **72**, 471–476.
Baguhl, M., Grün, E., and Landgraf, M.: 1996, *Space Sci. Rev.*, this issue.
Brandt, J. C., A'Hearn, M. F., Randall, C. E., Schleicher, D. G., Shoemaker, E. M., and Stewart, A. I. F.: 1996, *Earth, Moon, and Planets*, in press.
Cheng, A. F.: 1986, *J. Geophys. Res.* **91**, 4524–4530.
Cummings, A. C., and Stone, E. C.: 1990, *Proc. 21st Int. Cosmic Ray Conf.*, vol. 6, pp. 202–205.
Cummings, A. C., and Stone, E. C.: 1995, *Proc. 24th Int. Cosmic Ray Conf.*, vol. 4, pp. 497–500.
Cummings, A. C., and Stone, E. C.: 1996, *Space Sci. Rev.*, this issue.
Cummings, A. C., Stone, E. C., and Webber, W. R.: 1993, *J. Geophys. Res.* **98**, 15,165–15,168.
Eberhardt, P., Krankowski, D., Schulte, W., Dolder, U., Lämmerzahl, P., Berthelier, J. J., Woweries, J., Stubbemann, U., Hodges, R. R., Hoffmann, J. H., and Illiano, J. M.: 1987, *Astron. Astrophys.* **187**, 481–484.
Fisk, L. A., Kozlovski, B., and Ramaty, R.: 1974, *Astrophys. J.* **190**, L35–L37.
Fränz, M., Keppler, E., Krupp, N., Rouss, M. K., and Blake, J. B.: 1995, *Space Sci. Rev.* **72**, 339–342.
Geiss, J.: 1987, *Astron. Astrophys.* **187**, 859–866.
Geiss, J., Gloeckler, G., Balsiger, H., Fisk, L. A., Galvin, A. B., Gliem, F., Hamilton, D. C., Ipavich, F. M., Livi, S., Mall, U., Ogilvie, K. W., von Steiger, R., and Wilken, B.: 1992, *Science* **257**, 1535–1539.
Geiss, J., Gloeckler, G., Mall, U., von Steiger, R., Galvin, A. B., and Ogilvie, K. W.: 1994a, *Astron. Astrophys.* **282**, 924–933.

Geiss, J., Gloeckler, G., and Mall, U.: 1994b, *Astron. Astrophys.* **289**, 933–936.
Geiss, J., Gloeckler, G., Fisk, L. A., and von Steiger, R.: 1995, *J. Geophys. Res.* **100**, 23,373–23,377.
Gloeckler, G.: 1996, *Space Sci. Rev.*, this issue.
Gloeckler, G., and Geiss, J.: 1994, *Eos Trans. AGU* **75**(44), 512.
Gloeckler, G., Hovestadt, D., and Fisk, L. A.: 1979, *Astrophys. J.* **230**, L191–L195.
Gloeckler, G., Wilken, B., Stüdemann, W., Ipavich, F. M., Hovestadt, D., Hamilton, D. C., and Kremser, G.: 1985, *Geophys. Res. Lett.* **12**, 325–328.
Gloeckler, G., Geiss, J., Balsiger, H., Bedini, P., Cain, J. C., Fischer, J., Fisk, L. A., Galvin, A. B., Gliem, F., Hamilton, D. C., Hollweg, J. V., Ipavich, F. M., Joos, R., Livi, S., Lundgren, R., Mall, U., McKenzie, J. F., Ogilvie, K. W., Ottens, F., Rieck, W., Tums, E. O., von Steiger, R., Weiss, W., and Wilken, B.: 1992, *Astron. Astrophys. Suppl.* **92**, 267–289.
Gloeckler, G., Geiss, J., Balsiger, H., Fisk, L. A., Galvin, A. B., Ipavich, F. M., Ogilvie, K. W., von Steiger, R., and Wilken, B.: 1993, *Science* **261**, 70–73.
Gloeckler, G., Geiss, J., Roelof, E. C., Fisk, L. A., Ipavich, F. M., Ogilvie, K. W., Lanzerotti, L. J., von Steiger, R., and Wilken, B.: 1994, *J. Geophys. Res.* **99**, 17,637–17,643.
Gloeckler, G., Schwadron, N. A., Fisk, L. A., and Geiss, J.: 1995, *Geophys. Res. Lett.* **22**, 2665–2668.
Grün, E.: 1996, personal communication.
Grün, E., Zook, H. A., Baguhl, M., Balogh, A., Bame, S. J., Fechtig, H., Forsyth, R., Hanner, M. S., Horanyi, M., Kissel, J., *et al.*: 1993, *Nature* **362**, 428–430.
Gurnett, D. A.: 1995, *Space Sci. Rev.* **72**, 243–254.
Jenkins, E. B.: 1987, Hollenbach, D. J., and Thronson, H. A. (eds), *Interstellar Processes*. Dordrecht: Reidel, p. 375.
Jessberger, E. K., and Kissel, J.: 1991, Newburn, Jr. R. L., *et al.* (eds), *Comets in the Post-Halley Era*, vol. 2. Kluwer Academic Publishers, pp. 1075–1092.
Krankowski, D., Lämmerzahl, P., Herrwerth, I., Woweries, J., Eberhardt, P., Dolder, U., Herrmann, U., Schulte, W., Berthelier, J. J., Illiano, J. M., Hodges, R. R., and Hoffmann, J. H.: 1986, *Nature* **321**, 326–329.
Luhmann, J. G.: 1994, *J. Geophys. Res.* **99**, 13,285–13,305.
Lundin, R., Zakharov, A., Pellinen, R., Barabasj, S. W., Borg, H., Dubinin, E. M., Hultqvist, B., Koskinen, H., Liede, I., and Pissarenko, N.: 1990, *Geophys. Res. Lett.* **17**, 873–876.
McNutt, Jr. R. L., Belcher, J. W., and Bridge, H. S.: 1981, *J. Geophys. Res.* **86**, 8319–8342.
Möbius, E., Hovestadt, D., Klecker, B., Scholer, M., Gloeckler, G., and Ipavich, F. M.: 1985, *Nature* **318**, 426–429.
Neugebauer, M., Lazarus, A. J., Altwegg, K., Balsiger, H., Schwenn, R., Shelley, E. G., and Ungstrup, E.: 1987, *Astron. Astrophys.* **187**, 21–24.
Ogilvie, K. W., Gloeckler, G., and Geiss, J.: 1995, *Astron. Astrophys.* **299**, 925–928.
Rucinski, D., and Bzowski, M.: 1996, *Space Sci. Rev.*, this issue.
Young, D. T., Geiss, J., Balsiger, H., Eberhardt, P., Ghielmetti, A., and Rosenbauer, H.: 1977, *Geophys. Res. Lett.* **4**, 561–564.

RADIO EMISSIONS FROM THE OUTER HELIOSPHERE

D. A. GURNETT and W. S. KURTH
Dept. of Physics and Astronomy, The University of Iowa, Iowa City, IA, 52242, USA

Abstract. For nearly fifteen years the Voyager 1 and 2 spacecraft have been detecting an unusual radio emission in the outer heliosphere in the frequency range from about 2 to 3 kHz. Two major events have been observed, the first in 1983-84 and the second in 1992-93. In both cases the onset of the radio emission occurred about 400 days after a period of intense solar activity, the first in mid-July 1982, and the second in May-June 1991. These two periods of solar activity produced the two deepest cosmic ray Forbush decreases ever observed. Forbush decreases are indicative of a system of strong shocks and associated disturbances propagating outward through the heliosphere. The radio emission is believed to have been produced when this system of shocks and disturbances interacted with one of the outer boundaries of the heliosphere, most likely in the vicinity of the the heliopause. The emission is believed to be generated by the shock-driven Langmuir-wave mode conversion mechanism, which produces radiation at the plasma frequency (f_p) and at twice the plasma frequency ($2f_p$). From the 400-day travel time and the known speed of the shocks, the distance to the interaction region can be computed, and is estimated to be in the range from about 110 to 160 AU.

Key words: Heliospheric Radio Emissions, Heliosphere, Heliopause, Termination Shock

Abbreviations: PWS–Plasma Wave Subsystem, AU–Astronomical Unit, DSN–Deep Space Network, NASA–National Aeronautics and Space Administration, GMIR–Global Merged Interaction Region, MHD–Magnetohydrodynamic, CME–coronal mass ejection, f_p–plasma frequency, R–radial distance, AGC–automatic gain control

1. Introduction

The Voyagers 1 and 2 spacecraft both include an instrument called the Plasma Wave Subsystem (PWS) that can detect the electric field of plasma waves and low-frequency radio emissions. The Voyager plasma wave instrument was originally designed to study plasma waves in the magnetospheres of the outer planets (Scarf and Gurnett, 1977). However, because of an unexpected discovery, the plasma wave instrument is also providing important information on the structure of the outer heliosphere. On August 30, 1983, the PWS on Voyager 1 began detecting a weak radio emission in the frequency range around 3 kHz, slightly above the local solar wind plasma frequency (Kurth et al., 1984). At that time Voyager 1 was at a heliocentric radial distance of 17.9 Astronomical Units (AU). Subsequent investigations revealed that the radio emission could also be detected by Voyager 2, which was closer to the Sun, at a heliocentric radial distance of 12.7 AU. Since the same emission was being detected at two widely different locations, it was immediately recognized that the signal was most likely of heliospheric origin. However, it was difficult to rule out other potential sources, such as Jupiter. Subsequent observations over a period of nearly fifteen years now show that the radio emission is almost certainly generated in the outer regions of the heliosphere. The purpose of this paper is to summarize the current state of understanding of the heliospheric 2–3 kHz radio emissions. The presentation is organized into five sections. Section 2 discusses an overview of the radio spectrum observations, Section 3 discusses the relationship to intense solar events, Section 4 discusses the radio emission mechanism, Section 5 discusses the radial variation of the plasma frequency, and Section 6 discusses the interpretation of the spectrum.

Space Science Reviews **78**: 53–66, 1996.

2. Overview of the Radio Spectrum Observations

To give an overview of the observations, Figures 1 and 2 show spectrograms of the radio emission intensities detected by Voyagers 1 and 2 over a fourteen-year period, from January 1, 1982, to December 31, 1995. These spectrograms were produced using data from the PWS wideband receiver. The wideband receiver provides 4-bit samples of the electric field waveform at a sample rate of 28,800 s^{-1}. To produce the spectrograms, the waveform is Fourier transformed with a resolution of approximately 28 Hz. The color in each vertical line of the spectrogram represents a 15-second average of the intensities from the Fourier transforms. The time resolution is determined by the receiving capability of the NASA Deep Space Network (DSN) and varies from as low as one spectrum per month, to as high as two spectrums per week. The short vertical bars at the top of each spectrogram indicate when the individual spectrums were obtained. A color bar indicating the relative intensity is shown above each spectrogram. The dynamic range is 6 dB from the lowest intensity (blue) to the highest intensity (red). Relative intensity is shown because the wideband receiver utilizes an automatic gain control (AGC), which makes it difficult to determine absolute intensities.

Two periods of unusually intense radio emission activity can be seen in Figures 1 and 2; the first in 1983-84 (Kurth et al., 1984) and the second in 1992-93 (Gurnett et al., 1993). In addition, several somewhat weaker events have been observed, one in late 1985, one in 1989, three in 1990-91 (Kurth et al., 1987; Kurth and Gurnett, 1991), and several immediately after the 1992-93 event. By comparing the two spectrograms, it is evident that the radio emission spectrums detected by the two spacecraft are nearly identical, particularly during the later years when the spacecraft are farther from the Sun. During the 1983-84 event, one can see that the spectrum extends to lower frequencies at Voyager 1 than at Voyager 2. This difference in the spectrum is believed to be due to the propagation cutoff at the local electron plasma frequency, $f_p = 9\sqrt{n}$ kHz, where n is the electron number density in cm^{-3}. Since the electron density in the solar wind varies as $1/R^2$, where R is the radial distance from the Sun, the solar wind plasma frequency is expected to vary as 1/R. Based on the plasma measurements of Belcher et al. (1993), the plasma frequency at Voyager 2, which was at R $\simeq$ 13 AU, has an average value of 1.6 kHz, with peaks extending up to about 3.7 kHz. In contrast, the plasma frequency at Voyager 1, which was farther from the Sun (R $\simeq$ 18 AU), is estimated to have an average value of only 1.1 kHz, with peaks extending up to about 2.5 kHz. Thus, the radiation could reach Voyager 1 at a lower frequencies than for Voyager 2, which explains the difference in the low-frequency cutoffs for the two events.

Simple inspection of the spectrums during both the 1983-84 and 1992-93 events shows that the radio emission has two components, the first consisting of narrow bands that drift upwards in frequency from about 2.0 to 3.6 kHz at rates that vary from about 1 to 3 kHz/year, and the second consisting of a band of emission around 2 kHz that persists for time scales on the order of one year or more. The onset of an upward drifting band is often associated with an intensification of the 2-kHz band. Diffuse structures can also be seen in the 2-kHz band, sometimes drifting slowly upward in frequency. During the 1992-95 period, the 2-kHz band has a very sharp low-frequency cutoff at 1.8 kHz. This cutoff cannot be due to a local propagation cutoff, since the 1/R dependence shows that the local plasma frequency during the 1992-95 period should be only a few hundred Hz. It is also evident that the 1992-93 event is much more intense than the 1983-84 event, particularly for the 2-kHz band. The higher intensities during the 1992-93 event are attributed to the lower plasma frequencies compared to the 1983-84 event, which allows a broader range of frequencies to reach the spacecraft. By using absolute intensity measurements from the

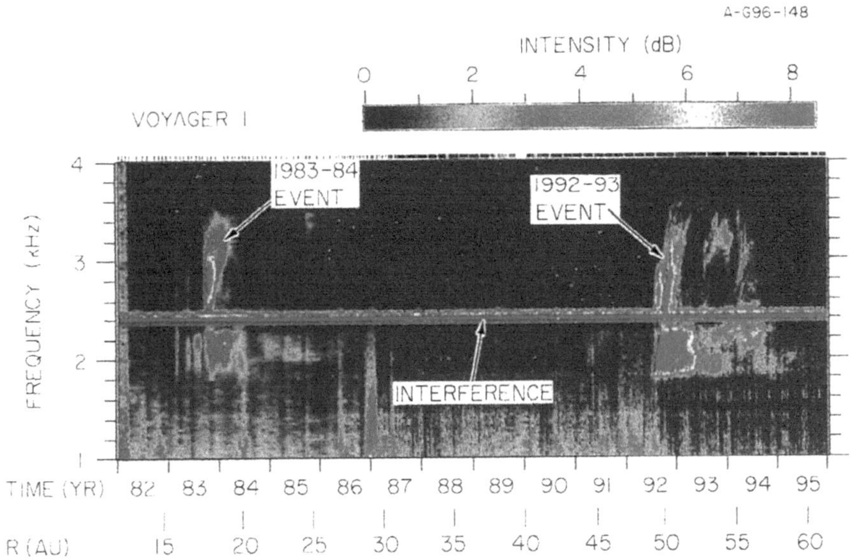

Figure 1. A 14-year frequency-time spectrogram showing the heliospheric radio emissions detected by Voyager 1. Two unusually strong events have been observed, the first in 1983-84 and the second in 1992-93, as well as a number of weaker events. The strong line at 2.4 kHz is interference from the spacecraft power system.

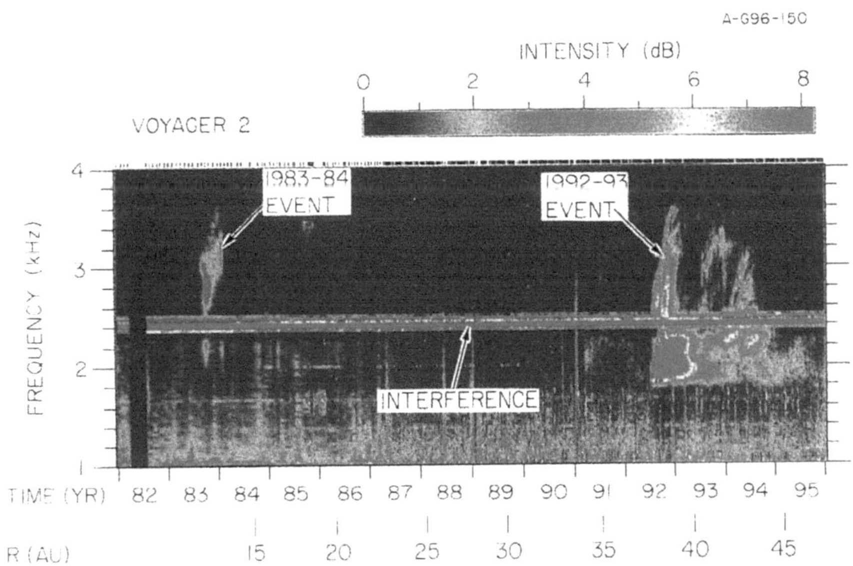

Figure 2. A 14-year spectrogram from Voyager 2 similar to Figure 1. Although separated by distances as large as 50 AU, the radio emission spectrums from the two spacecraft are remarkably similar, which shows that the source is at a considerable distance from the spacecraft, probably more than 50 AU.

PWS on-board spectrum analyzer, which has discrete channels at 1.78 and 3.11 kHz, Gurnett et al. (1993) estimated that the total radiated power for the 1992-93 event was at least 10^{13} Watts. This power level ranks the heliospheric 2–3 kHz radio emissions as one of the strongest radio sources in the solar system, considerably more intense than the strongest planetary radio sources, and comparable to the most intense solar radio bursts.

3. Relationship to Intense Solar Events

Since the heliospheric 2–3 kHz radio emission clearly occurs in distinct bursts, typically lasting a fraction of a year or longer, the question naturally arises as to what "triggers" these bursts. McNutt (1988) was the first to suggest that solar wind disturbances caused by transient activity at the Sun could trigger a burst of 2–3 kHz radio emission. In particular, he proposed that the 1983-84 radio emission event was produced by the interaction of a high-speed solar wind stream with the termination shock. This idea was further explored by Grzedzielski and Lazarus (1993), who identified a series of dynamic pressure increases in the solar wind that they believed were responsible for the 1983-84, 1985, and 1989 events, again assuming an interaction with the termination shock. Despite the merit of these ideas, the time delay between candidate solar wind transients and the onset of the radio bursts varied over such a wide range that the hypothesized cause-effect relationship was not convincing.

With the onset of the intense 1992-93 radio emission event, the evidence for a solar wind trigger improved significantly. This event was so intense that one would expect that it should be associated with an extraordinary solar event. Indeed, an extraordinary event was soon found, namely, the great cosmic ray Forbush decrease of 1991 (Gurnett et al., 1993), which occurred about 400 days before the onset of the radio emission. In the process of reviewing the earlier data, it was soon discovered that the intense 1983-84 radio emission event was also preceded in 1982 by an extraordinarily large Forbush decrease, again about 400 days before the onset of the radio emission. These relationships are illustrated in Figure 3, which shows the cosmic ray intensity from the Deep River neutron monitor in the top panel and the radio emission intensity from the 3.1-kHz channel of the Voyager 1 spectrum analyzer in the bottom panel. The two large Forbush decreases, labelled A and B, were produced by periods of intense solar activity in mid-July 1982 and May-June 1991, during the declining phases of solar cycles number 21 and 22. These two Forbush decreases are the two deepest Forbush decreases ever observed (21% and 30%, respectively). As can be seen, the 1983-84 radio emission event (labelled A′) started about 412 days after Forbush decrease A, and the 1992-93 radio emission event (labelled B′) started about 419 days after Forbush decrease B.

The current view is that large Forbush decreases, such as events A and B, are caused by a series of outward propagating solar wind disturbances that merge in the outer heliosphere to form a shell of compressed plasma and magnetic field called a Global Merged Interaction Region (GMIR). For a discussion of GMIRs, see Burlaga et al. (1993) and McDonald and Burlaga (1996). GMIRs are usually preceded by a strong leading shock that is formed by the coalescence of several shocks, each of which originates from a specific event at the Sun. The strong leading shock is typically followed by a region of turbulent plasma with numerous magnetohydrodynamic (MHD) discontinuities. Other shocks may also be imbedded within the GMIR. The strong turbulent magnetic fields in the GMIR scatter and impede the transmission of cosmic rays, thereby causing the transient cosmic ray intensity decreases known as Forbush decreases.

The Forbush decreases associated with events A and B were observed by a number of interplanetary spacecraft (Van Allen and Randall, 1985; Webber et al., 1986; Cliver et al.,

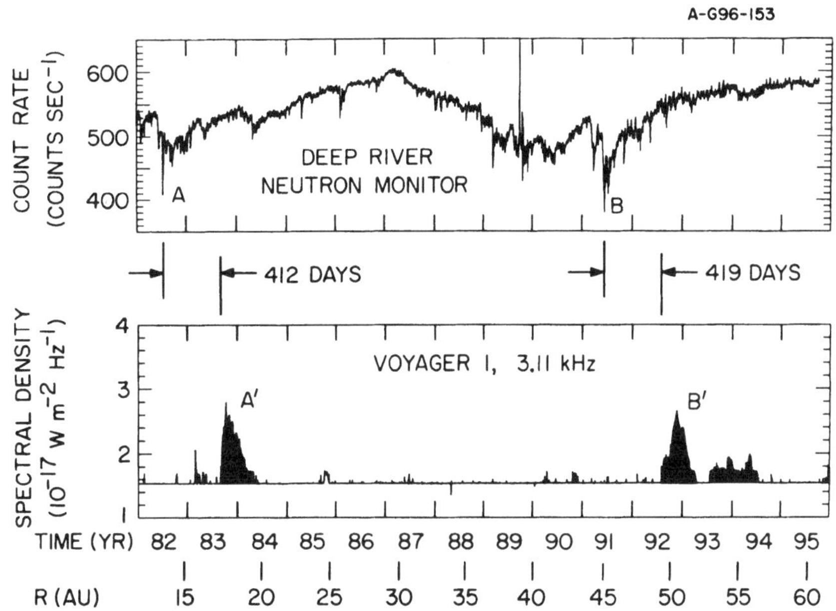

Figure 3. A 14-year plot showing the radio emission intensity at 3.11 kHz from Voyager 1, and the corresponding cosmic ray intensity from the Deep River neutron monitor. The Forbush decreases marked A and B are the two deepest depressions ever observed. The two strong heliospheric radio emission events marked A′ and B′, occurred about 400 days after these large Forbush decreases.

1987; Van Allen and Fillius, 1992; Webber and Lockwood, 1993; McDonald et al., 1994). Since both events are rather similar, we focus our attention on event B, which has the best coverage at large heliocentric distances. The top panel of Figure 4 shows an expanded plot of the count rate from the Deep River neutron monitor around the time of event B. The next four panels show the cosmic ray intensities from Pioneers 10 and 11, and from Voyagers 1 and 2. Clearly defined Forbush decreases can be seen at all four spacecraft, first by Pioneer 10 on day 233 at 34 AU, then by Voyager 2 on day 250 at 35 AU, then by Voyager 1 on day 257 at 46 AU, and finally by Pioneer 10 on day 273 at 53 AU. A strong leading shock was also observed coincident with these Forbush decreases by the Pioneer 10 magnetometer on day 232 (Winterhalter et al., 1992), by the Voyager 2 plasma instrument on day 251 (Belcher et al., 1993), and by the Voyager 1 plasma wave instrument on day 257 (Kurth and Gurnett, 1993).

Two periods of solar activity potentially contributed to the interplanetary shocks and associated cosmic ray decreases illustrated in Figure 4. The first, labelled the March solar events, consisted of 35 solar flares of classification M-5, or higher, that occurred in the southern hemisphere of the Sun from March 1 through March 31, 1991 (McDonald et al., 1994). These events caused the well-defined Forbush decrease in the Deep River neutron monitor on day 83 (March 24, 1991). The second, labelled the May-June solar events, consisted of an even more intense period of solar activity in the northern hemisphere about 6 weeks later, from May 25 to June 15, 1991. During this period, a total of 70 M-class and 6 X-class solar flares were observed (McDonald et al., 1994). This extraordinarily intense activity produced a sharp decline in the Deep River neutron monitor, causing the deep (30%) Forbush decrease labelled event B, which occurred on day 163 (June 12), 1991.

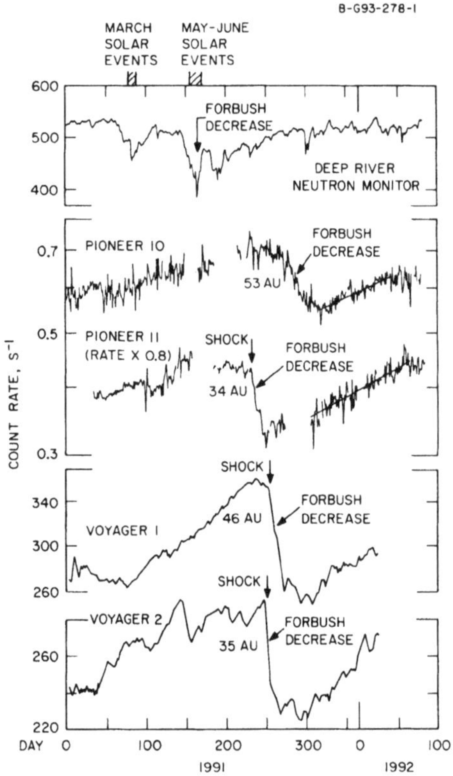

Figure 4. The cosmic ray intensities from Pioneers 10 and 11, and Voyagers 1 and 2, showing the large Forbush decreases produced by the May-June 1991 solar activity. This illustration was adapted from Van Allen and Fillius (1992) and Webber and Lockwood (1993).

The radial propagation velocity that one infers from these observations depends on which spacecraft are used and on which period of solar activity is assumed to be the causative event. Van Allen and Fillius (1992) and Webber and Lockwood (1993) both conclude that the May-June solar activity was primarily responsible for the cosmic ray decreases observed by Pioneers 10 and 11 and Voyagers 1 and 2. Initially, Van Allen and Fillius estimated the propagation speed of this event to be 820 km/s, but later, based on a more extensive analysis, Van Allen (1993) revised the best fit propagation speed to 865 $\pm$ 75 km/s. Webber and Lockwood quote two speeds, 580 and 740 km/s, depending on whether the time delay is computed between Earth and Voyager 2 or Earth and Voyager 1. McDonald et al. (1994), on the other hand, assumed that the March solar activity was responsible for the Forbush decreases observed in the outer heliosphere. Their estimate of the propagation speed was somewhat lower, 572 km/s.

In order to visualize how these events relate to the 1992-93 heliospheric radio emission event, it is useful to show all of these events on the same time scale. Such a plot is given in Figure 5 (from Gurnett and Kurth (1995) who also show a similar plot for event A). The top panel shows the count rate from the Deep River neutron monitor, and the bottom two panels show the radio emission intensity from the 1.78- and 3.11-kHz channels of the Voyager 1 spectrum analyzer. The times of the March and May-June solar events, and the

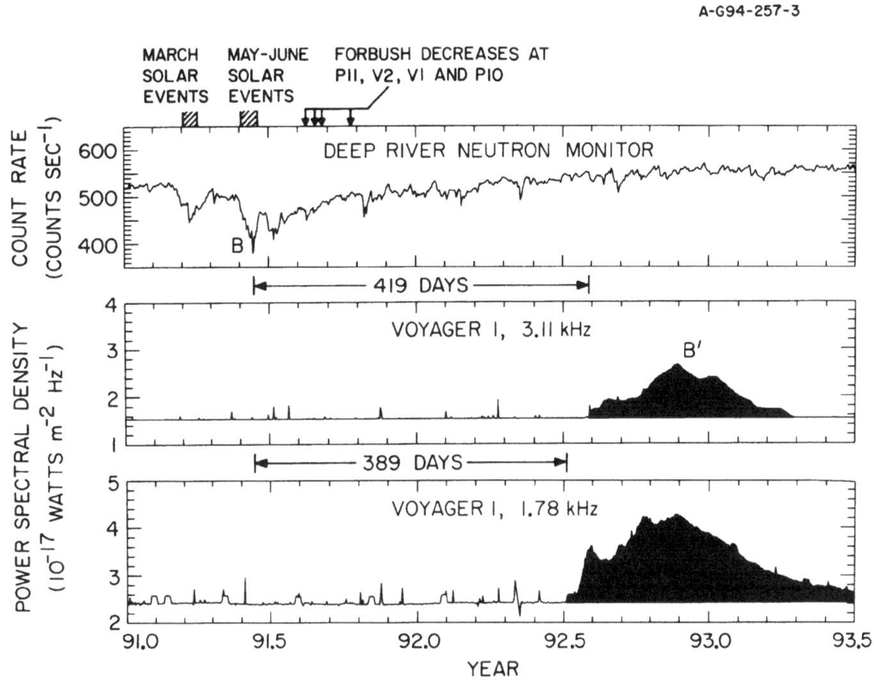

Figure 5. An expanded time scale plot showing further details of events B and B′. The deep Forbush decrease in 1991 was caused by two periods of intense solar activity, the first in March, and the second in May-June. The times at which this Forbush decrease was detected by Pioneer 11, Voyager 2, Voyager 1, and Pioneer 10 (see Figure 4) are shown at the top of the plot. The bottom two panels show the radio emission intensities at 1.78 and 3.11 kHz.

onset times of the Forbush decreases at Pioneers 10 and 11, and Voyagers 1 and 2 are shown at the top of the plot. As one can see, the onset time of the radio emission event in the 1.78-kHz channel occurred 389 days after Forbush decrease B, just a few days before the onset in the 3.11-kHz channel. If the propagation speed is assumed to be independent of distance from the Sun, then the radial distance to the radio emission source, R_S, can be computed by using the simple proportion, $R/\Delta t = R_S/389$, where R is the radial distance to the spacecraft, and Δt is the time delay from the event on the Sun to the onset of the Forbush decrease at that spacecraft. Using this formula, the distances to the source using the onset times for the Forbush decreases at Pioneers 10 and 11 and Voyagers 1 and 2, work out to be 187, 188, 190, and 156 AU. This simple calculation shows that the radio emission source is at a considerable distance from the Sun, well beyond 100 AU. Of course, the propagation speed most likely decreases in the outer regions of the heliosphere. The effect of the decreasing propagation speed with increasing radial distance has been studied by several investigators, including Gurnett et al. (1993), Steinolfson and Gurnett (1995), Gurnett and Kurth (1995), McNutt et al. (1995), and Liewer et al. (1996). These studies show that the distance to the source is most likely in the range from about 110 to 160 AU. Similar results have been obtained for the 1983-84 radio emission event (see Gurnett and Kurth (1995)).

4. Radio Emission Mechanism

Before one can attempt to interpret the radio emission spectrum, it is first necessary to understand the mechanism by which the radio emission is generated. Most of the well-known astrophysical radio emission mechanisms, such as synchrotron radiation and cyclotron maser radiation, require relatively strong magnetic fields. However, the magnetic field in the outer heliosphere is believed to be extremely weak, 1 nT or less (Axford, 1990). In the weak magnetic field regime only one mechanism is known that could account for the heliospheric radio emissions, namely, the shock-driven Langmuir-wave mode conversion mechanism (Cairns and Gurnett, 1992).

To explain how this radio emission mechanism works, it is useful to discuss type II solar radio bursts, which are believed to be generated by the same mechanism. Type II solar radio bursts are produced by coronal mass ejections (CMEs), which are transient ejections of material from the Sun. CMEs are typically accompanied by solar flares and other forms of solar activity (Gosling, 1993). As the ejected material from the CME moves outward from the Sun, it acts as a piston, which drives a shock ahead of the CME. Strong electric fields within the shock accelerate a beam of electrons outward along magnetic field lines ahead of the shock as shown in Figure 6. This beam generates electrostatic oscillations called Langmuir waves, which in turn decay into electromagnetic radiation via a nonlinear mode coupling process. Typically electromagnetic radiation is generated at two frequencies: the plasma frequency, f_p, and twice the plasma frequency, $2f_p$. The radiation at f_p is called the fundamental, and the radiation at $2f_p$ is called the harmonic. The decrease in the plasma frequency with increasing distance from the Sun causes the emission frequency to decrease as the shock propagates outward from the Sun, thereby producing the characteristic signature of a type II radio burst, which is an emission frequency that decreases with increasing time. The reason that the radiation is generated at both the fundamental and the harmonic is complicated, and has been the subject of numerous theoretical analyses. For a review of the mechanisms involved, see Melrose (1985).

5. Radial Variation of the Plasma Frequency

Since strong shocks are known to be propagating through the outer regions of the heliosphere in response to both the July 1982, and May-June 1991 events, it is clear that the proper conditions are present for generating the heliospheric radio emissions via the shock-driven Langmuir-wave mode conversion mechanism. The basic mechanism would then be similar to a type II solar radio burst. However, there are several important differences. Whereas the intensity of a type II radio burst varies in a smooth and continuous manner as the shock propagates outward from the Sun, the heliospheric radio emission events all have sudden onsets. Since no radio emission was detected as the shock propagated past the spacecraft, the sudden onset strongly suggests that the shock has encountered a plasma with significantly different properties, such as would occur at one of the outer boundaries of the heliosphere. There are two such boundaries, the termination shock and the heliopause. To decide which boundary is involved, we must consider the radial variation of the plasma frequency, since the emission must be generated at either f_p or $2f_p$.

As discussed earlier, in the region of supersonic flow the solar wind plasma frequency varies inversely with distance from the Sun. On a log-log plot, the f_p versus R profile is then a straight line with a slope of minus one, as shown in Figure 7. Given the large distances to the interaction region implied by the delay times and shock propagation speeds discussed earlier, it is reasonable to assume that the termination shock is located at about 100 AU.

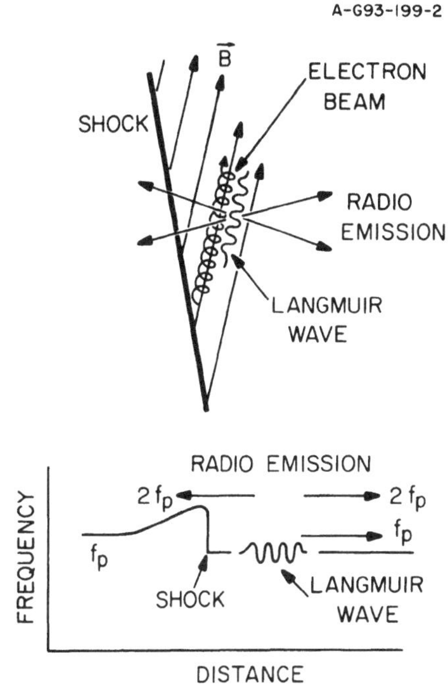

Figure 6. An illustration showing the generation of radio emission by an interplanetary shock via the Langmuir-wave mode conversion mechanism. An electron beam accelerated at the shock excites Langmuir waves, which in turn decay into electromagnetic radiation at the plasma frequency (f_p) and at twice the plasma frequency ($2f_p$).

This location is consistent with other current estimates of the distance to the termination shock (Axford, 1996). Since the average value of f_p at 1 AU is about 20 kHz, the average plasma frequency at 100 AU, just ahead of the termination shock, should be about 200 Hz. According to conventional MHD theory (Anderson, 1963), for a strong shock, which is what is expected at the termination shock, the plasma density should increase by a factor of four. Since a factor of four jump in the plasma density corresponds to a factor of two jump in the plasma frequency, the plasma frequency in the region immediately downstream of the termination shock should be about 400 Hz. Beyond the termination shock the plasma density is expected to remain roughly constant out to the heliopause. Plasma flow simulations (Baranov and Malama, 1993; Washimi, 1993; Steinolfson et al., 1994) show that the heliopause should be located at a heliocentric distance that is about 130 to 150% of the distance to the termination shock. In Figure 7 the heliopause is therefore assumed to be at a radial distance of 140 AU. At the heliopause, a second upward jump in the plasma density is expected to occur. Since the heliopause is a contact discontinuity, the density increases by whatever factor is necessary to maintain pressure balance with the interstellar medium. Present estimates are that the electron density in the interstellar medium near the Earth is in the range from about 0.06 to 0.1 cm^{-3} (Lallement et al., 1993), which corresponds to a plasma frequency of about 2.2 to 2.8 kHz. Since the heliospheric bow shock is expected to be a weak shock, or may not exist at all, the plasma density immediately beyond the heliopause is expected to be very similar to the plasma density in the interstellar medium.

A number of the early papers dealing with the origin of the heliospheric 2–3 kHz radio emission suggested that the radiation is generated upstream of the termination shock (Kurth

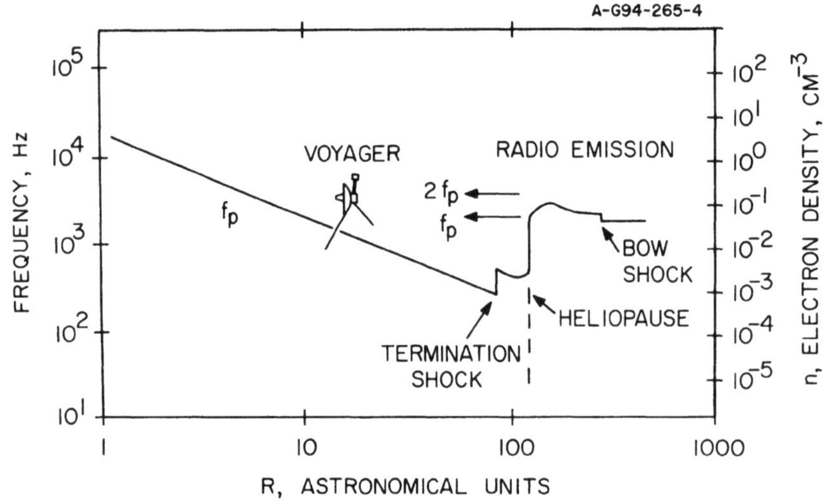

Figure 7. A model showing the plasma frequency (f_p) as a function of the radial distance (R) from the Sun. In the supersonic solar wind the plasma frequency varies inversely with the radial distance from the Sun. At the termination shock, according to conventional MHD theory, the plasma frequency is expected to increase by a factor of two. At the heliopause, the plasma frequency increases to whatever level is necessary to maintain pressure balance with the interstellar medium.

et al., 1984; McNutt, 1988; Grzedzielski and Lazarus, 1993). From simple inspection of Figure 7, one can see that the plasma frequency upstream of the termination shock (~200 Hz) is much too low to account for the 2–3 kHz radio emission, even if the emission is generated at $2f_p$. At the heliopause, the situation is much better. Since the plasma frequency on the interstellar side of the heliopause is most likely around a few kHz, there should be no problem generating frequencies of 2–3 kHz in the vicinity of the heliopause. For this reason, Gurnett et al. (1993) proposed that the heliospheric 2–3 kHz radio emission is generated in the vicinity of the heliopause.

Although the above model shows that the average plasma density in the region between the termination shock and the heliopause is too small to account for radio emission at 2–3 kHz, a number of ideas have been suggested that could raise the emission frequency in this region. For example, Cairns et al. (1992) have pointed out that the solar wind plasma density is often strongly enhanced by transient events at the Sun. Once these density enhancements enter the region beyond the termination shock, they would not only raise the local emission frequency, but they could also potentially block the escape of radiation from regions of lower density, thereby raising the average emission frequency. In a variation of this basic idea, Zank et al. (1994) suggested that the radio emission is produced by the interaction of an interplanetary shock with density enhancements that were previously injected into the region downstream of the termination shock. Donohue and Zank (1993) have also suggested that anomalous cosmic ray pressure could alter the structure of the termination shock in such a way that the density jump exceeds the factor of four limit imposed by the usual MHD jump conditions. A larger density jump at the termination shock would raise the plasma density throughout the region downstream of the shock, thereby raising the emission frequency. Whether any of these mechanisms prove to be viable remains to be determined.

6. Interpretation of the Radio Emission Spectrum

Since the emission frequency is controlled by the plasma density, the radio emission spectrum almost certainly contains valuable information on the structure of the outer heliosphere. Unfortunately, the spectrum is also quite difficult to interpret. At least three features must be explained: (1) the narrow bands that drift upward in frequency from about 2.0 to 3.6 kHz, (2) the persistent nearly constant frequency band at about 2 kHz, and (3) the sharp low-frequency cutoff at 1.8 kHz. If the radio emission is produced by an outward propagating shock, the upward drifting bands strongly suggest that the shock is propagating through a ramp of increasing plasma density. In this interpretation each upward drifting band would be caused by a different shock propagating through the same density ramp. The relatively small spread in the observed drift rates, from about 1 to 3 kHz/year, could be easily caused by variations in the shock speeds. Using this basic idea, Gurnett et al. (1993) suggested that the upward drifting bands were produced by shocks propagating through the density pileup region that is expected to exist near the nose of the heliosphere (Steinolfson et al., 1994). In a variation of this basic mechanism, Czechowski et al. (1995) have attributed the upward drifting bands to shocks propagating through a layer of cold high-density plasma that is formed by charge exchange cooling just inside the heliopause (also see Czechowski and Grzedzielski (1995)). Zank et al. (1994) have also attributed the upward drifting bands to a shock propagating through a density ramp. However, in their model the density ramp is caused by a transient structure in the region between the termination shock and the heliopause. Suitable transient density structures could be formed either by convection of a solar wind density enhancement through the termination shock, or by another slower moving shock.

During spacecraft roll maneuvers, a well-defined intensity modulation has been observed in the upward drifting bands at 3.11 kHz (Kurth, 1988; Gurnett et al., 1995), sometimes with intensity modulation factors as large as 60%. (Due to a data system failure on Voyager 2, roll measurements can only be made on Voyager 1). Although these one-dimensional roll modulation measurements cannot provide a unique direction to the source, the large modulation factors indicate that the angular size of the source must be relatively small. The exact size is difficult to estimate due to the unknown direction of arrival relative to the roll axis. A small source size is consistent with the relatively narrow bandwidths of the upward drifting bands, which indicate that the radio emission is generated in small isolated regions. The most likely explanation for a small isolated source is the magnetic field orientation relative to the shock front. From studies of the Earth's bow shock, it is known that the upstream electron beam is most intense for quasi-perpendicular shocks (Anderson et al., 1979). Since higher electron beam intensities most likely generate higher Langmuir wave intensities, the highest radio emission intensities are expected from regions where the magnetic field is nearly perpendicular to the shock normal. This type of magnetic control is believed to be the reason that type II radio emissions are not generated uniformly over the shock front, but rather originate from small isolated regions (Nelson and Robinson, 1975).

The current view regarding the origin of the 2-kHz band is that this radiation is also produced by a shock wave or a system of shock waves interacting with the outer heliosphere. To maintain the nearly constant emission frequency, the plasma frequency must be nearly constant throughout the interaction region. In contrast to the upward drifting bands, no roll modulation has ever been observed in the 2-kHz band. The absence of roll modulation most likely means that the source subtends a very large solid angle (i.e., approaching 4π). These observations support the view that the radiation is generated by a quasi-spherical shock or system of shocks propagating through a region of nearly constant plasma density. Two candidate source regions are immediately evident: (1) the region between the termination shock and the heliopause, and (2) the region beyond the heliopause. Pursuing the first possibility, Zank et al. (1994) and Whang and Burlaga (1995) both proposed that the 2-kHz band is produced in the region between the termination shock and the heliopause. In the model of Zank et al., the shock is propagating outward from the Sun, whereas in the model of Whang and Burlaga the shock is propagating inward toward the Sun (after reflected from the heliopause). These models both suffer from the previously mentioned difficulty that the plasma frequency in the region between the termination shock and the heliopause is probably too low to account for the observed emission frequency. However, if the density jump at the termination shock exceeds the MHD limit by a substantial factor, as suggested by Donohue and Zank (1993), or if the density is enhanced by the passage of a shock, as in the model of Whang and Burlaga, then it may be possible to account for the observed emission frequency. Pursuing the second possibility, Gurnett et al. (1993) proposed that the 2-kHz band is produced by a shock propagating through the region immediately beyond the heliopause. The plausibility of this model depends critically on the interstellar plasma density. If the plasma frequency in the interstellar medium is either about 1 kHz (if the emission is at $2f_p$) or about 2 kHz (if the emission is at f_p), then it is a plausible model. However, if the interstellar plasma frequency is substantially above 2 kHz (i.e., n $\gtrsim$ 0.05 cm^{-3}), as indicated by some recent measurements, then the radiation cannot be generated in the interstellar plasma. It would then be possible for the radiation to be trapped in the low-density cavity formed by the heliosphere, as has been suggested by Czechowski and Grzedzielski (1990). Repeated reflections within the heliospheric cavity rapidly leads to a nearly isotropic intensity distribution, which would explain the absence of roll modulation. If the Q-factor of the cavity is sufficiently high (i.e., high enough to give several hundred reflections), it may also be possible to explain the long persistence time of the 2-kHz band, as well as the slow frequency drifts that are sometimes observed (Czechowski et al., 1995).

An important feature of the 2-kHz band is the sharp, almost completely constant low-frequency cutoff at 1.8 kHz. This cutoff could be either due to a propagation cutoff at some point between the source and the spacecraft, or to an intrinsic characteristic of the source. Because of the large density fluctuations that exist in the region between the termination shock and the heliopause, one possibility is that the density peaks act to provide an effective propagation cutoff at a frequency that is several times the average local plasma frequency (Cairns et al., 1992), thereby accounting for the cutoff at ~1.8 kHz. If the propagation cutoff is controlled by the emission process, which would almost certainly be at $2f_p$, then the cutoff would correspond to the minimum plasma frequency, which implies a minimum plasma frequency of 900 Hz (i.e., n = 0.01 cm^{-3}) if the source is between the termination shock and the heliopause. If the source is located beyond the heliopause, then the cutoff would most likely be caused by the propagation cutoff at the plasma frequency in the interstellar plasma. In this case the plasma density in the interstellar medium would be 0.04 cm^{-3}, since n = 0.04 cm^{-3} corresponds to a plasma frequency of 1.8 kHz.

7. Conclusions

Considerable progress has been made since the original discovery of the 2–3 kHz radio emissions by Kurth et al. (1984). It is now known that the radio emission is caused by the interaction of an outward propagating shock, or a system of shocks, with one of the outer regions of the heliosphere, most likely in the region between the termination shock and the heliopause, or in the region near and slightly beyond the heliopause. The emissions are probably generated at the plasma frequency (f_p) or its harmonic ($2f_p$) by the shock-driven Langmuir-wave mode conversion mechanism. A strong case can be made that the intense bursts observed in 1983-84 and in 1992-93 were associated with periods of intense solar activity in 1982 and 1991. The propagation delay in both cases is about 400 days. From direct measurements of the propagation speed of the interplanetary shocks produced by these solar events, and knowledge of the time delay to the onset of the radio burst, the distance to the source can be estimated. These distances are quite large, well beyond 100 AU. Various calculations that take into account the decreasing speed of the shock with increasing distance from the Sun indicate that the radial distance to the interaction region is probably in the range from 110 to 160 AU. The radio emission spectrum is complicated and consists of three main features: (1) a series of narrow bands that drift upward in frequency from about 2.0 to 3.6 kHz at rates ranging from about 1 to 3 kHz/year, (2) a persistent band of emission around 2 kHz, and (3) a sharp low-frequency cutoff at 1.8 kHz. The upward drifting bands are most likely generated by a shock propagating through a ramp of rising density, either in the vicinity of the heliopause, or in association with transients density enhancements between the termination shock and the heliopause. The 2-kHz band is probably generated by the same shock as it propagates through an extended region of nearly constant plasma frequency, either between the termination shock and the heliopause, or beyond the heliopause.

Acknowledgements

This research was supported by NASA through contract 959193 with the Jet Propulsion Laboratory.

References

Anderson, J.E.: 1963, *Magnetohydrodynamic Shock Waves*, MIT Press, Cambridge, 23.

Anderson, K.A., Lin, R.P., Martel, F., Lin, C.S., Park, G.K., and Reme, H.: 1979, *Geophys. Res. Lett.* **6**, 401–404.

Axford, W.I.: 1990, in S. Grzedzielski and D.E. Page (eds.), *Physics of the Outer Heliosphere*, ed. by Pergamon Press, Oxford, 7–15.

Axford, W.I.: 1996, chapter in this book.

Baranov, V.B. and Malama, Yu.G.: 1993, *J. Geophys. Res.* **98**, 15,157–15,163.

Belcher, J.W., Lazarus, A.J., McNutt Jr., R.L. and Gordon Jr., G.S. Gordon: 1993, *J. Geophys. Res.* **98**, 15,177–15,183.

Burlaga, L.F., McDonald, F.B. and Ness, N.F.: 1993, *J. Geophys. Res.* **98**, 1–11.

Cairns, I.H. and Gurnett, D.A.: 1992, *J. Geophys. Res.* **97**, 6235–6244.

Cairns, I.H., Kurth, W.S. and Gurnett, D.A.: 1992, *J. Geophys. Res.* **97**, 6245–6259.

Cliver, E.W., Mihalov, J.D., Sheeley Jr., N.R., Howard, R.A., Koomen, M.J., and Schwenn, R.: 1987, *J. Geophys. Res.* **92**, 8487–8501.

Czechowski, A. and Grzedzielski, S.: 1993, *Nature* **344**, 640–641.

Czechowski, A. and Grzedzielski, S.: 1995, *Adv. Space Res.* **16**, (9)321–(9)325.

Czechowski, A. Grzedzielski, S. and Macek, W.M.: 1995, *Adv. Space Res.* **16**, (9)297–(9)301.

Donohue, D.J. and Zank, G.P.: 1993, *J. Geophys. Res.* **98**, 19,005–19,025.

Dunckel, N.: 1974, *Tech. Rep. 3469-2*, Radioscience Laboratory, Stanford Univ., Stanford, 114–131.

Gosling, J.T.: 1993, *J. Geophys. Res.* **98**, 18,937–18,949.

Grzedzielski, S. and Lazarus, A.J.: 1993, *J. Geophys. Res.* **98**, 5551–5558.

Gurnett, D.A. and Kurth, W.S.: 1995, *Adv. Space Res.* **16**, (9)279–(9)290.

Gurnett, D.A., Kurth, W.S., Allendorf, S.C. and Poynter, R.L.: 1993, *Science* **262**, 199–203.

Kurth, W.S.: 1988, in V.J. Pizzo, T.E. Holzer, and D.G. Sime (eds.), *Proc. of the Sixth Internat. Solar Wind Conf.*, 667.

Kurth, W.S. and Gurnett, D.A.: 1991, *Geophys. Res. Lett.* **18**, 1801–1804.
Kurth, W.S. and Gurnett, D.A.: 1993, *J. Geophys. Res.* **98**, 15,129–15,136.
Kurth, W.S., Gurnett, D.A., Scarf, F.L. and Poynter, R.L.: 1984, *Nature* **312**, 27–31.
Kurth, W.S., Gurnett, D.A., Scarf, F.L. and Poynter, R.L.: 1987, *Geophys. Res. Lett.* **14**, 49–52.
Lallement, R., Bertaux, J.-L. and Clark, J.T.: 1993, *Science* **260**, 1095–1098.
Liewer, P.C., Karmesin, S. Roy and Brackbill, J.U.: 1996, in D. Winterhalter, J. Gosling, S. Habbal, W. Kurth and M. Neugebauer (eds.), *Proceedings of Solar Wind Eight*, in press.
McDonald, F.B. and Burlaga, L.F.: 1996, in J.R. Jokipii, C.P. Sonnett and M.S. Giampapa (eds.),*Cosmic Winds and the Heliosphere*, Univ. of Arizona Press, in press.
McDonald, F.B., Barnes, A., Burlaga, L.F., Gazis, P., Mihalov, J. and Selesnick, R.S.: 1994, *J. Geophys. Res.* **99**, 14,705–14,715.
McNutt, R.L., Jr.: 1988, *Geophys. Res. Lett.* **15**, 1307–1310.
McNutt, R.L., Lazarus, A.J., Belcher, J.W., Lyon, J., Goodrich, C.C. and Kulkarni, R.: 1995, *Adv. Space Res.* **16**, (9)303–(9)306.
Melrose, D.B.: 1985, in D.J. McLean and N.R. Labrum, *Solar Radiophysics*, Cambridge Univ. Press, Cambridge, 177–210.
Nelson, G.J. and Robinson, R.D.: 1975, *Proc. Astron. Soc. Aust.* **2**, 370.
Scarf, F.L. and Gurnett, D.A.: 1977, *Space Sci. Rev.* **21**, 289–308.
Steinolfson, R.S. and Gurnett, D.A.: 1995, *Geophys. Res. Lett.* **22** 651–654.
Steinolfson, R.S., Pizzo, V.J. and Holzer, T.: 1994, *Geophys. Res. Lett.* **21**, 245–248.
Van Allen, J.A.: 1993, *Geophys. Res. Lett.* **20**, 2797–2800.
Van Allen, J.A. and Fillius, R.W.: 1992, *Geophys. Res. Lett.* **19**, 1423–1426.
Van Allen, J.A. and Randall, B.A.: 1985, *J. Geophys. Res.* **90**, 1399–1412.
Washimi, H.: 1993, *Adv. Space Res.* **6**, 227–236.
Webber, W.R. and Lockwood, J.A.: 1993, *J. Geophys. Res.* **98**, 7821–7825.
Webber, W.R., Lockwood, J.A. and Jokipii, J.R.: 1986, *J. Geophys. Res.* bf 91, 4103–4110.
Whang, Y.C. and Burlaga, L.F.: 1995, *Adv. Space Res.* (9)291–(9)295.
Winterhalter, D., Smith, E.J. and Klein, L.W.: 1992, *EOS Trans. AGU* **73(4)**, 237.
Zank, G.P., Cairns, I.H., Donohue, D.J. and Matthaeus, W.H.: 1994, *J. Geophys. Res.* **99**, 14,729–14,735.

Address for correspondence: D. Gurnett, Dept. of Physics and Astronomy, University of Iowa, Iowa City, IA, 52242, USA

A SUMMARY OF SOLAR WIND OBSERVATIONS AT HIGH LATITUDES: ULYSSES

R.G. MARSDEN
Space Science Department of ESA, Solar System Division, Estec, Noordwijk

Abstract. The Ulysses mission has provided the first in-situ observations of the solar wind covering all solar latitudes from the equator to the poles in both hemispheres. The measurements from the first polar passes, made at near-minimum solar activity conditions, have confirmed the basic picture established on the basis of remote sensing techniques: the high-latitude wind is fast, and originates in the polar coronal holes. The detailed in-situ observations have, however, revealed a number of features related to the global solar wind structure that were not expected: the transition between slow and fast wind was relatively abrupt, followed by a slight increase in speed toward the poles; the mass flux is almost independent of latitude, with only a modest increase at the equator; the momentum flux is significantly higher over the poles than near the equator, suggesting a non-circular cross-section for the flanks of the heliosphere.

1. Introduction

Launched in 1990, Ulysses is an exploratory mission carried out jointly by ESA and NASA and has as its primary objective the study of the interplanetary medium and solar wind as a function of heliographic latitude (Wenzel et al., 1992). There is good reason to believe that the conditions found in the narrow band of heliographic latitudes sampled by spacecraft confined to the ecliptic plane are not representative of the inner heliosphere as a whole, and yet attempts to understand the basic physical processes occurring within this environment have so far been based essentially on observations made in the ecliptic plane. Ulysses has, for the first time, permitted measurements to be made in situ away from the plane of the ecliptic and over the poles of the Sun (Marsden and Smith, 1996; Smith and Marsden, 1995; Smith et al., 1995). Its unique trajectory has taken the spacecraft into the uncharted third dimension of the heliosphere.

The phenomena being studied by the Ulysses mission are strongly influenced by the 11-year solar activity cycle. The polar passes of the prime mission occurred in 1994 and 1995 during the descending phase of the current solar cycle (no. 22), close to solar minimum. At this time, the Sun was in its most simple state: large coronal holes extended from the polar regions in both hemispheres; the Sun's surface magnetic field was largely dipolar, being positive (outward) in the north and negative (inward) in the south; there was a relative paucity of energetic flare and other transient events. Prior to Ulysses, it

Space Science Reviews **78**: 67–72, 1996.

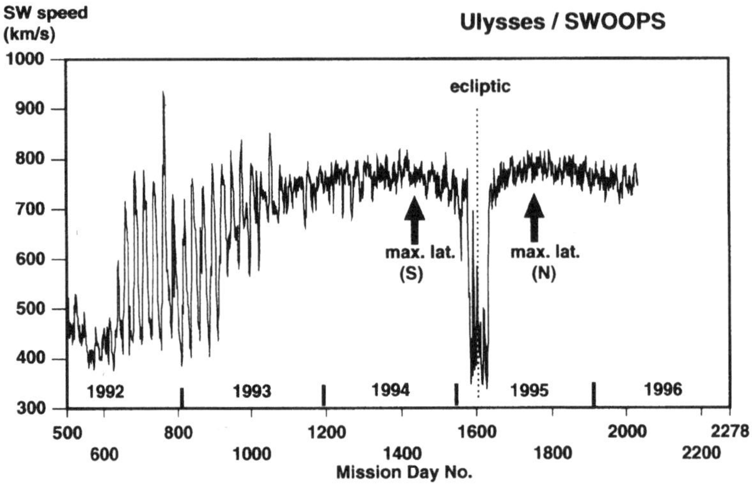

Fig. 1. Solar wind bulk flow speed measured by the SWOOPS plasma experiment on board Ulysses between 18 February 1992 and 1 May 1996.

was generally expected that the well-developed polar coronal holes would give rise to an increase in solar wind speed with increasing latitude. The precise nature of the latitude profile was, however, uncertain. Another open question concerned the variation of solar wind mass flux with latitude, since Lyman-alpha resonance glow and other measurements indicated a possible decrease in mass flux over the poles (Lallement et al., 1995). Now, with the excellent in-situ observations made by Ulysses at high latitudes, these and many other uncertainties have been removed.

The goal of this paper is provide an concise overview of the solar wind plasma observations made by Ulysses during the out-of-ecliptic phase of its mission (e.g. Phillips et al., 1995a,b). Of particular interest in the context of a workshop devoted to 'The Heliosphere in the Local Interstellar Medium' are those bulk flow parameters that are of importance in determining the global structure of the heliosphere. Measurements more directly related to the questions of solar wind origin and heating, including ion composition, are not considered in detail here. Following this introduction, in section 2 of this paper, we present a summary of the observations. The implications for the global structure of the heliosphere are briefly discussed in section 3.

2. Solar Wind Observations

Figure 1 shows an overview of the solar wind bulk flow speed recorded at Ulysses. Beginning in July 1992, the solar wind flow was dominated by the appearance of a single high-speed stream once per solar rotation, with slower

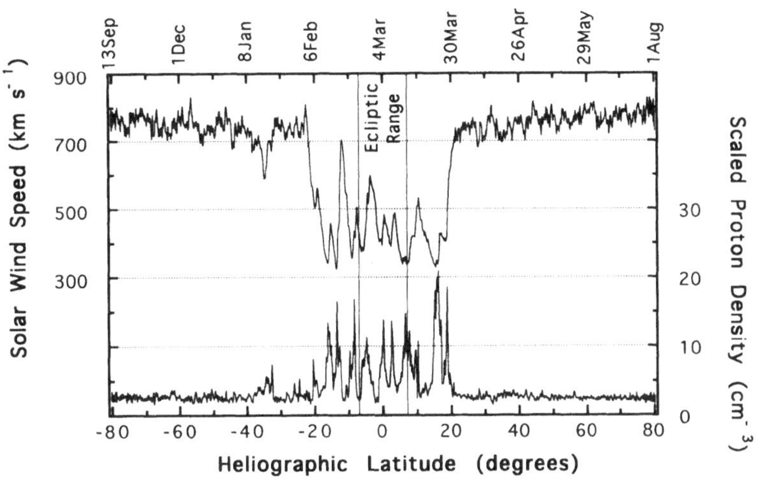

Fig. 2. Hourly averages of the solar wind bulk flow speed (upper trace) and proton density scaled to 1 AU (lower trace) measured by the SWOOPS experiment on Ulysses during the transit from the south to the north pole (from Phillips et al., 1995b).

solar wind in between. The source of the fast stream has been traced back to an equatorward extension of the southern polar coronal hole (Bame et al., 1993). The slower wind originated in the coronal streamer belt that encircles the Sun's magnetic equator. Starting in May 1993, this recurrent pattern underwent a change. While the dominant high-speed stream remained visible in the data, the speed of the wind in the inter-stream regions increased, significantly reducing the peak-to-valley excursions. As a consequence of its increasingly southern position at this time, Ulysses was no longer exposed to solar wind from inside the streamer belt, only to wind from the boundary region between the belt and the coronal hole and to fast wind from the hole itself.

Once above 40° southern latitude, Ulysses became totally immersed in fast solar wind from the south polar coronal hole flowing continuously at an average speed of 750 km/s (Phillips et al., 1995a). These conditions persisted throughout the south polar pass, continuing up to mid-January 1995, when the spacecraft's southerly latitude had decreased to 35°. At this point, the wind speed at Ulysses dropped briefly below 600 km/s, becoming fast again until the spacecraft crossed into the region of slow to medium solar wind from the coronal streamer belt.

The streamer belt had undergone significant changes since Ulysses left it in 1993. As shown in Figure 2, the band of slow to medium wind had become narrower, covering a range of 43 degrees in latitude at Ulysses (22° south to 21° north). Although the polar coronal holes typically had an equatorward extent of no more than 60° at the solar surface at this time, non-radial expansion of the solar wind had caused the boundaries to move significantly equatorward within 1.74 solar radii (Phillips et al., 1995), consistent with the

TABLE 1. Bulk Flow Parameters of the High-Latitude Solar Wind (from Feldman et al., 1996)

Parameter (scaled to 1 AU)	Mean	Median	5% Level	95% Level
N_pR^2 (cm^{-3})	2.47	2.42	1.82	3.26
V_{SW}(km s^{-1})	773	774	729	812
T_p(10^5 K)	1.86	1.84	1.39	2.42
T_e(10^5 K)	0.844	0.779	0.565	1.222
$T\alpha/Tp$	4.76	4.73	3.98	5.61
$N\alpha/Np$	0.0444	0.0439	0.0330	0.0574
Mass flux density (10^{12}amu $m^{-2}s^{-1}$)	2.28	2.21	1.69	3.08
Momentum flux density (nPa)	2.87	2.84	2.12	3.72
Kinetic energy flux density (10^{-3} joules m^{-2} s^{-1})	1.12	1.11	0.81	1.47

Ulysses results. Since early April 1995 to the present, Ulysses has remained immersed in fast solar wind, this time from the northern polar regions.

A major contribution made by Ulysses to the study of the solar wind has been the first ever in-situ determination of the latitudinal variation of its fluid parameters. Figure 3, adapted from Phillips et al. (1995b), shows the speed, density, mass flux, and momentum flux (ρu^2, where ρ is the mass density and u the speed), binned by solar rotation, for the latitude range covered by Ulysses (± 80.2°). Immediately apparent is the relatively small variation in all parameters poleward of 45° in both hemispheres, and the effect, particularly on the 95th-percentiles, of high-density, slow solar wind from the streamer belt at low latitudes. By restricting the dataset to the latitude range poleward of ± 60° , it is possible to derive average bulk flow parameters that may be considered to apply to the pure high-speed solar wind flow. This has been done by Feldman et al. (1996), and their results are reproduced in Table 1. Note that quantities involving density have been scaled to 1 AU using an R^{-2} scaling law.

3. Discussion and Conclusions

As pointed out by Feldman et al. (1996), the scaled Ulysses high-latitude mass flux and momentum flux show excellent agreement with in-ecliptic high-speed solar wind data (e.g., Schwenn, 1991). However, due to the mix of different wind conditions encountered at low latitudes, the solar rotation-averaged momentum flux density measured by Ulysses during the fast latitude scan

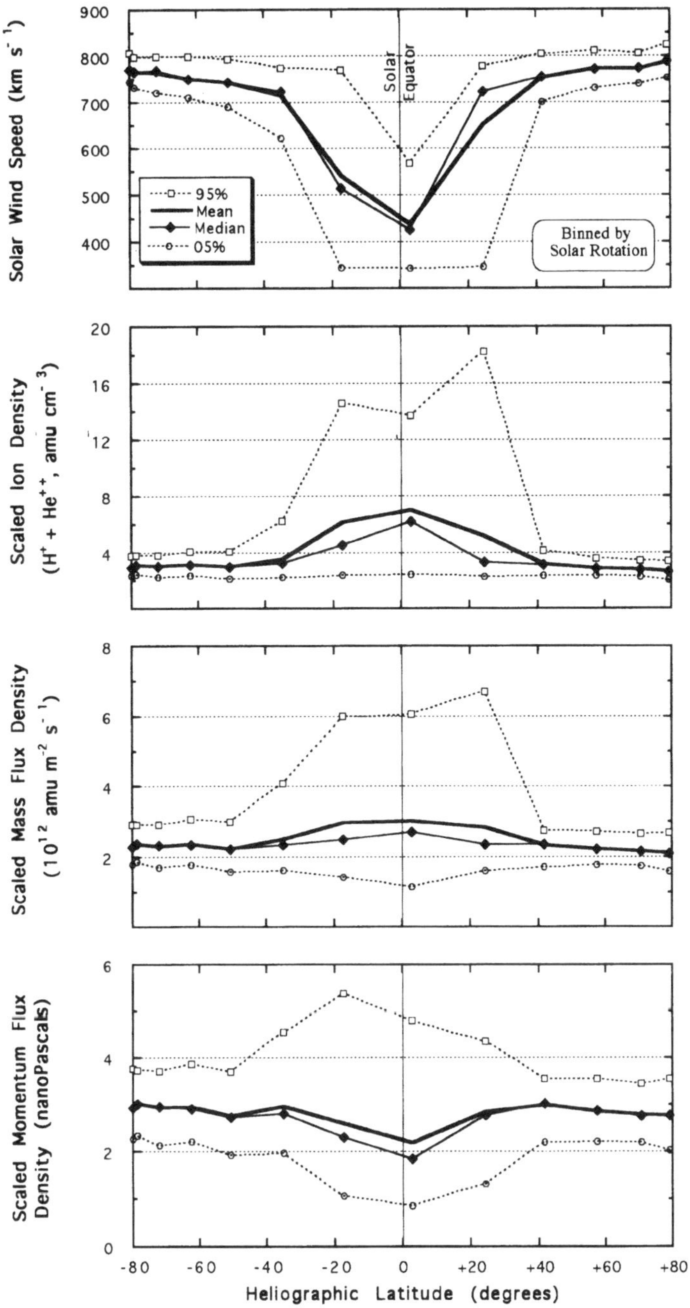

Fig. 3. Mean, median, 5th and 95th percentiles of solar wind fluid parameters measured by the solar wind plasma experiment on board Ulysses, binned by solar rotation, as a function of solar latitude (from Phillips et al., 1995b).

showed some dependence on latitude, the median being 37% lower at the equator than at high latitudes (Phillips et al., 1995b). Based on these median values, Phillips et al. (1995b) suggest that the heliopause and termination shock cross-sections in a plane normal to the interstellar wind direction are pinched in at the equator by ~20%. On the other hand, Goldstein et al. (1996) propose that the mean momentum flux is perhaps a better parameter than the median when considering the position of the heliopause and termination shock. In this case, the observed latitudinal variation would result in a somewhat less pronounced "pinching" (~10%). It should also be noted that the above configuration only applies to solar minimum conditions. CMEs and other transient phenomena give rise to large variations in momentum flux over the course of a solar cycle.

The mass flux density showed a modest increase (~15% in the median, ~25% in the mean value) at low latitudes compared with the polar regions. Lallement et al. (1995) have suggested that the equatorial "groove" observed in Lyman-alpha resonance glow patterns near solar minimum is best explained by an equatorial enhancement in the ionization rate of neutral interstellar hydrogen by charge exchange with solar wind protons. This in turn requires an increase in the equatorial solar wind proton mass flux, qualitively consistent with the Ulysses findings. Finally, the magnitude of the mass flux measured in situ by Ulysses, 2.28 10^8 $cm^{-2}s^{-1}$, as well as the absence of a decrease in flux at high latitudes, clearly rules out acceleration of the solar wind by classical thermal conduction alone (Barnes et al., 1995; Lallement et al., 1995).

Acknowledgements

The author would like to thank J.L. Phillips, Principal Investigator of the SWOOPS experiment on Ulysses, for helpful discussions and for providing several of the figures used in this paper.

References

Bame, S.J., et al.: 1993, *Geophys. Res. Lett.* **20**, 2323.
Barnes, A., Gazis, P.R. and Phillips, J.L.: 1995, *Geophys. Res. Lett.* **22**, 3309.
Feldman, W.C., et al.,: 1996, *Astron. Astrophys.*, in press.
Geiss, J., et al.: 1995, *Science* **268**, 1033.
Goldstein, B.E., et al.: 1996, *Astron. Astrophys.*, in press.
Lallement, R., Kyrola, E. and Summanen, T.: 1995, *Space Sci. Rev.* **72**, 455.
Marsden, R.G. and Smith, E.J.: 1996, *Adv. Space Res.* **17**, (4/5)293.
Phillips, J.L., et al.: 1995a, *Science* **268**, 1030.
Phillips, J.L., et al.: 1995b, *Geophys. Res. Lett.* **22**, 3301.
Schwenn, R.: 1991, in Schwenn, R., Marsch, E. (Eds.), *Physics of the Inner Heliosphere 1: Large Scale Phenomena*, Springer-Verlag, Berlin, 99.
Smith, E.J., Marsden, R.G. and Page, D.E.: 1995, *Science* **268**, 1005.
Smith, E.J. and Marsden, R.G.: 1995, *Geophys. Res. Lett.* **22**, 3297.
Wenzel, K.-P., et al.: 1992, *Astron. Astrophys. Suppl. Ser.* **92**, 207.

IONIZATION PROCESSES IN THE HELIOSPHERE – RATES AND METHODS OF THEIR DETERMINATION

D. RUCIŃSKI[1], A.C. CUMMINGS[2], G. GLOECKLER[3], A.J. LAZARUS[4], E. MÖBIUS[5] and M. WITTE[6]

[1] *Space Research Centre of the Polish Academy of Sciences, Bartycka 18A, 00-716 Warsaw, Poland*

[2] *Space Radiation Laboratory, California Institute of Technology, Pasadena CA 91125, USA*

[3] *Department of Physics, University of Maryland, College Park MD 20742, USA*

[4] *Center for Space Research, Massachusetts Institute of Technology, Cambridge MA 02139, USA*

[5] *Space Science Center, University of New Hampshire, Durham NH 03824, USA*

[6] *Max-Planck-Institut für Aeronomie, D-37191 Katlenburg-Lindau, Germany*

Abstract. The rates of the most important ionization processes acting in interplanetary space on interstellar H, He, C, O, Ne and Ar atoms are critically reviewed in the paper. Their long-term modulations in the period 1974 – 1994 are reexamined using updated information on relevant cross-sections as well as direct or indirect data on variations of the solar wind/solar EUV fluxes based on IMP 8 measurements and monitoring of the solar 10.7 cm radio emission. It is shown that solar cycle related variations are pronounced (factor of ~ 3 between maximum and minimum) especially for species such as He, Ne, C for which photoionization is the dominant loss process. Species sensitive primarily to the charge-exchange (as H) show only moderate fluctuations $\sim 20\%$ around average. It is also demonstrated that new techniques that make use of simultaneous observations of neutral He atoms on direct and indirect orbits, or simultaneous measurements of He^{+} and He^{++} pickup ions and solar wind particles can be useful tools for narrowing the uncertainties of the He photoionization rate caused by insufficient knowledge of the solar EUV flux and its variations.

1. Introduction

Interstellar neutral atoms penetrating the heliosphere are subjected to several ionization processes that reflect their interaction with the solar wind plasma and solar EUV radiation. These processes are commonly recognized to play a dual role. Firstly, they are responsible for the shaping of the local distribution of neutrals inside the Solar System. Secondly, they lead to the creation and determine the amount of the new, secondary components, as for example pickup ions. These ions while convected outwards with the solar wind may be preaccelerated at the travelling interplanetary shocks (Gloeckler et al., 1994) and later accelerated at the termination shock and thus are thought to be the source of the anomalous cosmic ray (ACR) component. To determine the properties of the gas in the Very Local Interstellar Medium (VLISM), or at least in the outer heliosphere, various experimental techniques based on the observations of the original neutral gas or the secondary products inside the Solar System are used. They include observations of the backscattered

Space Science Reviews **78**: 73–84, 1996.

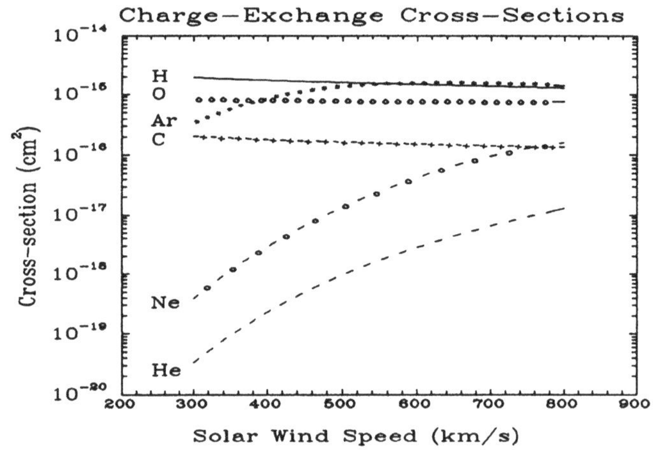

Figure 1. The dependence of the charge-exchange cross-section on the solar wind speed for H, He, C, O, Ne, and Ar.

solar EUV radiation (see e.g. Lallement et al., 1991; Quèmerais et al. 1994 and references therein), the most recent direct He gas measurements (Witte et al., 1993), pickup ion measurements (see e.g. Möbius et al., 1985, 1995; Gloeckler et al., 1993; Geiss et al., 1994), and observations of the ACR (Cummings and Stone, 1988, 1990, 1996). The reliability of the quantitative conclusions inferred from these studies is to large extent dependent on the accuracy of our knowledge of the ionization rates. Thus, the goal of this paper is to reexamine the efficiency of ionization processes for several interstellar species (H, He, C, O, Ne, Ar) that have been identified in the heliosphere. This includes in particular the variability of the ionization rates over the solar cycle. Though the study is mainly focused on photoionization and charge-exchange – the most important processes – we also refer briefly to other reactions of local importance, such as electron impact ionization, and double charge exchange with solar wind alpha particles which only recently could be traced to observable effects.

2. Efficiency of the ionization processes

2.1. Charge-exchange with solar wind protons

Revisiting charge-exchange between the solar wind protons and aforementioned neutral interstellar species, we adopt in our analysis the most recent data on the relevant cross-sections. The updated relations between the cross-sections and the solar wind speed based on studies by Barnett et al. (1990) for H and He; Ehrhardt and Langer (1987) for C; Stebbings et al. (1964) for O; and Nakai et al. (1987) for Ne and Ar, are shown in Fig.1 for the range 300 – 800 km/s. The experimental data can be represented either by simple analytical formulae or by fitted Chebyshev polynomials. While for H the modern values of Barnett et al. (1990) remain in good

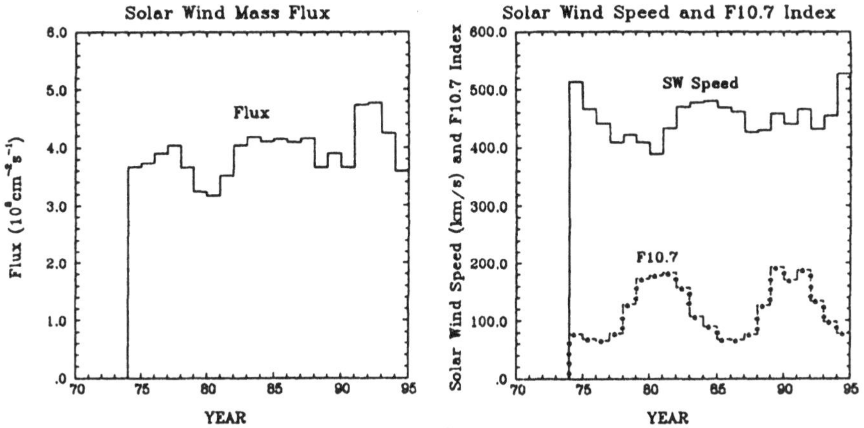

Figure 2. Yearly averages during 1974 – 1994 of the solar wind mass flux (left panel), solar wind speed (solid curve, right panel) from IMP 8 measurements and the yearly averages of the absolute 10.7 cm radio flux (in $10^{-22} Wm^{-2} Hz^{-1}$) (dash-circled curve, right panel).

agreement ($\sim$ 10 %) with the classical formula of Maher and Tinsley (1977), recent studies of Nakai et al. (1987) indicate much lower cross-sections for noble gases (mainly He and Ne), than reported earlier (Afrosimov et al., 1967; Tawara, 1978). To evaluate the charge-exchange rate (at 1 AU) for these species and to demonstrate their long-term variations, we use a comprehensive data set of yearly averages of the solar wind speed and mass flux from the IMP 8 measurements for the period 1974 – 1994. They are shown in Fig.2 together with the absolute F10.7 cm radio flux as an indicator for the solar cycle variation. The resulting charge-exchange rates are displayed in Fig.3. The temporal variation of the H rate is relatively low and its amplitude does not exceed $\sim$ 20% , what is consistent with observations from Helios (Schwenn, 1983). Similar behavior is found also for O and C. In contrast to that, the charge-exchange rates for Ne and He (see Fig.3, lower panel) may vary strongly (even by factor of $\sim$ 5) mainly due to the high sensitivity of the relevant cross-sections to the changes of the solar wind velocity. However, these drastic variations do not cause any practical consequences, since their contribution to the total ionization does not exceed $\sim$ 3% for Ne and is below 1% for He. In essence charge-exchange is much less effective for He and Ne than previously thought (Axford, 1970; Holzer, 1977).

2.2. Photoionization

Another important process of the interaction of neutral gases in interplanetary space is photoionization. The relevant dependencies of the cross-sections on wavelength taken from Allen (1973) for H; Marr and West (1976) for He, Ne and Ar; and from a compilation of M. Allen (1995, private communication) for C and O are shown in Fig.4. In contrast to the solar wind flux and velocity, for which continuous monitoring is available from several spacecraft (even over two decades), there are only very few direct measurements of the solar EUV flux, not even over any extended

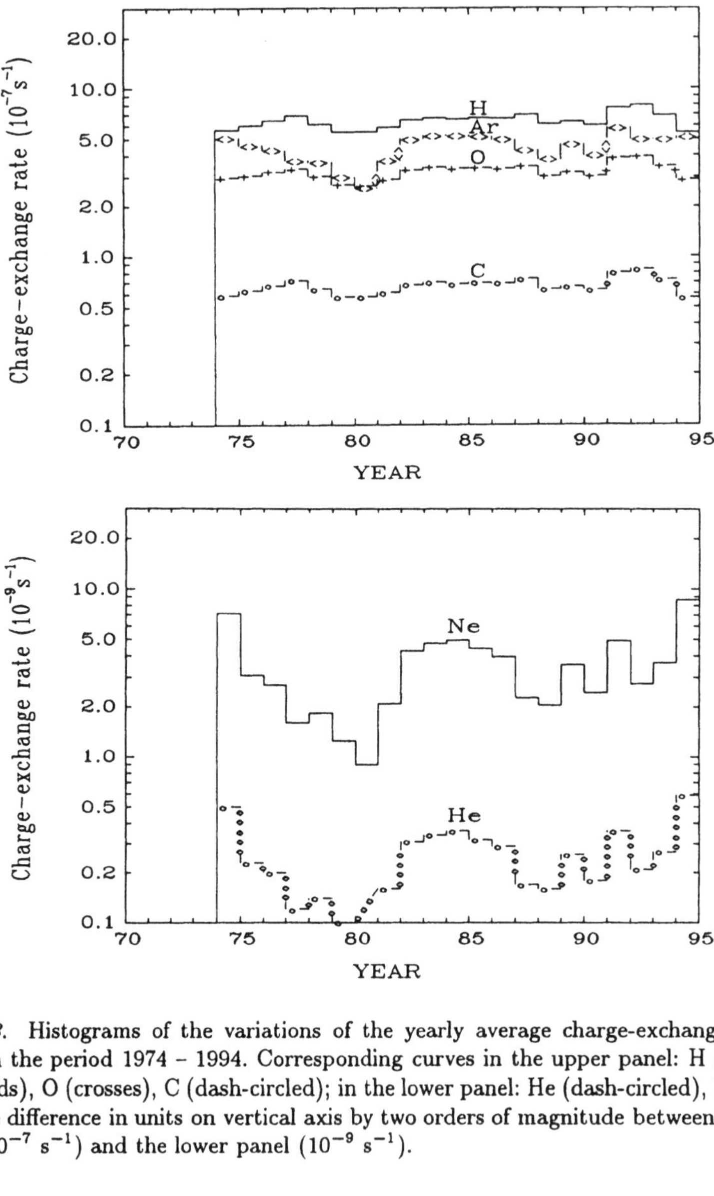

Figure 3. Histograms of the variations of the yearly average charge-exchange rates at 1 AU in the period 1974 - 1994. Corresponding curves in the upper panel: H (solid), Ar (diamonds), O (crosses), C (dash-circled); in the lower panel: He (dash-circled), Ne (solid). Note the difference in units on vertical axis by two orders of magnitude between the upper panel (10^{-7} s^{-1}) and the lower panel (10^{-9} s^{-1}).

period of time. The lack of continuous data requires the use in the current analysis of an indirect method that is based on the correlation between the solar EUV flux and the F10.7 cm radio flux. The existence of such a correlation has been pointed out for example by Feng et al. (1989), and Ogawa et al. (1990), who compiled results from various rocket/spacecraft measurements of the EUV flux. Recently, it was frequently used in more refined modeling of the solar EUV output (see e.g. Tobiska, 1993). Adopting the aforementioned cross-sections (see Fig.4) and using

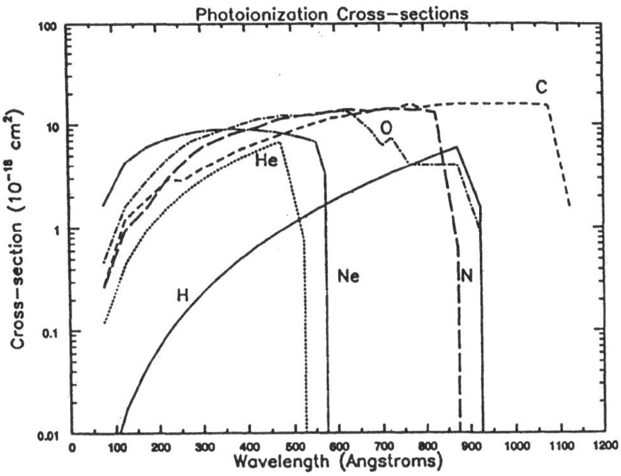

Figure 4. The dependence of the photoionization cross-section on wavelength for H, He, C, O, Ne and Ar.

the Atmospheric Explorer and rocket measurements of the solar EUV flux (Torr et al., 1979), with doubling of the flux below 250 Å, as suggested by Richards et al. (1994), we have calculated the photoionization rates of various species for two dates in April 1974 and February 1979 (74DOY113 and 79DOY050, respectively), which represent extreme (low and high) conditions of solar activity. Assuming a linear relation between the photoionization rate and the F10.7 cm flux which is in good agreement with the compilation of the experimental results by Ogawa et al. (1993), we have estimated the yearly averages of all rates for the period 1974 1994. Histograms of the photoionization rates calculated in this way are shown in Fig.5. As a consequence of the correlation of the photoionization rates with the F10.7 cm flux, they vary with the level of the solar activity. Relatively strong changes (typically by factor of $\sim$ 3) between the solar minimum and maximum are expected. However, we should stress here that our estimates are not based on direct absolute solar EUV flux measurements, but employ an indirect method. Therefore, the results may still contain significant uncertainties. Another reason for relatively large uncertainty are serious discrepancies between existing measurements as well as differences between the most commonly used EUV flux models, pronounced especially at wavelengths below 250 Å (Richards and Torr, 1984; Ogawa and Judge, 1986; Richards et al., 1994). Since the total photon flux in this interval may be comparable to the flux at 304 Å, dominating the solar EUV spectrum, a possible uncertainty in its determination by a factor of $\sim$ 2 may cause an error in the determination of the photoionisation rate of $\sim$ 10 – 15% for species with the high ionization potential, such as He and Ne. Comparing results shown in Fig.3 and in Fig.5 one can conclude that the photoionization is the absolutely dominant loss process for He, Ne, and C. It is comparable with charge-exchange for O and Ar. Only in the case of hydrogen charge-exchange is dominant with photoionization contributing only $\sim 10-20\%$ to the total rate. These rates and the range of their variability in the period 1974 – 1994 are compiled in Table 1.

Table 1a
Modulation of basic ionization rates at 1 AU over the period 1974 – 1994

	Charge Exchange $[10^{-7}s^{-1}]$			Photoionization $[10^{-7}s^{-1}]$			Ch- ex + photoionization* $[10^{-7}s^{-1}]$		
	Min.	Max.	Average**	Min.	Max.	Average**	Min.	Max	Average**
H	5.5	8.0	6.4±0.14	0.6	1.7	1.04±0.09	6.2	9.3	7.44±0.16
He	0.001	0.006	0.0025 ±0.0003	0.5	1.8	1.07±0.12	0.5	1.8	1.08±0.11
C	0.57	0.83	0.67±0.015	5.6	15.3	9.68±0.78	6.2	16.0	10.35±0.79
O	2.65	3.93	3.21±0.07	2.3	7.5	4.54±0.41	5.5	11.2	7.75±0.42
Ne	0.02	0.09	0.035 ±0.004	1.7	5.9	3.49±0.33	1.7	5.9	3.52±0.34
Ar	2.5	5.8	4.46±0.18	3.2	9.7	5.97±0.51	7.6	15.3	10.44±0.48
N***				2.3	7.2	4.38±0.38			

*Note that due to the time shift between the minimum and maximum of the two effects the total rate may not reproduce a simple sum of the min./max. rates of both processes.
**Uncertainty of averages represents the standard deviation of the mean value. The standard deviation for the sample of the 21 yearly averages is higher by factor $\sqrt{21}$.
***No charge exchange data found for nitrogen.

Table 1b
Variations of yearly averages of other quantities at 1 AU during 1974 – 1994

	Minimum	Maximum	Average
Solar wind flux $[10^{8}\ cm^{-2}\ s^{-1}]$	3.18	4.77	3.92±0.09
Solar wind speed $[km\ s^{-1}]$	389	527	451±7
10.7 cm radio flux $[10^{-22}\ W\ m^{-1}\ Hz^{-1}]$	66.0	192.1	119.4±10

2.3. Other ionization processes

Although charge-exchange and photoionization are the most significant reactions for the ionization of interstellar atoms, a few minor processes can be locally important. The studies of Askew and Kunc (1984) and Ruciński and Fahr (1989, 1991) showed that electron impact ionization plays a non-negligible role for H and He inside ~ 3 AU from the Sun, with a relatively strong increase of its rates at smaller distances. To calculate the characteristic rates for the process, we adopt analytical relations for the energy dependence of the cross-sections as given by Lotz (1967). The variations of the "core" and "halo" solar wind electron populations with the heliocentric distance are taken from the recent determinations of Scime et al. (1994) from Ulysses observations at $\sim 1-5$ AU: $T_c = 1.3 \cdot 10^5 R^{-0.85}$ [K], and $T_h = 9.2 \cdot 10^5 R^{-0.38}$ [K]. The relevant rates (at 1 AU) calculated for the solar wind density of 9 cm^{-3} (96% "core" + 4% "halo") are shown in Table 2. The characteristic contribution of the process to the total rate (at 1 AU) is of the order of $\sim$ 10% or even more. The process was recently recognized by Isenberg and Feldman (1995) as an important loss process for H and He at interplanetary shocks.

The significance of the effect near the Sun was also quite convincingly confirmed during recent direct measurements of the neutral helium gas on Ulysses (Witte et. al., 1996). It was demonstrated that to fit the observed signals referring separately to the He atoms on the 'direct' and 'indirect' orbits (i.e. those which swept out angle less or greater than 180° along their trajectory, respectively), one may either include the electron impact and adopt relatively low ($6.0 \cdot 10^{-8}$ s^{-1}) photoionization rate, adequate for the period of observation (March 1995), or if treating the photoionization as the only important effect – its rate must be much higher ($\approx 1.1 \cdot 10^{-7}$ s^{-1}) and thus irrelevant for that time. Another minor effect of obser-

Table 2
Typical electron impact ionization rate at 1 AU

Element	H	He	O	Ne	Ar
Electron impact ionization rate [10^{-7} s^{-1}]	0.64	0.16	1.30	0.88	2.20

vational importance is the double-charge exchange between neutral He and solar wind α-particles ($\sigma = 2.4 \cdot 10^{-16}$ cm^2 for $V_{SW} = 500$ km/s, see Gruntman (1994)). As discussed by Ratkiewicz et al. (1990), the process should lead to the creation of a non-negligible population of the He^{++} pickup ions. This was recently confirmed by Ulysses measurements (Gloeckler and Geiss, 1994). However, no independent information on the related charge-exchange rate is available from this result. In order to derive the interstellar He density from the simultaneously measured pickup ion and solar wind flux a charge-exchange cross section has to be adopted from published values. It should be noted that this is probably the largest uncertainty in this method, because the ratio of pickup and solar wind flux is measured with the same instrument and therefore no absolute calibration factors influence the result.

3. In-situ methods to determine ionization rates

Direct measurements of interstellar gas and its ionization products also provide a new handle on ionization rates. This carries the promise to substantially improve the accuracy of these rates in the near future in a cross-calibration between complementary techniques. As can be easily seen in the following discussion, the new in-situ

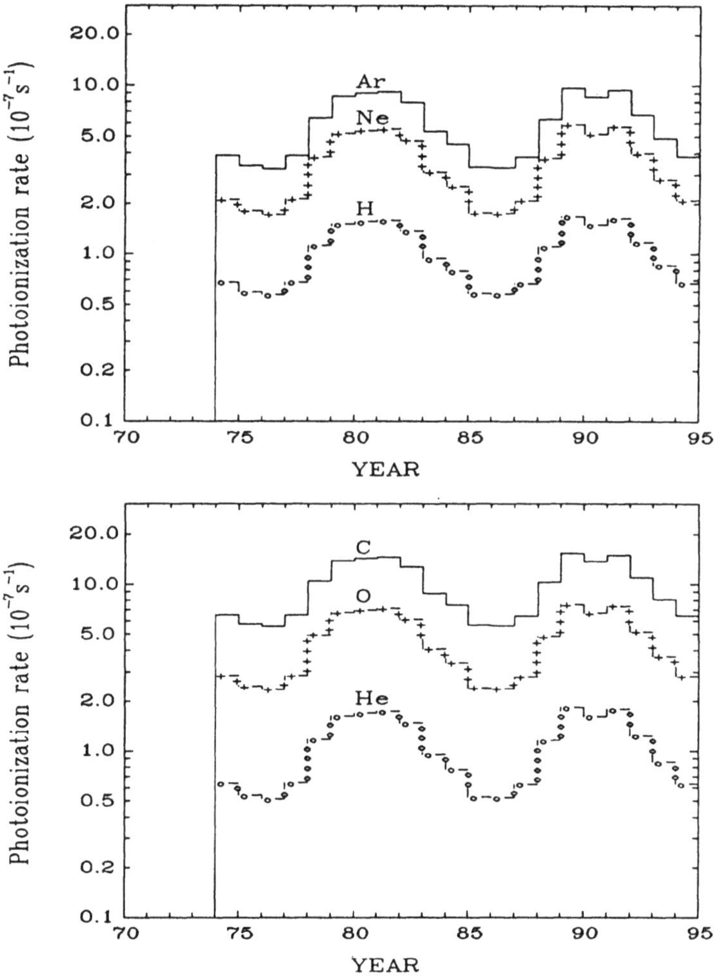

Figure 5. Histograms of the variations of the yearly average photoionization rates at 1 AU (in 10^{-7} s^{-1}) in the period 1974 – 1994. Corresponding curves in the upper panel: H (dash-circled), Ar (solid), Ne (crosses); in the lower panel: He (dash-circled), O (crosses), C (solid).

measurements are sensitive to the total ionization rate, i.e. an integral determination of all ionization processes combined is made. A compilation of current results from in-situ measurements in comparison with the contribution from most important ionization processes is shown in Table 3. Due to the adiabatic cooling of the pickup ion distribution during the outward convection with the radially expanding solar wind the velocity distribution is a direct image of the neutral gas distribution along the Sun-spacecraft line upstream of the observer. Therefore, it is possible to derive radial density gradient from the energy spectra of pickup ions taken in the

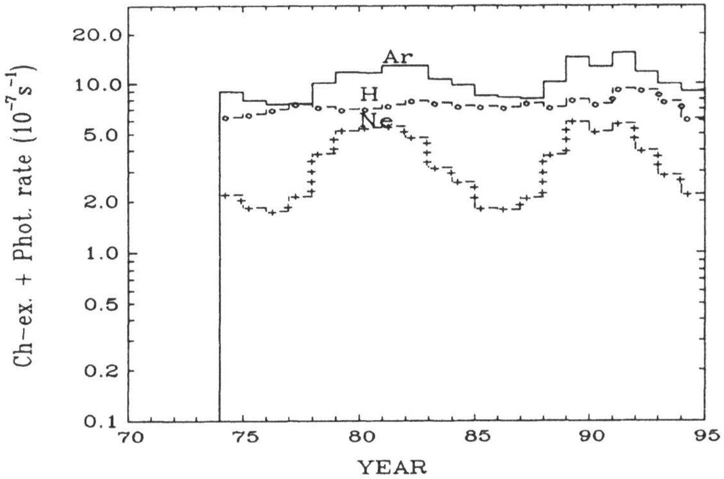

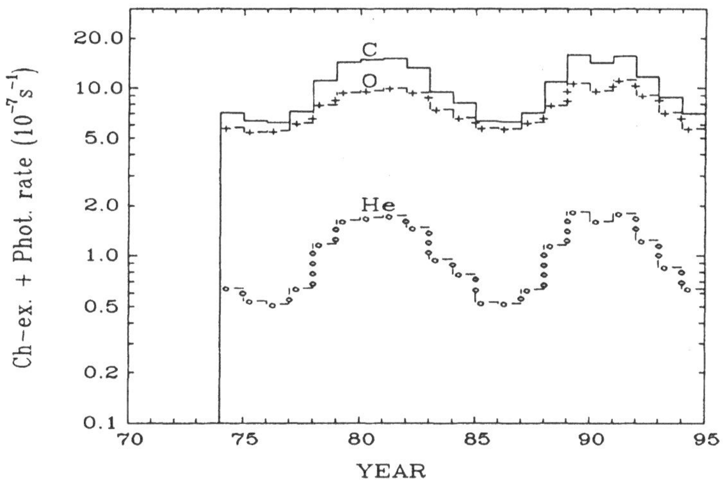

Figure 6. Histograms of the variations of the total (i.e. charge-exchange + photoionization) rates at 1 AU (in 10^{-7} s^{-1}) in the period 1974 - 1994. The curves in upper panel correspond to: H (dash-circled), Ne (crosses), Ar (solid); whereas in lower panel to: He (dash-circled), O (crosses) and C (solid).

solar wind direction and compare this with model distributions of the interstellar gas in the inner Solar System (e.g. Möbius et al., 1988). The radial gradient is a sensitive indicator of the total destruction rate of neutral gas, as long as the observations are made within the typical penetration distance of the species from the Sun ($\leq$ 1 AU for He and $\leq$ 5 AU for H, Gloeckler et al. (1994)). The resulting energy spectra have led to constraints on the ionization rates which are compatible with other observations. It should be noted that the resulting ionization rate represents

Table 3
Methods of determination of ionization rates applied for helium and hydrogen

Species Method	Photo-ionization [s^{-1}]	Related 10.7 flux [$10^{-22}\cdot$ $Wm^{-2}Hz^{-1}$]	Charge exchange [s^{-1}]	Electron impact [s^{-1}]	Total [s^{-1}]	Uncertainties/ comments
Helium						
Compilation of EUV measurements and literature studies*	**$5.0\cdot10^{-7}$** (min.) **$1.8\cdot10^{-7}$** (max)*	66 (min.) 192 (max)	**$1.0\cdot10^{-10}$** (min.) **$6.0\cdot10^{-10}$** (max.)*		**$5.0\cdot10^{-8}$** (min.) **$1.8\cdot10^{-7}$** (max.)* av. **$1.08\cdot10^{-7}$**	EUV results vary greatly (±25%), EUV flux at λ< 250 Â poorly known.
Fits to He^+ pickup ion spectra + EUV (*Möbius et al.1995*)		75	4% of the total rate**	~10% of the total rate	**5.5 – $6.5\cdot10^{-8}$** (1985)	Pickup spectrum varies with transport effects: EUV used as add. constraint
Fits to He^+ pickup ion spectra (*Gloeckler 1996*)		77			**6.5 – $8.5\cdot10^{-8}$** (1994)	Pickup spectrum varies with transport effects
He on direct & indirect orbits (*Witte et al.1996*)	**$6\cdot10^{-8}$** (1994–95)			rest	**9 – $11\cdot10^{-8}$** (1994 – 95)	Large effect from e-impact & elastic collisions at 0.2 AU
He^{++} pickup ions; SW flux+σ_{ch-ex} ** (*Gloeckler 1996*)	$3\cdot10^{-10}$		**$1.7\ 10^{-9}$**		**$2.0\ 10^{-9}$**	Accuracy of σ_{ch-ex} **
Hydrogen						
Compilation of SW measurements and literature studies	**$6.0\cdot10^{-8}$** (min.) **$1.7\cdot10^{-7}$** (max.)*	66 (min.) 192 (max.)*	**$5.5\ 10^{-7}$** (min.) **$8.0\cdot10^{-7}$** (max.)*		**$6.2\ 10^{-7}$** (min.) **$9.3\ 10^{-7}$** (max.)* av. **$7.44\ 10^{-7}$**	Solar wind measurements,accuracy of σ_{ch-ex}, highly variable photo contribution
Fits to H^+ pick-up ion spectra (*Gloeckler 1996*)					**5 – 6 10^{-7}** (1994) high latitude	Anisotropy of pickup spectrum

Values in boldface print have been determined or compiled with appropriate method.
* Compilation over period 1974 – 1994; total rates do not include electron impact.
** Rates refer to the production of He^{++} pickup ions; charge exchange rate denotes here double charge exchange with solar wind alpha particles.

an integral value over several months prior to the measurement, i.e. the travel time of the interstellar neutral gas through the inner Solar System.

Another practical way to determine the total destruction rate of the neutral gas, applied already for the case of He is the comparison of the fluxes of atoms on direct and indirect orbits reaching the neutral gas detector able to differentiate these components (e.g. GAS experiment on Ulysses). As it can be easily seen, the atoms that have passed the point of closest approach (indirect orbits) have been exposed to ionization processes for a longer time. Therefore the flux density on indirect orbits is reduced over that on direct ones. From the difference of the related neutral atom fluxes Witte et al. (1996) have recently computed the combined ionization rate for interstellar helium (see also Section 2.3).

4. Summary

In summary, we can conclude that the contributions of the different ionization processes to the total ionization rate for the most abundant interstellar species are basically known. The ionization of the noble gases He and Ne is almost completely dominated by photoionization, whereas for H charge-exchange with the solar wind is most important. For other species, such as O and Ar, both processes contribute significantly. Electron impact ionization can typically contribute by $\sim$ 10% to the total rate in the inner Solar System. Because direct measurements of the solar EUV flux are not yet continuously available, the variation of the ionization rate over the solar cycle still contains a relatively large uncertainty. The recent measurements of pickup ion distributions and of the neutral helium gas provide an independent tool to determine the total ionization rate that can be used to cross calibrate with the results obtained for the individual ionization processes.

Acknowledgements

The authors are grateful to M.Allen for supplying us with new data on photoionization cross-sections compiled by him. We thank also M.Gruntman for drawing our attention to and support in collecting the most recent data on charge-exchange cross-sections. D.R. was supported by grant No.2 P03C.004.09 from the Committee for Scientific Research (Poland). This work was also supported in part through NASA contract NAS7-918, NSF Grant INT-911637, NASA Grant NAGW-2579. The SWICS/Ulysses work contributing to the paper was supported by the NASA/JPL contract 955460 and the Swiss National Science Foundation.

References

Afrosimov, V.V, Mamaev, Yu. A., Panov, M.N., and Fedorenko, N.V.: 1969, *Sov. Phys. Tech. Phys.* **14**, 109

Allen, C.W. 1973: in *Astrophysical Quantities*, University of London, The Athlone Press

Askew, S.D., and Judge, D.L.: 1984, *Planet. Space Sci.* **32**, 779

Axford W.I.: 1972, in *Solar Wind*, NASA SP-308, p.689

Barnett, C.F., Hunter, H.T., Kirkpatrick, M.I., Alvarez, I, Cisneros, C. and Phaneuf, R.A.: 1990, *Atomic data for fusion*, Oak Ridge Natl. Lab., ORNL-6086/VI

Cummings, A.C., and Stone, E.C.: 1996, *Space Sci. Rev.*, this issue

Cummings, A.C., and Stone, E.C.: 1990, *Proc. of 21-st Internat. Cosmic Ray Conf.* **6**, 202

Cummings, A.C., and Stone, E.C.: 1988, *Proc. of the Sixth Internat. Solar Wind Conf.* **2**, 599

Ehrhardt, A.B., Langer, W.D.: 1987, *Report PPPL-2477*, Princeton University

Feng, W., Ogawa, H.S. and Judge, D.L.: 1989, *J.Geophys. Res.* **94**, 9125

Geiss, J., Gloeckler, G., Mall, U., von Steiger, R., Galvin, A.B., and Ogilvie, K.W.: 1994, *Astron. Astrophys.* **282**, 924

Gloeckler, G.: 1996, *Space Sci. Rev.*, this issue

Gloeckler, G., and Geiss, J.: 1994, *EOS Transactions* **75** No 44, 512

Gloeckler, G., Geiss, J., Balsiger, H., Fisk, L.A., Galvin, A.B., Ipavich, F.M., Ogilvie, K.W., von Steiger, R., and Wilken, B.: 1993, *Science* **261**, 70
Gloeckler, G., Geiss, J., Roelof, E.C., Fisk, L.A., Ipavich, F.M., Ogilvie, K.W., Lanzerotti, L.J., von Steiger, R., and Wilken, B.: 1994, *J. Geophys. Res.* **99**, 17637
Gruntman, M.: 1994, *J. Geophys. Res.* **99**, 19213
Holzer, T.E.: 1977, *Revs. Geophys. Space Phys.* **15**, 467
Isenberg, P.A., and Feldman, W.C.: 1995, *Geophys. Res. Lett.* **22**, 873
Lallement, R., Bertaux, J.-L., Chassefière, E., and Sandel, B.R.: 1991, *Astron. Astrophys.* **252**, 385
Lotz, W.: 1967, *Zeitschrift f. Physik*, **206**, 205
Maher, L.J., and Tinsley, B.A.: 1977, *J. Geophys. Res.* **82**, 689
Marr, G.V., and West, J.B.: 1976, *Atomic Data and Nucl. Data Tables* **18**, 497
Möbius, E., Hovestadt, D., Klecker, B., Scholer, M., Gloeckler, G., and Ipavich, M.: 1985, *Nature* **318**, 426
Möbius, E., Klecker, B., Hovestadt, D., and Scholer, M.: 1988, *Astrophys. Space Sci.* **144**, 487
Möbius, E., Ruciński, D., Hovestadt, D., and Klecker, B.: 1995, *Astron. Astrophys.* **304**, 505
Nakai, Y., Shirai, T., Tabata, T., Ito, R.: 1987, *Atomic Data and Nucl. Data Tables* **37**, 69
Ogawa, H.S., Canfield, L.R., McMullin, D., and Judge, D.L.: 1990, *J.Geophys. Res.* **95**, 4291
Ogawa, H.S., and Judge, D.L.: 1986, *J. Geophys. Res.* **91**, 7089
Ogawa, H.S., Phillips, E., and Judge, D.L.: 1993, AGU Spring Meeting - poster
Quèmerais , E., Bertaux, J.-L., Sandel, B.R., and Lallement, R.: 1994, *Astron. Astrophys.* **290**, 941
Ratkiewicz, R., Ruciński, D., and Ip, W.-H: 1990, *Astron. Astrophys.* **230**, 227
Richards, P.G., Fennelly, J.A., and Torr, D.G.: 1994, *J. Geophys. Res.* **99**, 8981
Richards, P.G., and Torr, D.G.: 1984, *J. Geophys. Res.* **89**, 5625
Ruciński, D., and Fahr, H.J.: 1991, *Ann. Geophys.* **9**, 102
Ruciński, D., and Fahr, H.J.: 1989, *Astron. Astrophys.* **224**, 290
Schwenn, R.: 1983, in *Solar Wind Five*, NASA CP-2280, p.489
Scime, E.E., Bame, S.J., Feldman, W.C., Gary, S.P., Phillips, J.L., and Balogh, A.: 1994, *J. Geophys. Res.* **99**, 23401
Stebbings, R.F., Smith, A.C.H., and Ehrhardt, H.: 1964, *J.Geophys. Res.* **69**, 2349
Tawara, H.: 1978, *Atomic Data and Nucl. Data Tables* **22**, 491
Tobiska, W.K.: 1993, *J. Geophys. Res.* **98**, 18,879
Torr, M.G., Torr, D.G., Ong, R.A., Hinteregger, H.E.: 1979, *Geophys. Res. Lett.* **6**, 771
Witte, M., Rosenbauer, H., Banaszkiewicz, M., and Fahr, H.J.: 1993, *Adv. Space Res.* **13(6)**, 121
Witte, M., Banaszkiewicz, M., and Rosenbauer, H.: 1996, this issue

Address for correspondence: Dr. Daniel Ruciński, Space Research Centre of the Polish Academy of Sciences, Bartycka 18A, 00-716 Warsaw, Poland

3-D MAGNETIC FIELD AND CURRENT SYSTEM IN THE HELIOSPHERE

HARUICHI WASHIMI
Shonan Institute of Technology, Tsujido, Fujisawa 251, Japan

TAKASHI TANAKA
Communications Research Laboratory, Nukui-kitamachi, Koganei, Tokyo 184, Japan

Abstract. In this paper a global system of the magnetic field and current from the interaction of the solar wind plasma and the interstellar medium is modeled using a 3-D MHD simulation. The terminal shock, the heliopause and the outer shock are clearly determined in our simulation. In the heliosheath the toroidal magnetic field is found to increase with the distance from the sun. The magnetic field increases rapidly in the upstream region of the heliosheath and becomes maximum between the terminal shock and the heliopause. Hence a shell-type magnetic wall is found to be formed in the heliosheath. Because of this magnetic wall the radially expanding solar wind plasma changes its direction tailward in all latitudes except the equatorial region. Only the equatorial disk-like plasma flow is found to extend to the heliopause through the weak magnetic-field region around the equator. Two kinds of global current loops which sustain the toroidal magnetic field in the heliosphere are found in our simulation.

The influence of the 11-year solar cycle variation of the magnetic polarity is also examined. It is found that the polarity of the toroidal magnetic field in the heliosheath switches at every solar cycle change. Hence the heliosheath is found to consist of such magnetized plasma bubbles. The neutral sheets are found to extend between such magnetized plasma bubbles in the 3-D heliosheath in a complicated form. The magnetic-pressure effect on the heliosheath plasma structure is also examined.

1. Introduction

In a simplified model there are two kinds of fundamental magnetic components in interplanetary space. One, B_R, has a split-monopole (poloidal) configuration. That is, it is a radial magnetic field centered on the sun but its polarity reverses between the northern and southern hemispheres by forming a neutral magnetic sheet on the solar equatorial plane. This split-monopole component is formed by an interaction between the seed solar magnetic-dipole field and the radial outflow of the solar wind plasma. The other fundamental component is a toroidal magnetic field B_ϕ, formed by the solar rotation: $B_\phi = (R \sin\theta \cdot \Omega/V) \cdot B_R$, where Ω is the angular speed of the solar rotation, R the distance from the sun, θ the colatitude angle, and V the velocity of the solar wind. The strength of the split-monopole field B_R decreases with the square of R while that of B_ϕ is simply inversely proportional to R. These components have almost the same intensity at $R = 1$ AU, and B_R decreases with R very quickly in deep interplanetary space. Thus we may consider only B_ϕ in discussions of the magnetic field in the deep interplanetary space and heliosheath.

The poloidal current, which is associated with B_ϕ, flows from the outer heliosphere into the solar corona along the solar equatorial plane and flows out of the

Space Science Reviews **78**: 85–94, 1996.

corona toward the outer heliosphere in middle and high latitudes. How do these out-going and in-coming currents connect in the outer heliosphere?

It has been suggested that the toroidal magnetic field is amplified at the terminal shock and in the heliosheath in association with the deceleration of the solar wind (Cranfill 1971, Axford 1972, Lee 1986, Holzer 1988, Suess 1990). What is the structure of the poloidal current that sustains this amplified toroidal magnetic field?

To answer these questions we perform a 3-D MHD computer simulation of the interaction between the solar wind plasma and the interstellar medium by using a high-resolution, the total variation diminishing (TVD) method (Tanaka, 1994). Focusing our discussion on MHD processes, we neglect the effects of the neutral particles in the interstellar medium and of cosmic rays. We first study the magnetic-field and current system in the outer heliosphere in detail when the polarity of the toroidal magnetic field is fixed on the inner boundary of our simulation. In the next step we examine the response of this system when the polarity of the the toroidal magnetic field on the inner boundary switches at every 11-year solar-cycle change.

The magnetic-pressure effect in the heliosheath, where the toroidal magnetic field is stimulated and amplified, is one of the interesting physical problems in heliospheric physics. B_ϕ increases with the colatitude angle θ, becomes maximum at $\theta_0(\approx \pi/2)$ very near the equator, falls to zero on the equator, and then changes its polarity in the southern hemisphere. Thus the magnetic pressure pushes plasma both poleward and equatorward from the region $\theta = \theta_0$ in the northern hemisphere, and from the region $\theta = \pi - \theta_0$ in the southern hemisphere. These kinds of magnetic-pressure effects have also been evaluated by simulation in some other cosmic plasmas (e.g., Washimi and Shibata, 1993; Washimi et al., 1996). The magnetic pressure also confines the inside solar wind because the conservation relation $V \cdot R \cdot B_\phi = const$ indicates that the magnetic intensity is stronger at larger R in the heliosheath if the subsonic solar wind decelerates faster than R^{-1}. The contraction of the scale of the terminal shock due to the magnetic pressure was discussed theoretically in a spherically symmetric system by Cranfill (1971), Axford (1972), Lee (1986), Holzer (1988) and Suess (1990). A 2-D MHD simulation showed the poleward flow due to the magnetic collimation by the toroidal field and the contraction of the scale of the terminal shock (Washimi, 1993). This contraction was also analyzed in some detail recently in a 3-D MHD simulation performed by using the Lax-Wendroff scheme (Nozawa and Washimi, 1996). We expect that our new simulation using a high-resolution TVD scheme can further clarify these magnetic-pressure effects.

2. Numerical Method

Our numerical simulation is based on the ideal time-dependent MHD equations. An unstructured grid system is used for the calculation in order to obtain a high-resolution near the inner boundary and the solar rotation axis. Because the ordinary finite-difference method (FDM) used to solve differential equations numerically is not applicable in an unstructured grid system, we instead use the finite-volume method (FVD) based on the flux conservation law. As with the FDM, some viscous terms have to be included in order to solve hyperbolic equation systems stably. In the present method, a TVD viscous term is introduced through an upwinding method after the rotation of the equation system. We use a third-order TVD scheme based on the monotonic upstream scheme for the conservation law (MUSCL) approach with a linearized Riemann solver (Tanaka, 1994). This scheme is useful for shock-capturing problems and also enables us to perform stable computations with high-order accuracy even in subsonic regions.

The inner and outer boundaries for the simulation are 50 and 1,000 AU from the sun. The sun is at the center of the simulation region. The x axis is in the equatorial plane pointing toward the upstream direction of the interstellar wind, the z axis is parallel to the solar rotation axis pointing the north, and the y axis is set as to form a right-hand coordinate system. The grid system for the calculation is not aligned to the xyz coordinate system but is generated from the spherical coordinates (r, θ, ϕ), with the number of grid points being $(62 \times 60 \times 64)$: the grid size ΔR is $\sim$ 6 AU near the inner boundary, increases with R up to $\sim$ 20 AU near the outer boundary, while $\Delta\theta$ and $\Delta\phi$ are 3 and 5.6 degrees, respectively. On the inner boundary, a radial solar wind flow and interplanetary toroidal magnetic field are given. On the outer boundary, a uniform interstellar medium is prescribed at the upstream side and the condition of zero-gradients is assigned at the downstream side. We assume that the solar wind speed is 400 km/s, that the density and the interplanetary toroidal magnetic field at 1 AU are 5×10^6 m^{-3} and 2.8 nT, and that the interstellar plasma speed, plasma density and magnetic field are 25 km/s, 10^5 m^{-3}, and 0.15 nT. The interstellar plasma flow is assumed to be parallel to the solar equatorial plane and the interstellar magnetic field is parallel to the solar rotation axis. The temperature of both of the solar wind plasma at the inner boundary and of the interstellar plasma is assumed to be 10^4 K. The computation is performed step by step until a quasi-steady state is obtained.

3. Magnetic Wall and Global Loop Current

The global structure of the plasma in the outer heliosphere is shown in Fig. 1. The terminal shock with the Mach disk (Wallis and Dryer, 1976) and the heliopause are clearly evident in panels P and T, and the outer shock is evident in panel P. The magnetic field intensity is shown in Fig. 2. It is seen that B_y, the toroidal-field component on the meridional plane, is dominant in the solar wind plasma (in the interplanetary space and the heliosheath), whereas B_z is the main field in the interstellar plasma. It is evident that B_y is enhanced in the heliosheath, and this enhancement is due to the accumulation of the magnetic field associated with the deceleration of the solar wind at the terminal shock and beyond. B_y should disappear on the neutral plasma sheet, but its amplitude is high on the equator (the lower-half of the B_y panel). This means that the neutral sheet in the heliosheath bends toward

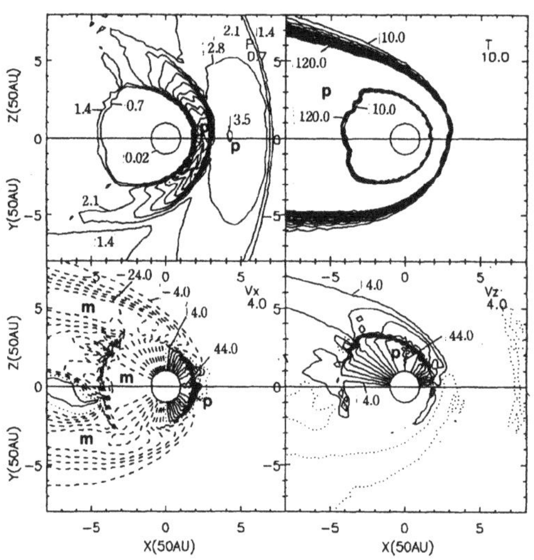

Figure 1. Global structure of the plasma in outer heliosphere. The upper and lower halves of each panel,respectively, correspond to the meridional plane (x-z plane) and the equatorial plane (x-y). The upper-left, upper-right, lower-left, and lower-right panels respectively show the equipressure contours ($\Delta P = 0.7$), the equitemperature contours ($\Delta T = 10.0$, up to 120.0), the equi-velocity contours of the x component ($\Delta V_x = 4.0$), and the equi-velocity contours of the z component ($\Delta V_z = 4.0$). The normalized values of pressure, temperature, and velocity in these panels are 1.44×10^{-14} N m^{-2}, 10^4 K, and 9.1×10^3 m s^{-1}. Solid and dashed lines respectively indicate positive and negative values, while dotted lines correspond to zero value. Regions of plus maximum and minus minimum are shown by p and m, respectively.

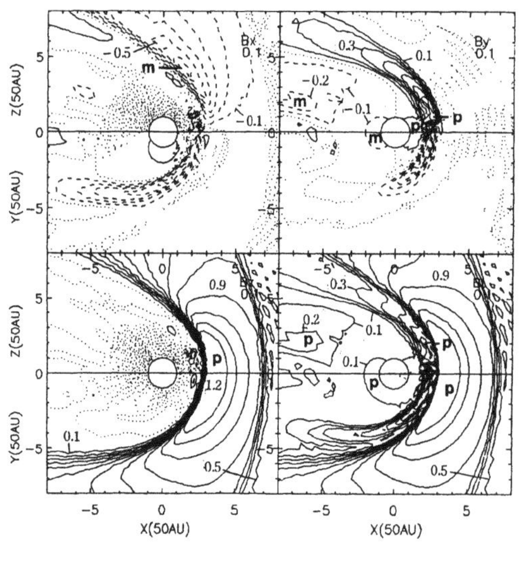

Figure 2. Global structure of the magnetic field in the outer heliosphere. The upper and lower halves of each panel respectively correspond to the meridional plane (x-z plane) and the equatorial plane (x-y). The upper-left, upper-right, lower-left and lower-right panels respectively show B_x contours ($\Delta B_x = 0.1$), B_y ($\Delta B_y = 0.1$), B_z ($\Delta B_z = 0.1$), and $|B|$ ($\Delta |B| = 0.1$). The normalized values of the magnetic intensity in these panels is 0.3 nT. Solid and dashed lines respectively indicate positive and negative values, while dotted lines correspond to zero value. Regions of plus maximum and minus minimum are shown by p and m, respectively.

the upper side under the influence of the interstellar medium, and hence the amplitude of B_y is greater on the equatorial plane than on the meridional plane. This also means that our system is not a complete north-south symmetric system: the magnetic field in the LISM is oriented z-direction while the toroidal field in the heliosheath is north-south anti-symmetric at the initial time of the simulation. This magnetic-field configuration breaks the symmetry of the plasma configuration. In addition, there is no dominant restoration force to keep symmetry of the plasma in the outer heliosphere.

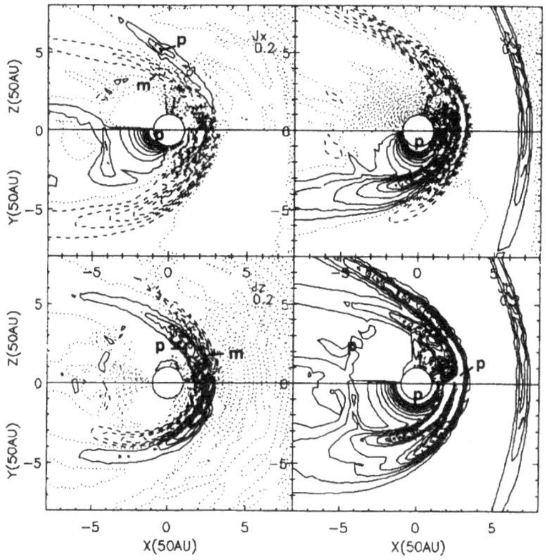

Figure 3. Global structure of the electric current in the outer heliosphere. The upper and lower halves of each panel respectively correspond to the meridional plane (x-z plane) and the equatorial plane (x-y). The upper-left, upper-right, lower-left, and lower-right panels respectively show contours of J_x ($\Delta J_x = 0.1$), J_y ($\Delta J_y = 0.1$), J_z ($\Delta J_z = 0.1$), and $|J|$ ($\Delta |J| = 0.1$). The normalized values of the current intensity in these panels is 3.18×10^{-17} Ampere$\cdot$ m^{-2}. Solid and dashed lines respectively indicate positive and negative values, while dotted lines correspond to zero value. Regions of plus maximum and minus minimum are shown by p and m, respectively.

It is interesting that the equipressure contours indicate a pressure minimum in the middle region of the heliosheath on the upstream side, and that the pressure depression is more clear on the equatorial plane than on the meridional plane. By taking into account that B_y is not zero but is stronger on the equatorial plane than on the meridional plane due to the upward bending of the equatorial sheet, it is recognized that these pressure-depressions must be due to a magnetic-pressure effect. In fact, the region of this series of the pressure-depressions in Fig. 1 is identical to that of the maximum-B_y in Fig. 2, and it extends in a shell-like form in 3-D space. Thus a magnetic wall with a shell-like structure is found in the middle of the heliosheath on the upstream side.

It is interesting that the region of the positive V_x (solid lines) in the heliosheath just outside the terminal shock in the panel V_x of Fig. 1 is very thin. This indicates that the radially expanding solar wind in the middle and high latitudes on the upstream side is stopped by the magnetic wall and changes its direction downstream (dashed lines). Only the sheet plasma flow near the equator (it is bent upward) is found to keep a positive V_x (series of projections of solid lines) up to near the heliopause and then change its direction tailward along the heliopause surface. Thus, it is clear that the region between the terminal shock and the magnetic-wall in the heliosheath is occupied by the solar wind in the middle and high latitudes and that the region between the magnetic wall and the heliopause is occupied by the solar wind which comes along the equatorial neutral sheet.

The poloidal current intensity is shown in Fig. 3. In the panels J_x and J_z we see that the poloidal current in the positive-x direction comes from tail region along the surface of the heliopause (solid lines in panel J_x and dashed lines in panel J_z). This current nears the equatorial region on the heliopause surface and then goes into the inner region of the heliosheath along the (bending) neutral sheet. Some of

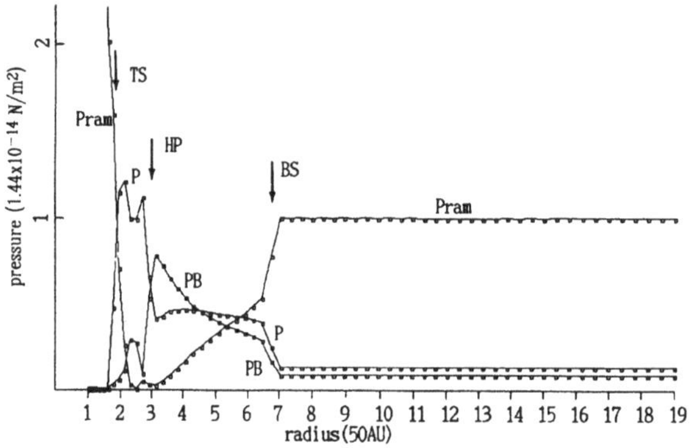

Figure 4. Spatial variation of the ram pressure (Pram), thermal pressure (P), and magnetic pressure (PB) along the line θ(colatitude) $= 80°$ on the meridional (x-z) plane on the upstream side. TS, HP, and BS respectively mean the terminal shock, the heliopause and the bow shock (outer shock).

the poloidal current goes up to the polar region along the surface of the terminal shock and then goes back to the tail from the polar region of the terminal shock (dashed lines in panel J_x and solid lines in panel J_z) along the axis of the toroidal magnetic field (dotted line in the panel B_y in Fig. 2) and will connect step-by-step to the original current in the positive-x direction on the heliopause surface in the tail region. This current thus constitutes a global 3-D loop structure in the heliosheath.

The other current goes into interplanetary space across the terminal shock along the equatorial sheet and reaches the solar corona. From the solar corona the poloidal current goes out again at all latitudes except at the equator and reaches the terminal shock surface. This current goes up along the terminal shock surface to the pole region, and then goes back to the tail together with the first current along the axis of the toroidal magnetic field and will finally connect to the original current in the positive-x direction. This second current thus also constitutes a global 3-D loop structure through the heliosheath, interplanetary space, and solar corona, and it sustains the interplanetary toroidal magnetic field.

The variation of the ram pressure, thermal pressure and magnetic pressure is shown in Fig. 4. The terminal shock is near $R = 1.8$ (in units of 50 AU), where the ram pressure falls and the thermal pressure and the magnetic field increase sharply, and the heliopause is near $R = 2.9$, where the magnetic pressure is on the second maximum and the ram pressure is minimum (stagnation point). As the authors are aware of, since the heliopause is a tangential discontinuity, the total pressure (thermal + magnetic) is conserved over the heliopause, which follows quite nicely from Fig. 4.

It is interesting that the magnetic wall is found near $R = 2.3$, where the thermal pressure becomes minimum in the heliosheath. The ram pressure also becomes minimum (≈ 0) at the magnetic wall which indicates that the magnetic wall stops the radial outgoing flow of the solar wind. It is interesting that the ram pressure has a finite value again between the minimum (magnetic wall) and the heliopause, which shows that this solar wind plasma does not come directly across the magnetic wall, but does along the equatorial sheet where the magnetic field is zero or very weak.

Thus we can conclude that the solar wind plasma in the heliosheath is separated into two parts: the first part is in the region between the terminal shock and the magnetic wall, and the second part is in the region between the magnetic wall and the heliopause. These two parts of the plasma are connected each other at the equatorial sheet. The first kind of the current loop discussed above flows through these two parts of the plasma and sustains the magnetic field of the wall.

4. Solar-cycle Variation

The above analysis has clarified the fundamental magnetic field and current system, but to get a more realistic configuration the 11-year solar cycle variation of the magnetic field polarity should be taken into account. The response of the magnetic field in the outer heliosphere to the solar-cycle variation, when the polarity of B_ϕ on the inner boundary switches at every 11-year solar cycle variation, is shown in Fig. 5. It is found that in association with the magnetic-polarity change of the sun the polarity of the carried magnetic field in the heliosheath also switch-

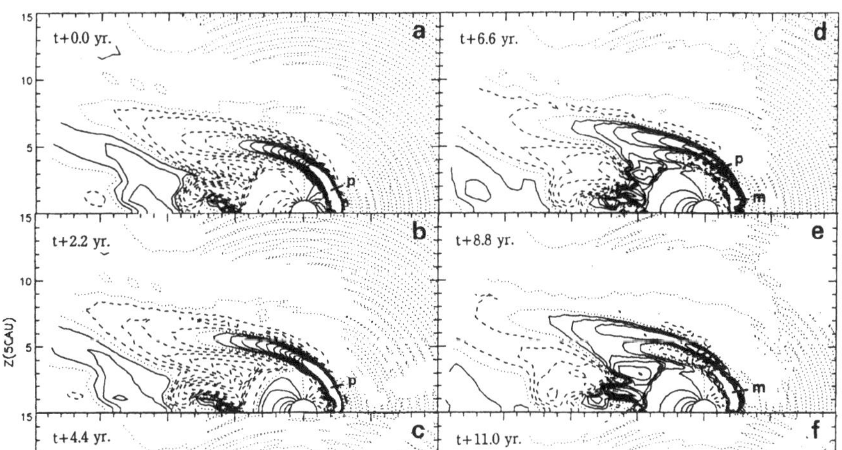

Figure 5. Solar-cycle variation of the toroidal magnetic field B_y. The global structure of the toroidal magnetic field at every 2.2 year (0.2 cycle) period on the meridional plane is shown. Solid and dashed lines respectively indicate positive and negative values of B_y, while dotted lines correspond to zero value. Regions of plus maximum and minus minimum are shown by p and m, respectively. It is apparent that the polarity of B_y of (f) is just opposite of that of (a).

es. The heliosheath is thus found to consist of a series of such magnetized plasma bubbles. It is interesting to note that, as seen in panels a, b, and c, the negative B_y of the previous cycle (near the heliopause) and the new cycle (outside of the Mach disk) seem to merge continuously. The positive B_y region of the previous and new cycles should be between these two negative B_y regions, but this channel is too thin and disappears in our simulation due to the viscosity. We should note, however, that between the two negative B_y regions the plasma sheet still exists in the off-meridional plane because the negative B_y of the previous and new cycles connect to the positive B_y of the previous and new cycles, respectively, through the off-meridional space in the 3-D heliosheath. The same situation is found to be reproduced in the positive B_y region of the new and the next-new cycles in panels d, e, and f. Thus the neutral sheet is produced between every plasma bubble, and is constructed in a complicated form in the 3-D heliosheath.

5. Magnetic-pressure Effect

Figure 6 shows the global structures of P and B_y when B_ϕ is 0.5, 1, and 2 (in units of 2.8 nT) at 1 AU. It is clear that the pressure depression, which corresponds to the magnetic wall in the middle region of the heliosheath, is found even for the weakest

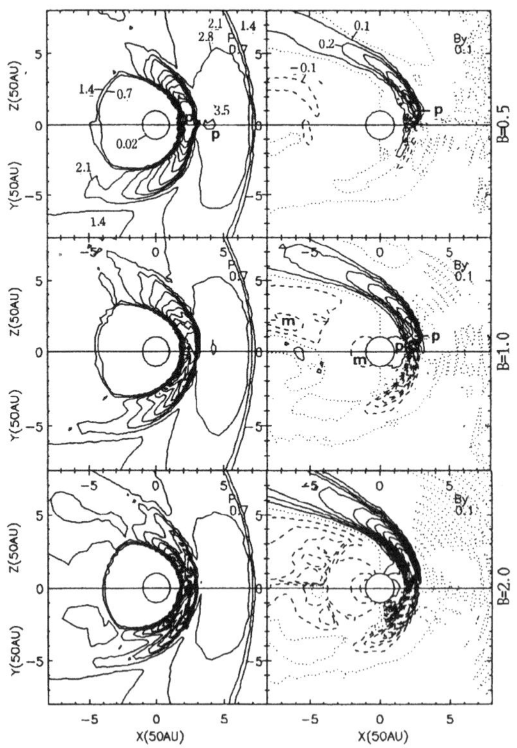

Figure 6. Magnetic pressure. The left and right panels respectively show the equipressure contours (P) and equitoroidal magnetic field contours (B_y) on the meridional plane. The upper and lower halves of each panel, respectively, correspond to the meridional and equatorial planes. The upper, middle and lower panels respectively correspond to $B_\phi = 0.5$, 1 and 2 (in units of 2.8 nT) at $R = 1$ AU. Regions of plus maximum and minus minimum are shown by p and m, respectively.

B_ϕ and is more evident for stronger B_ϕ. This means that the magnetic wall is one of the fundamental structures defining the global configuration of the heliosphere.

The contraction of the scale of the terminal shock due to the magnetic-pressure is also confirmed: as seen in Fig. 6, the scale is smaller for stronger B_ϕ. It is interesting that the scale of the heliopause seems to be the same for the B_ϕ intensities in this figure; hence the volume of the heliosheath seems greater for stronger B_ϕ.

We note that the magnetic-pressure pushes the heliosheath plasma poleward in middle and high latitudes, and simultaneously confines the plasma at low latitudes along the equatorial plane. Such phenomena are clearly evident in 3-D colour graphics which will be published in another paper, but can also be discerned here: the equatorial neutral sheet (which is bent upward) in panels P and B_y is more evident for stronger B_y.

6. Conclusion

A global system of the magnetic field and current resulting from the interaction of the solar wind plasma and the interstellar medium has been examined by using a 3-D MHD simulation. We have used a high-resolution finite-volume TVD scheme and for the solar wind, and interstellar plasmas and magnetic field we have used some standard set of parameters described in Section 2. The toroidal magnetic field in the heliosheath is found to increase with R. It increases rapidly in the upstream region of the heliosheath because the solar wind speed decreases rapidly and the toroidal field thus accumulates there, and it becomes maximum between the terminal shock and the heliopause. Hence a shell-type magnetic wall is found to be formed in the heliosheath. Because of this magnetic wall the radially expanding solar wind plasma changes its direction tailward in all latitudes except the equatorial region. Only the equatorial disk-like plasma flow can extend to the heliopause through the valley of the magnetic wall around the equator. After nearing the heliopause surface, this plasma turns and flows tailward through the region between the magnetic wall and the heliopause.

We find two kinds of global poloidal current loops in the 3-D heliosphere. One of them comes from the tail region toward the equatorial plane along the heliopause surface, goes into the heliosheath along the equatorial neutral sheet up to the terminal shock, goes up to the pole region along the surface of the terminal shock, and then goes back to the tail along the axis of the toroidal magnetic field. Hence this current surrounds the magnetic wall and sustains the enhanced magnetic field in the heliosheath.

The second kind of loop also comes from the tail region toward the equatorial plane along the heliopause surface, and then it goes straight into the solar corona along the equatorial neutral sheet through the heliosheath and interplanetary space. This current goes out of the corona to the terminal shock at middle and high latitudes. After reaching the terminal shock, this current goes up to the pole region along the terminal shock surface and then goes back to the tail along the axis of the toroidal magnetic field. Hence, this current sustains the interplanetary toroidal magnetic field and also the enhanced field in the heliosheath.

The influence of the 11-year solar cycle variation of the magnetic polarity has also been examined. The two kinds of current loops switch their orientation at every solar-cycle change. It has been found that in associating with the magnetic-polarity change of the sun, the polarity of the carried magnetic field in the heliosheath also switches. The heliosheath is thus found to consist of a series of such magnetized plasma bubbles. The neutral sheets extend between such magnetized plasma bubbles in 3-D space in a complicated form. More detailed analysis of these sheets will be especially important for the study of the penetration of the galactic cosmic rays into the heliosphere.

The magnetic-pressure effect in the heliosheath seems weaker than that inferred from a 2-D simulation (Washimi, 1993). This difference is due to the difference in geometry: The deceleration of the solar wind in the heliosheath is not as sharp in 3-D case, because the solar wind on the upstream side of the heliosheath can change its direction and escape tailward through the lateral side without a strong deceleration. This results in the magnetic-pressure effect being weaker in the 3-D case. The effect is nonetheless confirmed to be important in the 3-D case for a standard intensity of B_ϕ such as 2.8 nT at 1 AU.

Acknowledgements

This work was supported by the Interministry Fundamental Research Program of the Science and Technology Agency and by a grant-in-Aid for Scientific Research from the Ministry of Education, Science, Sports and Culture of Japan (06640569). The computations were performed on the vector-parallel supercomputer Fujitsu VPP500.

References

Axford, W.I.: 1972, *Solar Wind, NASA Spec. Publ., NASA SP-308*, 609.
Cranfill, C.W.: 1971 *Ph.D. dissertation*, Univ. Calif. San Diego.
Holzer, T.E.: 1989, *Ann. Rev. Astron. Astrophys.* **27**, 199–234.
Lee, M.A.: 1988 *Proc. VI Internat. Solar Wind Conf.*, NCAR, Boulder, Colo.
Nozawa, S. and Washimi, H.: 1996, submitted for publication.
Suess, T.S.: 1990, *Reviews Geophys.* **28**, 97–115.
Tanaka, T.: 1994, *Comput. Phys.* **111**, 381–389.
Wallis, M.K., and Dryer, M.: 1976, *Ap. J.* **205**, 895–899.
Washimi, H.: 1993 *Adv. Space Res.* **13**(6), 227–236.
Washimi, H., and Shibata, S.: 1993, *Mon. Not. R. Astr. Soc.* **262**, 936–944.
Washimi, H., Shibata, S., and Mori, M.: 1996, *PASJ* **48**, 23–28.

Address for correspondence: Shonan Institute of Technology, Tsujido, Fujisawa 251, Japan

MODELLING THE HELIOSPHERE

G.P. ZANK and H.L. PAULS
Bartol Research Institute
University of Delaware

Abstract. An overview of our present efforts at the Bartol Research Institute in modelling the large-scale interaction of the solar wind with the local interstellar medium is presented. Particular stress is placed on the self-consistent inclusion of neutral hydrogen in the models and both 2D and 3D structure is discussed. Observational implications are noted.

Key words: interstellar neutrals, numerical modelling, solar wind

1. Introduction

The ionized and neutral components of the local interstellar medium (LISM) determine the global nature and structure of the heliosphere. Neutral interstellar gas flows into the heliosphere, experiencing some deceleration and "filtration" on its passage across the heliospheric boundaries before it acts to decelerate the supersonic outflowing solar wind. The weak coupling of neutral gas and plasma affects both distributions in important ways, and their self-consistent coupling is crucial in modelling the solar wind-LISM interaction. *Baranov and Malama* [1993] developed such a coupled model, using a Monte Carlo algorithm to evaluate the neutral hydrogen (H) distribution and a 2D steady-state fluid description of the plasma. We have since developed an alternative series of models [*Pauls et al.* 1995, 1996; *Zank et al.* 1996a,b,c; *Williams et al.* 1996a,b; *Pauls and Zank* 1996a,b] which are still self-consistent but based on a multi-fluid description of the neutrals. The need for modelling the neutrals as a multi-fluid stems from the variation in the charge exchange cross-section mean free path for H in different regions of the solar wind and LISM. Large anisotropies are introduced in the neutral gas distribution by charge exchange with the solar wind plasma (both shocked and unshocked) and these are well captured on the basis of multi-fluid hydrodynamics.

In this review, we present a synopsis of results from our modelling efforts and refer the reader to the article by *Baranov and Malama* [1996] for a discussion of the Monte Carlo model. Although our multi-fluid approach cannot provide neutral spectra, it possesses, nonetheless, several distinct advantages over the Monte Carlo model. (i) It provides a physically intuitive description of the neutrals (ii) Unlike the present Monte Carlo model, difficulties with particle statistics are absent in the multi-fluid approach, so enabling us to determine accurately the neutral H distribution well inside the termination shock. (iii) Our multi-fluid model is fully time-dependent, so allowing us to address the intrinsic variability of the solar wind and the possibility that the heliospheric boundaries are time dependent. (iv) The model

Space Science Reviews **78**: 95–106, 1996.

is computationally efficient and easily extended to 3D. The issue of computational efficiency is of particular importance when additional processes associated with cosmic rays [*Lee and Axford* 1988; *Donohue and Zank* 1993; *Zank et al.* 1993, 1994], magnetic fields [*Washimi and Tanaka*, 1996] and a non-symmetric solar wind [*Pauls and Zank* 1996a,b] are included.

Limitations of space preclude our discussing the important topics of cosmic ray mediation of the termination shock [e.g., *Donohue and Zank* 1993], the interaction of the heliospheric boundaries with interplanetary disturbances [*Story and Zank* 1995, 1996], the nature of shock propagation in the outer heliosphere [*Zank and Pauls* 1996] and turbulence generated by pick-up ions [*Zank et al.* 1996d,e].

2. Models

Instead of attempting to solve the neutral Boltzmann equation directly [although see *Zank et al.* 1996b], we recognize that, to a good approximation, there exist essentially three distinct neutral H components [*Holzer* 1972; *Hall* 1992] corresponding to three physically distinct regions of origin. Neutral H atoms whose source lies beyond the heliosphere (region 1) are component 1. This "thermal" neutral component is thus interstellar in origin although dynamical changes to the distribution result from charge exchange with the weakly compressed interstellar plasma upstream of the heliopause. By contrast, neutral component 2, which is produced in the solar wind shock heated heliosheath and heliotail (region 2) is suprathermal, the high temperatures reflecting charge exchange with a 10^6K plasma. Although of low density, component 2 can be dynamically important. The third or "splash" component is produced in the cold supersonic solar wind itself (region 3) and this very tenuous component is characterized by high radially outward velocities. Each of these three neutral components is well represented by a distinct Maxwellian distribution function appropriate to the characteristics of the source distribution. The production and loss terms corresponding to each neutral component then assume simple forms [*Ripkin and Fahr* 1983; *Zank et al* 1996a; *Williams et al.* 1996a]. The complete highly non-Maxwellian H distribution function is then the sum over the three components, *i.e.*, $f(\mathbf{x}, \mathbf{v}, t) = \sum_{i=1}^{3} f_i(\mathbf{x}, \mathbf{v}, t)$. Under the assumption that each of the neutral component distributions is approximated adequately by a Maxwellian, one obtains immediately an isotropic hydrodynamic description for each neutral component. Nonetheless, it must be emphasized that the full neutral description f corresponds to a highly anisotropic neutral distribution.

Our multi-fluid description comprises all three neutral fluids coupled to a hydrodynamic plasma. Even this model is computationally demanding on todays fastest supercomputers. However, an important simplification to the multi-fluid model which captures most of the essential features of both the multi-fluid and Monte Carlo models is that used by *Pauls et al.* [1995]. In this simplified approach, only

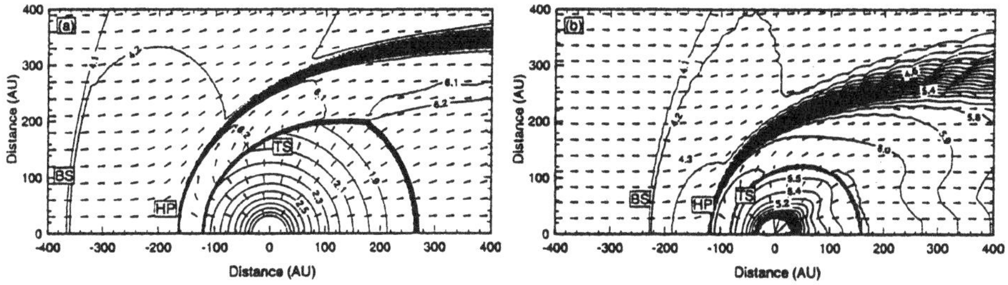

Figure 1. (**a**) Normalized flow vectors and $\log(T)$ contours depicting the interaction of the solar wind and the LISM when charge exchange is ignored. (**b**) The same as (a) except interstellar hydrogen is now included self-consistently [*Pauls et al.* 1995, 1996].

component 1 is computed. Thus, production and loss of component 1 occurs in region 1 and losses elsewhere, as before, but components 2 and 3 are neglected everywhere. This allows for considerable savings in computational time and the multi-fluid and the *Pauls et al.* two-fluid model are compared in detail in *Zank et al.* [1996a].

In concluding this section, we note that the charge-exchange cross-sections quoted by *Maher and Tinsley* [1977] and *Fite et al.* [1962] differ by $\sim$ 40% at 1eV energies and this gives rise to similar uncertainties in the interplanetary neutral number densities. Secondly, additional collisional processes (H-H, H-P) may be important in the heliotail [*Williams et al.* 1996a].

3. 2D Heliospheric Structure

3.1. Gas Dynamic Versus PZW Model

The basic features of the solar wind-LISM interaction can be understood on the basis of the computationally simpler PZW model. The steady-state plasma structure of the gas dynamic model is illustrated in Figure 1a [*Baranov and Malama* 1993; *Steinolfson et al.* 1994; *Pauls et al.* 1995], and the characteristic bullet shape and elongation of the termination shock is clearly exhibited. Figure 1b, by contrast, uses precisely the same boundary conditions but includes neutrals self-consistently in the manner prescribed by PZW. The PZW heliosphere is reduced noticeably in size and the heliosheath flow remains subsonic (see *Pauls et al.* [1995, 1996] for further discussion). The neutral distribution is discussed below.

3.2. Multi-fluid Two-shock Model

Here we summarize the two-shock results obtained using the full multi-fluid model. Our model has been compared to a Monte Carlo model run and, in the vicinity of

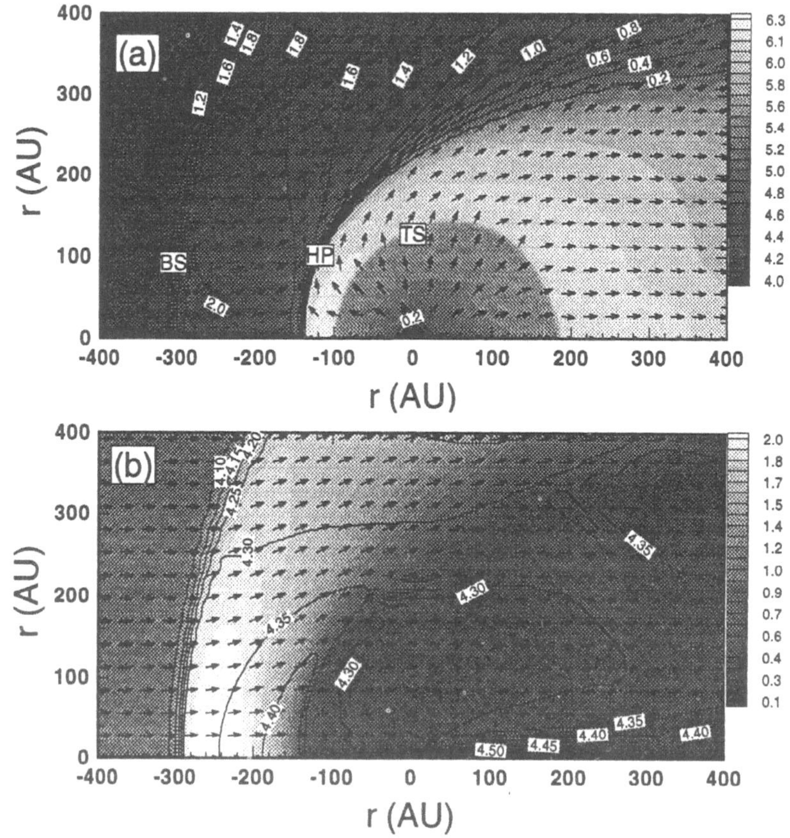

Figure 2. (**a**) A 2D plot of the global plasma structure for the two-shock model. The shading refers to the Log(plasma temperature), the contours show the normalized (to the upstream LISM value) density and the arrows depict flow direction. The bow shock (BS), heliopause (HP) and termination shock (TS) are marked. (**b**) A 2D plot of the component 1 neutral distribution for the two-shock model. The shading here corresponds to density and the contours to Log(temperature) [*Zank et al.* 1996a].

the heliospheric boundaries, very reasonable agreement between the two approaches is found. Inside the heliospheric termination shock, however, the neutral statistics of the *Baranov and Malama* [1995] run become so poor that we cannot compare the two simulations [*Williams et al.* 1996b].

In view of the limitations on space, we provide little discussion about the basic results (see instead *Zank et al.* [1996a] and *Williams et al.* [1996b]). At 1AU, the solar wind number density, velocity and temperature are taken to be 5cm^{-3}, 400km/s and 10^5K respectively. For the LISM, the plasma and neutral number density are 0.07 and 0.14cm^{-3} respectively, the velocity is 26km/s and the temperature is 10 900K.

In Figures 2a and 2b, 2D plots of the plasma and component 1 (*i.e.*, neutrals

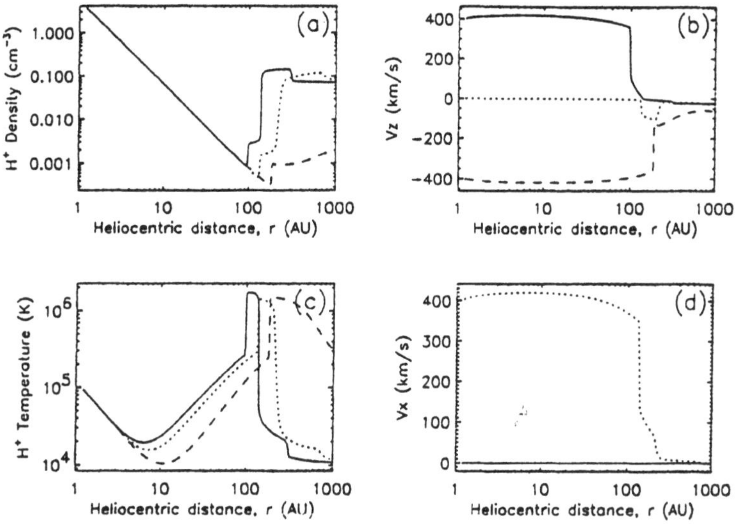

Figure 3. Lines of sight profiles for the plasma in the upstream direction (—), sidestream direction (· · ·) and downstream direction (- - -). Shown from left to right and top to bottom are **(a)** the density profiles; **(b)** the v_z plasma velocity profiles; **(c)** the temperature profiles, and (d) the velocity v_x profiles [*Zank et al.* 1996a].

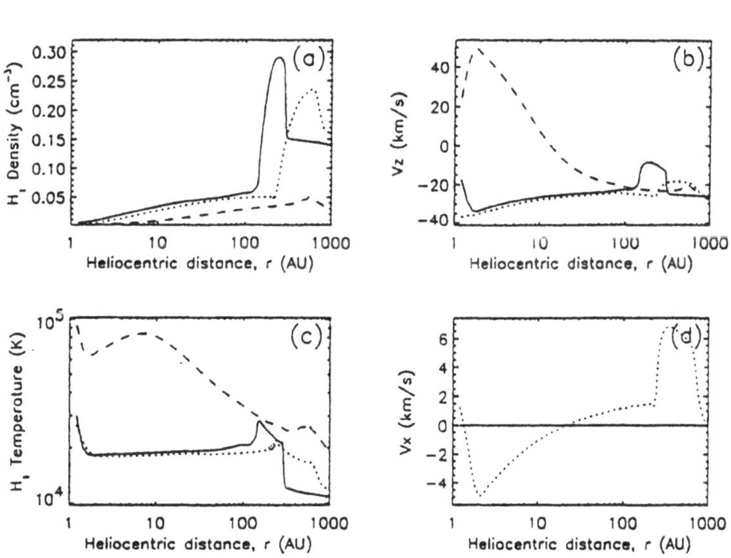

Figure 4. Profiles for component 1 neutrals (format as in Figure 3) [*Zank et al.* 1996a].

of interstellar origin) are presented. A large hydrogen wall is formed between the BS and HP in the upstream direction. In Figures 3 and 4, we show line of sight profiles for the plasma and component 1 neutrals respectively. The following points can be made. (i) The TS is located at $\sim$ 95AU in the upstream direction. (ii) Interstellar component 1 is decelerated, filtered and heated between the BS and HP in the upstream direction, so leading to the formation of a hydrogen wall with a maximum density $\sim$ 0.3cm^{-3}. Component 1 neutrals cross the TS with speeds of $\sim$ 21km/s, densities $\sim$ 0.06cm^{-3} and temperatures of 20 000K. (iii) Leakage of component 2, produced via charge exchange with the hot shocked solar wind plasma, into the cooler shocked LISM heats the plasma through secondary charge exchange, thus leading to the formation of an extended thermal foot abutting the HP. The heating affects the BS location and component 1 temperature. (iv) Upstream and downstream heliospheric temperatures of component 1 are highly asymmetric, as are the neutral gradients ($\sim R^{0.25}$ and $R^{0.35}$ respectively, R the heliospheric radius). (vii) *Zank et al.* [1996a] point out the possibility that the HP is weakly time dependent on long time scales ($\sim$ 180 years).

Two points related to the current status of observations need to be made. (1) *Bertaux et al.* [1985] claim to observe more modest temperatures for interstellar hydrogen in the heliosphere in the upstream direction (8000K) than we predict for a two-shock model, much less than the 20 000K upper limit suggested by *Adams and Frisch* [1977]. In the downwind direction, *Clarke et al.* [1995] present observations which are consistent with neutral temperatures $\sim$ 20 – 30 000K, somewhat low when compared to the simulations. The hydrogen wall temperature, column density and velocity appear to be consistent with the observations presented by *Linsky and Wood* [1996]. (2) H-H collisions in the heliotail suggest that the neutral temperature along that line of sight should be $\sim 10^5$K with an associated column density of $\sim 10^{14}$cm^{-2}. Observations by *Gry et al.* [1995] and *Bertin et al.* [1995] in the direction of Sirius may provide preliminary evidence for such hot interstellar neutrals.

3.3. Multi-fluid One-shock Model

Although most models assume that the interstellar wind impinging on the heliosphere is supersonic, thus necessitating a two-shock model [*Baranov et al.* 1970], it is by no means clear that such an assumption is completely warranted. Our knowledge of both the local interstellar magnetic field strength and orientation and the energy density in cosmic rays is somewhat rudimentary. The canonical interstellar pressure contributed by cosmic rays is $\sim 10^{-12}$dynes/cm^2 [*Ip and Axford* 1985] with perhaps $\sim (3 \pm 2) \times 10^{-13}$dynes/cm^2 contributed by cosmic rays of energy 300MeV/nuc. and less. These estimates are very uncertain, particularly at MeV energies. By adopting a cosmic ray pressure of 3×10^{-13}dynes/cm^2 together with the local interstellar parameters used above, then the LISM sound speed is $\sim$ 27km/s, implying that the LISM wind is subsonic. It is sensible therefore to consider the possibility of a subsonic interstellar wind. The global heliosphere should more closely resemble the *Parker* [1963] model. We now choose a larger "effective" temperature to account for the added contribution from cosmic rays (and per-

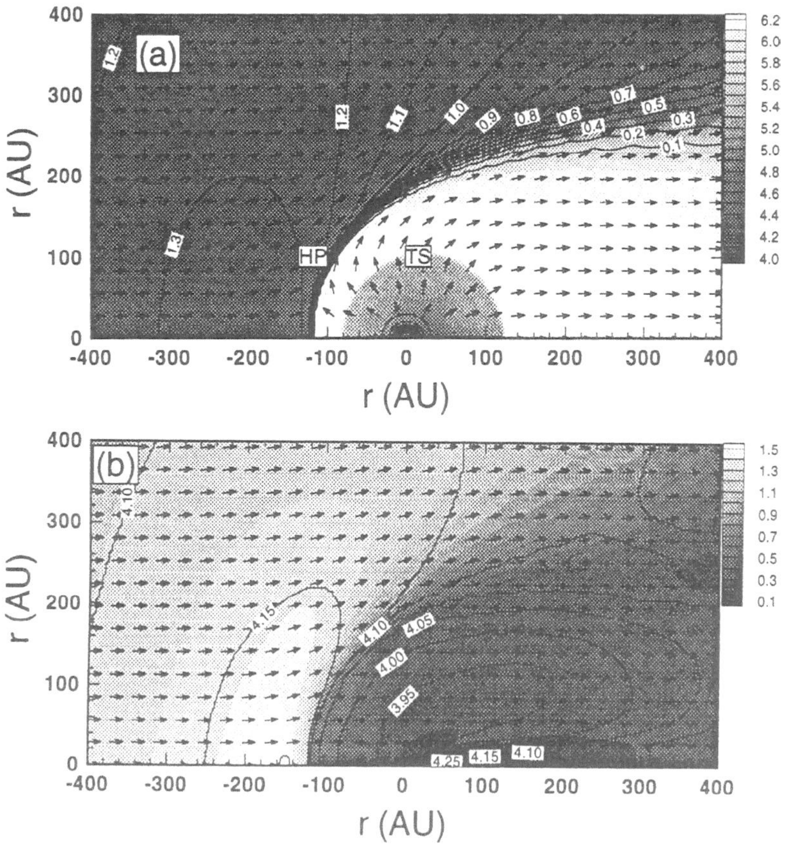

Figure 5. (**a**) A 2D plot of the global plasma structure and (**b**) the component 1 neutral distribution for the one-shock model (format as in Figure 2) [*Zank et al.* 1996a].

haps the magnetic field). The electron temperature is maintained at 10 900K, equal to the assumed thermal proton and neutral interstellar temperatures. No charge exchange is assumed to occur between the cosmic rays and neutrals due to the former's very low number density and the cosmic ray contribution is neglected within the heliosphere.

Global plots of the simulations are presented in Figure 5. Line of sight profiles for the plasma, and H1 components are illustrated in Figures 6 and 7. Again, we simply summarize the main points. (i) Although a bow shock is absent, some adiabatic compression of the incident interstellar flow is evident, which results in the formation of a more extended, lower amplitude hydrogen wall. The hydrogen wall is far more localized than its two-shock counterpart and the amplitude of the wall in the upstream direction is $\sim 0.21\mathrm{cm}^{-3}$ (still larger than the LISM value). (ii) The heliosphere is less distorted along the axis of symmetry than the two-shock case and smaller (due to the higher assumed LISM pressure). (iii) The number density

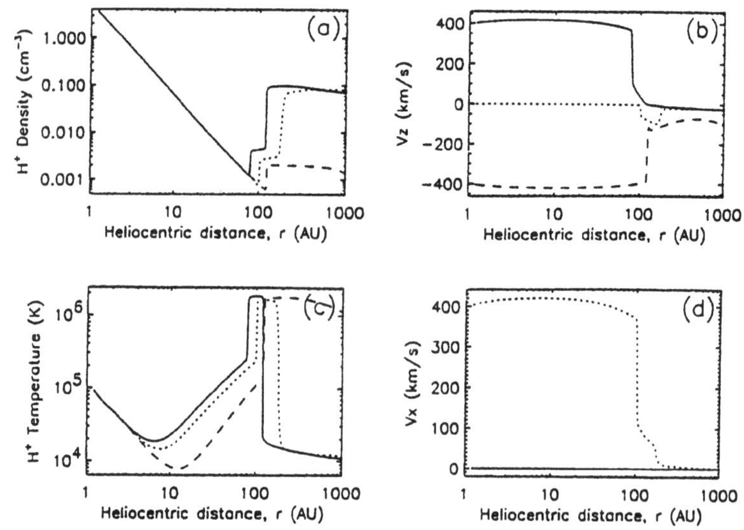

Figure 6. Lines of sight profiles for the one-shock model plasma. (Format as in Figure 3.)

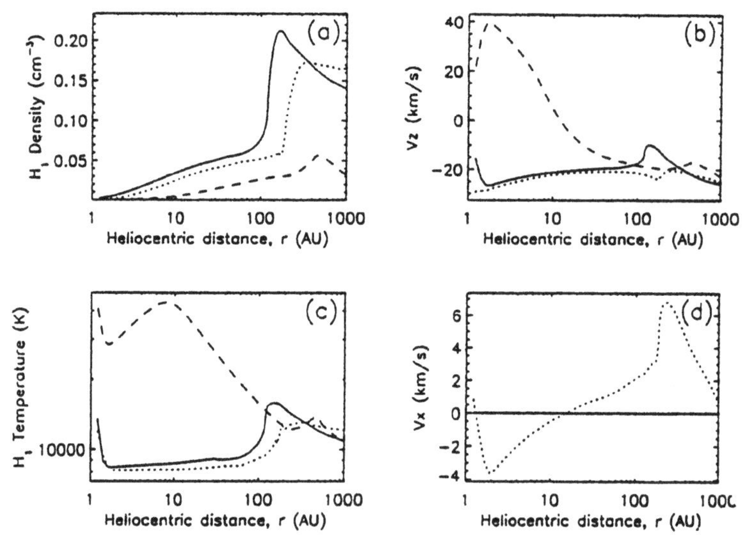

Figure 7. Lines of sight profiles for the one-shock component 1 neutrals.

of H1 flowing across the TS is $\sim$ 0.06cm^{-3} with a velocity of $\sim$ 20km/s, almost identical to the two-shock model. (iv) The upstream and downstream temperature characteristics of the heliospheric component 1 differ significantly between the one- and two-shock models. In the upstream direction, a slight cooling of the neutrals is predicted ($\sim$ 9000K from the 10 900K assumed LISM temperature). A temperature asymmetry between upstream and downstream heliospheric neutral temperatures is again present. The one-shock upstream temperature is consistent with the observations of *Bertaux et al.* [1985] ($\sim$ 8000K).

4. 3D Heliospheric Structure

All approaches to modelling the solar wind - LISM interaction have so far assumed that the solar wind is isotropic. Ulysses, in its pass over the south solar pole [*Smith et al.* 1995], found that the wind speed increased from $\sim$ 400km/s in the ecliptic to $\sim$ 700km/s over the pole and the proton number density decreased similarly from $\sim$ 8cm^{-3} to 3cm^{-3}. The proton temperature also increased from $\sim$ 50 000K to 200 000K [*Phillips et al.* 1995]. The increased velocity implies a factor 1.5 increase in momentum flux from ecliptic to polar regions [*Phillips et al.* 1996]. During solar minimum, the solar wind may be regarded as possessing two components – a steady, long-lived, hot, low-density high speed wind emanating from two large polar coronal holes, one in each hemisphere and extending down to $\sim$ 35° heliolatitude, and bounding a cool, sluggish, high density, somewhat turbulent solar wind. The first results of time-dependent, fully 3D modelling have been presented by *Pauls and Zank* [1996a,b] and are summarized below.

4.1. Gas Dynamic Models

The results of modelling an anisotropic solar wind for both a sub- and supersonic interstellar wind are described here and we assume that no neutrals are present [*Pauls and Zank* 1996a,b]. In Figure 8, we show $\log T$ and normalized flow vectors for a cut through the ecliptic plane (a) and polar plane (b). In this figure, a supersonic interstellar wind is assumed, hence the presence of the bow shock. Figure 9 is similar to Figure 8 except that the impinging interstellar wind is now subsonic.

The main results of the 3D gas dynamic models can be summarized as follows. (i) The solar wind ram pressure increase with heliolatitude results in increased distances to the TS and HP over the poles of the Sun compared to these distances in the ecliptic plane, so leading to an elongation of the TS and HP along the Sun's polar axis. (ii) Due to the polar elongation, the shocked solar wind and LISM material at the heliospheric nose flows to the heliotail in the ecliptic plane, rather than following the longer path over the solar poles. (iii) The heliotail flow is time dependent and turbulent and a pair of large counter-rotating vortices are present. (iv) The ecliptic heliosheath flow in the one-shock simulation is, like the two-shock model, accelerated to supersonic speeds, and consequently at low latitudes, the global TS structure for the one-shock model acquires a bullet shape.

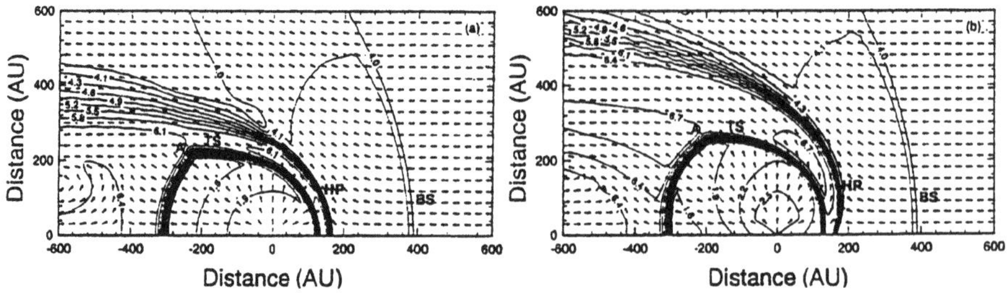

Figure 8. **(a)** Log(temperature) and normalized flow vectors in the ecliptic plane for the two-shock anisotropic solar wind case. TS, HP, and BS refer to the termination shock, heliopause and bow shock respectively. The position of a triple point in the flow is shown by A. **(b)** The same as (a) except now in the polar plane.

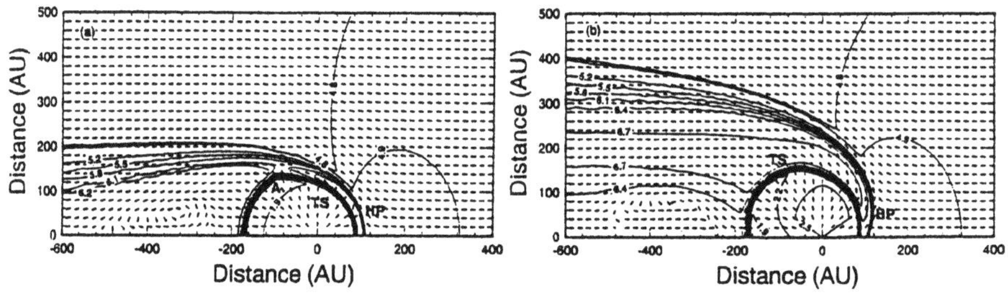

Figure 9. One-shock 3D gas dynamic heliospheric structure (format as in Figure 8).

4.2. The 3D Neutral Model

The self-consistent inclusion of interstellar neutrals in the two shock non-uniform solar wind model described above [see *Pauls and Zank* 1996b] modifies the global heliospheric structure in important ways. Only charge exchange with interstellar neutral hydrogen is modelled [*Pauls et al.* 1995]. Figure 10 illustrates polar and ecliptic plasma structure and Figure 11 that of the neutral distribution. We find the following results. (i) As before, charge exchange in the solar wind reduces the extent of the heliosphere and its boundaries. (ii) The polar elongation is less pronounced than in the no charge exchange case. (iii) The turbulent heliotail of the no charge exchange simulation is absent when charge exchange is included. (iv) A hydrogen wall is formed with densities in excess of 2.8 times the interstellar neutral hydrogen density. Due to filtering by the hydrogen wall, the H density inside the heliosphere is $\sim 1/2$ that of the interstellar H density.

5. Conclusions

Unfortunately, owing to limitations of space, important topics related to the structure of the termination shock, its mediation by cosmic rays and pick-up ions, the

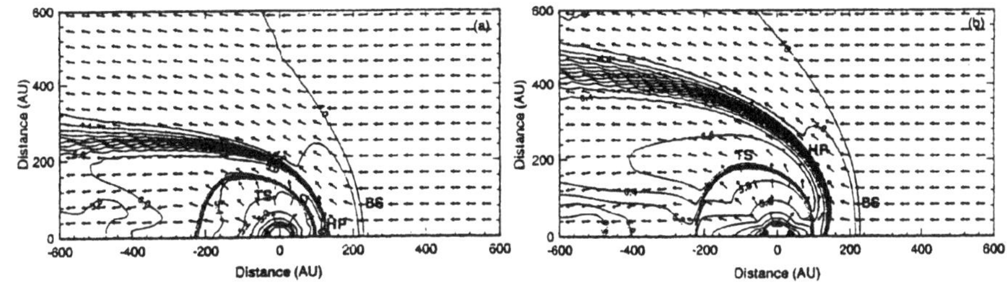

Figure 10. (**a**) Log(plasma temperature) and normalized flow vectors in the ecliptic plane for the two-shock anisotropic two fluid simulation. (**b**) The same as (a) but in the polar plane.

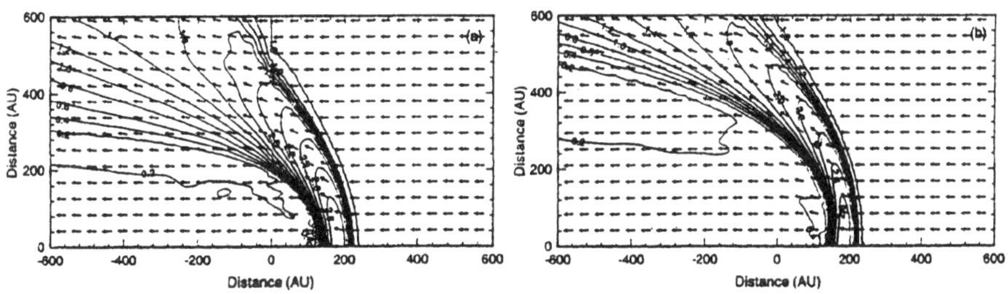

Figure 11. (**a**) Normalized (to interstellar value) neutral hydrogen density contours and flow vectors in the ecliptic plane for the two-shock anisotropic two fluid simulation. (**b**) The same as (a) but in the polar plane.

interaction of the heliospheric boundaries with interplanetary disturbances, turbulence and shock propagation in the outer heliosphere, etc. were not addressed in this review. Instead, a synopsis of our current efforts at the Bartol Research Institute in modelling the large-scale interaction of the solar wind with the LISM was presented and the importance of including neutral hydrogen was emphasized. These models, as detailed as they are, should not be regarded as the last word on the subject and additional surprises are sure to arise with the inclusion of hitherto neglected physical processes.

Acknowledgements

This work was supported in part by NASA grants NAGW-3450 and NAGW-2076 and an NSF Young Investigator Award ATM-9357861. The computations were performed in part on the CRAY-YMP at the San Diego Supercomputer Center. GPZ thanks Prof. J. Geiss for his kind hospitality, for organizing a stimulating meeting and the ISSI for their financial assistance.

References

Adams, T.F., and P.C. Frisch, *Astrophys. J., 212,* 300, 1977.
Baranov, V.B., K. Krasnobaev, and A. Kulikovsky, *Dovkl. Akad. Nauk SSSR, 194,* 41, 1970.
Baranov, V.B., and Y.G. Malama, *J. Geophys. Res., 98,* 15,157, 1993.
Baranov, V.B., and Y.G. Malama, *J. Geophys. Res., 100,* 14,755, 1995.
Baranov, V.B., and Y.G. Malama, *Space Sci. Rev.*, this issue, 1996.
Bertaux, J.-L., R. Lallement, V.G. Kurt, and E.N. Mironova, *Astron. Astrophys., 150,* 1, 1985.
Bertin, P., A. Vidal-Madjar, R. Lallement, R. Ferlet, and M. Lemoine, *Astron. Astrophys., 302,* 889, 1995.
Clarke, J.T., R. Lallement, J-L. Bertaux, and E. Quémerais, *Astrophys. J., 448,* 893, 1995.
Donohue, D.J., and G.P. Zank, *J. Geophys. Res., 98,* 19,005, 1993.
Fite, W.L., A.C.H. Smith, and R.F. Stebbings, *Proc. R. Soc. London, A., 268,* 527, 1962.
Gry, C., L. Lemonon, A. Vidal-Madjar, M. Lemoine, and R, Ferlet, *Astron. Astrophys., 302,* 497, 1995.
Hall, D.T., *Ultraviolet resonance radiation and the structure of the heliosphere,* Ph.D. thesis, Univ. of Arizona, Tucson, 1992.
Holzer, T.E., *J. Geophys. Res., 77,* 5407, 1972.
Ip, W.-H. and W.I. Axford, *Astron. Astrophys., 149,* 7, 1985.
Lee, M.A. and W.I. Axford, *Astron. Astrophys., 194,* 297, 1988.
Linsky, J.L. and B.E. Wood, *Astrophys. J.,* to appear, 1996.
Maher, L., and B. Tinsley, *J. Geophys. Res., 82,* 689, 1977.
Parker, E.N., *Interplanetary Dynamical Processes,* Interscience, New York, 1963.
Pauls, H.L., G.P. Zank and L.L. Williams, *J. Geophys. Res., 100,* 21,595, 1995.
Pauls, H.L., G.P. Zank and L.L. Williams, in press *Proc. Solar Wind 8,* 1996.
Pauls, H.L., and G.P. Zank, *J. Geophys. Res.,* in press, 1996a.
Pauls, H.L., and G.P. Zank, *J. Geophys. Res.,* submitted, 1996b.
Phillips, J.L., S.J. Bame, W.C. Feldman, B.E. Goldstein, J.T. Gosling, C.M. Hammond, D.J. McComas, M. Neugebauer, E.E. Scime, and S.T. Suess, *Science, 268,* 1030, 1995.
Phillips, J.L., S.J. Bame, W.C. Feldman, B.E. Goldstein, J.T. Gosling, C.M. Hammond, D.J. McComas, and M. Neugebauer, in press, *Proc. Solar Wind 8,* 1996.
Ripkin, H.W. and H.J. Fahr, *Astron. Astrophys., 122,* 181, 1983.
Smith, E.J., R.G. Marsden, and D.E. Page, *Science, 268,* 1005, 1995.
Steinolfson, R.S., V.J. Pizzo and T. Holzer, *Geophys. Res. Lett., 21,* 245, 1994.
Story, T.R., and G.P. Zank, *J. Geophys. Res., 100,* 9849, 1995.
Story, T.R., and G.P. Zank, *J. Geophys. Res.*, submitted, 1996.
Washimi, H., and T. Tanaka, *Space Sci. Rev.*, this issue, 1996.
Williams, L.L., D.T. Hall, H.L. Pauls, and G.P. Zank, *Astrophys. J.,*, submitted, 1996a.
Williams, L.L., H.L. Pauls, G.P. Zank, and D.T. Hall, in press *Proc. Solar Wind 8,* 1996b.
Zank, G.P., G.M. Webb and D.J. Donohue, *Astrophys. J., 406,* 67, 1993.
Zank, G.P., I.H. Cairns, D.J. Donohue, and W.H. Matthaeus, *J. Geophys. Res.*, *99,* 14,729, 1994.
Zank, G.P., and H.L. Pauls, in preparation, *J. Geophys. Res.*, 1996.
Zank, G.P., H.L. Pauls, L.L. Williams and D.T. Hall, *J. Geophys. Res.*, submitted, 1996a.
Zank, G.P., H.L. Pauls, L.L. Williams and D.T. Hall, in press *Proc. Solar Wind 8,* 1996b.
Zank, G.P., H.L. Pauls and L.L. Williams, in press *Proc. Solar Wind 8,* 1996c.
Zank, G.P., H.L. Pauls, I.H. Cairns and G.M. Webb, *J. Geophys. Res., 101,* 457, 1996d.
Zank, G.P., W.H. Matthaeus and C.W. Smith, *J. Geophys. Res.*, in press, 1996e.

Address for correspondence: G.P. Zank and H.L. Pauls, Bartol Research Institute,
The University of Delaware, Newark, DE 19716, U.S.A.
(email: zank@bartol.udel.edu, louis@bartol.udel.edu)

THE TERMINATION SHOCK

THE TERMINATION SHOCK OF THE SOLAR WIND

MARTIN A. LEE
Institute for the Study of Earth, Oceans and Space,
University of New Hampshire, Durham, NH 03824 USA

Abstract. An overview of the solar wind termination shock is presented including: its place in the heliosphere and its origin; its structure including the role of interstellar pickup ions and galactic and anomalous cosmic rays; its inferred location based on Lyman-α backscatter, Voyager radio signals, and anomalous cosmic rays; its shape and movement.

1. Introduction

It is a particular pleasure for me to present this lecture of the First ISSI Workshop on the Heliosphere in the Local Interstellar Medium at the Physikalisches Institut of the University of Bern. It gives me an opportunity to visit the Institut for the first time, to visit friends and colleagues, and to enjoy the wonderful view of Bern and the Bernese Alps from the rooftop. I wish I could greet you appropriately in the local dialect, Swiss German!

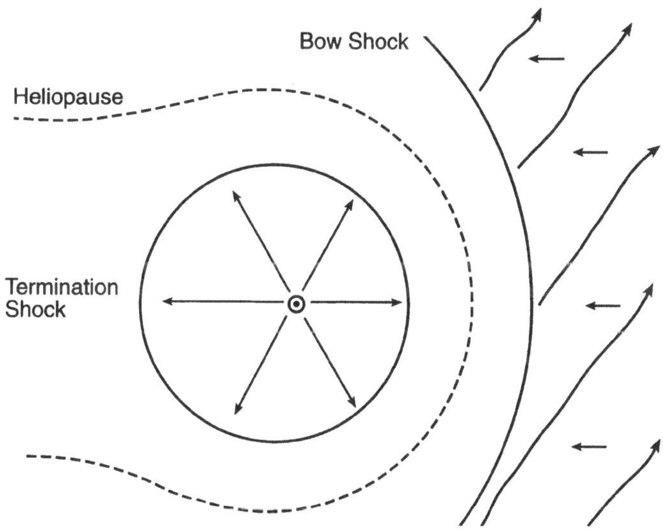

Fig. 1. Schematic view of the heliosphere with the Sun (and the orbit of Jupiter) at the center, the radial solar wind flow, and the flow of the local interstellar medium (with embedded magnetic field) from right to left. Important boundaries are indicated.

The topic of this Workshop has been the structure of the heliosphere, that volume of space created by the solar wind and constrained and shaped by the very local interstellar medium (VLISM). Figure 1 shows an oversimplified schematic view of the heliosphere. At the center is the Sun and the orbit of Jupiter to indicate the scale. The solar wind flows radially outwards from the Sun as

Space Science Reviews **78**: 109–116, 1996.

shown. At the right is shown the plasma of the VLISM flowing toward the Sun with its embedded interstellar magnetic field with some unknown orientation. Since the solar wind is supersonic (technically super-fast-magnetosonic) and the solar wind and VLISM plasmas cannot mix, the interaction between the two necessarily creates two separate boundaries. The boundary between plasma of solar and interstellar origin is the heliopause, denoted by the dashed line. The second boundary is the solar wind termination shock, at which the solar wind is heated and decelerated to subsonic flow. There may also be a bow shock if the flow of the VLISM relative to the Sun is supersonic, where the interstellar gas is initially deflected to flow around the heliopause. The interaction is complicated by asymmetry introduced by the solar wind and interstellar magnetic fields and by variations in the wind ram pressure. It is also complicated by the interstellar neutral gas which penetrates the heliopause and affects the solar wind plasma (I shall return to the neutral gas later). Nevertheless, Figure 1 provides a useful conceptualization of the heliosphere and its boundaries.

The major feature of the heliosphere is the solar wind termination shock. It is the outer boundary of our space plasma environment as we know it, and it signals the presence of the interstellar medium beyond. The termination shock is the primary treasure sought by the newly confirmed NASA Interstellar Mission consisting of Voyagers 1 and 2 and Pioneer 10. This lecture is a termination shock primer!

2. The Morphology of the Termination Shock

How does one view the formation of the termination shock? It can be viewed as the required deceleration of the supersonic solar wind in response to the finite pressure of the VLISM. However, this steady-state view does not help conceptualize its formation.

Alternatively, one can imagine a star which suddenly launches a wind, akin perhaps to the explosion of a supernova. As shown in Figure 2, the new wind expels and accelerates the surrounding gas which creates a high-pressure interaction region moving away from the star, in which is embedded the boundary (a contact discontinuity) between stellar wind and surrounding gas. The over-pressure of the interaction region drives two shocks: a "reverse" shock, attempting to propagate against the flow of the wind back toward the star, and a "forward" shock, propagating into the surrounding gas. A similar interaction region within the solar wind bounded by forward and reverse shocks is created when a fast stream of wind from a coronal hole overtakes slow wind ahead of it due to solar rotation.

For a supernova the interaction region would be expelled to great distance. For a stellar wind which becomes steady the configuration shown in Figure 2 would relax to a stationary configuration. For a static surrounding medium the reverse shock would stabilize at a fixed distance as the wind termination shock, while the contact discontinuity and the forward shock would recede to great distance from the star. For supersonic motion of the surrounding medium from right to left in Figure 2, the interaction region would stabilize at the position of total pressure balance between the approaching flows of the wind and the surrounding medium. The configuration shown in Figure 1 then results with the shocked subsonic flows deflected to the left in the figure to create a heliospheric tail. The orientation and strength of the interstellar magnetic field may affect the shape and extent of the heliospheric tail.

You can explore an instructive analog of the termination shock in two dimensions when you next run water into the sink in order to brush your teeth. The water from the tap flows smoothly and radially over the sink surface away from the localized area of initial impact. Along a circular boundary there is a transition of the flow at a hydraulic jump or bore (Whitham, 1974) to a deeper slower turbulent flow toward the drain. If the drain gets in the way, you can hold a plate under the tap and allow the turbulent flow to spill over the edge of the plate. The hydraulic jump corresponds to the termination shock, and the height of the plate rim (or the curvature of the sink) corresponds to the pressure of the VLISM. This analogy was discussed by Axford (1972) and by Jokipii and McDonald (1995).

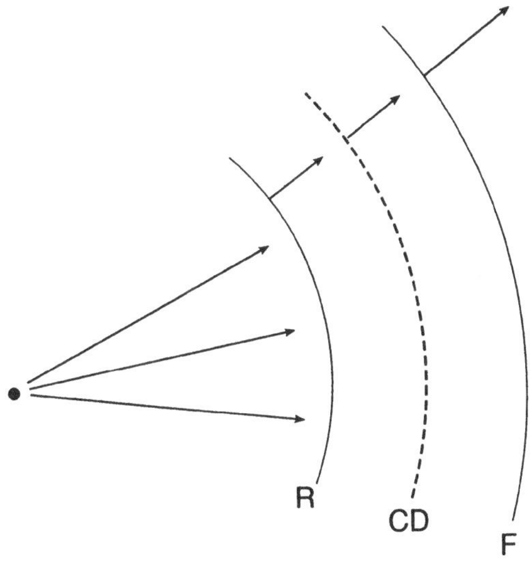

Fig. 2. Schematic view of a stellar wind expelling the local interstellar medium through the formation of a compression region bounded by a reverse (R) shock, and a forward (F) shock which propagates into the interstellar medium. Stellar and interstellar plasmas are separated by a contact discontinuity (CD).

With the plate in hand you can illustrate several features of termination shock behavior very simply. If you tilt the plate down to the left you will obtain an asymmetric shock with the turbulent flow channeled over the left plate rim, similar to the configuration shown schematically in Figure 1. If you increase the volume of water from the tap the "shock" moves out, and if you decrease the volume the "shock" moves in. Such oscillations in termination shock location occur in response to solar-cycle variations in solar wind ram pressure.

3. The Structure of the Termination Shock

Since we haven't yet visited the termination shock and made direct observations, there is an element of fantasy in what I shall say. Nevertheless I shall proceed with impunity!

What are the characteristics of the termination shock and how will we recognize it in the Voyager data? As a collisionless shock we expect the various plasma (and neutral) components to couple the upstream solar wind and downstream heliosheath plasmas on different spatial scales. This implies a broad transition with nested smaller-scale structures within. Such a shock could be even broader, but the tight winding of the Archimedes spiral magnetic field at large heliocentric radial distances implies a perpendicular shock on average, for which the magnetic field inhibits transport of ions and electrons back upstream from the shock.

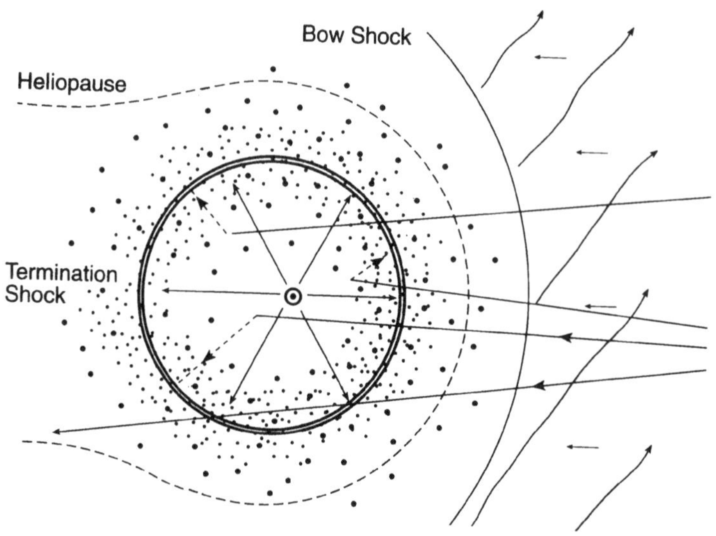

Fig. 3. Diagram of the heliosphere (from Fig. 1) with the particle populations indicated schematically which influence termination shock structure: galactic cosmic rays (large dots), anomalous cosmic rays (small dots), interstellar neutral gas (solid trajectories), and interstellar pickup ions (dashed trajectories). The inner circle indicates the transition at which pickup ions are injected at the termination shock to become low-energy anomalous cosmic rays.

Figure 3 shows superimposed on the heliospheric structure of Figure 1 the particle populations which influence the shock structure. The outer dark circle depicts the plasma "subshock", at which the solar wind plasma is decelerated and heated. The width of the subshock is approximately the proton gyroradius for a speed equal to the solar wind speed: $r_g = V_{SW}/\Omega$, where Ω is the cyclotron frequency. With V_{SW} = 800 km/s, representative of a fast stream, and $B = 5 \times 10^{-7}$ G, we obtain $r_g \cong 10^{-3}$ AU. The galactic cosmic rays (large dots) with a typical energy of 1 GeV diffuse and drift across the "subshock" into the heliosphere and couple to the solar wind plasma on the characteristic lengthscale of K/V_{SW}, where K is the radial spatial diffusion coefficient. The resulting decrease in cosmic ray intensity with decreasing heliocentric radial distance occurs over a lengthscale of tens of AU and is known as the solar modulation of galactic cosmic rays.

There also exists a complex chain of coupling processes between the upstream and downstream plasmas due to interstellar atoms. These neutral particles (denoted by the straight solid trajectories in Figure 3) flow into the solar wind with a bulk flow speed of about 20-25 km/s and a thermal speed of the same order. Some traverse the heliosphere without interacting. Others, particularly if they penetrate close to the Sun, are photoionized or electron-impact ionized, or suffer a charge-exchange interaction with a solar wind ion. Subsequently the newborn ion is "picked-up" by the solar wind. The pickup process, which heats and decelerates the wind, is part of the coupling process and therefore part of the termination shock structure. The pickup ions (denoted by straight dashed trajectories in Figure 3) are advected with the solar wind to the termination subshock. The pickup process causes them to have a suprathermal distribution with a characteristic thermal speed of V_{SW}, which accounts for the heating of the solar wind. This distribution apparently causes a fraction of them to be preaccelerated at the subshock. The mechanism for this preacceleration is unknown. It presumably occurs on the scale of an ion gyroradius in front of the subshock, as indicated by the circle adjacent to and inside the termination subshock in Figure 3. A possible mechanism for this preacceleration is shock surfing (Sagdeev, 1966; Zank et al., 1996; Lee et al., 1996) in which ions skip along the shock surface parallel to the motional electric field while they are trapped between the subshock potential ramp and the solar wind Lorentz force. Once these pickup ions are preaccelerated they become sufficiently mobile in spite of the quasi-perpendicular shock geometry to be further accelerated by the process of diffusive shock acceleration. These shock accelerated pickup ions are accelerated to energies of ~ 200 MeV/Q to become the cosmic ray anomalous component (denoted by small dots in Figure 3) which is also part of the termination shock structure (Fisk et al., 1974; Pesses et al., 1981; Jokipii, 1986; Lee and Axford, 1988). The most energetic anomalous cosmic rays diffuse into the heliosphere with a lengthscale, K'/V_{SW}, on the order of a few AU, smaller than that of galactic cosmic rays since $K' < K$. Actually, the pickup ions which are directly, or after preacceleration, transmitted at the subshock could provide much of the required shock dissipation and allow most of the solar wind plasma to be transmitted and compressed adiabatically at the termination subshock.

Thus, the termination shock as shown schematically in Figure 3 has a complex extended spatial structure. Since we observe the interstellar neutral gas, interstellar pickup ions, modulated galactic cosmic rays, and anomalous cosmic rays even in the inner heliosphere, we are already observing the upstream precursors of the termination shock. We are already within the shock! Of course most of the variation in solar wind density and speed occurs at the subshock where the solar wind plasma and the interstellar pickup ions are heated. It is the subshock, its location and structure, which is the major observational goal of NASA's Interstellar Mission.

4. The Location of the Termination Shock

The distance from the Sun to the termination shock (or subshock) is determined primarily by pressure balance between the solar wind ram pressure and the total pressure of the VLISM, which includes the magnetic field pressure, plasma thermal pressure, and ram pressure. Since the solar wind ram pressure decreases with heliocentric radial distance r as r^{-2}, a distance can be found at which the pressures balance.

A simple and illustrative calculation of the shock location for a configuration with spherical symmetry was presented by Parker (1963). He assumed a cold radial solar wind with constant speed V_0 and mass density ρ_0 at distance $r = r_0$, a strong hydrodynamic termination shock, and an isotropic interstellar pressure P_∞. By imposing the Rankine-Hugoniot relations at the shock, requiring that Bernoulli's Law be satisfied beyond the termination shock, and insisting that the pressure equals P_∞ as $r \rightarrow \infty$, he obtained for the distance r_S to the shock

$$(r_s/r_0)^2 = \rho_0 V_0^2 P_\infty^{-1} \left\{ 2(\gamma+1)^{-1} \left[(\gamma+1)^2/(4\gamma) \right]^{\gamma/(\gamma-1)} \right\} \tag{1}$$

where γ is the ratio of specific heats. Equation (1) is a statement of pressure balance. The factor in curly brackets accounts for the fact that there is a flow in the heliosheath and that the pressure just beyond the shock is less than P_∞. For $\gamma = 5/3$ this factor has the value 0.88. For a reasonable choice of solar wind parameters and P_∞, equation (1) yields shock locations in the range 70-140 AU (e.g., Holzer, 1989). The uncertainty arises from uncertainties in the interstellar magnetic field and the pressure of the ionized component of the interstellar medium, which have been a focus of this Workshop.

In view of the uncertainties inherent in a "first-principles" prediction of r_S, it is worthwhile asking if there are any indirect determinations of the distance to the termination shock. First of all we should ask the "messengers" who have been there. The interstellar neutral gas in the inner heliosphere has traversed the outer heliosphere, and its spatial distribution depends on heliospheric structure. The neutral gas charge exchanges with the interstellar plasma, which is deflected around the heliopause, to create a neutral hydrogen enhancement in the interface region between the termination shock and bow shock, known as the "hydrogen wall". The positive radial gradient in neutral hydrogen density in the direction of the "wall" is larger beyond the termination shock since the plasma density, and therefore the charge exchange rate, is higher. Similarly the negative radial gradient in neutral hydrogen density is larger beyond the shock in the opposite direction toward the heliospheric tail. By comparing the Lyman-α backscatter measured by Pioneer 10 and Voyager, Hall et al. (1993) exploit this asymmetry to infer a distance of 70-105 AU to the termination shock.

The 2-3 kHz radio signals observed by Voyager in the outer heliosphere (Gurnett et al., 1993) also provide an estimate of distance to the termination shock. The most plausible origin of the radio signals is via mode conversion from electrostatic waves at the heliopause, generated by the large shocks beyond the termination shock associated with global merged interaction regions (GMIR) in the solar wind. The 400-day delay between the cosmic ray decreases associated with the GMIR as observed by Voyager and the radio bursts for both the 1983 and 1992 bursts, implies a distance of about 110-160 AU to the heliopause, and a distance of 70-120 AU to the termination shock.

The most compelling estimate of the distance to the termination shock is based on the anomalous cosmic rays, which are accelerated at the shock and observed by Voyager in the outer heliosphere (Stone et al., 1996). Using a radial diffusion coefficient consistent with theoretical models, the observed intensities of the various species, and a transport model, the cosmic ray energy spectra and gradients must be matched. Requiring that the low-energy spectra of all species have the same power law at the shock yields a most probable heliocentric distance to the shock in the upwind direction of Voyager of about 85 AU. This value is consistent with the previous estimates and is probably the best bet. An interesting byproduct of this analysis is the shock compression ratio X, which follows from the energy spectral index. The derived value X = 2.6 is remarkably low, since the termination shock has traditionally been assumed to be strong. The recognition that interstellar pickup hydrogen dominates the pressure in the outer heliosphere has reduced the expected value of X, but observed pickup ion densities are still unable to yield such a low compression ratio. This discrepancy is unresolved.

In principle the solar-cycle variation of galactic cosmic rays, which roam around the outer heliosphere partially reaccelerated at the termination shock, could also provide an additional estimate of the distance to the shock, but current models are probably not adequate to the task. If you don't believe these indirect estimates, you could consult the experts! An assemblage of experts met at the University of New Hampshire in 1989 and cast secret ballots with their informed estimates of shock distance. The resulting histogram was published in Lee (1990). The mean distance was 61 AU (within the current location of Voyager 1!). Three distinguished scientists cast ballots for no shock, and one placed the shock at 1000 AU. The absence of a termination shock, while in principle possible for a wind continually heated and decelerated by the pickup of ionized interstellar gas, would present a challenge for the origin of the anomalous component.

5. Shape and Movement of the Termination Shock

The above estimates of shock location have referred to the upwind direction, to the right in Figures 1 and 3. The shock distance is larger in the downwind direction, as shown in Figures 1 and 3, due to the left to right pressure gradient in the subsonic flow. According to equation (1) the shock location also depends on wind ram pressure which depends on time and is not uniform in all directions. Indeed the ram pressure of the high-speed wind over the poles of the Sun during solar minimum activity exceeds that of the slow wind adjacent to the streamer belt at low heliographic latitudes (Phillips et al., 1995). We may therefore expect the shock to be further from the Sun over the poles during solar minimum. Local variations in solar wind ram pressure should also produce a shock with a warped surface (Suess, 1993).

Time variations of wind ram pressure will also cause the shock location to vary with time. The shock location should respond to solar-cycle variations in ram pressure, as well as to variations on a shorter timescale due to solar wind interaction regions or traveling shocks. Since the timescale for the global response of the heliosphere, R/V_{SW}, where R is a characteristic lengthscale of the heliosphere, is about 2 years, the variation in shock location on the solar-cycle timescale can be calculated as a sequence of equilibrium values based on the variable ram pressure. On the shorter timescale the shock moves as an infinite planar shock in response to the transmission of upstream sound waves (Landau and Lifshitz, 1959), entropy waves (Grzedzielski, 1993) or shock waves (Barnes, 1993). As a result the termination shock is dynamic on a large range of timescales, so that Voyager may expect several traversals of the shock.

Acknowledgements

I would like to thank Johannes Geiss, Rudi von Steiger, and the entire staff of ISSI for a very enjoyable well-organized Workshop. The generosity of ISSI is gratefully acknowledged, as is the patience of R vS in awaiting this paper. I acknowledge several valuable discussions with W.I. Axford, A. Cummings, L. Fisk, P. Isenberg, I. Mann, E. Möbius, D. Rucinski, and G. Zank. This work was supported in part by NSF grant ATM-9633366, and NASA grants NAGW-2579 and NAG5-1479.

References

Axford, W.I.: 1972, in C.P. Sonett, P.J. Coleman, Jr., and J.M. Wilcox (eds.), 'The Interaction of the Solar Wind with the Interstellar Medium', *Solar Wind*, NASA SP 308, p. 609.

Barnes, A.: 1993, 'Motion of the Heliospheric Termination Shock: A Gasdynamic Model', *J. Geophys. Res.* **98**, 15137.

Fisk, L.A., Kozlovsky, B., and Ramaty, R.: 1974, 'An Interpretation of the Observed Oxygen and Nitrogen Enhancements in Low-energy Cosmic Rays', *Astrophys. J.* **190**, L35.

Gurnett, D.A., Kurth, W.S., Allendorf, S.C., and Poynter, R.L.: 1993, 'Radio Emission from the Heliopause Triggered by an Interplanetary Shock', *Science* **262**, 199.

Grzedzielski, S.: 1993, 'The Heliospheric Boundary Regions', *Adv. Space Res.* **13** (6), 147.

Hall, D.T., Shemansky, D.E., Judge, D.L., Gangopadhyay, P., and Gruntman, M.A.: 1993, 'Heliospheric Hydrogen Beyond 15 AU: Evidence for a Termination Shock', *J. Geophys. Res.* **98**, 15185.

Holzer, T.E.: 1989, 'Interaction Between the Solar Wind and the Interstellar Medium', *Ann. Rev. Astron. Astrophys.* **27**, 199.

Jokipii, J.R.: 1986, 'Particle Acceleration at a Termination Shock, 1, Application to the Solar Wind and the Anomalous Component, *J. Geophys. Res.*, **91**, 2929.

Jokipii, J.R., and McDonald, F.B.: 1995, 'Quests for the Limits of the Heliosphere', *Scientific American* **272**, 62.

Landau, L.D., and Lifshitz, E.M.: 1959, *Fluid Mechanics*, Pergamon Press, New York, p. 332.

Lee, M.A., Shapiro, V.D., and Sagdeev, R.Z.: 1996, 'Pickup Ion Energization by Shock Surfing', *J. Geophys. Res.* **101**, 4777.

Lee, M.A.: 1990, in S. Grzedzielski and D.E. Page (eds.), 'Ion Acceleration to Cosmic Ray Energies', *Physics of the Outer Heliosphere*, COSPAR, Pergamon Press, New York, p. 157.

Lee, M.A., and Axford, W.I.: 1988, 'Model Structure of a Cosmic-ray Mediated Stellar or Solar Wind', *Astron. Astrophys.* **194**, 297.

Parker, E.N.: 1963, *Interplanetary Dynamical Processes*, Wiley-Interscience, New York, pp. 113-128.

Pesses, M.E., Jokipii, J.R., and Eichler, D.:1981, 'Cosmic Ray Drift, Shock Wave Acceleration, and the Anomalous Component of Cosmic Rays', *Astrophys. J.* **246**, L85.

Phillips, J.L. et al.: 1995, 'Ulysses Solar Wind Plasma Observations From Pole to Pole', *Geophys. Res. Lett.* **22**, 3301.

Sagdeev, R.Z.: 1966, in M.A. Leontovich (ed.), 'Cooperative Phenomena and Shock Waves in Collisionless Plasmas', *Reviews of Plasma Physics* **4**, Consultants Bur., New York, p. 23.

Stone, E.C., Cummings, A.C., and Webber, W.R.: 1996, 'The Distance to the Solar Wind Termination Shock in 1993 and 1994 from Observations of Anomalous Cosmic Rays', *J. Geophys. Res.* **101**, 11017.

Suess, S.T.: 1993, 'Temporal Variations in the Termination Shock Distance', *J. Geophys. Res.* **98**, 15147.

Whitham, G.B.: 1974, *Linear and Nonlinear Waves*, John Wiley & Sons, New York, p. 458.

Zank, G.P., Pauls, H.L., Cairns, I.H., and Webb, G.M.: 1996, 'Interstellar Pickup Ions and Quasi-Perpendicular Shocks: Implications for the Termination Shock and Interplanetary Shocks', *J. Geophys. Res.* **101**, 457.

COMPOSITION OF ANOMALOUS COSMIC RAYS AND IMPLICATIONS FOR THE HELIOSPHERE

A. C. CUMMINGS and E. C. STONE
California Institute of Technology, Pasadena, CA 91125

Abstract. We use energy spectra of anomalous cosmic rays (ACRs) measured with the Cosmic Ray instrument on the Voyager 1 and 2 spacecraft during the period 1994/157-313 to determine several parameters of interest to heliospheric studies. We estimate that the strength of the solar wind termination shock is 2.42 (-0.08, +0.04). We determine the composition of ACRs by estimating their differential energy spectra at the shock and find the following abundance ratios: H/He = 5.6 (-0.5, +0.6), C/He = 0.00048 ± 0.00011, N/He = 0.011 ± 0.001, O/He = 0.075 ± 0.006, and Ne/He = 0.0050 ± 0.0004. We correlate our observations with those of pickup ions to deduce that the long-term ionization rate of neutral nitrogen at 1 AU is $\sim 8.3 \times 10^{-7}$ s^{-1} and that the charge-exchange cross section for neutral N and solar wind protons is $\sim 1.0 \times 10^{-15}$ cm^2 at 1.1 keV. We estimate that the neutral C/He ratio in the outer heliosphere is $1.8(-0.7, +0.9) \times 10^{-5}$. We also find that heavy ions are preferentially injected into the acceleration process at the termination shock.

Key words: abundances, anomalous cosmic rays, *Voyager*, interstellar medium, heliosphere, solar wind termination shock

1. Introduction

Anomalous cosmic rays (ACRs) are energetic particles which are thought to be accelerated pickup ions. The main acceleration is thought to take place at the solar wind termination shock (Pesses et al., 1981). The source of the pickup ions that become ACRs is believed to be neutral gas from the interstellar medium (Fisk et al., 1974). The ACR component currently consists of seven elements: H, He, C, N, O, Ne, and Ar (Garcia-Munoz et al., 1973; McDonald et al., 1974; Hovestadt et al., 1973; Cummings and Stone, 1988, 1990; Christian et al., 1988, 1995; McDonald et al., 1995). Previous studies of their composition have led to estimates of the abundances of neutral gas in the interstellar medium (Cummings and Stone, 1987, 1990). However, those studies were made before the pickup ion observations became available and the fractionation in the acceleration and propagation processes could only be roughly estimated.

In this study we adopt a new approach that makes use of the pickup ion observations and the ACR observations to gain information on the injection of the pickup ions into the acceleration process at the termination shock. This approach has only become viable as solar modulation has lessened to such an extent that the observed ACR energy spectra are less modulated than at any time in the past. This decreased modulation results in ACR spectra which show signs of a power-law dependence at low energies, which makes possible estimates of the ACR spectra at the shock.

The ACR shock spectra are estimated for H, He, C, N, O, and Ne. Ar, which was included in previous studies, is observable only at energies above the roll-off energy of the shock spectrum. Thus we can gain no information on the low-energy power-law portion of the shock spectrum, which is what is required for this study. Estimates of the neutral densities of H, He, N, O, and Ne in the outer heliosphere (30 - 60 AU) are available from pickup ion studies (Geiss et al., 1994; Geiss and Witte, 1996). We use a model of their ionization and subsequent transport to estimate the fluxes of pickup ions incident on the termination

Space Science Reviews **78**: 117–128, 1996.

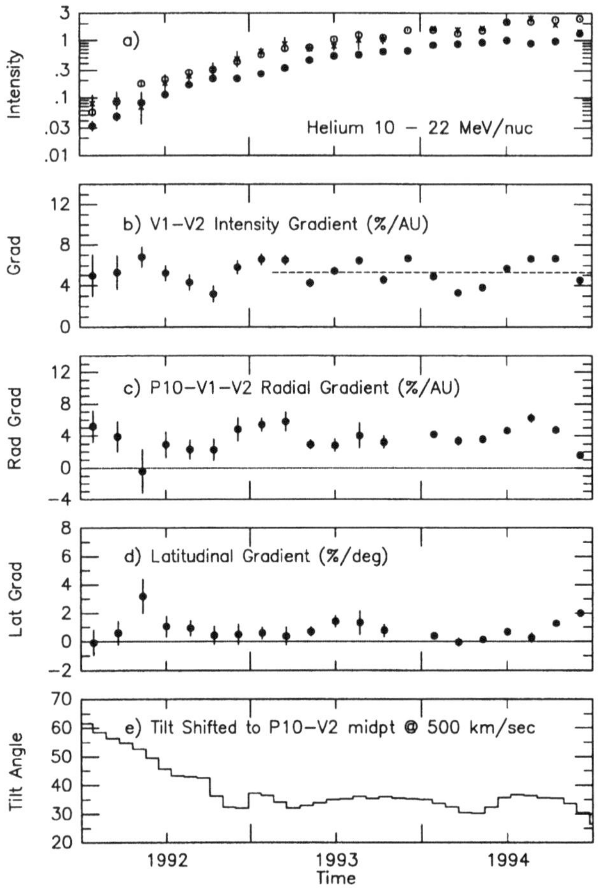

Figure 1. Intensity of helium (52-day averages) measured at P10 (crosses), V2 (solid circles), and V1 (open circles) versus time. b) Intensity gradient between V1 and V2 for He with 9.3 - 22.3 MeV/nuc. The dashed line is the mean for the period indicated. c) Radial gradient of ACR He with ~ 10 - 22 MeV/nuc. d) Latitudinal gradient of ACR He with ~ 10 - 22 MeV/nuc. e) Estimated tilt of the neutral current sheet shifted to the mid-point of V2 and P10. Each tilt observation (Hoeksema, private communication, 1995) covers a single solar rotation or ~ 26 days. We have performed a 3-solar-rotation moving average on the supplied tilt data set before plotting, in order to approximate the average conditions between V2 and P10 which are ~ 17 AU apart.

shock. We use a theory of particle acceleration at the termination shock to estimate the resulting power-law differential energy spectra for these elements. These spectra can be compared with our derived ACR shock spectra to estimate the relative injection efficiencies of pickup ion H^+, He^+, O^+, and Ne^+. We do not have independent information on the charge-exchange cross section of N, so in the case of N we can estimate this quantity and the ionization rate of N at 1 AU by using an estimate of the injection efficiency of pickup ion N^+ based on that of O^+ and Ne^+. Also, by using an estimate of the injection efficiency of C^+, we can estimate the neutral C abundance in the outer heliosphere.

2. ACR Shock Spectra

In order to estimate the ACR shock spectra, we compare our observations made with the Cosmic Ray instrument (Stone et al., 1977) on V1 and V2 during 1994/157-313 with a spherically-symmetric equilibrium model of the propagation of ACRs, including diffusion, convection, and adiabatic deceleration (Fisk, 1971). This period consists of three 52-day periods which were part of a recent study on the distance to the termination shock (Stone et al., 1996). The average latitudinal gradient for ACR He with 10 - 22 MeV/nuc was ~ 0.8%/deg for the period considered in that study, 1993/52 - 1994/365. This latitudinal

Table I
V2 radial shifts from Equation 1

Element	$<T>$ (MeV/nuc)	$<G_{V1V2}>$ (%/AU)	G_{V1V2} (%/AU)	Δr_{V2} (AU)
He	20.1	4.49	5.32	-2.41
	26.0	2.69	3.37	-3.26
	34.8	3.33	3.87	-2.13
	43.9	2.07	2.40	-2.12
O	4.3	3.20	4.75	-6.34
	4.9	3.18	4.66	-6.10
	6.2	3.25	3.91	-2.64
	7.9	2.81	3.51	-3.29
	10.5	2.91	3.52	-2.71

gradient is small enough that we feel justified in using a spherically-symmetric model of propagation to make a first order estimate of the shock spectra, shock location, and other parameters. We will later make a correction to the estimated shock location by accounting for the finite latitudinal gradient.

Figure 1 is reproduced from Figure 2 of Stone et al. (1996) and displays the time history of He with 10 - 22 MeV/nuc, three gradients of these particles, and the tilt of the neutral current sheet. The time period we have chosen for this study, 1994/157-313, comprise the 4th through 6th 52-day periods of 1994 shown in panels a-d of Figure 1. In Figure 1b, intensity gradients between V1 and V2 are shown and it appears that the period 1994/157-313 chosen for this study has a larger than average gradient. The gradient for the period 1993/52 - 1994/365 is seen to wander about a mean value of 5.3%/AU showing no systematic increase or decrease. The non-statistical variations are caused by transient disturbances and ideally we would use a period that was unaffected by such disturbances, as was done in the Stone et al. (1996) study. However, in order to get good statistical precision for the heavier ACR species and to observe ACR H at both V1 and V2, it was necessary to use several of the most recent periods for the analysis.

By examining Figure 1a, we deduce that the transient disturbance during 1994/157-313 is affecting the V2 fluxes, making them lower than they would otherwise be. This atypical gradient can be accounted for in our model fits by moving the V2 spacecraft towards the Sun from its actual position and away from V1. Thus, the intensity we observe at V2 during this transient decrease is representative of the equilibrium intensity at a position some distance sunward of V2.

In order to estimate the effective radial location of V2, we examined the V1/V2 intensity gradients in four He energy intervals and in five O energy intervals for 11 52-day periods (1993/105 - 1994/313) and for the period 1994/157-313. We computed the distance, Δr_{V2}, to move V2 for each energy interval from the equation:

$$\Delta r_{V2} = \frac{< G_{V1V2} > - G_{V1V2}}{< G_{V1V2} >} (r_{V1} - r_{V2}) \tag{1}$$

where $< G_{V1V2} >$ is the average V1/V2 intensity gradient for the 11 time periods, G_{V1V2} is the observed V1/V2 intensity gradient for the period 1994/157-313, and r_{V1} and r_{V2} are the average radial locations of the V1 and V2 spacecraft, respectively. The average latitudes and radial locations of the spacecraft are 32.5°N and 56.8 AU for V1 and 12.3°S and 43.7

AU for V2. The gradients and radial shifts for V2 calculated from Equation 1 are displayed in Table I. We find that the average V2 correction for the 9 observations is -3.4 $\pm$ 0.5 AU. Thus the effective radial location of V2 for the purposes of the model fits is 40.3 AU.

In the model calculations we fit the ACR H, He, C, N, O, and Ne spectra at V1 and V2. These spectra are obtained by subtracting estimated spectra of galactic cosmic rays from the observed spectra. The estimated GCR spectra for each element were derived from the observed C and O spectra at high energies. For V1, a power-law fit, with the spectral index fixed at 1.0, was made to three C intensity values with energies between 36 and 106 MeV/nuc. Also in the fit was the highest energy O data point at 94 - 125 MeV/nuc, scaled in intensity by the factor 1.095 to represent C. The resulting V1 GCR C spectrum in units of $(\mathrm{m}^2\ \mathrm{s}\ \mathrm{sr}\ \mathrm{MeV/nuc})^{-1}$ was $j_{V1} = 6.51 \times 10^{-5}$ T, where T is energy in MeV/nuc. For V2, a similar fit was made, except a C data point with energies 20 - 36 MeV/nuc was added to the fit. The resulting V2 GCR C spectrum was $j_{V2} = 5.59 \times 10^{-5}$ T. The assumed GCR abundances of the other species, relative to C, are: 107, 35.6, 0.228, 0.913, and 0.132 for H, He, N, O, and Ne. For N, O, and Ne, these ratios are from Simpson (1983) and represent observed GCR ratios at 70 - 280 MeV/nuc at 1 AU. For He, we use the estimate from Simpson (1983) at 600 - 1000 MeV/nuc. For H, we estimate a ratio of GCR H/He = 3 from our observed V1 H and He energy spectra. This ratio is smaller than the value of 4.7 $\pm$ 0.5 derived by Simpson (1983).

The ACR He shock spectrum is assumed to be a power-law in energy per nucleon, $dJ/dT = j_{He} \cdot T^{\gamma}$. The shock spectra of the other elements are assumed to be a power-laws with the same index. For O this is an adequate approximation for energies up to $\sim$ 10 MeV/nuc and for ACR He this approximation is valid up to $\sim$ 60 MeV/nuc (see Stone et al. (1996) for more discussion). Above a total energy of $\sim$ 150 - 240 MeV the energy spectra of these elements exhibit an approximately exponential roll-off (Stone et al., 1996).

We assume the diffusion coefficient, κ (in $\mathrm{cm}^2\ \mathrm{s}^{-1}$), is given by

$$\kappa = \frac{\kappa_0 \beta r (1 + \kappa_S^2) R^2}{1 + (\kappa_S R)^2} \qquad (r \gg 1\ \mathrm{AU}) \tag{2}$$

where κ_0 and κ_S are scaling factors, β is particle speed, r is heliocentric radial distance in AU, and R is rigidity in GV. This form was used by Stone et al. (1996) and can be derived from the quasilinear formulation of Bieber et al. (1995). There are ten free parameters in the model: the shock location (r_S), the diffusion coefficient scaling factor (κ_0), the diffusion coefficient shape factor (κ_S), the power-law index (γ) of the energy spectrum at the shock, the intensity scaling factor (j_{He}) of the ACR He shock spectrum, and the ratios of the intensity scaling factors of the other elements to that of ACR He at the shock (H/He, C/He, N/He, O/He, and Ne/He). We assume that the solar wind velocity, V, is 500 km s^{-1}, which is close to the average value at V2 of 490 km s^{-1} for 1993/1 - 1994/365 (Richardson, private communication, 1995). The fits are not sensitive to the actual value of V, but a different V would result in a proportionally different κ_0.

The data and best-fit model curves for the period 1994/157-313 are shown in Figures 2a-f. The fits were made only in the energy regions shown by the solid lines, below the exponential roll-off. The χ^2 of the ten-parameter best fit to the 86 data points participating in the fits in Figure 2 is 51.2. Figure 3 shows the best-fit diffusion coefficient as a function of rigidity.

We investigated the confidence limits for each parameter in two ways. We first estimated the 68% confidence limits by iteratively changing and fixing the value of one parameter and re-fitting until we found the parameter value where the χ^2 had increased by 1 (see Press et al. (1992)). We did this in turn for all ten parameters. The best-fit parameter values and the 68% confidence limits are shown in Table II for the period 1994/157-313.

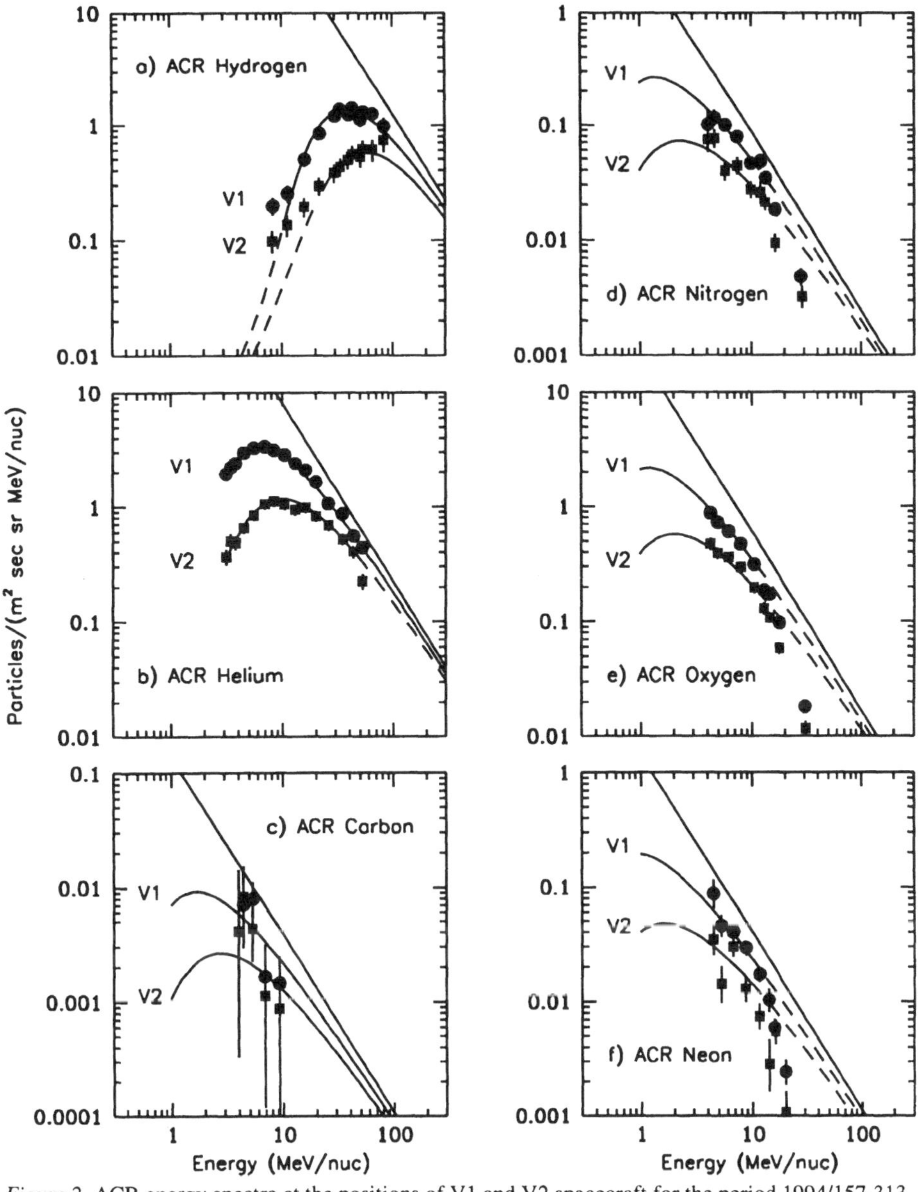

Figure 2. ACR energy spectra at the positions of V1 and V2 spacecraft for the period 1994/157-313. The curves represent the 10-parameter best-fit energy spectra at the solar wind termination shock, V1, and V2, as described in the text. a) ACR H, b) ACR He, c) ACR C, d) ACR N, e) ACR O, and f) ACR Ne.

In the second method we account for modelling uncertainties by using the estimated uncertainty in the effective radial position of V2 (0.5 AU). We performed two additional fits, one using the upper limit for the effective radial location for V2 and another using the lower limit. The resulting parameters are shown as the model limits in Table II.

For the shock distance, we need to apply an additional correction to account for the small but finite latitudinal gradient. In the study by Stone et al. (1996), the authors took

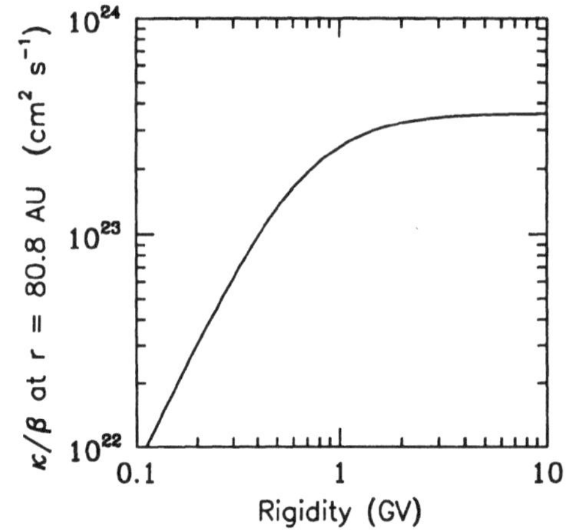

Figure 3. Best-fit diffusion coefficient divided by particle velocity versus rigidity at 80.8 AU. The form of the diffusion coefficient is described in the text.

Table II
Fit parameters for period 1994/157-313

Parameter*	Fit Value**	68% Lower Limit	68% Upper Limit	Model Lower Limit	Model Upper Limit
γ	-1.55 (-0.06, +0.03)	-1.62	-1.53	-1.56	-1.55
r_S (AU)	80.8 (-3.0, +2.4)***	77.9	82.9	81.7	79.8
κ_0 ($\times 10^{21}$)	3.10 (-0.22, +0.17)	2.92	3.22	3.23	2.98
κ_S	1.51 (-0.09, +0.11)	1.42	1.62	1.49	1.54
j_{He}	285 (-69, +81)	217	366	288	279
H/He	5.6 (-0.5, +0.6)	5.1	6.2	5.7	5.5
C/He ($\times 10^{-4}$)	4.8 ($\pm$ 1.1)	3.7	6.0	4.8	4.9
N/He ($\times 10^{-2}$)	1.1 ($\pm$ 0.1)	1.0	1.2	1.1	1.1
O/He ($\times 10^{-2}$)	7.5 ($\pm$ 0.6)	7.0	8.1	7.5	7.7
Ne/He ($\times 10^{-3}$)	5.0 ($\pm$ 0.4)	4.6	5.5	5.0	5.1

* See text for definition of symbols.
** Approximate uncertainty in parenthesis derived from the root mean square of the statistical (68%) uncertainty and the model uncertainty. The model uncertainty illustrates the change in the best-fit values resulting from an uncertainty of $\pm$ 0.5 AU in the effective radial location of V2.
*** Shock location estimate taking account of finite latitudinal gradient is 100 $\pm$ 6 AU (see text).

account of this effect by making a correction to the radial locations of V1 and V2 before using the spherically-symmetric model to fit the energy spectra. The correction they derived was

$$\ln(r'/r_{act}) = (G_\Theta \cdot \Theta / G_0) \tag{3}$$

where r' is the effective location, r_{act} is the actual location, G_Θ is the average latitudinal gradient, and Θ is the absolute value of the latitude of the spacecraft, and where a model was used in which the radial gradient in the intensity j is proportional to $1/r$, with

$(1/j)(\partial j/\partial r) = G_0/r$. In the current study, we make a correction to the radial location of V2 to account for the transient intensity decrease observed on V2. The fits are then made with the actual V1 radial location and the transient-decrease-corrected radial location of V2. Thus the resulting shock location, r_S, is a first-order estimate of the location if there were no latitudinal gradient. To derive a correction factor to apply to r_S, we assume that the ratio of intensities at two locations can be described by the following equation:

$$j_0 = j_1 \exp[-\int_{r_0}^{r_1} (CV/\kappa)dr] \tag{4}$$

where j_i is the intensity at radial location r_i, C is the Compton-Getting factor, V is the solar-wind speed, and κ is the diffusion coefficient. We assume that the diffusion coefficient is proportional to radial distance, $\kappa = \kappa_1 \cdot r$, so that Equation 4 can be written

$$\ln(j_0/j_1) = -(CV/\kappa_1)\ln(r_1/r_0) \tag{5}$$

where r_1 and r_0 are the radial locations where the intensities j_1 and j_0 are measured, respectively. Thus,

$$\frac{\ln(j_0/j_1)}{\ln(r_1/r_0)} = \text{constant} \tag{6}$$

By applying Equation 6 for two cases: 1) $r_1 = r_S$ and $r_0 = r_{V1}$ and 2) $r_1 = r_{V1}$ and $r_0 = r_{V2}$ we find

$$\frac{\ln(r_S/r_{V1})}{\ln(r_{V1}/r_{V2})} = \frac{\ln(j_{V1}/j_S)}{\ln(j_{V2}/j_{V1})} \tag{7}$$

If we substitute the latitude-corrected spacecraft positions for the actual positions, the best-fit shock position should change such that the same shock flux j_S and the fluxes j_{V1} and j_{V2} are recovered. Since the right side of Equation 7 will remain unchanged, so must the left side. That is,

$$\frac{\ln(r_S'/r_{V1}')}{\ln(r_{V1}'/r_{V2}')} = \frac{\ln(j_{V1}/j_S)}{\ln(j_{V2}/j_{V1})} \tag{8}$$

and thus

$$\ln(r_S'/r_{V1}') = \ln(r_S/r_{V1})\frac{\ln(r_{V1}'/r_{V2}')}{\ln(r_{V1}/r_{V2})} \tag{9}$$

From Equation 3,

$$\ln(r_{V1}'/r_{V1}) = (G_\Theta \cdot \Theta_{V1}/G_0) \tag{10}$$

and

$$\ln(r_{V2}'/r_{V2}) = (G_\Theta \cdot \Theta_{V2}/G_0) \tag{11}$$

By using these two equations in Equation 9, it can be shown that

$$\ln(r_S'/r_{V1}') = a\ln(r_S/r_{V1}) \tag{12}$$

where

$$a = 1 + \frac{G_\Theta}{G_0} \frac{(\Theta_{V1} - \Theta_{V2})}{\ln(r_{V1}/r_{V2})} \tag{13}$$

Using Equations 10, 12, and 13, it can be shown that

$$\frac{r_S'}{r_S} = (\frac{r_S}{r_{V1}})^b \exp(G_\Theta \cdot \Theta_{V1}/G_0) \tag{14}$$

where $b = a - 1$. Using the values of $G_\Theta = 0.78 \pm 0.18$ %/deg, and $G_0 = 1.96 \pm 0.19$ from Stone et al. (1996), and $r_S = 80.8$ AU from Table II, we find $r_S'/r_S = 1.24 \pm 0.07$. Applying this correction to the derived fit shock location from Table II, we find $r_S' = 100\pm6$ AU. We caution, however, that this estimate is based on Equation 5 which is only an approximation to the solution of the full transport equation. No systematic error to account for this source of uncertainty has been added.

The shock strength (see, e.g., Potgieter and Moraal (1988)), s, is related to the spectral index by: $s = (2\gamma - 2)/(2\gamma + 1)$. From the values of γ in Table II, the inferred strength of the shock is 2.42 (-0.08, +0.04). The shock is not a strong shock (s = 4; γ = -1), a finding in agreement with the results from the study by Stone et al. (1996).

3. Injection Efficiency of Pickup Ions at Shock

To estimate the efficiency ϵ_i for the injection of pickup ion species i into the acceleration process, we follow the calculations of Lee (1983) from which it can be shown that the accelerated spectrum is given in units of (cm^2 s sr MeV/nuc)$^{-1}$ by

$$j = p^2 f(p) = \frac{\nu \epsilon_i F_i}{4\pi} E_{0i}^{(\nu-4)/2} E^{(-\nu+2)/2} \tag{15}$$

where $\nu \equiv 3/(1 - \rho_u/\rho_d) = 5.11$ for a downstream/upstream density ratio of 2.42, F_i(cm^{-2} s^{-1}) is the pickup ion flux at the shock, and E_{0i} is the injection energy corresponding to $2V_{SW}$, taken here to be 5.2×10^{-3} MeV/nuc. The pickup ion fluxes at the shock can be estimated from the observations of Geiss et al. (1994), who inferred the densities of neutral H, He, C, N, O, and Ne in the outer heliosphere shown in Table III. Using these values, the model of Vasyliunas and Siscoe (1976) for the distribution of these neutrals throughout the heliosphere, and the long-term ionization rates at 1 AU from Rucinski et al. (1996) (also shown in Table III), we estimate the fluxes of H^+, He^+, O^+, and Ne^+ pickup ions at the nose of the heliosphere shown in Table III and displayed in Figure 4. We do not include the pickup ion fluxes of C^+ in Table III because the neutral density of C from the pickup ion studies is only an upper limit. The N^+ pickup ion flux is also missing from Table III because we could not find the charge-exchange cross section for N and hence we could not independently estimate the long-term ionization rate of N at 1 AU. Later, we will use our observations to make estimates of these parameters for N and of the neutral C density in the outer heliosphere.

Equation 15 can be written:

$$j_i = D_i \, \epsilon_i \, E^{-1.55} \text{ (m}^2 \text{ s sr MeV/nuc)}^{-1} \tag{16}$$

where the coefficients D_i are shown in Table IV. These are the calculated ACR spectra at the shock and they can be compared with the derived ACR shock spectra from the model

Table III
Pickup ion parameters

Element	Neutral abund. in outer heliosphere*	Total ioniz. rate at 1 AU** ($\times 10^{-7}$ s^{-1})	Pickup ion flux at nose of heliosphere (cm^{-2} s^{-1})
H	7.7 ± 1.3	7.44 ± 0.16	10240
He	1.00***	1.08 ± 0.10	228
C	$\leq 0.8 \times 10^{-3}$	10.35 ± 0.77	
N	0.5 (-0.3, +0.5) $\times 10^{-3}$		
O	3.5 (-1.4, +1.8) $\times 10^{-3}$	7.75 ± 0.41	5.3
Ne	0.7 (-0.5, +1.0) $\times 10^{-3}$	3.52 ± 0.33	0.53

* (Geiss and Witte, 1996; Gloeckler, 1996)
** (Rucinski et al., 1996)
*** Neutral density = 0.0155 ± 0.0025 cm^{-3} (Geiss and Witte, 1996)

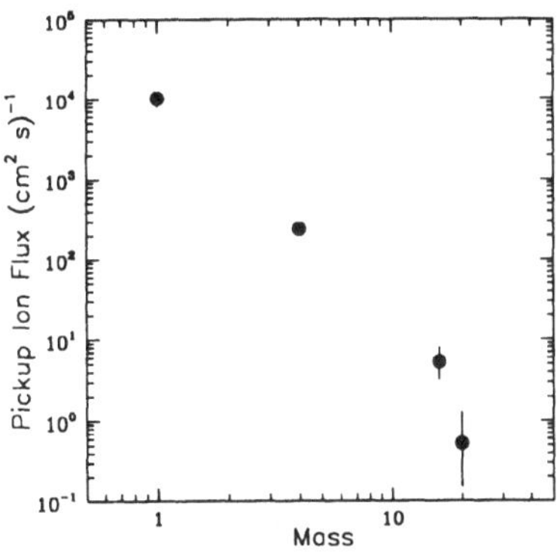

Figure 4. Estimated fluxes of pickup ions at the nose of the heliosphere at 100 AU, the estimated location of the solar wind termination shock.

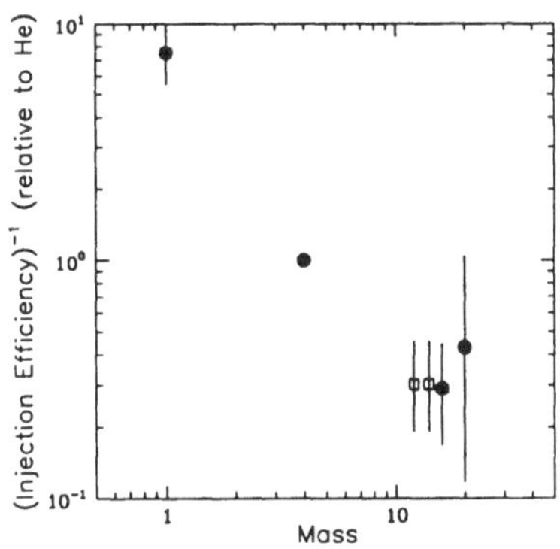

Figure 5. Estimated injection efficiencies of pickup ions into the acceleration process at the solar wind termination shock.

fits (Figure 2) to derive the relative injection efficiencies shown in Table IV. The relative efficiencies are much more reliably determined than the absolute efficiencies because the uncertainty in the spectral index is removed when the ratio is taken. The estimated uncertainties in the relative injection efficiencies include the uncertainties in the ACR intensity scaling factors from Table II and the uncertainties in the pickup ion fluxes, which are assumed to be dominated by the uncertainties in the neutral abundances from the pickup ion studies shown in Table III. It does not include any uncertainties associated with the simplifications inherent in the acceleration model in Equation 15. The relative injection efficiencies are shown as a function of particle mass in Figure 5. The observed preferential injection for the heavier particles is qualitatively consistent with Monte Carlo studies of shock acceleration by Ellison et al. (1981) but is in disagreement with the results of Kucharek and Scholer (1995).

Table IV
Injection efficiencies

Element	Coeff. D_i from Eq. 16	j_i from Table II	Inj. eff.	$(\text{Inj.eff.})^{-1}$ relative to He
H	2.25×10^6	1600	7.04×10^{-4}	7.5 (-2.5, +2.4)
He	5.34×10^4	285	5.29×10^{-3}	1.00
O	1.16×10^3	21.4	1.82×10^{-2}	0.29 (-0.12, +0.15)
Ne	1.16×10^2	1.43	1.23×10^{-2}	0.43 (-0.31, +0.62)

4. Ionization Rate of N at 1 AU

The open squares in Figure 5 represent weighted averages of the inverse injection efficiencies for O^+ and Ne^+. They are plotted at mass numbers 12 and 14 to represent the estimated injection efficiencies of C^+ and N^+. For N^+, the value plotted, 0.30 (-0.11, +0.15) implies an absolute injection efficiency of 0.0174. From Table II we estimate that the ACR N shock spectral intensity coefficient is $j_N = 3.1$ which implies that the coefficient D_i for N (D_N) in Equation 16 is 3.1/0.0174 = 178. D_N is proportional to the flux of pickup ions at the shock. By scaling from the flux of He^+ pickup ions at the shock from Table III, we find that the expected flux of N^+ pickup ions at the shock is 0.81 cm^{-2} s^{-1}. Using the model of Vasyliunas and Siscoe (1976) for the distribution of neutrals in the heliosphere and the estimated neutral N density in the outer heliosphere from Table III (7.8×10^{-6} cm^{-3}), we estimate that the long-term ionization rate at 1 AU of neutral N must be $\sim 8.3 \times 10^{-7}$ s^{-1}. This value is similar to that derived by Rucinski et al. (1996) for O (7.8×10^{-7} s^{-1}).

5. C Abundance

For C, only an upper limit to the neutral abundance in the outer heliosphere is available from the pickup ion observations. We can supply an estimate of this abundance by using a similar technique to that described above to deduce the ionization rate of N at 1 AU, except that in this case we have the long-term ionization rate of C from Table III and the unknown is the neutral abundance of C in the outer heliosphere. If we assume the injection efficiency for C^+ is as plotted in Figure 5, it can be shown that the neutral C density in the outer heliosphere must be $2.8\ (-1.0, +1.4) \times 10^{-7}$ cm^{-3}, which implies the C/He ratio in the outer heliosphere is $1.8(-0.7, +0.9) \times 10^{-5}$. The uncertainty is estimated from the uncertainty computed for the injection efficiency for C^+ plotted in Figure 5 and the uncertainty in the neutral density of He in Table III. The neutral abundances of all the elements are shown in Figure 6. The abundances are plotted relative to He and for H, N, O, and Ne are from the pickup ion observations (Gloeckler, 1996; Geiss et al., 1994). For C, we show two abundances which are in good agreement, one from this study and one from Cummings and Stone (1990) derived from observations made during a solar minimum period in 1987.

6. Discussion

We consider the shock location derived from this study, 100 ± 6 AU, to be less accurate than the shock location of 85 ± 5 AU derived for a slightly different period (1994/157-209)

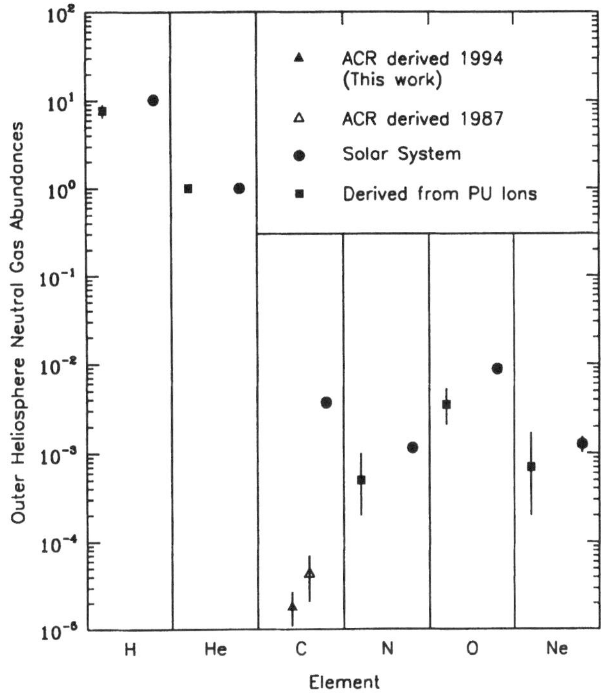

Figure 6. Estimated abundances of neutral gases in the outer heliosphere. The solid squares are from pickup ion studies (Geiss et al., 1996), the solid circles are solar system abundances (Grevesse and Anders, 1988), and the solid triangle is derived in this study.

by Stone et al. (1996). The reason is that the V1/V2 intensity gradients for the longer period considered in this study are typically larger than the average. Therefore, this period is likely affected by a transient or transients that have decreased the intensity at V2. While we have tried to take this into account by changing the location of V2 for the fits, the previous study has an advantage in that it used periods when there were apparently no transients present. We feel that the shock spectral intensity ratios are relatively insensitive to this effect and should be accurate. As evidence of this we find that the O/He ratios derived in the two studies are in good agreement. We also note that in Table II the model upper and lower limits for the intensity ratios and the spectral index are essentially the same as their respective nominal fit values.

The long-term ionization rate of neutral N derived in this study can be used to estimate the charge-exchange cross-section for the reaction $H^+ + N \rightarrow H + N^+$ for which we could find no reference in the literature. The study of ionization rates by Rucinski et al. (1996) did not address the charge-exchange process for N for that reason. However, that study did result in the estimate of 4.4×10^{-7} s^{-1} for the long-term photoionization rate of N at 1 AU. If we assume that the total ionization rate is due to photoionization and charge-exchange with the solar wind, then we estimate that the charge-exchange rate for N is $\sim 3.9 \times 10^{-7}$ s^{-1}. The long-term average solar wind flux is 3.9×10^8 cm^{-2} s^{-1} (Rucinski et al., 1996), which implies that the cross-section for charge exchange is $\sim 1.0 \times 10^{-15}$ cm^{-2} at the average solar wind velocity of 450 km s^{-1} (1.1 keV).

The neutral C abundance derived in this study may be useful in estimating the ionization state of the very local interstellar medium (VLISM). Recently, Frisch (1995) used a previous estimate of the neutral C/O ratio = 0.0039 (-0.0020, +0.0039) from Cummings and Stone (1990), assuming the ratio in the outer heliosphere is the same as it is in the VLISM, to help deduce that the VLISM was likely highly ionized ($\sim$ 69 - 81%). From this study, we estimate that the neutral C/O ratio is 0.005 ± 0.003. The major reason for the change from the previous study is that the O in this study is from the pickup ion observations, whereas it was from the solar system abundances (Grevesse and Anders, 1988) in the previous study. We believe our current technique results in a more accurate estimate of the neutral C/O ratio. We caution, however, that our ratio represents an estimate for the outer heliosphere, just sunward of the termination shock, and filtering through the heliospheric interface has not been considered and may alter the ratio in the VLISM (see, e.g., Fahr et al. (1995)).

Acknowledgements

We are grateful to J. T. Hoeksema for providing the tilt observations prior to publication. We thank J. Richardson and J. Belcher for providing the Voyager 2 solar wind speed data. This work was supported by NASA under contract NAS7-918.

References

Bieber, J. W., Burger, R. A., and Matthaeus, W. H.: 1995, *Proc. Int. Conf. Cosmic Ray 24th*, **SH**, 694
Christian, E. R., Cummings, A. C., and Stone, E. C.: 1988, *Astrophys. J. Lett.*, **334**, L77
Christian, E. R., Cummings, A. C., and Stone, E. C.: 1995, *Astrophys. J. Lett.*, **446**, L105
Cummings, A. C., and Stone, E. C.: 1987, *Proc. Int. Conf. Cosmic Ray 20th*, **3**, 413
Cummings, A. C., and Stone, E. C.: 1988, Proc. Sixth Internat. Solar Wind Conference (Boulder), *NCAR Technical Note 306* 2, 599
Cummings, A. C., and Stone, E. C.: 1990, *Proc. Int. Conf. Cosmic Ray 21st*, **6**, 202
Ellison, D. C., Jones, F. C., and Eichler, D.: 1981, *J. Geophys.*, **50**, 110
Fahr, H. J., Osterbart, R., and Rucinski, D.: 1995, *Astron. Astrophys.*, **294**, 587
Fisk, L. A.: 1971, *J. Geophys. Res.*, **76**, 221
Fisk, L., Kozlovsky, B., and Ramaty, R.: 1974, *Astrophys. J. Lett.*, **190**, L35
Frisch, P. C.: 1995, *Space Sci. Rev.*, **72**, 499
Garcia-Munoz, M., Mason, G. M., and Simpson, J. A.: 1973, *Astrophys. J. Lett.*, **182**, L81
Geiss, J., Gloeckler, G., Mall, U., von Steiger, R., Galvin, A. B., and Ogilvie, K. W.: 1994, *Astron. Astrophys.*, **282**, 924
Geiss, J., and Witte, M.: 1996, *Space Sci. Rev.*, this issue
Gloeckler, G.: 1996, *Space Sci. Rev.*, this issue
Grevesse, N., and Anders, E.: 1988, *AIP Conference Proceedings*, **183**, 1
Hovestadt, D., Vollmer, O., Gloeckler, G., and Fan, C. Y.: 1973, *Phys. Rev. Lett.*, **31**, 650
Kucharek, H., and Scholer, M.: 1995, *J. Geophys. Res.*, **100**, 1745
Lee, M. A.: 1983, *J. Geophys. Res.*, **88**, 6109
McDonald, F. B., Teegarden, B. J., Trainor, J. H., and Webber, W. R.: 1974, *Astrophys. J. Lett.*, **187**, L105
McDonald, F. B., Lukasiak, A., and Webber, W. R.: 1995, *Astrophys. J. Lett.*, **446**, L101
Pesses, M. E., Jokipii, J. R., and Eichler, D.: 1981, *Astrophys. J. Lett.*, **246**, L85
Potgieter, M. S., and Moraal, H.: 1988, *Astrophys. J.*, **330**, 445
Press, W. H., Teukolsky, S. A., Vettering, W. T., and Flannery, B. P.: 1992, Numerical Recipes in C The Art of Scientific Computing, *Cambridge U. Press, 2nd Ed.*, p551
Rucinski, D., Cummings, A. C., Gloeckler, G., Lazarus, A. J., Möbius, E., and Witte, M.: 1996, *Space Sci. Rev.*, this issue
Simpson, J. A.: 1983, *Ann. Rev. Nucl. Sci.*, **33**, 323
Stone, E. C.,Vogt, R. E., McDonald, F. B., Teegarden, B. J., Trainor, J. H., Jokipii, J. R., and Webber, W. R.: 1977, *Space Sci. Rev.*, **21**, 355
Stone, E. C., Cummings, A. C., and Webber, W. R.: 1996, *J. Geophys. Res.*, **101**, 11017
Vasyliunas, V. M., and Siscoe, G. L.: 1976, *J. Geophys. Res.*, **81**, 1247

Address for correspondence: A. C. Cummings, Mail Code 220–47, California Institute of Technology, Pasadena, CA 91125

IMPLICATIONS OF A WEAK TERMINATION SHOCK

L. A. FISK
Department of Atmospheric, Oceanic, and Space Sciences
University of Michigan, Ann Arbor, MI 48109-2143, U.S.A.

Abstract. Recent observations from the Voyager spacecraft have suggested that the spectrum of the anomalous cosmic ray component is relatively steep at the termination shock, which is believed to be responsible for accelerating these particles. This conclusion argues that the termination shock must be weak, which in turn requires that the upstream Mach number in the solar wind must be quite low, ~2.4. It is pointed out that such conditions are unlikely to prevail at all locations along the shock front. However, it is possible for such conditions to exist at the interface between high speed streams at high heliographic latitudes and the region at low latitudes where high and low speed streams have interacted and come into equilibrium. This discussion suggests a preferred location for the injection of the anomalous component into the shock acceleration process.

Key words: anomalous cosmic rays – solar wind – termination shock

1. Introduction

The purpose of this paper is to discuss the conditions in the solar wind immediately upstream from the termination shock. These conditions, of course, define the inner boundary condition for the interaction of the solar wind with the local interstellar medium – the subsequent shock transition, the subsequent downstream, subsonic flow, and the merging with the interstellar medium. We will undertake this discussion in the context of a recent set of observations and analysis by Stone *et al.* (1996) of the spectrum of the anomalous cosmic ray component at the termination shock, from which we can infer that the shock strength is weak, and therefore that the upstream Mach number is low. We will conclude that such conditions cannot prevail all along the shock front, but rather they may be unique to the interface region where high-speed streams at high heliographic latitudes interact with lower speed flow near the solar equatorial plane.

2. Implications of the Anomalous Component Spectrum

We begin by reviewing briefly the analysis of Stone *et al.* (1996) and the resulting implications of the observed spectra of anomalous cosmic rays for the strength of the termination shock. As was postulated by Fisk *et al.* (1974), the anomalous component is considered to originate as interstellar neutral gas that is swept into the heliosphere by the motion of the Sun relative to the local interstellar medium; this accounts for its anomalous composition of mainly helium, nitrogen, oxygen, and neon. Once here, it is ionized by charge-exchange and photoionization and becomes the pick-up ions, which have energies of ~1 keV/nucleon and which are now seen by the SWICS instrument on Ulysses (Gloeckler *et al.*, 1993). The pick-up ions are then convected outward by the solar wind into the outer heliosphere, and the prevailing theory, as postulated by Pesses et al. (1981), is that when they encounter the termination shock they are accelerated to high energies to form the anomalous cosmic ray component which is observed at energies of ~ 10's of MeV/nucleon, by, for example, the Voyager spacecraft (Stone *et al.*, 1996).

Space Science Reviews **78**: 129–136, 1996.

The Voyager 1 spacecraft is now more than 50 AU from the Sun. It is a fairly straightforward calculation to estimate the spectrum of the accelerated anomalous particles, as it would appear at the termination shock, the distance of which is now estimated to be 70-90 AU from the Sun. The anomalous particles are singly charged – there is time to ionize the particles only once. They are thus of quite high rigidities, comparable to galactic cosmic rays with energies of ~ GeV/nucleon which do not experience much modulation in the solar wind. Equivalently, the extrapolation from the observed spectra at Voyager to the expected spectra at the termination shock is not large, and Stone *et al.* (1996) find the interesting result that the expected spectrum is quite steep – the intensity j varies with energy E as $j \propto E^{-\beta}$, with $\beta = 1.42 \pm 0.08$.

The theory for the acceleration of energetic particles at shocks is quite simple. Particles bounce back and forth across the shock front. They make head-on collisions with the magnetic irregularities on the upstream side, and they gain energy. They make over-taking collisions on the downstream side, and they lose energy. The upstream velocity is larger than the downstream one, and so the particles experience a net energy gain. Indeed, there is a simple formula, which relates the spectral index to the jump in flow speed of the plasma:

$$\frac{u_u}{u_d} = \frac{(2+2\beta)}{(2\beta-1)}. \tag{1}$$

Here, u_u and u_d is the upstream and downstream flow speed, respectively. This formula assumes only that the accelerated particles are in steady state, i.e. the number of particles and the flux of these particles, in a given velocity range, are conserved. This approach is also formally equivalent to assuming the particles gain energy by drifting through the electric field at the shock front. The work done by the shock electric field is identical to the work done by the solar wind on the pressure gradient of the pick-up ions, which is set up as they attempt to diffuse upstream from the shock front. It is quite unreasonable to assume that any one particle drifts undisturbed along the full length of the termination shock front. Rather, particles will be scattered off of and return to the shock many times where they experience diffusive or equivalently electric field acceleration, and the average behavior at any one location on the shock front should reasonably be described by (1).

With the value of $\beta = 1.42 \pm 0.08$ for the spectrum of the anomalous particles at the termination shock, the resulting flow speed jump is then found by Stone *et al.* (1996) from (1) to be $u_u/u_d = 2.63 \pm 0.14$. The termination shock can reasonably be assumed to behave as a simple gas dynamic shock since the expected ratio of the thermal pressure, due to the pick-up ions, to the magnetic pressure, is of order 10. The velocity jump is then related to the upstream Mach number by (e.g., Gombosi, 1994).

$$\frac{u_u}{u_d} = \frac{(\gamma+1)M_u^2}{2+(\gamma-1)M_u^2}. \tag{2}$$

where γ is the ratio of specific heats, or $\gamma = 5/3$. Thus, with $u_u/u_d = 2.63 \pm 0.14$, the upstream Mach number is found to be $M_u \approx 2.4$.

The termination shock could, of course, be a fairly complicated shock. As the anomalous particles are accelerated from low to high energies – from keV/nucleon to MeV/nucleon energies – they could exert a pressure force on the upstream solar wind, particularly the lower energy particles could exert such a force, which will slow down the solar wind before it reaches the shock front. There could be a fore shock region where the solar wind slows down adiabatically, followed by a fairly weak shock.

For the high energy anomalous particles, however, the detailed shock structure does not matter. These particles bounce back and forth across the shock front on relatively large spatial scales. The particles are quite mobile in the solar wind, otherwise they would not be seen with relatively little modulation at Voyager. The flow speed jump that these higher energy particles experience, then, from which they gain their energy and form their spectrum, is the total velocity jump. It does not matter whether the jump occurs in a fore shock region followed by a weak shock, or in a single shock jump.

If we stand back far enough from a shock, we can apply the Rankine-Hugoniot relations – mass, momentum, and energy must be conserved. Thus, if we are not interested in the details of the shock transition, just in the total speed jump, then (2), which is derived from the Rankine-Hugoniot relations, applies. Equivalently, if the total speed jump is only 2.6, the Mach number of the upstream solar wind, unaffected by the shock, must be only 2.4. Thus, if we believe the Stone *et al.* (1996) observations that the spectrum of the anomalous particles is steep ($\beta = 1.42$), and we believe that these particles are accelerated in a traditional shock acceleration model at the termination shock, then we are forced to conclude that the solar wind upstream from the shock front, and unaffected by the shock, has a very low Mach number.

There have been suggestions that galactic cosmic rays, which have fairly large pressures in the outer heliosphere, could affect the solar wind termination shock (Donohue and Zank, 1993). To have an influence on the solar wind, the cosmic rays must exert a pressure gradient. However, the pressure resides primarily in cosmic rays with energies of about 1 GeV/nucleon, which are quite mobile in the solar wind, and which typically have gradients of only a few percent per AU (e.g., Lockwood and Webber, 1993). There are, in fact, no significant pressure gradients in the galactic cosmic rays, and they exert little force on the solar wind.

3. Mechanisms for Yielding a Low Mach Number in the Outer Heliosphere

If the Mach number of the solar wind in the outer heliosphere is only ~2.4, we have to find a mechanism or mechanisms to develop a large internal pressure in the solar wind, perhaps through the use of the interstellar pick-up ions themselves, through some acceleration mechanism or instability, or by some proportional reduction in the solar wind flow energy. Consider several possibilities.

Left to its own devices, the steady solar wind is not much help. In a constant solar wind flow the internal pressure will cool adiabatically, the ram pressure will decline in proportion to density, and the Mach Number will go up dramatically. In a simple solar wind model, then, with no other processes involved, the Mach number of the solar wind in the outer heliosphere will be very large. Of course, there are other processes involved; in particular, the interstellar pick-up ions are continuously injected into the solar wind through the ionization of interstellar neutrals, and in the process of being picked up by the solar wind and subsequently isotropized, acquire a thermal energy equal to the solar wind flow energy. Indeed, the pick-up ions beyond about 10 AU from the Sun are the dominant internal pressure force in the solar wind (e.g., Holzer, 1977).

We can estimate how much pressure P should reside in the pick-up ions with the following equation:

$$u\frac{dP}{dr} + \frac{10}{3}\frac{u}{r}P = \beta_p n_n(r,\Theta)\frac{mu^2}{3}\frac{r_o^2}{r^2}. \tag{3}$$

We assume that pick-up ions are injected into the solar wind with an ionization rate, β_p, at heliocentric radial distance $r_o = 1AU$. The pick-up ions are primarily hydrogen, so the ionization is due mainly to charge-exchange with the solar wind, the density of which varies as r^{-2}. The number injected is proportional to the interstellar neutral density n_n, which is a function of r and Θ, the angle between the solar wind velocity and the upstream direction of interstellar neutrals. We take n_n to be constant, which is a reasonable approximation in the outer heliosphere. The solar wind flow speed is u and the proton mass m.

Equation (3) is a direct consequence of the equation derived by Vasyliunas and Siscoe (1976) to calculate the expected distribution function of pick-up ions; (3) can be formed by taking the second moment of the Vasyliunas and Siscoe equation with respect to particle velocity, to form an equation for the pressure. The particles are assumed to be convected outward with the solar wind; they suffer adiabatic deceleration in the expanding solar wind; and there is continuous injection of new ions.

Equation (3) can be readily solved to yield:

$$P(r) = \frac{3}{7}\frac{r_o^2}{ur}\beta_p n_n \frac{mu^2}{3}. \tag{4}$$

The average production rate of pick-up ions which is required to account for the observations of pick-up hydrogen as seen by Ulysses in the inner heliosphere is β_p = 7.5 x 10^{-7} sec^{-1} (Gloeckler *et al.*, 1993). The expected density of interstellar hydrogen is $n_n = 0.08\ cm^{-3}$. A reasonable speed for the solar wind, near the equatorial plane in the outer heliosphere, is u = 455 km/s. With these values in (4), the average pressure of the pick-up ions at 80 AU, and thus the internal pressure in the solar wind at this distance is $P = 0.075$ eV/cm^3, consistent with more detailed calculations by Holzer (1977). The average solar wind density at $\sim$ 5 AU is observed to be $\sim$ 0.3 cm^3. Thus, the expected Mach number at 80 AU is $\sim$ 4.6. With pick-up ions alone, with no further acceleration, the average Mach number is a factor of two higher than the required value of $\sim$ 2.4. The Mach number goes as the square root of the pressure, or for the pick-up ions with no further acceleration to be responsible for the inferred low Mach number in the solar wind in the outer heliosphere, we would have to be off by a factor of 4 in their pressure, which appears to be unlikely.

Consider the possibility that the pick-up ions are accelerated through some interaction with the magnetic field or turbulence in the solar wind, and their pressure is increased. Indeed, the spectrum of pick-up ions seen near the equatorial plane in the inner heliosphere has a pronounced tail at speeds above twice the solar wind speed, the speed the particles can acquire through the pick-up process (Gloeckler *et al.*, 1994). This tail results, presumably, from acceleration of the pick-up ions in Co-rotating Interaction Regions, where high and slow solar wind flows interact, and which contain both shocks and extensive turbulence. Indeed, the pressure in the tail of the pick-up ion distribution can be comparable to or greater than the pressure in the initial pick-up ion distribution. Thus, should such acceleration continue into the outer heliosphere, then perhaps the pick-up ion pressure could be increased by the required factor of $\sim$ 4, and the Mach number lowered to $\sim$ 2.4.

The question arises, however, as to whether there is a sufficient energy source for such acceleration. The tail of the pick-up ion distribution in the inner heliosphere is produced by stream-stream interactions in the solar wind – high speed streams overtaking slower ones and producing shocks and turbulence. Such processes, however, diminish with increasing heliocentric distance. The solar wind near the equatorial plane tends to a nearly constant speed, with considerable variations in density and in the magnetic field strength (e.g., Belcher et al., 1987).

Indeed, the largest free energy source that is potentially available after stream-stream interactions come into equilibrium is the large scale variations in the magnitude of magnetic field which result from the piling up of solar wind material as the high speed flows overtake slower flows (e.g., Burlaga *et al.*, 1995). Magnitude variations in the magnetic field tend to be small in the solar wind on small spatial scales; the variations are Alfvénic in nature.

However, on large spatial scales the field magnitude can vary by a factor of order ~ 2, or the energy density varies by ~ 4.

Suppose, then, that it is possible to extract energy from the large magnitude variations in the solar wind magnetic field; perhaps the variations are not static, but propagating, and they can be damped by a transit-time damping mechanism (Fisk, 1976). If we consider that the energy in the fluctuations can be extracted with a characteristic damping length of 80 AU, then there is an additional term on the right side of (3), or

$$u\frac{dP}{dr} + \frac{10}{3}\frac{u}{r}P = \beta_p n_n(r,\Theta)\frac{mu^2}{3}\frac{r_o^2}{r^2} + 4\frac{u}{R}\frac{B_e^2}{8\pi}\frac{r_o^2}{r^2}. \tag{5}$$

where $R = 80$ AU and B_e is the magnetic field strength at Earth. We assume that the field magnitude varies by a factor of 2, and the field is primarily azimuthal, i.e. that the energy in the field declines as r^{-2}.

Equation (5) can also be readily solved, and with $B_e = 3.5$ nT we find that the increase in the pressure in the pick-up ions due to the damping of the largest variations in the magnetic field that are available in the solar wind – the large-scale variations in the field magnitude – is only $\sim 10\%$. That is, acceleration of the pick-up ions is unlikely to be significant. This result is not surprising. The main source of energy for the pick-up ions is the flow energy of the solar wind. The particles, when they are first ionized, acquire the solar wind flow energy. Since the flow energy in the solar wind is so much larger than any internal energy in the solar wind (e.g., the magnetic field energy), it is not surprising that the pick-up ions acquire most of their energy through the initial pick-up process, and not by subsequent acceleration from the damping of waves or turbulence.

Equivalently, if we are to find a way to increase the internal pressure in the solar wind to where the Mach number is only ~ 2.4, it will be necessary to tap the flow energy of the solar wind. No other source is sufficient. Near the equatorial plane, access to the flow energy is difficult. High and low speed streams certainly interact near the equatorial plane and this converts flow energy into internal pressure. However, such interaction occurs in the inner heliosphere, and once completed, the flow stabilizes to a relatively constant flow speed, after which the pressure should tend to cool adiabatically and the Mach number tends to increase again.

There are, however, some possibilities at higher heliographic latitudes. Recall near solar minimum conditions the solar wind is fast and steady at high latitudes, flowing outward from the polar coronal holes with speeds of 700-800 km/s (e.g., McComas *et al.*, 1995). Near the equatorial plane, high and low speed streams interact and yield a steady flow at a lower speed of ~ 450 km/s. At some latitude, then, there must be an interface between this high speed flow and the lower speed flow. The interaction region near the equatorial plane may expand in latitude with increasing heliocentric distance – the conversion of flow energy into thermal pressure will drive an expansion in latitude – so that at each heliocentric radius there is a latitude where undisturbed high speed flow is encountering the lower speed flow. At this location the flow energy of the high speed flow is converted into internal pressure, and the Mach number of the high speed flow drops substantially. If this conversion of flow energy to internal pressure occurs relatively close to the termination shock, then in this location the Mach number of the solar wind upstream from the termination shock could have the required low value.

Consider a simple calculation. If we take the enthalpy per unit mass plus $u^2/2$ to be constant along a streamline, then a high speed flow with an initial speed of u_i and Mach number M_i will slow down to a final velocity and Mach number of u_f and M_f, subject to:

$$\frac{u_f^2}{u_i^2} = \frac{(3/M_i^2 + 1)}{(3/M_f^2 + 1)}. \tag{6}$$

Thus, if the initial Mach number is M_i = 4.6, as is expected in the outer heliosphere with the large pick-up ion pressure, and the required final Mach number is M_f =2.4, then, from (6) the speed of the high speed flow must be reduced to u_f/u_i = 0.86. Such a reduction is not unreasonable in that a high speed flow of 750 km/s, colliding with a slower 450 km/s flow and coming to rest at the speed of the center of mass of these two flows, would have its speed reduced by approximately this amount.

It would appear then that it is possible to have the required Mach number upstream from the termination shock of $\sim$ 2.4 at a specific location, viz. at latitudes above the equatorial plane where high and low speed streams are actively interacting in the outer heliosphere. At lower latitudes, the interaction occurs in the inner heliosphere, and although the Mach number is reduced by the interaction, the subsequent expansion of the solar wind will tend to increase the Mach number again. At higher latitudes, the high speed flow persists all the way out to the termination shock, and the Mach number remains high. This suggestion that there is an intermediate latitude range where the Mach number of the solar wind is small requires, of course, that the interaction region of high and low speed flows near the equatorial plane does not expand rapidly with latitude, and enclose the heliosphere at relatively small heliocentric distances. It is necessary to preserve the high speed flow at these intermediate latitudes to large heliocentric distances, $\sim$ 80 AU, and to have the interaction with the lower speed flow occur at this location, immediately upstream from the termination shock.

We should ask, of course, whether the anomalous cosmic rays seen by Voyager could in fact have originated from this intermediate latitude range, where we might expect a low upstream Mach number and a weak termination shock. Indeed, that could be the case. Gradient and curvature drifts, and diffusion, during this portion of the solar magnetic cycle are expected to be inward in radial distance, and downward in latitude (e.g., Jokipii et al., 1977). The anomalous particles seen by Voyager should thus have originated at a higher latitude and a larger heliocentric distance, which is not inconsistent with a location on the termination shock where the upstream Mach number could be low.

It is also interesting to note that the interface between the high speed flow at higher latitudes, and the lower speed flow near the equatorial plane has a large velocity shear, $\sim$ 350 km/s, and could be Kelvin-Helmholtz unstable. The magnetic fields in the high and lower speed flows have different origins at the Sun, and thus do not cross the interface, nor tend to suppress the instability. This instability could be responsible for reducing the speed of the high speed flow, and thus for converting flow energy into internal pressure. It could also be responsible for accelerating the pick-up ions in advance of the termination shock. One of the difficult problems for accelerating the pick-up ions at the termination shock is to provide a mechanism for their injection into the shock acceleration process. The pick-up ions have thermal speeds comparable to the solar wind speed. It is difficult for such particles to propagate upstream in the solar wind, particularly when the magnetic field is highly azimuthal as it is in the outer heliosphere. Such particles might thus be expected simply to be convected through the shock front, and not to bounce back and forth and experience the energization envisioned in (1). However, if such particles are pre-accelerated, in advance of the shock, the injection is then straight-forward. A Kelvin-Helmholtz instability is a Venturi effect. It creates pressure variations in the flow, which could result in variations in the magnetic field strength. Such variations could provide a natural acceleration of the pick-up ions through a magnetic pump mechanism.

Finally, we note that if this explanation is correct, there are some interesting implications for the solar cycle dependence of the Mach number at the termination shock. The polar coronal holes expand as the solar cycle evolves towards solar minimum. The interface, then, where high and slower speed flows interact in the outer heliosphere, upstream from the shock, will then move downward in latitude as we approach solar minimum, and retreat as conditions evolve towards solar maximum. With regard to acceleration at the termination shock, it is reasonable to expect that particles are accelerated everywhere along the shock front since drift motions will carry the particles to all latitudes, following their injection even from a preferred location near the high and slower speed flow interface. However, the behavior of the spectral shape of the accelerated particles at the shock front with latitude should vary during the solar cycle, which may be reflected in the intensity of anomalous particles seen near the equatorial plane, or by Ulysses at high latitudes.

4. Summary

In summary, there are interesting observations by Stone *et al.* (1996) which suggest that the spectrum of the anomalous cosmic rays appears to be steep at the termination shock, which, in turn, suggests that the total jump in velocity at the shock is relatively small. This can occur only if the upstream Mach number is only of order 2.4. When we consider how the Mach number could be so low, the only viable mechanism is the conversion of solar wind flow energy into internal pressure. Near the equatorial plane, such conversion occurs at relatively small heliocentric radial distances (within $\sim$ 20 AU of the Sun), as high and low speed solar wind flows interact, after which the Mach number will tend to increase again. The only possibility appears to be to perform this conversion at higher latitudes where the high speed streams, at least near solar minimum conditions, may be preserved into the outer heliosphere, and then slow down by interacting with slower speed flow that has expanded upward in latitude from the equatorial region. It may also be possible that the interface between the lower speed flow near the equatorial region, and the high speed flow at high latitudes is Kelvin-Helmholtz unstable. The termination shock that follows this region of low Mach number would be an ideal candidate for the acceleration of the anomalous component because the effects of the Kelvin-Helmholtz instability could be to pre-accelerate the pick-up ions, and readily inject them into the shock acceleration process. These arguments suggest that the entire termination shock is not weak, but rather it can be weak only in the mid-latitude region, with the remaining shock conforming to the usual expectations that it is strong and embedded in a relatively high Mach number flow.

Acknowledgements

This work was supported, in part, by NASA/JPL contract 955460.

References

Belcher, J.W., Lazarus, A.J., McNutt, R.L, Jr., and Gordon, G.S., Jr.: 1993, *J. Geophys. Res.* **98**, 15177.

Burlaga, L.F., Ness, N.F., and McDonald, F.B.: 1987, *J. Geophys. Res.* **92**, 13647.

Donohue, D.J., and Zank, G.P.: 1993, *J. Geophys. Res.* **98**, 19005.

Fisk, L.A., Kozlovski, B., and Ramaty, R.: 1974, *Astrophys. J.* **190**, L35.

Fisk, L.A.: 1976, *J. Geophys. Res.* **81**, 4633.

Gloeckler, G., Geiss, J., Balsiger, H., Fisk, L.A., Galvin, A.B., Ipavich, F.M., Ogilvie, K.W., von Steiger, R., and Wilken, B.: 1993, *Science* **261**, 70.

Gloeckler, G., Geiss, J., Roelof, E. C., Fisk, L. A., Ipavich, F. M., Ogilvie, K.W., Lanzerotti, L.J., von Steiger, R., and Wilken, B.: 1994, *J. Geophys. Res.* **99**, 17637.

Gombosi, T.I.: 1994, in *Gaskinetic Theory*, Cambridge University Press, 266.
Holzer, T.E.: 1977, *Rev. Geophys. Space Phys.* **15**, 467.
Jokipii, J.R., Levy, E.H., and Hubbard, W.B.: 1977, *Astrophys. J.* **213**, 861.
Lockwood, J.A., and Webber, W.R.: 1993, in *Proc. 23rd Int. Cosm. Ray Conf.* (Calgary) **3**, 469.
McComas, D.J., Phillips, J:L., Bame, S.J., Gosling, J.T., Goldstein, B.E., and Neugebauer, M.: 1995, *Space Sci. Rev.* **72**, 93.
Pesses, M.E., Jokipii, J.R., and Eichler, D.: 1981, *Astrophys. J.* **246**, L85.
Stone, E.C., Cummings, A.C., and Webber, W.R.: 1996, *J. Geophys. Res.* **101**, 11017.
Vasyliunas, V.M., and Siscoe, G.L.: 1976, *J. Geophys. Res.* **81**, 1247.

THE ACCELERATION OF PICKUP IONS

J.R. JOKIPII AND J. GIACALONE
The University of Arizona, Tucson, AZ, USA

Abstract. The well-established association of pickup ions with anomalous cosmic rays shows that acceleration of pickup ions to energies above 1 GeV occurs. At present, diffusive shock acceleration of the pickup ions at the termination shock of the solar wind seems to be the best candidate for acceleration to the high energies of anomalous cosmic rays, accounting well for many of their observed properties. However, it is shown that acceleration of pickup ions from their initial energies by this process appears to be difficult at very strong, nearly perpendicular shocks such as the termination shock. This "injection" problem remains without a clear solution. A number of alternatives have been proposed for the initial acceleration of pickup ions to the point where diffusive acceleration at the termination shock can take over, but none of these processes has yet emerged as a clear favorite.

Key words: cosmic rays - solar wind - acceleration

1. Introduction

Pickup ions, the result of the ionization of interstellar neutral particles in the heliosphere and their subsequent pickup by the solar wind's convected magnetic field, are accelerated in the heliosphere. This is documented by the well-established association of anomalous cosmic rays (herein ACR) with the pickup ions and by the Ulysses observations of direct acceleration out of the pickup-ion distribution. This paper reviews our current understanding of the acceleration process, from the keV energies of pickup ions to the highest energies observed, more than 1 GeV. Intimately connected with this acceleration is spatial transport of the particles throughout the heliosphere.

In this process, a basic parameter such as the gyro-radius, for example, varies by more than a factor of a million in in going from the lowest to the highest energies and from the inner parts of the heliosphere to its outer boundary. Hence, very different approximations must be utilized for different parts of the problem. Historically, because our understanding of high-energy charged-particle acceleration and transport is more-developed, the first problem to be studied quantitatively was that of the acceleration of already-energetic ($\approx$ 100 keV), ACR particles to the highest energies observed. The initial acceleration of the pickup ions has been studied more-recently and here our understanding is less developed. For this reason, we will begin with ACR having energies of the order of 100 keV and higher. These are apparently accelerated primarily at the termination shock of the solar wind, and behave much as ordinary cosmic rays. The latter part of this paper will discuss the acceleration of the lowest-energy particles, and point out the difficulties of getting pickup ions accelerated.

2. The Anomalous Component

The anomalous component of cosmic rays is composed of fluxes of helium, nitrogen, oxygen, neon, protons and low levels of carbon (for a recent review, see Klecker, 1995), which are observed, in the inner heliosphere, to be enhanced in a region of the energy spectrum ranging from a kinetic energy of 20 MeV to perhaps 300 MeV. The observed radial

Space Science Reviews **78**: 137–148, 1996.

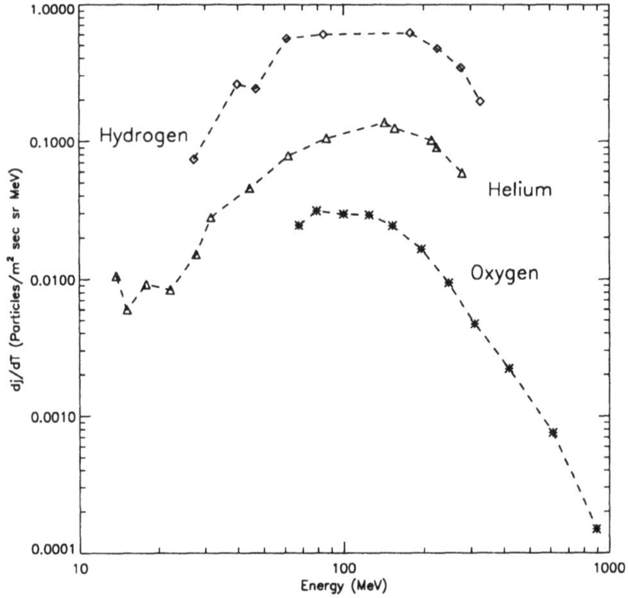

Figure 1. Energy spectra of anomalous oxygen, helium and hydrogen observed near 21 AU in 1985 (Cummings and Stone, 1987, 1988).

intensity gradient is positive out to the maximum distance reached by current spacecraft, indicating that this component probably originates in the interaction of the solar wind with the interstellar medium. Observed spectra of ACR oxygen, helium and protons are shown in Figure 1.

The first significant advance in our understanding of the origin of the anomalous component was due to Fisk, Kozlovsky and Ramaty (1974) (herein FKR). They pointed out that interstellar neutral atoms, streaming into the solar system, have a probability of being ionized, either by charge exchange with the solar wind or by solar UV radiation. These newly-ionized, singly-charged particles are then swept out of the inner solar system by the solar wind and subsequently accelerated. This explains very nicely the composition of the anomalous component since only initially-neutral particles can participate in the process.

Pesses, Jokipii and Eichler (1981) subsequently pointed out that many features of the anomalous component could be explained if the acceleration of the freshly-ionized particles occurs at the termination shock of the solar wind, by the mechanism of diffusive shock acceleration. Jokipii (1986) presented results from a quantitative two-dimensional numerical simulation, in which the full transport equation was solved. It was found that the basic observed features of the spectrum and spatial gradients could be explained very naturally in terms of this model. Because of the small charge and consequent high magnetic rigidity of the ACR, gradient and curvature drifts of the particles, both along the face of the shock and in the solar wind play a major role. Steenberg and Moraal (1995) presented results from a simulation neglecting drifts.

It has been established that the acceleration of anomalous oxygen to ≈ 200 MeV must occur in less than a few years. This is because if acceleration took significantly longer, further electron loss would occur and the ACR would be multiply-charged rather than singly-charged at this energy (Jokipii, 1992). Diffusive shock acceleration can accelerate particles to these energies in a time less than or on the order of a year, whereas other mechanisms in the weak fields of the outer heliosphere take significantly longer. Therefore, diffusive shock acceleration at the quasi-perpendicular termination shock of the solar wind seems to be the acceleration mechanism of choice for the acceleration of ACR to high energies.

3. The Model For Acceleration of ACR to High Energies

The transport of charged particles in the solar wind is governed by the ambient electric and magnetic fields, because particle-particle collisions are totally negligible. The fast charged particles are subject to four distinct transport effects. Because of the magnetic field, they are convected with the fluid flow at velocity $\mathbf{V}_w$. In addition, because the magnetic field varies systematically over large scales, the curvature and gradient drifts (velocity $\mathbf{V}_d$) are coherent over large distances. The expansion or compression of the wind causes associated energy changes. Finally, there is an anisotropic random walk or spatial diffusion (diffusion tensor κ_{ij}) which is caused by scattering due to random magnetic irregularities carried with the flow, which also maintains a nearly-isotropic angular distribution in the fluid frame.

The resulting superposition of coherent and random effects were combined first by Parker (1965) to obtain the generally accepted transport equation, which may be written for the quasi-isotropic distribution function $f(\mathbf{r}, p, t)$ of cosmic rays of momentum p at position $\mathbf{r}$ and time t:

$$\frac{\partial f}{\partial t} = \frac{\partial}{\partial x_i}\left[\kappa_{ij}\frac{\partial f}{\partial x_j}\right] - V_{w,i}\frac{\partial f}{\partial x_i} - V_{d,i}\frac{\partial f}{\partial x_i} + \frac{1}{3}\frac{V_{w,i}}{\partial x_i}\left[\frac{\partial f}{\partial \ell np}\right] + Q(\mathbf{r}, p, t) \quad (1)$$

The guiding-center drift velocity is given in terms of the local magnetic field $\mathbf{B}$ and the particle charge q and speed w by $\mathbf{V}_d = (pcw/3q)\,\nabla \times (\mathbf{B}/B^2)$, where c is the speed of light. This transport equation is remarkably general, and is a good approximation if there is enough scattering by the magnetic irregularities to keep the distribution function nearly isotropic, and if the particles have random speeds substantially larger than the background fluid convection speed. In particular, since the fluid velocity need not be a continuous function of position, **all** of the standard theory of diffusive shock acceleration is contained in this equation.

3.1. Heliospheric Configuration

The general configuration of the inner heliosphere is well understood, and extrapolation to the termination shock probably does not introduce significant uncertainties. The configuration beyond the shock is uncertain, and we adopt a simple configuration which contains the basic features expected.

The solar wind velocity is taken to be radial out to a spherical termination shock at a radius R_{ts} at which it drops by a factor of r_{sh} (which is 4 for a strong shock), and then decreases as $1/r^2$ (small Mach number flow) out to an outer boundary R_b, where the energetic particle are presumed to escape. Typically, R_{ts} is taken to be some 70–100 AU and R_b some 30%–50% larger.

During the years around each solar minimum, the interplanetary magnetic field is generally organized into two hemispheres separated by a thin, nearly equatorial current sheet, across which the field reverses direction. In each hemisphere the field is generally assumed to be a classical Archimedean spiral, with the sense of the field being outward in one hemisphere and inward in the other. The field direction alternates with each 11-year sunspot cycle, so that during the 1996 sunspot minimum, the northern field was directed outward from the sun (conventionally denoted as $A > 0$), but in 1986 the northern field pointed inward ($A < 0$). This field is assumed to continue beyond the termination shock, with the spiral angle reflecting the local V_w.

There is now evidence that the **polar** magnetic field differs considerably from this spiral (Jokipii and Kóta, 1989, and Jokipii *et al.*, 1995), so the polar field is modified in our simulations. The magnetic-field structure for the years near sunspot maximum is not simple, so the following discussion is most-relevant during the several years around sunspot minimum.

3.2. General Considerations

Diffusive shock acceleration is contained within the cosmic-ray transport equation (1), if one allows the fluid flow to have a compressive discontinuity corresponding to the shock wave. The acceleration can occur for any angle between the shock normal and the magnetic field and is in general due to both compression and drift. The process at quasi-perpendicular shocks is more closely related to drift in the $\mathbf{V_w} \times \mathbf{B}$ electric field than to compressive Fermi acceleration. If the scattering frequency is significantly less than the gyro-frequency, the energy ΔT gained by a particle having electric charge Ze, at a quasi-perpendicular shock, is approximately Ze times the electrostatic potential $\Delta\phi$ gained in drifting along the shock face (Jokipii, 1990): $\Delta T \approx Ze\Delta\phi$. Note that the scattering must be sufficient to maintain near isotropy. For billiard-ball scattering this requires $\lambda_{\|}/r_g \ll w/V_w$. It follows, then, that there is in general a characteristic energy $T_c \approx Ze(\Delta\phi)_{max}$ above which the spectrum begins to decrease rapidly. Below this energy the spectrum is often a power law. Particles will "drift off" of the shock before gaining more energy. The cutoff need not be sharp, since some particles will be scattered back and forth, receiving more or less energy than T_c.

We may write for the solar wind

$$Ze(\Delta\phi)_{max} = ZeB_r r^2 \Omega_{\odot}/c \approx 240Z \text{ MeV} \tag{2}$$

where the numerical value results from using a radial magnetic field of 3.5γ at a radius of 1 AU and a solar rotational angular velocity $\Omega_{\odot} = 2.9 \times 10^{-6}$ (corresponding to a magnetic field magnitude of 5γ at 1 AU). Singly-charged particles accelerated at the termination shock would then have a spectrum which exhibits a decrease above an energy between 200 and 300 MeV.

3.3. Acceleration Rate

We next consider the acceleration **rate**. The particles gain only a small amount of energy (a fraction of order V_w/w) each time they cross the shock. The time for acceleration of particles from an initial momentum p_o to a momentum p may be written as (Forman and Morfill, 1979; Drury, 1983, Jokipii, 1987)

$$\tau_a \simeq \frac{3r_{sh}}{V_w^2(r_{sh}-1)} \int_{p_o}^{p} \{\kappa_1 + r_{sh}\kappa_2\} \frac{dp'}{p'} \tag{3}$$

where r_{sh} is the shock ratio and if θ is the angle between the shock normal and the magnetic field, κ_1 and κ_2 are the upstream and downstream values of $\kappa_{xx} = \kappa_{\|}\sin^2(\theta) + \kappa_{\perp}\cos^2(\theta)$,

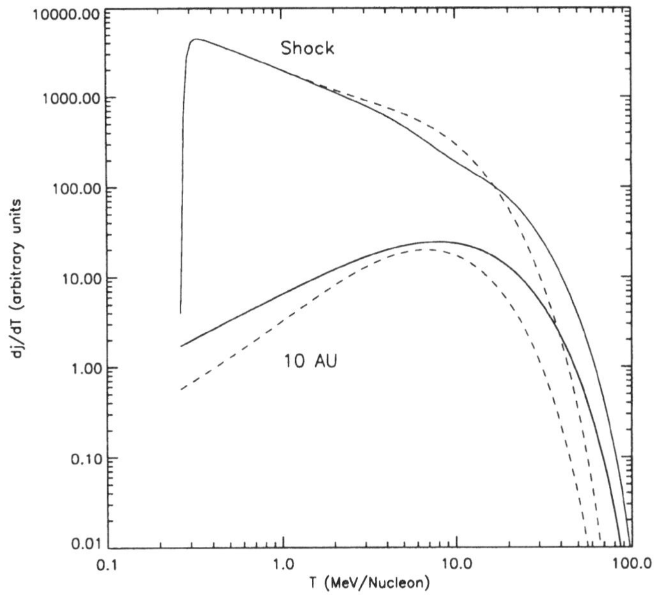

Figure 2. Plot of computed energy spectra near the heliospheric equator, for two radii, and for the two signs of the heliospheric magnetic field. The solid lines correspond to the case where the northern heliospheric magnetic field is directed outward $A > 0$), corresponding to the present sunspot minimum. The dashed lines are for $A < 0$, corresponding to the 1986 sunspot minimum. The parameters are as in the text.

and V_1 is the shock velocity normal to the upstream gas. For a constant, planar shocks, the solution at any time is a power law up to a gradual cutoff at a momentum near that defined in equation (2). In other cases the spectrum is often a power law with a cutoff determined by something other than time. Since generally $\kappa_\perp \ll \kappa_\parallel$, acceleration at a perpendicular shock is significantly faster than at a parallel shock. This makes the termination shock an excellent site for acceleration to high energies. Reasonable parameters yield a time for acceleration to a few hundred MeV in of the order of a year, whereas parallel shocks or other acceleration mechanisms take much longer.

3.4. Simulation Results

Our basic numerical model (including two spatial dimensions, heliocentric radius and polar angle) follows the acceleration of low-energy, singly-charged particles injected into the solar wind *a la* FKR and then accelerated at the termination shock. The injection energy used is typically of the order of 100 keV, because of the limited dynamical range available in numerical solutions. The method of solution is then to follow these particles in time as they are accelerated at the shock and propagate throughout the heliosphere, until the distribution reaches a steady steady state. The characteristic time to approach a steady state is found to be 2–3 years or less, except for the highest-energy ACR.

The theoretical spectra from a simulation, in which the injection of low-energy particles was spatially uniform over the shock, are shown in Figure 2. Here nominal values for the

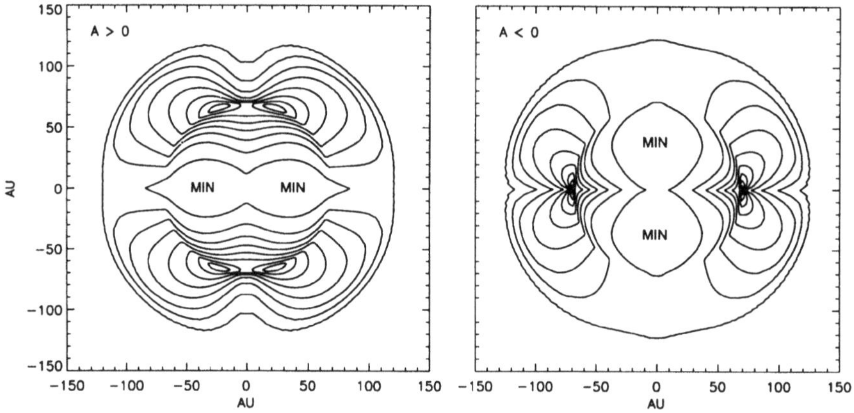

Figure 3. Equal intensity contours computed for 8 MeV/nucleon anomalous oxygen, as a function of position in a meridional plane, as computed for the model parameters discussed in the text. The left panel is for $A > 0$ and the right is for $A < 0$.

parameters were used. $V_w = 400$ km/sec, $\kappa_{\|} = 1.5 \times 10^{22} P^{.5} \beta$ cm^2/sec, $\kappa_{\perp} = .1\ \kappa_{\|}$, $R_{ts} = 70$ AU, and $R_b = 130$ AU (here P is the particle rigidity in GV, and β is the ratio of the particle speed to the speed of light). A modified polar field (Jokipii and Kóta, 1989) was used. Note that the relative normalization (amplitude) of the two spectra are not significant and may be changed by small variation of the parameters. The computed energy spectra of anomalous protons, helium, neon, etc. are in good agreement with the observed spectra.

Figure 3 illustrates typical contour plots of the intensity of the modeled anomalous oxygen as a function of heliocentric radius and polar angle, in a solar meridional plane. We see that the intensity increases with radius, much as does that of the galactic cosmic rays, out to the termination shock (at a radius of 60 AU). Beyond the shock, the intensity decreases out to the outer boundary of the heliosphere. Along the shock, the **maximum** intensity occurs at a latitude which shifts as the sign of the magnetic field changes. If $A > 0$, the particles drift toward the pole along the shock face and then inward and down from the poles to the current sheet, the intensity maximum is near the poles. On the other hand, when $A < 0$ the maximum at the shock shifts toward the equator.

Measurements of the radial and latitudinal gradients of the ACR have been reported by a number of authors (McKibben *et al.*, 1979; McDonald and Lal, 1986; Cummings *et al.*, 1987, 1995). It is found that in the inner heliosphere the radial gradient of the anomalous oxygen is significantly higher than in the outer heliosphere. The magnitude of the radial gradient computed in the model is quite consistent with that observed, both in absolute magnitude and in radial variation.

A robust consequence of the models is that the latitudinal gradient near the current sheet, near sunspot minimum, should change sign in alternate sunspot minima. In particular it is predicted that, during the present and 1975 sunspot minima, the intensity of both galactic cosmic rays and the anomalous component should **increase** away from the current sheet, whereas during the 1986 sunspot minimum the sign should change and the cosmic rays would **decrease** away from the current sheet. Near sunspot minimum, when the current sheet is nearly flat, the effects of these drifts are expected to be the most important. These consequences are robust in that they are true for the entire range of reasonable parameters considered, and hence cannot be easily changed.

In 1977, Pioneer observed a positive gradient of anomalous helium away from the current sheet (McKibben *et al.*, 1979), as predicted. Observations carried out during the last sunspot minimum during a period in 1984 and 1985, when the current sheet tilt went below the latitude of the Voyager I spacecraft, Cummings and Stone (1987) reported that the sign of the latitudinal gradient changed from being positive to negative, again as predicted by the theory.

4. The Charge State of Anomalous Cosmic Rays

The charge state of anomalous cosmic-ray oxygen at energies of $\approx$ 10 MeV/n has been shown to be essentially one (Adams *et al.*, 1991). Adams and Leising (1991) then showed that this implied that the source was less than .2 pc away, if they came on a straight line. Jokipii (1992) pointed that this result also implied a significant upper limit of about 4 years on the time for acceleration of these particles. Very-recent work (Mewaldt et al., 1996) reports the existence of multiply-charged ACR oxygen at energies above some 300 MeV, and Jokipii (1996) showed that these observations agree very well with the model discussed here, indicating that singly-charged ACR decrease quickly at energies higher than T_c as discussed above in Section 3.2.

5. The Acceleration of Pickup Ions

The above discussion has summarized the present status of the acceleration of ACR's to the highest energies observed, in excess of 1 GeV. Diffusive acceleration at the termination shock of the solar wind accounts for the observations very well. The physics of the initial stage of this acceleration process from the low energies of the pickup ions ($\sim$ 1 keV/N in the plasma frame immediately after ionization) is not yet fully resolved. This "injection problem" stems from the fact the the pickup ions have initial speeds that are comparable to the convection speed of the solar wind and are not fast enough to encounter the nearly-perpendicular shock several times. Charged particles move along the magnetic field much more effectively than across it and since the magnetic field is convecting through the shock, particles which have speeds comparable to the flow speed, tend to convect through the shock. On the other hand, high-energy particles can encounter the shock several times due to the fact that their velocities are much larger than the convection speed.

Because propagating *interplanetary* shocks only move $\approx$ 100 km/s faster than the solar wind, interstellar pickup ions, moving several times faster, can more easily encounter these shocks many times. Hence, the propagating shocks are much more natural "injectors" of energetic particles, which can then undergo further acceleration at the termination shock. This is especially true in the inner heliosphere where the orientation of the magnetic field is more favorable for injection to occur.

5.1. Scatter-Free Limit

As was discussed in Section 3.4, diffusive shock acceleration at a perpendicular shock is applicable only when the criterion $\lambda_{\|}/r_g \ll w/V_1$ is satisfied. Here, V_1 is the upstream flow speed relative to the shock. For the terminations shock and other standing shocks in the solar wind, unaccelerated pickup ions do not meet this criterion. Consequently, a boost in energy is required. This boost may be achieved at interplanetary shocks even though these shocks are expected to be considerably weaker than the termination shock.

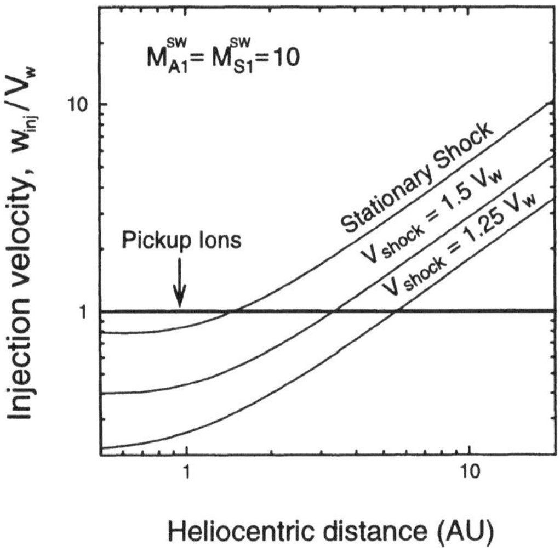

Figure 4. Injection velocity as a function of heliocentric distance for two radially propagating shocks and a stationary shock in a standard Parker-spiral magnetic field.

It is illustrative to consider the interaction between the low-energy pickup ions and the shock in the limit of no scattering. In this case, it is straightforward to demonstrate (Decker, 1988) that a fraction of an isotropic ensemble of charged particles with plasma-frame speed $w > V_1 \sec\theta b^{-\frac{1}{2}}$ (b is the ratio of the down stream to upstream magnetic field intensity), released upstream of a shock, is reflected by the shock back upstream and gain a small amount of energy (a factor of $\approx b$) in the process. These particles will propagate upstream where they will eventually scatter (weak scattering limit) and return to the shock at some later time. Particles that go through the shock on their first encounter are not likely to return to the shock. It is reasonable to regard this critical speed to be that at which particles are are capable of diffusive shock acceleration to higher energies.

Consider a shock propagating radially from the sun at a speed $V_{shock} = V_w + V_1$ measured relative to the inertial (sun) frame. The injection speed can be written

$$w_{inj} = V_1 \sec\theta b^{-\frac{1}{2}} \tag{4}$$

Particles with speed $w > w_{inj}$ are reflected by the shock (and we loosely refer to these as being "injected") whereas those with $w < w_{inj}$ are convected downstream from the shock. b is determined by the Rankine-Hugoniot shock-jump relations and is a function of the upstream Alfv̀en and sound Mach numbers measured relative to the shock frame, and the shock-normal angle.

Shown in Figure 4 is the injection speed (obtained from equation 4) as a function of heliocentric distance for a nominal Parker-spiral IMF plotted for three different shocks: two propagating shocks with speeds indicated and a stationary shock ($V_1 = V_w$). It is clear that propagating shocks can "inject" pickup ions at larger heliocentric radii than stationary shocks.

5.2. Scattering by Magnetic Irregularities

In order for charged particles to be efficiently accelerated there must be sufficient scattering to keep the particles near the shock. It was recently shown by Giacalone and Jokipii (1996a,b) that simple scattering off of random magnetic irregularities, convecting with the fluid, cannot readily account for the acceleration of low-energy pickup ions. Using an ad hoc scattering model (Giacalone *et al.*, 1994) it was shown that some acceleration does take place, but that this depends on the scattering law that is chosen. Even in the presence of large-scale magnetic fluctuations that lead to large shock-intersecting loops of the magnetic-field lines (Decker, 1993; Giacalone and Jokipii, 1996a,b), the low-energy pickup ions do not have enough speed along the field lines to encounter the shock at these places and simply convect with them downstream. Kucharek and Scholer (1995) and Liewer *et al.* (1995) have shown that low-energy pickup ions can be efficiently accelerated at a stationary termination shock by utilizing the one-dimensional hybrid simulation program. However, the shock must have a local shock-normal angle less than about 60° for this to happen. Although large-scale fluctuations of the interplanetary magnetic field will lead to fluctuations in the local shock-normal angle of the termination shock, it is unclear how long the shock will remain in this geometry. For the most part, the termination shock is expected to be nearly perpendicular.

5.3. Using the Internal Structure of Shocks

Some of the details of the internal structure of collisionless shocks may aid in surmounting the injection problem, however, this is still not resolved. One aspect that has received some attention is the effect of the cross-shock electric potential which is due to the inertia of the protons. Zank *et al.* (1996) and Lee *et al.* (1996) have suggested that a fraction of the pickup ions have a small velocity normal to the shock front, are unable to penetrate this barrier and are specularly reflected back upstream. At each reflection, since the transverse velocity is essentially conserved, the particle skims along the shock face in the same direction as the motional electric field and gains more energy. Once the particle's velocity is large enough that the Lorentz force in the direction normal to the shock exceeds the electrostatic force, the particle will convect through the layer. This energy is determined by the shock thickness, and to be significant, the shock must be as thin as the electron inertial length, which is much smaller than expected for the termination shock.

5.4. Acceleration By Turbulence Inside CIR's

Recently, Gloeckler *et al.* (1994) reported observational evidence for the acceleration of pickup ions inside CIR's, some distance away from the shocks bounding the CIR's. They suggested that pre-acceleration of the pickup ions to several times their initial energy is done by waves and turbulence in the CIR's. In this picture, the CIR-associated shocks would still be required to further-accelerate the particles to the $\approx$ MeV energies observed in association with CIR's.

6. A New Model for the Source and Acceleration of ACR

One possible solution to the problem of accelerating pickup ions at the termination shock was proposed recently by Giacalone and Jokipii (1995). They suggested, that the pickup ions are initially accelerated to energies of some 5–10 MeV by propagating shocks in the inner solar system, and these particles are subsequently accelerated at the termination shock. The scenario proposed by Gloeckler *et al.* (1994), for the initial acceleration of the pickup ions, would clearly be compatible with this general picture. Here we summarize briefly their analysis.

They assume an interplanetary-shock-accelerated source of energetic particles in the inner heliosphere, of the form

$$Q(r,\theta,p,t) = p^{-\gamma} \exp\left[-\frac{p}{p_{cut}}\right] \exp\left[-\frac{(r-r_0)^2}{L_c^2}\right] \exp\left[-\frac{(\theta-\pi/2)^2}{\theta_0^2}\right] \quad (5)$$

where r, θ are spherical coordinates. Equation (5) is solved numerically with (6) as a source until a steady-state is reached. The resulting spectra are normalized using Voyager 2 observations as discussed in the next section.

The parameters of the source are $\gamma = 5$, p_{cut} corresponds to an energy of 6 MeV, $r_0 = 10$ AU, $L_c = 6$ AU, and $\theta_0 = 0.4$ rad. The remaining simulation parameters are: $R_{ts} = 100$ AU, $R_b = 140$ AU $\kappa_{\parallel} = 1.5 \times 10^{22} P^{.5} \beta$ cm^2/sec and $\kappa_{\perp} = 0.015\ \kappa_{\parallel}$. The particle energy range considered in the simulation were from 50 keV to 1.7 GeV. These are entirely plausible parameters.

6.1. Normalizing the Spectra

To normalize the spectra obtained from our model, we use inner-heliosphere observations from Voyager 2. We use 6-month averages of ions with $Z \geq 1$ observed by the LECP experiment on board Voyager 2 from 1992 through 1994. This time period is dominated by recurrent flux enhancements with periods of ≈ 26 days. We arbitrarily chose the first 6-month average of 1993 (day 1 through day 183) in which Voyager 2 was located at about 40 AU from the sun. This spectrum is shown as the stars in Figure 5. The solid line shows the simulated spectrum at 40 AU near the heliographic equatorial plane. The dot-dashed line shows the simulated spectrum at 20 AU. The square data points at higher energies are Voyager observations reported by Cummings and Stone (1987) (see also fig. 1). We see that the model, normalized to the 40 AU observations at ~ 5 MeV, passes quite close to the higher-energy observations of anomalous cosmic rays obtained near 20 AU. This agreement leads us to conclude that the model scenario is plausible. This must be viewed with some caution, however, since the 40 AU lower-energy data was taken in a different time epoch from the 20 AU ACR data and the sign of the magnetic field was different. In addition, the present comparison applies to anomalous hydrogen only. A further analysis using low-energy helium or oxygen is necessary.

6.2. Constraints

There are two additional observational constraints relevant to this model which should be discussed. First, observations of singly-charged carbon (Geiss *et al.*, 1995) indicate that C^+ is about as abundant as O^+ in the inner heliosphere (1–3 AU). While there is no anomalous cosmic-ray carbon there is clearly anomalous oxygen leading to an apparent paradox in our theory. However, it is important to point out that the source of C^+ is within a few AU, and therefore the flux will fall off as r^{-2} beyond this distance, whereas O^+ comes from interstellar neutrals. Moreover, O^+ does not peak until 5–10 AU (Geiss *et al.*, 1994) and falls off less rapidly than r^{-2} due to the fact that oxygen is continually being pickup up throughout the heliosphere. For distances greater than ~ 3 AU oxygen is far more abundant than carbon and, as shown in Section 5, interplanetary shock acceleration, occurring from

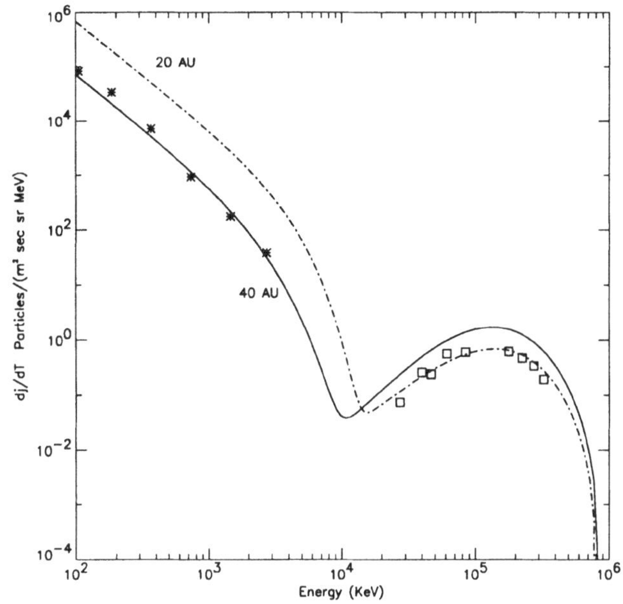

Figure 5. Comparison of the computed energy spectra of anomalous hydrogen, near the equator and at 20 and 40 AU, for the parameters discussed in the text. The overall normalization is chosen to fit the low-energy points. This then fixes the higher energy points, which are in good agreement with the data.

perhaps $\sim 1 - 10$ AU, can provide the initial boost in energy for oxygen to distances beyond 3 AU. Therefore, the lack of observed anomalous carbon is simply due to the fact that the galactic component exceeds that due to accelerated pickup carbon in the heliosphere (accelerated first by interplanetary shocks and ultimately at the termination shock).

The second constraint is imposed by the recent observation of multiply-charged ACR oxygen (Mewaldt *et al.*, 1996 and Jokipii, 1996). The ACR which are accelerated to some 5–10 MeV in the inner heliosphere may be further ionized before they reach the termination shock to be accelerated to high energies. Preliminary calculations suggest that this process is consistent with the picture of acceleration and transport of the higher-energy singly- and multiply-charged ACR given by Mewaldt *et al.* (ibid) and Jokipii (ibid).

7. Discussion and Summary

Anomalous cosmic rays can be understood as the consequence of acceleration at the termination shock of the solar wind. Such acceleration can account quite naturally for the observed composition, charge state, energy spectrum and spatial distribution. However, the direct acceleration of pickup ions at the termination shock seems to be very difficult. Acceleration of these low-energy particles at propagation shocks in the inner heliosphere is found to be much easier. We have shown that a recently-proposed scenario of initial acceleration (or "pre-acceleration") at propagating or co-rotating shocks in the inner solar system is consistent with observations of both low-energy particles and $\approx$ 100 MeV anomalous cosmic rays in the inner heliosphere.

Acknowledgements

This work was supported, in part, by NASA/JPL as part of an interdisciplinary investigation on the Ulysses mission, JPL contract 959376. It was also supported, in part by the US NSF under Grant ATM 9307046 and by NASA under Grants NAGW 50552251 and NAGW 1931.

References

Adams, J.H. Jr., *et al.*: 1991, *Astrophys. J.* **375**, L45.
Adams J.H., and Leising, M.D.: 1991, *Proc. 22nd Int. Cos. Ray Conf.*, Dublin, **3**, 304.
Cummings, A.C., and Stone, E.C.: 1987, *Proc. 20th Int. Cos. Ray Conf.*, Moscow, **3**, 421.
Cummings, A.C., and Stone,E.C.: 1988, *Astrophys J.* **334**, L77.
Cummings, A.C., *et al.*: 1987, *Geophys. Res. Lett.* **14**, 174.
Cummings, A.C., *et al.*: 1995, *Geophys. Res. Lett.* **22**, 341.
Decker, R.B.: 1988, *Space Sci. Rev.* **48**, 195.
Decker, R.B.: 1993, *J. Geophys. Res.* **98**, 33.
Drury, L. O'C.: 1983, *Rep. Prog. Phys.* **46**, 973.
Fisk, L.A., Kozlovsky, B. and Ramaty, R.: 1974 *Astrophys. J.* **190**, L35.
Forman, M.A., and Morfill, G.: 1979 in *Proc. 16th Intl. Cos. Ray Conf.* Kyoto, **5**, 328.
Geiss, J., *et al.*: 1994 , *Astron. & Astrophys.* **289**, 924.
Geiss, J., *et al.*: 1995, *J. Geophys. Res.* **100**, 23,373.
Giacalone, J., and Jokipii, J.R.: 1995, *Eos Trans.* AGU, **76** (46), Fall Meet. Suppl.
Giacalone, J., and Jokipii, J.R.: 1996a, *J. Geophys. Res.* **101**, 11,095.
Giacalone, J., and Jokipii, J.R.: 1996b, *Proc. of the Solar Wind 8 conference*, in press.
Giacalone, J., Jokipii, J.R., and Kóta, J.: 1994, *J. Geophys. Res.* **99**, 19,351.
Gloeckler, G., *et al.*: 1994, *J. Geophys. Res.* **99**, 17637.
Jokipii, J.R.: 1986, *J. Geophys. Res.* 91, 2929.
Jokipii, J.R.: 1987, *Astrophys. J.* **313**, 842.
Jokipii, J.R.: 1990, in *Physics of the Outer Heliosphere*, ed S. Grzedzielski and D.E. Page (Oxford: Pergamon), 169.
Jokipii, J.R.: 1992, *Astrophys. J.* 393, L41.
Jokipii, J.R.: 1996, *Astrophys. J.* **466**, L47.
Jokipii, J.R., and Kóta, J.: 1989, *Geophys. Res. Lett.* **16**, 1.
Jokipii, J.R., *et al.*: 1995, *Geophys. Res. Lett.* **22**, 3385.
Klecker, B.: 1995, *Space Sci. Rev.* **72**, 419.
Kucharek, H., and Scholer, M.: 1995, *J. Geophys. Res.* **100**, 1745.
Lee, M.A., Shapiro, V., and Sagdeev, R.: 1996, *J. Geophys. Res.* **101**, 4777.
Liewer, P.C., Rath, S., and Goldstein, B.E.: 1995, *J. Geophys. Res.* **100**, 19,809.
McDonald, F.B, and Lal, N.: 1986, *Geophys. Res. Lett.* **13**, 781.
McKibben, R.B, Pyle, K.R. and Simpson, J.A.: 1979, *Astrophys. J.* **227**, L147.
Mewaldt, R., *et al.*: 1996, *Astrophys. J.* **466**, L43.
Parker, E.N.: 1965, *Planet. Space Sci.* **13**, 9.
Pesses, M.E., Jokipii, J.R. and Eichler, D.: 1981 *Astrophys. J.* **246**, L85.
Steenberg, C.D. and Moraal, H.: 1995, in *Proc. 24th Intl. Cosmic Ray Conf.* Rome, **3**, 739.
Zank, G.P., Pauls, H.L., Cairns, I.H., and Webb, G.M.: 1989, *J. Geophys. Res.* **101**, 457.

THE ISOTOPIC COMPOSITION OF ANOMALOUS COSMIC RAYS FROM SAMPEX

R. A. LESKE, R. A. MEWALDT, A. C. CUMMINGS, J. R. CUMMINGS and E. C. STONE
California Institute of Technology, Pasadena, CA 91125

T. T. VON ROSENVINGE
NASA/Goddard Space Flight Center, Code 661, Greenbelt, MD 20771

Abstract. Measurements of the anomalous cosmic ray (ACR) isotopic composition have been made in three regions of the magnetosphere accessible from the polar Earth orbit of *SAMPEX*, including the interplanetary medium at high latitudes and geomagnetically trapped ACRs. At those latitudes where ACRs can penetrate the Earth's magnetic field while fully stripped galactic cosmic rays (GCRs) of similar energies are excluded, a pure ACR sample is observed to have the following composition: ^{15}N/N < 0.023, ^{18}O/^{16}O < 0.0034, and ^{22}Ne/^{20}Ne $= 0.077(+0.085, -0.023)$. We compare our values with those found by previous investigators and with those measured in other samples of solar and galactic material. In particular, a comparison of ^{22}Ne/^{20}Ne measurements from various sources implies that GCRs are not simply an accelerated sample of the local interstellar medium.

Key words: abundances, isotopes, anomalous cosmic rays, *SAMPEX*, neon, interstellar medium, heliosphere, trapped heavy ions

1. Introduction

Anomalous cosmic rays (ACRs) originate from neutral interstellar atoms that have been swept into the heliosphere, ionized by solar UV or charge exchange with the solar wind, convected into the outer heliosphere, and then accelerated to energies of $\sim$ 1 to $>$ 50 MeV/nuc (Fisk et al., 1974). They are mainly singly–charged, and include H, He, C, N, O, Ne, and Ar (see review by Klecker 1995). When impinging on the upper atmosphere, ACRs may become stripped of additional electrons and trapped in the Earth's magnetosphere by the mechanism of Blake and Friesen (1977), forming a radiation belt composed of interstellar material (Cummings et al., 1993). Since its launch in 1992, the Mass Spectrometer Telescope (MAST; Cook et al., 1993) on the *Solar, Anomalous, and Magnetospheric Particle Explorer* (*SAMPEX*) has been measuring the composition and energy spectra of ACRs in interplanetary space and in the magnetosphere, providing a new source of information on interstellar matter.

This paper discusses ACR isotope measurements, which are important for studying the evolution of the local interstellar medium (ISM) since the formation of the solar system and are relevant to galactic cosmic ray (GCR) isotope measurements (Mewaldt et al., 1984). We present measurements of N, O, and Ne isotopes (updated and expanded from Leske et al., 1995a and Mewaldt et al., 1996a) from three ACR samples in the near–Earth environment, and compare these with each other and with the composition of other solar system and galactic material.

Space Science Reviews **78**: 149–154, 1996.

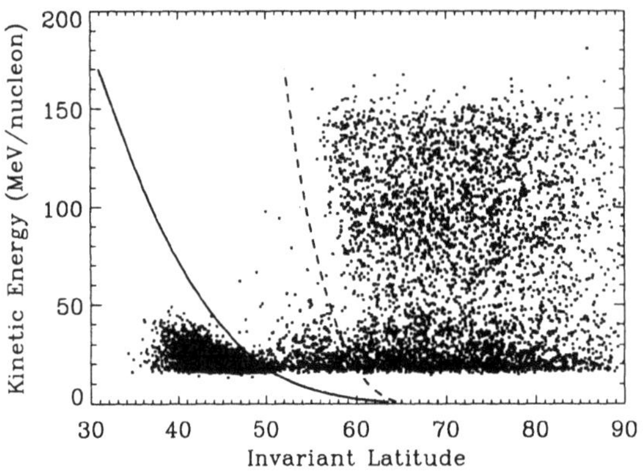

Figure 1. Kinetic energy vs invariant latitude for quiet time O events observed by MAST between July 1992 and November 1995, showing cuts used to select a pure ACR sample at mid–latitudes (see text). Trapped ACRs make up the dense concentration at low latitudes, while GCRs are restricted to the higher latitudes. (The drop in density between $\sim$ 45 and 70 MeV/nuc is due in part to the loss of an intermediate–range detector in July 1994).

2. Anomalous Cosmic Ray Isotopic Composition

At least three distinct populations of ACR nuclei are accessible in the polar Earth orbit of *SAMPEX*, each occupying a different region of the orbit, as illustrated in the plot of kinetic energy vs invariant latitude for quiet time O shown in Figure 1. At high latitudes, the energetic particle population of interplanetary space at 1 AU is sampled; GCRs extend to $\sim$ 150 MeV/nuc in the plot, with a pronounced enhancement visible at low energies due to ACRs. In the mid–latitude interval, GCRs with $\leq$ 150 MeV/nuc fall below the local geomagnetic cutoff rigidity and are excluded, while singly–charged (and even doubly– or higher–charged; Mewaldt et al., 1996a) ACRs have a higher rigidity than fully stripped GCRs at the same energy per nucleon and can penetrate to these latitudes. Located at still lower latitudes is a trapped ACR belt (Cummings et al., 1993) with the highest particle fluxes. The solid curve in Figure 1, which bounds the trapped population, corresponds to the product $\epsilon Q = 0.9$ (Selesnick et al., 1995), where the adiabaticity parameter ϵ is the ratio of the particle gyroradius to the local scale length of the magnetic field and Q is the ionic charge. The dashed curve is an empirically derived cutoff for fully stripped particles (Leske et al., 1995b), conservatively adjusted down 10% in rigidity to reduce potential GCR contamination from cutoff suppression during geomagnetically active periods. Events between the two curves represent a “pure” sample of ACR O; similar cuts were made for N and Ne to select geomagnetically filtered ACRs. (See Mewaldt et al., 1996b for the elemental composition of ACRs measured in this way.)

Figure 2 shows the mass distributions of N, O, and Ne from all three regions, taken from MAST data during solar quiet times between July 1992 and November 1995. The data here have been further restricted to times when the instrument trigger rate was < 15000 s^{-1}, which helps to minimize resolution–degrading effects of

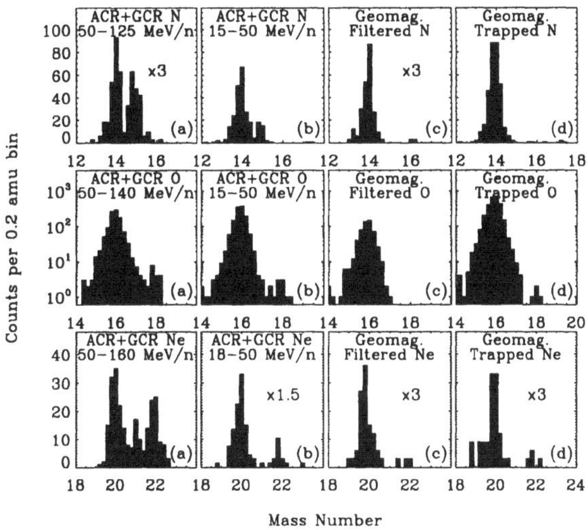

Figure 2. N, O (note logarithmic scales), and Ne isotope distributions from the MAST instrument on SAMPEX, including: a) > 50 MeV/nuc GCRs from latitudes $\Lambda > 65°$; b) < 50 MeV/nuc cosmic rays (mainly ACRs) from $\Lambda > 65°$; c) geomagnetically filtered, "pure" ACRs from $\sim 50° < \Lambda < 60°$ (see Figure 1); and d) ACRs trapped in the magnetosphere ($\Lambda \sim 45°$).

trapped proton and alpha particle pileup in the position sensing detectors (as well as periods of sporadically noisy detectors).

Over the geomagnetic poles (latitudes $\Lambda > 60°$), the isotopic ratios of these elements vary with energy. At energies > 50 MeV/nuc, where GCRs dominate, most of the observed ^{15}N and ^{18}O is produced by cosmic ray spallation during transport through the galaxy and is not directly indicative of either the GCR source or the local ISM abundances (see e.g., Gibner et al., 1992). However, most of the observed ^{22}Ne originates in the cosmic ray source. Below 50 MeV/nuc, where ACRs dominate, the ^{15}N/N, ^{18}O/^{16}O, and ^{22}Ne/^{20}Ne ratios suddenly drop by factors of more than ~ 2 to 5 from those observed at higher energies, as illustrated in more detail in the right–hand panels of Figure 3 (following Mewaldt et al., 1984). Curves shown in the figure are preliminary estimates of the expected energy dependence of the isotopic ratios. They represent a weighted average of the observed GCR isotopic ratio and the assumed ACR ratio indicated on each curve (taken to be either solar or GCR source values), using energy–dependent weighting factors obtained from power–law fits to the ACR and GCR spectra. Our measurements provide the first clear evidence of the expected energy dependence of the ^{18}O/^{16}O ratio, and a much improved measure of the ^{15}N/N energy dependence.

A comparison of the isotopic ratios for the geomagnetically filtered ACRs in the mid–latitudes with those of the interplanetary fluxes at high latitudes reveals that GCRs make a significant contribution to interplanetary ^{15}N, ^{18}O, and ^{22}Ne even at the lowest energy interval in Figures 2 and 3. It is possible to subtract the GCR contributions to obtain a corrected ACR ratio, but this analysis is not yet complete. Instruments to be flown on the *Advanced Composition Explorer* (*ACE*) in 1997 will extend isotope measurements to lower energies (e.g., 5 to 15 MeV/nuc), where the ACR flux is greater and GCR contamination is minimized, with a collecting

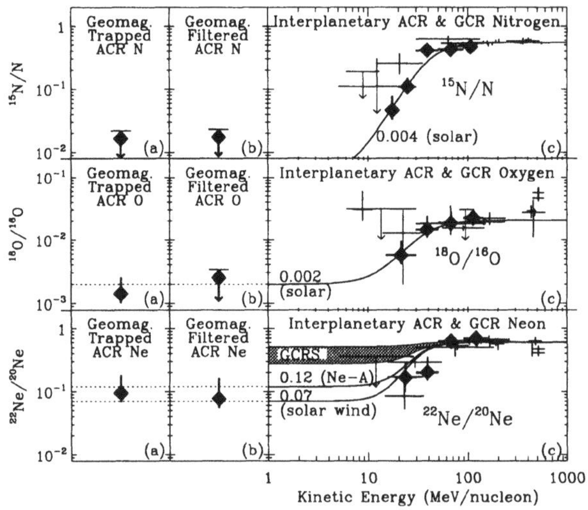

Figure 3. Observed isotopic composition of N, O, and Ne from MAST (*diamonds*), for a) trapped ACRs, b) geomagnetically filtered ACRs, and c) interplanetary particles (plotted vs energy) compared to previous measurements (*pluses*) and expected values (curves and shaded region; see text). Previous data are from the compilation by Mewaldt et al. (1984), and from Krombel and Wiedenbeck (1988), Garcia-Munoz et al. (1993), Connell and Simpson (1993a,b), DuVernois et al. (1993), Gibner et al. (1992), Lukasiak et al. (1994), Webber et al. (1996), and Cummings et al. (1991).

power > 30 times that on *SAMPEX*. However, *ACE* will not be able to make use of geomagnetic filtering as the satellite will be stationed outside the magnetosphere. Only upper limits are possible on the filtered ^{15}N and ^{18}O abundances at present due to limited statistics. The geomagnetically filtered ^{22}Ne/^{20}Ne ratio is $\simeq 0.1$, with a sizable statistical uncertainty.

To obtain ACR isotopic abundances from the geomagnetically trapped population, corrections must be applied for mass–dependent processes such as trapping efficiency and lifetime. These corrections have not yet been applied, but it appears that their effects are of opposing sign and small compared to the statistical uncertainties of our present measurements. Note that the observed ^{22}Ne/^{20}Ne ratio is again ~ 0.1, and the ^{18}O/^{16}O ratio appears to be consistent with the solar value of 0.002. The relatively intense flux of ACRs in this region offers the potential for excellent ACR isotope measurements, once the details of the trapping process are better understood.

3. Discussion

Although limited in statistical accuracy, the geomagnetically filtered ACR sample is far less subject to GCR contamination than the high latitude, low energy interplanetary sample, and less prone to selection processes than the trapped sample. In this region, we find preliminary values of the arriving ACR isotopic abundances (at the 84% confidence level) to be: ^{15}N/N < 0.023, ^{18}O/^{16}O < 0.0034, and ^{22}Ne/^{20}Ne $= 0.077^{+0.085}_{-0.023}$. After applying corrections for ACR acceleration and transport effects (Cummings et al., 1984, 1991) appropriate for measurements at 1 AU at this point in the solar cycle, these values become: ^{15}N/N < 0.023, ^{18}O/^{16}O < 0.0035,

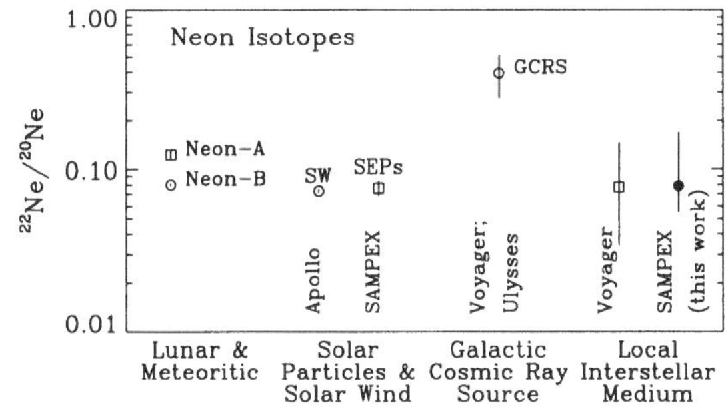

Figure 4. Comparison of ^{22}Ne/^{20}Ne ratios (references in text).

and ^{22}Ne/^{20}Ne $= 0.078^{+0.090}_{-0.024}$. While the ^{15}N and ^{18}O upper limits still exceed the solar system values for these species by factors of $\sim$ 6 and 1.8 respectively, both are significant improvements over previous attempts to measure these ratios in ACRs (Figure 3). The ^{18}O/^{16}O upper limit is becoming comparable to values measured in other parts of the galaxy (Wilson and Rood, 1994).

To place the Ne measurements in context, Figure 4 compares ^{22}Ne/^{20}Ne measurements from several sources. There is some disagreement regarding the isotopic composition of solar Ne. In the Cameron (1982) table of solar system abundances, the meteoritic component "Neon–A" (with ^{22}Ne/^{20}Ne = 0.122) was used. Anders and Grevesse (1989), on the other hand, chose the solar wind value of ^{22}Ne/^{20}Ne = 0.076 (Geiss et al., 1972), which is close to the lunar/meteoritic component "Neon–B", thought to be implanted solar wind. Solar energetic particle (SEP) measurements from *SAMPEX* (Selesnick et al., 1993) find a ^{22}Ne/^{20}Ne ratio very close to the solar wind value. Determination of the GCR source (GCRS) ratio for ^{22}Ne/^{20}Ne depends on the details of transport through the galaxy, particularly the interaction cross sections for producing secondary ^{22}Ne from heavier species in spallation reactions with the gas of the interstellar medium. Various recent calculations of the GCRS ^{22}Ne/^{20}Ne ratio range from 0.322 (Connell and Simpson, 1993a) to 0.448 (Lukasiak et al., 1994). This range of values, along with the reported 1σ uncertainties, is represented by the shaded region in Figure 3 and the error range of the GCR data point in Figure 4. It is clear that the GCR source ratio greatly exceeds any of the solar system components. The *SAMPEX* and *Voyager* (Cummings et al., 1991) values for the local ISM are both $\sim$ 0.1 and are not sufficiently accurate to differentiate Neon–A and Neon–B, but are clearly much less than the GCR source ratio. Although the values obtained from interplanetary ACRs (Figures 2b and 3c) and from trapped ACRs (Figures 2d and 3a) require further analysis to estimate possible systematic corrections and uncertainties, they also support a ratio of $\sim$ 0.1 rather than a value as high as $\sim$ 0.4.

These ^{22}Ne/^{20}Ne results indicate that GCRs are not simply a sample of local ISM that has been accelerated to high energies (e.g., Olive and Schramm, 1982), but rather suggest that GCRs include contributions from sources especially rich in ^{22}Ne, such as Wolf–Rayet stars (Prantzos et al., 1986). This work illustrates the potential of ACRs to provide unique information on the composition of the local ISM and to better understand the nature of GCRs. In the coming years we can expect improved statistical accuracy from *SAMPEX* as it continues to gather data under solar minimum conditions with increased ACR fluxes, and improved capability from *ACE*.

We appreciate contributions to this work by R. S. Selesnick. This work was supported by NASA under contract NAS5-30704 and grant NAGW-1919.

References

Anders, E. and Grevesse, N.: 1989, *Geochim. Cosmochim. Acta*, **53**, 197
Blake, J. B. and Friesen, L. M.: 1977, *Proc. 15th Internat. Cosmic Ray Conf. (Plovdiv)* **2**, 341
Cameron, A. G. W.: 1982, in *Essays in Nuclear Astrophysics*, C. A. Barnes, D. D. Clayton, and D. N. Schramm, Cambridge Univ. Press, p. 23
Connell, J. J. and Simpson, J. A.: 1993a, *Proc. 23rd Internat. Cosmic Ray Conf. (Calgary)*, **1**, 559
Connell, J. J. and Simpson, J. A.: 1993b, *Proc. 23rd Internat. Cosmic Ray Conf. (Calgary)*, **1**, 547
Cook, W. R. et al.: 1993, *IEEE Trans. Geosci. Remote Sensing*, **31**, 557
Cummings, A. C., Stone, E. C. and Webber, W. R.: 1984, *ApJL*, **287**, L99
Cummings, A. C., Stone, E. C. and Webber, W. R.: 1991, *Proc. 22nd Internat. Cosmic Ray Conf. (Dublin)*, **3**, 362
Cummings, J. R. et al.: 1993, *Geophys. Res. Letters*, **20**, 2003
DuVernois, M. A., Garcia–Munoz, M., Pyle, K. R. and Simpson, J. A.: 1993, *Proc. 23rd Internat. Cosmic Ray Conf. (Calgary)*, **1**, 563
Fisk, L. A., Kozlovsky, B. and Ramaty, R.: 1974, *ApJ*, **190**, L35
Garcia-Munoz, M., Pyle, K. R., Simpson, J. A. and Thayer, M.: 1993, *Proc. 23rd Internat. Cosmic Ray Conf. (Calgary)*, **1**, 543
Geiss, J., Buehler, F., Cerruti, H., Eberhardt, P. and Filleux, Ch.: 1972, *Apollo 16 Preliminary Science Report*, NASA SP–315, p. 14–1
Gibner, P. S., Mewaldt, R. A., Schindler, S. M., Stone, E. C. and Webber, W. R.: 1992, *ApJL*, **391**, L89
Klecker, B.: 1995, *Space Science Reviews*, **72**, 419
Krombel, K. E. and Wiedenbeck, M. E.: 1988, *ApJ*, **328**, 940
Leske, R. A. et al.: 1995a, *Proc. 24th Internat. Cosmic Ray Conf. (Rome)*, **2**, 606
Leske, R. A., Cummings, J. R., Mewaldt, R. A., Stone, E. C. and von Rosenvinge, T. T.: 1995b, *ApJL*, **452**, L149
Lukasiak, A., Ferrando, P., McDonald, F. B. and Webber, W. R.: 1994, *ApJ*, **426**, 366
Mewaldt, R. A., Spalding, J. D. and Stone, E. C.: 1984, *ApJ*, **283**, 450
Mewaldt, R. A., Leske, R. A. and Cummings, J. R.: 1996a, *Proc. Maryland Conf. on Cosmic Abundances*, in press
Mewaldt, R. A. et al.: 1996b, *Geophys. Res. Letters*, **23**, 617
Olive, K. A. and Schramm, D. N.: 1982, *ApJ*, **257**, 276
Prantzos, N., Doom, C., Arnould, M. and deLoore, C.: 1986, *ApJ*, **304**, 695
Selesnick, R. S. et al.: 1993, *ApJL*, **418**, L45
Selesnick, R. S. et al.: 1995, *J. Geophys. Res.*, **100**, 9503
Wilson, T. L. and Rood, R. T.: 1994, *Annual Rev. Astron. Astrophys.*, **32**, 191
Webber, W. R., Lukasiak, A., McDonald, F. B. and Ferrando, P.: 1996, *ApJ*, **457**, 435

Address for correspondence: R. A. Leske, Mail Code 220–47, California Institute of Technology, Pasadena, CA 91125

THE LOCAL INTERSTELLAR MEDIUM

GHRS OBSERVATIONS OF THE LISM

JEFFREY L. LINSKY
JILA, University of Colorado, Boulder, CO 80309-0440 USA

Abstract. The GHRS has obtained high-resolution spectra of interstellar gas toward 19 nearby stars. These excellent data show that the Sun is located inside the Local Interstellar Cloud (LIC) with other warm clouds nearby. I will summarize the physical properties of these clouds and the three-dimensional structure of this warm interstellar gas. There is now clear evidence that the Sun and other late-type stars are surrounded by hydrogen walls in the upwind direction. The D/H ratio probably has a constant value in the LIC, $(1.6 \pm 0.2) \times 10^{-5}$, consistent with the measured values for all LIC lines of sight.

Key words: LISM, interstellar medium, deuterium

1. Comparison of LISM Science Objectives with the Instrument Capabilities of the GHRS

To study the kinematics, physical properties, and chemical composition of the warm gas in the local interstellar medium (LISM) one must use an ultraviolet spectrograph with high spectral resolution and signal/noise (S/N). High spectral resolution is needed to identify individual clouds, a term that is often used to characterize gas moving with the same bulk velocity. Clouds with line of sight velocity separations as small as 2.6 km s^{-1} have already been identified for the Procyon line of sight and smaller velocity separations may exist. High spectral resolution is also needed to measure line widths, which depend on temperature (T) and turbulent velocity (ξ) according to the relation $0.60 \times FWHM \approx b = \sqrt{0.0166T/A + \xi^2}$. This relation is valid for optically thin lines, where A is the atomic weight, ξ is the most probable speed of the turbulent mass motions (km s^{-1}), and b is the line width parameter, also in km s^{-1}. Measurement of the widths of the Mg II and Fe II lines (see Table I), which are needed to infer the value of ξ in a cloud, requires the resolution of the Goddard High Resolution Spectrograph (GHRS) echelles. High S/N is essential to measure the shapes of deep absorption features and especially the shape of the highly saturated H I Lyman-α interstellar absorption line. UV spectroscopy is needed because the column densities for short path lengths are very small, and resonance lines, typically found in the UV, are the most opaque lines for each ion.

Table I summarizes the capabilities of past, present, and future high-resolution UV spectrographs in space. Neither the Copernicus nor IUE spectrographs could resolve line profiles, and their low sensitivities hindered studies of weak lines, such as the important Lyman-α resonance line of D I, toward nearby stars. The echelle spectrographs on the GHRS (EA for short wavelengths and EB for longer wavelengths) have provided most of what we now know about the LISM, and

Space Science Reviews **78**: 157–164, 1996.

Table I
Comparison of High Resolution Spectroscopic Capabilities.

Satellite/ Instrument	Years Operational	Spectral Range (Å)	Spectral Resolution (km s^{-1})
Copernicus	1975 – 1979	900 – 3000	15
IUE/echelles	1978 –	1170 – 3400	25 – 30
HST/GHRS/EA	1990 – 1996*	1150 – 1730	3.57
HST/GHRS/G160M	1990 – 1996	1150 – 2300	20
HST/GHRS/G140M	1990 – 1996*	1100 – 1900	15
HST/GHRS/EB	1990 – 1996	1700 – 3200	3.54
HST/STIS/1.4	1997 –	1150 – 1700	2.9
HST/STIS/2.4	1997 –	1650 – 3100	2.9
HST/STIS/1.3	1997 –	1150 – 1700	12.5
HST/STIS/2.3	1997 –	1650 – 3100	12.8
FUSE	1998 –	900 – 1180	10 – 12

* Not operational in 1992 – 1994.
Typical value of $b_{\mathrm{H\,I}} = 11$ km s^{-1}.
Typical value of $b_{\mathrm{D\,I}} = 8$ km s^{-1}.
Typical value of $b_{\mathrm{Mg\,II}} = 2.6$ km s^{-1}.
Typical value of $b_{\mathrm{Fe\,II}} = 2.4$ km s^{-1}.

we anticipate that the echelles on the new Space Telescope Imaging Spectrograph (STIS) will extend this work by virtue of its large simultaneous spectral coverage to study many lines at one time. The Far Ultraviolet Spectrograph Explorer (FUSE) will permit studies of resonance lines of other ions (e.g., O VI, C III) at shorter wavelengths.

2. Kinematics of the LISM

Lallement and Bertin (1992) proposed that the Sun lies inside a cloud, which they called the Local Interstellar Cloud (LIC), because the line of sight velocities toward 6 nearby stars observed in ground-based Ca II spectra and the velocity of interstellar He I into the solar system are consistent with a single flow vector. GHRS spectra of the Mg II and Fe II resonance lines (2796, 2803, and 2600 Å) formed in the lines of sight toward other nearby stars (Lallement *et al.*, 1995) confirmed this picture with the flow vector magnitude 26 ± 1 km s^{-1} from Galactic coordinates $l = 186° \pm 3°$ and $b = -16° \pm 3°$ in the heliocentric rest frame. In the local standard of rest (defined by the motion of nearby stars), the LIC flow is from Galactic coordinates $l = 331.9°$ and $b = +4.6°$. The direction of this flow suggests that it originates from the expansion of a large superbubble created by supernovae and stellar winds from the Scorpius-Centaurus OB Association (Crutcher, 1982; Frisch, 1995).

GHRS spectra are confirming that the kinematical structure of the LISM is indeed very complex. Most lines of sight show at least one velocity component in addition to the LIC, even for stars as close as Sirius (2.7 pc) and Procyon (3.5 pc), indicating that additional clouds lie outside of the LIC at short distances. Table II lists the number of clouds now identified on the lines of sight toward the stars observed with the GHRS, where L refers to the LIC. In the Galactic Center direction the Sun lies close to the edge of the G cloud as the α Cen stars show interstellar absorption only at the velocity of this cloud. The Sun likely lies very close to the edge of the LIC toward the North Galactic Pole as 31 Comae (Piskunov *et al.*, 1996) shows only one velocity component that is inconsistent with the LIC vector.

3. Physical Properties of the LISM

The temperature and nonthermal broadening of interstellar gas can be measured by comparing line widths of low mass elements (H and D) and high mass elements (e.g., Mg and Fe). For the line of sight to Capella, Linsky *et al.* (1995) derived $T = 7000 \pm 500 \pm 400$ K and $\xi = 1.6 \pm 0.4 \pm 0.2$ km s^{-1}, where the second uncertainty refers to the likely systematic errors in the uncertain intrinsic stellar emission lines. They also found $T = 6900 \pm 80 \pm 300$ K and $\xi = 1.21 \pm 0.27$ for the Procyon line of sight. Temperatures and turbulent velocities measured for other lines of sight through the LIC are consistent with these values. For example, Lallement *et al.* (1994) found that $T = 7600 \pm 3000$ K and $\xi = 1.4^{+0.6}_{-1.4}$ km s^{-1} for the LIC component toward Sirius, and Gry *et al.* (1995) found that $T = 7200 \pm 2000$ K and $\xi = 2.0 \pm 0.3$ km s^{-1} for the LIC component toward ϵ CMa. Recent analyses (Piskunov *et al.*, 1996) of the LIC components toward HR 1099, 31 Com, β Cet, and β Cas yield similar results, and *in situ* measurements of the LISM H I and He I atoms flowing through the heliosphere (cf. Lallement *et al.*, 1994) yield consistent values for the temperature. We therefore conclude that $T \approx 7000$ K and $\xi \approx 1.2$ km s^{-1} in the LIC.

Other clouds have different parameters. The G cloud, for example, is cooler with $T = 5400 \pm 500$ K and $\xi = 1.20 \pm 0.25$ km s^{-1} along the α Cen line of sight. Component 2 toward ϵ CMa is also cooler with $T = 3600 \pm 1500$ K and $\xi = 1.85 \pm 0.3$ km s^{-1} (Gry *et al.*, 1995) Hotter gas is inferred for several clouds toward ϵ CMa and for some of the gas toward Sirius (Bertin *et al.*, 1995).

Electron densities can be derived from the Mg II/Mg I column density ratio, assuming ionization equilibrium, but the derived densities are very temperature sensitive. For example, the ratio $R = N_{\rm Mg\ II}/N_{\rm Mg\ I} = 220^{+70}_{-40}$ for the Sirius line of sight (Lallement *et al.*, 1994) implies that $n_e = 0.3 - 0.7$ cm^{-3} if $T = 7000$ K and a much wider range for the plausible uncertainty in T. For the LIC component of the ϵ CMa line of sight, the measured ratio results in $n_e = 0.09^{+0.23}_{-0.07}$ cm^{-3} (Gry *et al.*, 1995) and a range in ionization fraction from nearly neutral to mostly ionized. For the LIC component of the Sirius line of sight, Frisch (1995) used both the Mg and C ionization fractions, which have opposite temperature dependencies, to derive $n_e = 0.22 - 0.44$ cm^{-3}. Thus for $n_{\mathrm{H\,I}} = 0.1$ cm^{-3}, hydrogen in the LIC is mostly ionized with the photoionizing flux primarily from the star ϵ CMa.

4. The Three-Dimensional Structure of the LISM

The first attempts to map the distribution of interstellar gas in the solar neighborhood were based on measurements of $N_{\mathrm{H\,I}}$ extracted from Copernicus and IUE spectra primarily toward OB stars located near the Galactic plane. These maps (Frisch & York, 1983; Paresce, 1984) show a very asymmetric distribution of neutral hydrogen gas with the third Galactic quadrant (centered at $l = 225°$) showing the smallest column ($N_{\mathrm{H\,I}} \approx 5 \times 10^{18}$ cm^{-2}). This small column extends to at least 200 pc toward the star ϵ CMa (Gry *et al.*, 1995). GHRS spectra of some 19 mostly late-type stars now permit us to study the region within 20 pc of the Sun in more detail. These spectra sample lines of sight at Galactic latitudes from the North Galactic Pole (31 Com) to near the South Galactic Pole (β Cet), and they permit us to determine the amount of gas in each cloud along each line of sight.

Linsky, Piskunov, & Wood (1996) have created a three-dimensional morphological model for the LISM using the 12 lines of sight with Lα spectra listed in Table II and 5 additional lines of sight toward hot white dwarfs for which the analysis of EUVE spectra (Dupuis *et al.*, 1995) provide accurate estimates of the total value of $N_{\mathrm{H\,I}}$ (but no information on the specific clouds contributing to the H I column). Linsky *et al.* (1996) found that the LIC is flattened in the Galactic plane with minimum and maximum dimensions of about 1.6 and 4.8 pc, respectively. On the other hand, the total amount of warm gas in all clouds in the LISM is elongated roughly perpendicular to the plane with minimum and maximum dimensions of about 5.1 and 8.2 pc, respectively. These dimensions are based on the assumptions that the observed warm gas is located close to the Sun, can be approximated by a triaxial ellipsoid, and has a mean density of $n_{\mathrm{H\,I}} = 0.10$ cm^{-3}. The dimensions of the ellipsoids will scale inversely with the mean density. The direction of minimum hydrogen absorption through the LISM is near Galactic coordinates $l = 262°$ and $b = +22°$. They also find that $N_{\mathrm{H\,I}}$ is similar for stars located in the sky within 12°, which indicates that 12° is a typical angular scale of the clouds close to the Sun.

5. Discovery of Hydrogen Walls Around the Sun and Stars

Whereas the resonance lines of Mg II and Fe II are useful for studying clouds with $N_{\mathrm{H\,I}} \geq 10^{17}$ cm^{-2}, only the Lyman-α line of H I has sufficient opacity to study warm gas with columns as small as $N_{\mathrm{H\,I}} \approx 10^{13}$ cm^{-2}. For this reason and because of the uncertain shape of the intrinsic stellar emission line against which the interstellar absorption is measured, one must obtain Lyman-α line profiles with

Table II
Summary of GHRS Observations of the LISM.

Star (References)	d (pc)	l (°)	b (°)	Grating† (line)	Clouds in LOS	D/H (10^{-5})
α Cen A* [a,b]	1.3	316	−01	E(Lα,MgI+II,FeII)	G	
α Cen B [b]	1.3	316	−01	E(Lα,MgII)	G	
Sirius [a,c,d]	2.7	227	−09	G(Lα), E(MgI+II,FeII)	L+1	
ϵ Ind [e]	3.4	336	−48	E(Lα)	G?	1.6 ± 0.4
Procyon [f]	3.5	214	+13	G(Lα), E(MgII,FeII)	L+1	
α Aql [a]	5.0	48	−09	E(FeII,MgII)	L+2	
α PsA [a]	6.7	21	−65	E(FeII)		
Vega [a]	7.5	68	+19	E(FeII)	L+2	
β Leo [a]	12.2	251	+71	E(FeII,MgII)	L+2	
Capella* [f,g]	12.5	163	+05	E(Lα,MgII,FeII), G(Lα)	L	$1.60^{+0.14}_{-0.19}$
β Cas [h]	14	118	−03	G140M(Lα)	L	1.6 ± 0.4
β Cet [h]	16	111	−81	G(Lα), E(MgII)	L+1	2.2 ± 1.1
β Pic [a]	16.5	258	−31	E(FeII)	L	
λ And [e]	24	110	−15	E(Lα)	L+?	1.7 ± 0.5
δ Cas [a]	27	127	−02	E(MgII)	L	
HR1099* [h]	33	185	−41	E(Lα,MgII,FeII), G(Lα)	L+2	1.46 ± 0.09
G191-B2B [i]	48	156	+07	G(Lα), E(MgII,FeII)	L+2	$1.4^{+0.1}_{-0.3}$
31 Com [h]	80	115	+89	G(Lα), E(MgII)	1	1.5 ± 0.4
ϵ CMa [j]	187	240	−11	G(Lα), E(MgI+II,FeII)	L+5	

* These stars were observed twice. Capella and HR 1099 were observed near opposite quadratures.
† Gratings: G = G160M, E = Echelle-A or Echelle-B.
References: [a] Lallement *et al.* 1995, [b] Linsky & Wood 1996, [c] Lallement *et al.* 1994, [d] Bertin *et al.* 1995, [e] Wood *et al.* 1996, [f] Linsky *et al.* 1995, [g] Linsky *et al.* 1993, [h] Piskunov *et al.* 1996, [i] Lemoine *et al.* 1995, [j] Gry *et al.* 1995.

very high S/N and then analyze the data very carefully. Linsky and Wood (1996) therefore obtained high S/N GHRS spectra of two of the very nearest stars, α Cen A and α Cen B. They anticipated that the analysis of the Lyman-α lines toward these stars would be straightforward, as the line of sight is short (1.3 pc) and presumably simple. The intrinsic Lyman-α profile of α Cen A should be very similar to the Sun, as the two stars have the same spectral type, and the lines of sight to α Cen A and α Cen B (separated by only 20″) should have the same properties and thus provide redundant information.

Figure 1 shows the observed Lyman-α profile toward α Cen B and the best model fit in which the velocity and temperature of the interstellar H I are constrained to be the same as that obtained from the D I, Mg II, and Fe II lines. Clearly there is missing opacity near zero flux on the red side of the interstellar absorption, indicating additional absorption by gas that (i) is redshifted compared to the interstellar flow velocity of 18.0 km s^{-1}, (ii) is hotter that the interstellar gas (required to

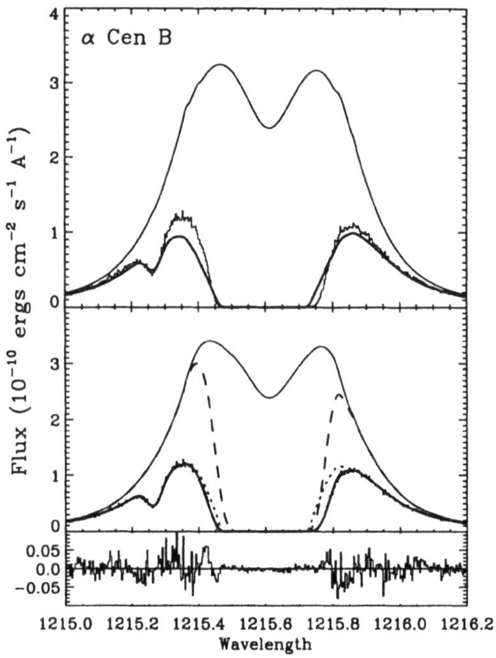

Figure 1. Upper panel: comparison of the observed Echelle-A spectrum (noisy line) of α Cen B with the assumed intrinsic stellar spectrum (smooth thin line) and the best one-component model fit (thick solid line). Middle panel: best two-component model with the absorption due to the ISM component only (dotted line), absorption due to the H wall component only (dashed line), and the total absorption (thick solid line). Lower panel: residuals between the observed profile and the two-component fit (from Linsky & Wood, 1996).

fit the gentler slope of the red side of the absorption compared to the blue side), and (iii) has a relatively low column density (the additional absorption has no Voigt wings). Because there is missing opacity at zero flux, no sensible change in the assumed intrinsic stellar profile can explain the discrepancy. Figure 1 also shows a least-squares fit to the observed profile by a two-component model. The second component gas is hot ($T = 29,000 \pm 5,000$ K), has a low column density ($\log N_{\mathrm{H\,I}} = 14.74 \pm 0.24$), and is redshifted by 2–4 km s^{-1} relative to the main component of the interstellar gas.

When Linsky and Wood (1996) derived these results they were unaware of the location of the second absorption component H I toward α Cen. The location became clear later at the 1995 July 12-13 meeting of the IUGG in Boulder, Colorado, where Baranov, Zank, and Williams presented their calculations of the interaction between the solar wind and the incoming interstellar flow. Their models (Baranov & Malama, 1993, 1995; Pauls *et al.*, 1995) which include charge exchange between the outflowing solar wind protons and the inflowing H I atoms, show that near the heliopause there is a region of decelerated, hot hydrogen with higher density than in the LIC. This H I pileup region located about 200 AU in the upstream direction (depending on the proton density in the LIC) has been called the "hydrogen wall." Because the column density, temperature, and flow velocity of the H I agree well with the parameters derived for the second component toward α Cen, Linsky and Wood (1996) concluded that the second component originates in the wall. Before this time the hydrogen wall was just an interesting theoretical concept with no observational confirmation, although Lyman-α backscattering observations (e.g., Quémerais *et al.*, 1995) indicated that $n_{\mathrm{H\,I}}$ increases outward toward the heliopause.

Are there hydrogen walls around other stars? A stellar hydrogen wall would be seen as a second absorption component shifted to shorter wavelengths compared to the interstellar gas flowing toward the star, because the stellar wall would have to be viewed from the upwind direction. Wood, Alexander, & Linsky (1996) found that a one-component model for the interstellar absorption toward ϵ Ind could not explain the absorption on the blue side of the interstellar Lyman-α line. They concluded that a second component blueshifted by 18 ± 6 km s^{-1} with respect to the interstellar flow was needed with $\log N_{\mathrm{H\,I}} = 14.2 \pm 0.2$ and $T = 100,000 \pm 20,000$ K. The high temperature and large blueshift are consistent with the higher inflow velocity of 64.0 km s^{-1} toward this rapidly moving star. They also found evidence for a hydrogen wall around λ And with a smaller blueshift and temperature. Because the presence of a hydrogen wall requires a solar-like wind, the discovery of hydrogen walls around these two stars provides the first clear evidence that stars similar to the Sun actually have winds.

6. The D/H Ratio in the LISM

An important objective of the GHRS studies of nearby stars is to derive the D/H ratios along a number of lines of sight and to determine whether this ratio is constant in the LISM. The primordial value of D/H measures the baryon density of the universe and the parameter Ω_{B}, which is the ratio of the baryon density to the density needed to halt the expansion of the universe (the so called "closure density"). The value of D/H in the LISM is smaller than the primordial ratio because nuclear processes in stellar interiors destroy deuterium and stellar winds and supernova explosions eject deuterium-poor gas into the ISM. Comparison of hydrogen and deuterium column densities obtained from the same GHRS Lyman-α spectrum is a particularly good way to determine the D/H ratio, since in the warm absorbing gas H I and D I are the dominant stages of ionization with little association into molecules or depletion on to grains.

Table II summarizes the D/H ratios measured by various authors in the LIC component along different lines of sight (LOS). The measurements are all consistent with a single value, $(1.6 \pm 0.2) \times 10^{-5}$. Thus the LISM appears to be well mixed on a time scale that is short compared to major perturbing events such as supernova explosions.

Acknowledgements

This work is supported by NASA through grant S-56460-D. The author thanks Brian Wood for creating the figure and ISSI for its hospitality and support.

References

Baranov, V. B., & Malama, Y. G.: 1993, *JGR*, **98**, 15,157.
Baranov, V. B., & Malama, Y. G.: 1995, *JGR*, **100**, A8, 14,755.
Bertin, P., Vidal-Madjar, A., Lallement, R., Ferlet, R., & Lemoine, M.: 1995, *A&A*, **302**, 889.
Crutcher, R.M.: 1982, *ApJ*, **254**, 82.
Dupuis, J., Vennes, S., Bowyer, S., Pradhan, A. K., & Thejll, P.: 1995, *Ap.J.*, **455**, 574.
Frisch, P. C.: 1995, *Science*, **265**, 1423.
Frisch, P. C., & York, D. G.: 1983, *Ap.J.*, **271**, L59.
Gry, C., Lemonon, L., Vidal-Madjar, A., Lemoine, M., & Ferlet, R., 1995, *A&A*, **302**, 497.
Lallement, R., & Bertin, P.: 1992, *A&A*, **266**, 479.
Lallement, R., Bertin, P., Ferlet, R., Vidal-Madjar, A., & Bertaux, J. L.: 1994, *A&A*, **286**, 898.
Lallement, R., Ferlet, R., Lagrange, A. M., Lemoine, M., & Vidal-Madjar, A.: 1995, *A&A*, **304**, 461.
Lemoine, M., Vidal-Madjar, A., Ferlet, R., Bertin, P., Gry, C., and Lallement, R.: 1995, in The Light Element Abundances, ed. P. Crane (Berlin; Springer), p. 233.
Linsky, J. L. & Wood, B. E.: 1996, *Ap.J.*, **462**, to appear May 20.
Linsky, J. L., Brown, A., Gayley, K., Diplas, A., Savage, B. D., Ayres, T. R., Landsman, W., Shore, S. N., & Heap, S. R.: 1993, *Ap.J.*, **402**, 694.
Linsky, J. L., Diplas, A., Wood, B. E., Brown, A., Ayres, T. R., & Savage, B. D.: 1995, *Ap.J.*, **451**, 335.
Linsky, J. L., Piskunov, N., & Wood, B. E.: 1996, submitted to *Science*.
Paresce, F.: 1984, *A.J.*, **89**, 1022.
Pauls, H. L., Zank, G. P., & Williams, L. L.: 1995, *JGR*, **100**, 21,595.
Piskunov, N., Wood, B., Linsky, J. L., Dempsey, R. C., & Ayres, T. R.: 1996, submitted to *Ap.J.*.
Quémerais, E., Sandel, B. R., Lallement, R., & Bertaux, J.-L.: 1995, *A&A*, **299**, 249.
Wood, B. E., Alexander, W. R., & Linsky, J. L.: 1996, *Ap.J.*, in press.

IN SITU MEASUREMENTS OF INTERSTELLAR DUST WITH THE ULYSSES AND GALILEO SPACEPROBES

MICHAEL BAGUHL, EBERHARD GRÜN and MARKUS LANDGRAF
Max-Planck-Institut für Kernphysik, Heidelberg, Germany

Abstract. Interstellar dust was first identified by the dust sensor onboard Ulysses after the Jupiter flyby in February 1992. These findings were confirmed by the Galileo experiment on its outbound orbit from Earth to Jupiter. Although modeling results show that interstellar dust is also present at the Earth orbit, a direct identification of interstellar grains from geometrical arguments is only possible outside of 2.5 AU. The flux of interstellar dust with masses greater than $6 \cdot 10^{-14} g$ is about $1 \cdot 10^{-4} m^{-2} s^{-1}$ at ecliptic latitudes and at heliocentric distances greater than $1 AU$. The mean mass of the interstellar particles is $3 \cdot 10^{-13} g$. The flux arrives from a direction which is compatible with the influx direction of the interstellar neutral Helium of 252° longitude and 5.2° latitude but it may deviate from this direction by 15 – 20°.

1. Introduction

Interstellar dust in the heliosphere was first detected by Ulysses at a distance of 5 AU (Grün et al., 1993). Dust measurements with the Pioneer 10 and 11 spacecraft at distances between 3 and 18 AU (Humes, 1980) could not be explained with orbits typical for classical zodiacal dust, but the results are compatible with a randomly inclined population of dust on bound orbits. Interstellar meteors were detected using the AMOR radar in New Zealand (Taylor et al., 1996). This detection refers to grains with sizes greater than $40 \mu m$ and only very fast ($v_\infty \approx 100 kms^{-1}$) meteors were considered as of interstellar origin. These interstellar grains contribute with 1% to the impact of dust grains larger than $40 \mu m$ to the Earth atmosphere. A possible explanation of the failure to measure submicron–sized grains at the Earth orbit was given by (Jokipii et al., 1976).

The Ulysses and Galileo spacecraft were launched in October 1990 and 1989 respectively. Galileo performed gravity–assist flybys on Venus and two times on Earth before heading out for Jupiter. Ulysses was launched on a direct trajectory to Jupiter and used its gravity to propel the spacecraft to an elliptical orbit almost perpendicular to the ecliptic plane. The spacecraft carry two identical dust detectors capable of measuring the dust impact speed, mass and impact direction and possibly particle charge. The setup and calibration are described in detail in (Grün et al., 1992a) and (Grün et al., 1992b). The measurable speeds range from 2 kms^{-1} to 70 kms^{-1}, the

Space Science Reviews **78**: 165–172, 1996.

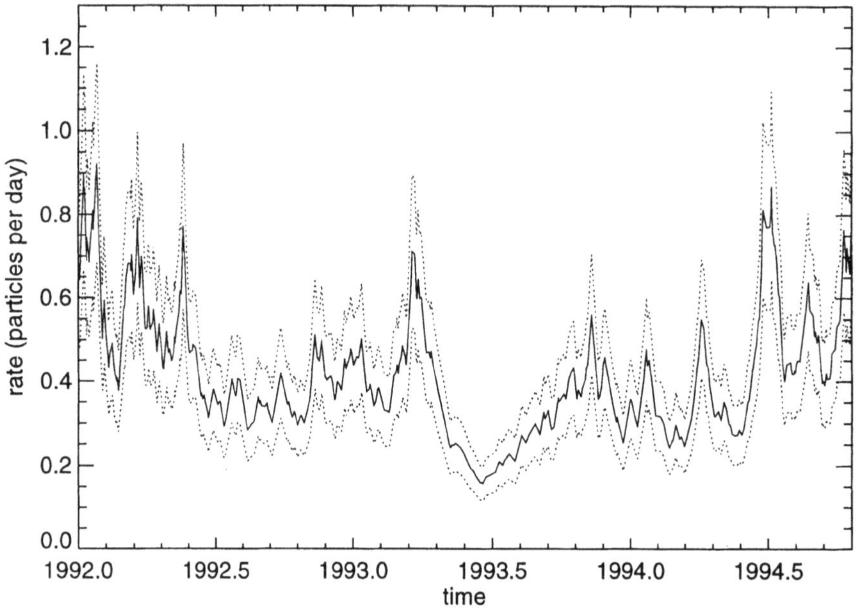

Figure 1. Ulysses flux rate. A sliding mean over 15 dust impacts was applied. The solid line shows the sliding impact rate, the broken lines denote the statistical 1σ–errors. Jovian stream particles, identified by their collimation in impact time and direction, were excluded from the displayed impact rate.

mass threshold is highly speed dependent and ranges from $10^{-15}g$ to $10^{-11}g$. The opening angle of the detector is 140°.

2. Ulysses measurements

In figure 1 we show the flux of dust particles observed by Ulysses during the first 2.8 years after Jupiter flyby. From this plots the Jupiter streams as described in (Grün et al., 1993) are excluded. The impact rate of the dust particles varies around its mean value of 0.45 day^{-1} with a small deviation of 0.16 day^{-1} over the complete range of heliocentric latitudes (nearly zero before Jupiter encounter in February 1992, then decreasing to $-80°$ while passing the Sun's south pole in October 1994, then increasing to $-55°$ in December 1994). This by itself suggests that measured dust population does not belong to the solar system. The zodiacal dust cloud in the inner solar system is on moderately inclined prograde orbits and extends only to 30° around the ecliptic plane. As (Grün et al., 1992c) show, the almost constant flux before and after the Jupiter flyby is a hint to retrograde orbits. Since almost all sources of zodiacal dust, namely asteroids and short period comets

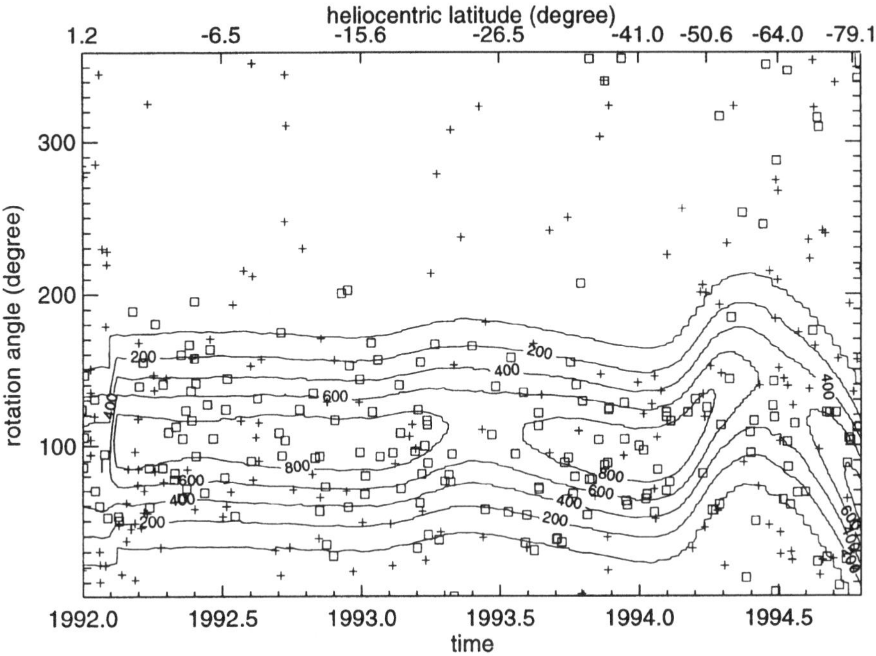

Figure 2. Impact direction of dust particles measured by Ulysses (For the definition of rotation angle see (Grün et al., 1993)). The contour lines give the differential geometrical sensitive area in cm^2 for impacts from a direction towards the upstream direction of the interstellar gas (Witte et al., 1993). The squares denote particles with signal amplitudes $> 1 \cdot 10^{-13} C$, plus signs denote impacts with smaller signal amplitudes. The contour lines included give the differential sensitive area in cm^2 for dust impacts from the flight direction of the neutral interstellar gas. The latitude of Ulysses in the heliocentric System is given at the top.

are on predominantly prograde, low inclined orbits, this is an additional reference to an interstellar origin.

Additional information on the dust particle origin can be collected from the impact directions. Figure 2 shows the sensor orientation and time for every impact, in comparison with the sensitive area of the detector for impacts from the upstream direction of interstellar gas. It can be seen that the direction of impacts of the fast dust particles (i.e. particles with high signal amplitude) is compatible with the direction of the gas. This becomes even more impressive by the fact that prograde dust on predominantly prograde orbits and low eccentricity should be sensed from rotation angles between 180° and 360°. The apparently worse fit at the highest latitudes may be explained by gravitational or electromagnetic deflection. Hamilton (pers. comm.) showed that this deflection may change the impact direction by up to 10°.

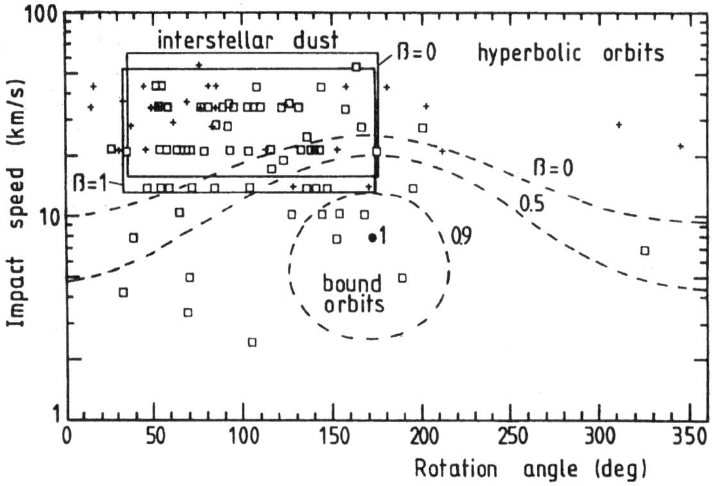

Figure 3. Measured impact speeds as a function of rotation angle for Ulysses data taken from one year after Jupiter flyby. The lines give the limits for particles on bound orbits for two different radiation pressure coefficients β. The boxes denote the opening cone around the interstellar direction of 140° and the 1σ uncertainty around the interstellar dust impact speed of $26 kms^{-1}$ and $27.2 kms^{-1}$. Interstellar dust impacts should lie within this box.

The measured impact speeds are also compatible with the assumption that the dust particles are predominantly on hyperbolic orbits. The limits for bound orbit speeds are given in figure 3. Taking into account that the measured speed is accurate within a factor 2, the plot clearly shows that the majority of the particles are on hyperbolic orbits entering the solar system.

3. Galileo measurements

We argued in the previous section that Ulysses detected interstellar dust after the Jupiter flyby. If this is true, interstellar particles should have been seen also by Ulysses before the Jupiter encounter as well as by Galileo enroute to Jupiter. In the former case, Ulysses' highly eccentric orbit caused the impact directions of classical interplanetary and interstellar dust particles to largely overlap. Therefore, both types are not distinguishable in the Ulysses data set on the basis of impact direction. In contrast, interstellar particles can be distinguished from dust on low–eccentricity low–inclination orbits over a large part of Galileo's orbit. This can be seen in figures 4a and 4b, which give the impact directions of all dust impacts registered after the second Earth encounter. The good correlation between measured rotation angles and interstellar sensitive area can be seen in figure 4a. However,

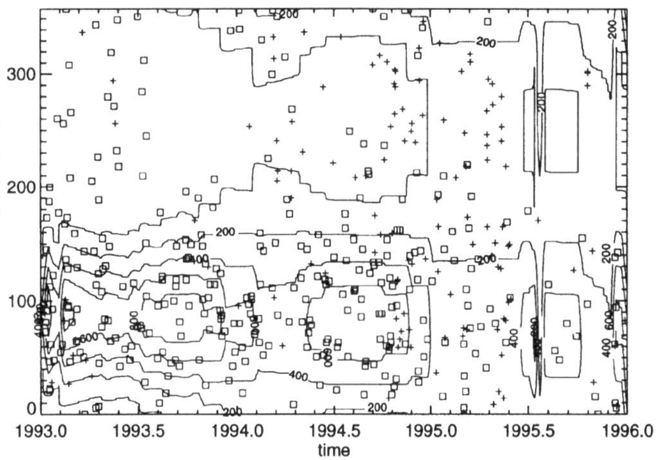

Figure 4a. Impact direction of dust particles versus time of Galileo. The contour lines give the differential geometrical area in cm^2 for impacts from the downstream direction of the interstellar gas. The squares denote particles with signal amplitudes $> 1 \cdot 10^{-13} C$, plus signs denote impacts with smaller signal amplitudes. Jupiter stream impacts have been excluded by their impact time as given in (Grün et al., 1996a). Note the almost empty space after 1995.0 caused by severe dead time due to Jupiter stream impacts.

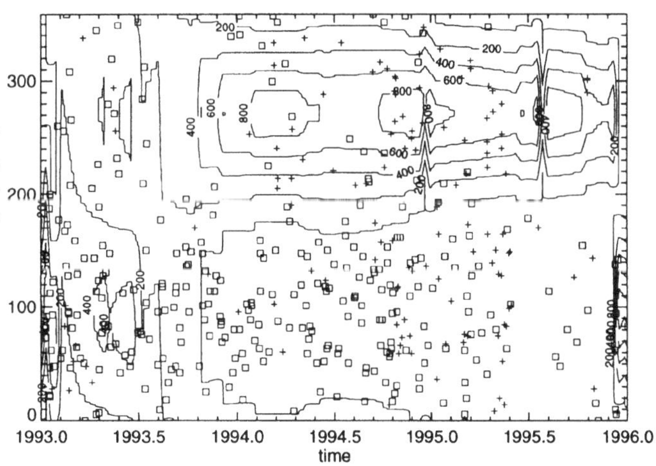

Figure 4b. The contour lines give the differential geometrical area in cm^2 for impacts of particles on prograde orbits with small inclinations.

in the Galileo case the geometry is more complicated than that of Ulysses. Since Galileo moves in the prograde direction, for part of the orbit, from distances between 1 and $2.6AU$, dust particles on prograde orbits with small inclinations are sensed from the same direction as the interstellar particles.

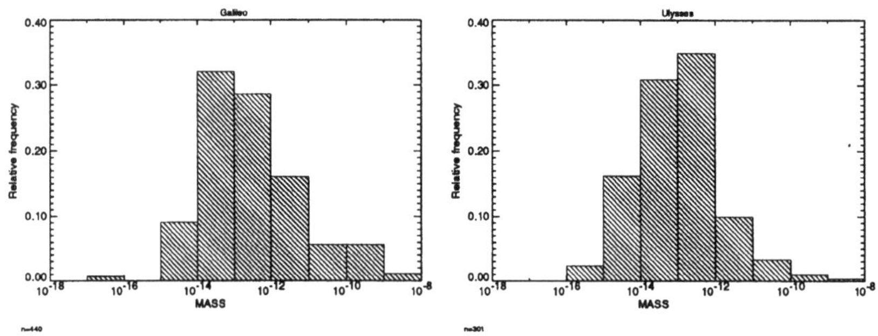

Figure 5. **Mass distribution of the Galileo (left) and Ulysses (right) dust impacts. Selected are all impacts compatible with an interstellar impact direction. Jupiter stream particles have been excluded.**

From directional arguments a clear identification of interstellar dust impacts is only possible outside the asteroid belt (fig. 4b).

The mass distribution of the identified interstellar particles is given in figure 5. The mass is given in g. Included are all particles with rotation angles compatible with the interstellar direction. To make the distributions comparable and to exclude contaminations of small Jupiter particles, only impacts with a signal amplitude (ion charge) greater than $1 \cdot 10^{-13} C$ are included. Before day 192 of 1994, Galileo was only sensitive to impacts above this threshold, other impacts were mostly overwritten by noise. See (Grün et al., 1995b) for a discussion of the Galileo effective threshold.

4. Discussion

From the data presented, it can be seen that dust of interstellar origin is the best explanation for the majority of the dust flux measured by Ulysses in the outer solar system. Galileo confirms this result. Although speed measurements in both cases indicate hyperbolic velocities for the impacting particles, the best argument for an interstellar origin are the impact directions of the dust particles. For Ulysses, the combined effects of increasing distance from the ecliptic plane and radial distance do not significantly influence the flux rate and the distribution of impact directions in agreement with the expected behaviour of interstellar particles. Whereas during the in–ecliptic leg of the Ulysses orbit no distinction between interstellar and bound orbits can be derived from the impact directions, the out–of–ecliptic leg and the Galileo measurements show that interstellar dust is definitely present in the ecliptic plane outside about $2.8 AU$. The situation within $2.8 AU$ is more complicated: Modeling shows (Grün et al., 1996b), that the data taken around the Ulysses ecliptic passage in March 1995 can be only explained by the assumption of interstellar dust contributing to the total flux by approximately 30% at the Earth orbit. Although a clear identification of interstellar dust from impact direction arguments can not be made inside 2.8AU the data don't contradict this assumption. (McDonnell et al., 1975) argue from Pioneer 8 and 9 measurements that the contribution of interstellar dust to the flux at terrestrial distances should be lower than 3%, a value derived from flux isotropy arguments. Further analysis of recent data is necessary to improve the reliability of this estimate.

Although the interstellar gas and dust directions are compatible, more analysis of the data is necessary to give the best fitting direction. Whether or not it deviates from the gas direction will allow important clues about the dust dynamic in the heliosphere and may give clues to the parameters of the surrounding local interstellar cloud itself.

The mass distribution of the measured interstellar dust particles indicate a dropoff at small masses that can not be explained by the sensor threshold. The dropoff value is at least one order of magnitude higher than the threshold. The dropoff indicates that smaller interstellar dust particles, known to exist in interstellar space, are kept out of the heliosphere by defocusing Lorentz forces. Modeling of these processes is underway (Grün et al., 1994). The mean mass of the distribution is $3 \cdot 10^{-13} g$. This is about a factor of 30 heavier than the mass expected from astronomical observations of the large scale interstellar dust component. Whereas astronomical measurements, however, are integrated over a large distance, our measurements are the first ever on the parameters of our local interstellar dust environment itself.

References

M. Baguhl, E. Grün, D.P. Hamilton, G. Linkert, R. Riemann, P. Staubach: 1994, 'The flux of interstellar dust observed by Ulysses and Galileo' *Space Sci. Rev.* **72** pp. 471-476

M. Baguhl, D.P. Hamilton, E. Grün, S.F. Dermott, H. Fechtig, M.S. Hanner, J. Kissel, B.A. Lindblad, D. Linkert, G. Linkert, I. Mann, J.A.M. McDonnell, G.E. Morfill, C. Polanskey, R. Riemann, G. Schwehm, P. Staubach and H.A. Zook: 1995, 'Dust measurements at high ecliptic latitudes' *Science* **268** pp. 1016-1019

E. Grün, H. Fechtig, R.H. Giese, J. Kissel, D. Linkert, D. Maas, J.A.M. McDonnell, G.E.Morfill, G. Schwehm, H.A. Zook: 1992, 'The Ulysses dust experiment' *Astron. Astrophys. Suppl. Ser.* **92** pp. 411-423

E. Grün, H. Fechtig, M.A. Hanner, J. Kissel, B.A. Lindblad, D. Linkert, G. Linkert, G.E. Morfill, H.A. Zook: 1992b, 'The Galileo Dust Detector' *Space Science Reviews* **60/1-4** p. 317

E. Grün, H.A. Zook, M. Baguhl, H. Fechtig, M.S. Hanner, J. Kissel, B.-A. Lindblad, D. Linkert, G. Linkert, I. Mann, J.A.M. McDonnell, G.E.Morfill, C. Polanskey, R.

Riemann, G. Schwehm, N.Siddique: 1992c, 'Ulysses dust measurements near Jupiter' *Science* **257** pp. 1550-1552

E. Grün, H.A. Zook, M. Baguhl, A. Balogh, S.J. Bame, H. Fechtig, R. Forsyth, M.S. Hanner, M. Horanyi, J. Kissel, B.-A. Lindblad, D. Linkert, G. Linkert, I. Mann, J.A.M. McDonnell, G.E.Morfill, J.L. Phillips, C. Polanskey, G. Schwehm, N.Siddique, P. Staubach, J. Sveska, A. Taylor: 1993, 'Discovery of jovian dust streams and interstellar grains by the Ulysses spacecraft' *Nature* **362** pp. 428-430

E. Grün, B. Gustafson, I. Mann, M. Baguhl, G.E.Morfill, P. Staubach, A. Taylor, H.A. Zook: 1994, 'Interstellar dust in the heliosphere' *Astron. Astrophys.* **286** pp. 915-924

E. Grün, M. Baguhl, H. Fechtig, J. Kissel, D. Linkert, G. Linkert, R. Riemann: 1995, 'Reduction of Galileo and Ulysses dust data' *Planet. Space Sci.* **43** pp. 941-951

E. Grün, M. Baguhl, N. Divine, H. Fechtig, M. S. Hanner, J. Kissel, B.-A. Lindblad, D. Linkert, G. Linkert, I. Mann, J. A. M. McDonnell, G. E. Morfill, C. Polanskey, R. Riemann, G. Schwehm, N. Siddique, P. Staubach and H. A. Zook: 1995, 'Three years of Galileo dust data' *Planet. Space Sci.* **43** pp. 953-969

E. Grün, M. Baguhl, R. Riemann, H.A. Zook, S. Dermott, H. Fechting, B.Å. Gustafson, D. Hamilton, M.S. Hanner, M. Horanyi, K.K. Khurana, J. Kissel, M. Kivelson, B.A. Lindbald, D. Linkert, G. Linkert, I. Mann, J.A.M. McDonnell, G.E. Morfill, C. Polanskey, R. Srama: 1996a, 'Dust Storms from Jupiter' *Nature* in press

E. Grün, M. Baguhl, P. Staubach, S. Dermott, H. Fechting, B.Å. Gustafson, D.P. Hamilton, M.S. Hanner, M. Horanyi, J. Kissel, B.A. Lindblad, D. Linkert, G. Linkert, I. Mann, J.A.M. McDonnell, G.E. Morfill, C. Polanskey, G.Schwehm, R. Srama, H.A. Zook: 1996b, 'South–North and Radial Traverses Through the Zodiacal Cloud' submitted to *Icarus*

D.P. Hamilton und J.A. Burns: 1993, 'Ejection of dust from Jupiter's gossamer ring' *Nature* **364** pp. 695-699

M. Horanyi, E. Grün, G.E. Morfill: 1993a, 'The dust skirt of Jupiter: a possible explanation of the Ulysses dust events' *Nature* **363** pp. 144-146

M. Horanyi, E. Grün, G.E. Morfill: 1993b, 'The dusty ballerina skirt of Jupiter' *JGR* **98/A12** pp. 21245-21251

D.H. Humes: 1980, 'Results of Pioneer 10 and 11 Meteoroid experiments: Interplanetary and near-Saturn' *JGR* **85/A11** p. 5841

J.R.Jokipii et al.: 1976 *Nat.* **264** p. 424

K.P. Wenzel, R.G. Marsden, D.E. Page, E.J. Smith: 1992, 'The Ulysses mission' *Astron. Astrophys. Suppl. Ser.* **92** pp. 411-423

J.A.M. McDonnell, O.E. Berg: 1975, 'Bounds for the interstellar to solar system microparticle flux ratio over the mass range $10^{-11} - 10^{-13}$ g' *Space Research* **XV** Academie Verlag: Berlin, pp.555-563.

M. Witte , H. Rosenbauer , M. Banaszkiewicz, H. Fahr: 1993, 'The Ulysses Neutral Gas Experiment: Determination of the Velocity and Temperature of the Interstellar Neutral Helium' *Advances in Space Res* **13** pp. (6)121-(6)130

A.D. Taylor, W.J. Baggaley, D.I. Steel: 1996, 'Discovery of interstellar dust entering the Earth's atmosphere' *Nature* **380** pp.323–325

THE LOCAL BUBBLE

Current state of observations and models

DIETER BREITSCHWERDT *
Max-Planck-Institut für extraterrestrische Physik, Postfach 1603, D-85740 Garching, Germany
Email: breitsch@rosat.mpe-garching.mpg.de

Abstract. Recently, observations with the ROSAT PSPC instrument and the spectrometers onboard the EUVE satellite have given new detailed information on the structure and physical conditions of the Local Bubble. From the early rocket experiments, and in particular from the WISCONSIN Survey, the existence of a diffuse hot gas in the vicinity of the solar system, extending out to about 100 pc, has been inferred in order to explain the emission below 0.3 keV. The higher angular resolution and sensitivity of ROSAT made it possible to use diffuse neutral clouds as targets for shadowing the soft X-ray background. Thus, in some directions, more than half of the flux in the 0.25 keV band appears to come from outside the Local Bubble. Further, measurements of the diffuse EUV in the LISM, show surprisingly few emission lines. These findings are in conflict with the standard LHB model, which assumes a local hot ($T \sim 10^6$ K) plasma in CIE. Model calculations, based on the non-equilibrium cooling of an expanding plasma, show a promising way of reconciling all available observations. Thus the present temperature within the LB may be as low as 4×10^4 K and its number density as large as 2×10^{-2} cm^{-3}, giving a total pressure that is roughly in agreement with the Local Cloud.

Key words: Local Interstellar Medium, Soft X-ray Background, diffuse emission, non-equilibrium plasma models

Abbreviations: CIE – collisional ionization equilibrium; ISM – Interstellar Medium; LHB – Local Hot Bubble; LB – Local Bubble; LISM – Local ISM; SB – superbubble; SXR – soft X-ray; SXRB – SXR Background; VLISM – Very Local ISM

1. Introduction

Generally, our intuitive concept of a background radiation is that of a diffuse emission from a more or less homogeneous distribution of very distant sources. This is certainly true for the first background radiation ever observed, which was in X-rays above 2 keV (Giacconi *et al.* , 1962), as well as for the Cosmic Microwave background. However, for a softer X-ray component below 2 keV (Bowyer *et al.* , 1968), due to photoelectric absorption by neutral hydrogen, the mean free path of the photons becomes less than the radius of the Galaxy. This emission extends down to less than 100 eV, and thus a local Galactic origin for the ultrasoft (0.08 - 0.3 keV) component seems compelling. In any case, a pure extragalactic origin was questionable, because the absolute intensity of the measured flux exceeded the downward

* Heisenberg Fellow

Space Science Reviews **78**: 173–182, 1996.

extrapolation of the spectrum above 2 keV (Henry *et al.*, 1968). After brightness distributions of SXRs for most of the sky were available in the early 70's, and with the recognition of a wide spread hot phase (Cox & Smith, 1974; McKee & Ostriker, 1977) of the ISM from ubiquitous OVI absorption lines (Jenkins & Meloy, 1974), the nexus between the soft X-ray emission and a supernova origin became suggestive.

However, it has been pointed out (McCammon & Sanders, 1990) that the hot intercloud medium, with a typical temperature of $\sim 5 \times 10^5\,K$ (McKee & Ostriker, 1977), fails to reproduce the correct ratios (c.f. McCammon & Sanders, 1990) of the WISCONSIN C-band (160 - 284 eV), B-band (130 - 188 eV) and Be-band (77 - 111 eV). Although the origin of the SXRB is still under debate, there is agreement that our LISM is a highly ionized region with low HI column density, i.e. $N_H \leq 10^{20}\,cm^{-2}$ within a radius of less than 100 pc. For this and other reasons that will be discussed in some detail in the next Section, the region is called the Local Bubble (LB).

Why is the LB, apart from its not yet fully understood relation to the SXRB, an interesting subject of investigation?

Firstly, there is a wealth of observational data available in different wave bands, which allow us to test current ISM models. Indeed, this may be the reason that too simple models are in conflict with some of the observations, and thus one is forced to give up convenient assumptions of a zero order model. In this spirit, that a model should be as simple as possible, but not simpler, I will present in Section 3 a first order model, in which the assumption of ionization equilibrium has been rejected in favour of a self-consistent dynamical and thermal evolution of the LB. It is shown that the results are consistent with all present data, but owing to the complexity of the ISM, it is clear that further modifications will be necessary as observations are progressing.

Secondly, if one would be able to explain the LISM observations by a more general model that is also appropriate, though with different boundary conditions, for other ISM regions, one could answer the question whether the LISM is really so distinctly unusual as it has been claimed (Cox & Reynolds, 1987).

Thirdly, knowing the physical conditions of the LB, helps us to provide input parameters for our very local environment, the heliosphere. This touches the problem of the interaction between the LB and the Local Cloud, that surrounds our solar system, extending out to $\sim 5\,pc$ with an average HI column density of $N_{HI} \approx 3 \times 10^{18}\,cm^{-2}$ (Chassefière *et al.*, 1988). This is discussed in detail in the papers by P. Frisch and R. Lallement (this volume).

Finally, we should bear in mind, how lucky we are that our solar system is located in a region of low density, that is transparent to optical photons. If we were sitting deep inside a dark cloud, our concept of the world would certainly be very different, let alone the consequences for astronomy and navigation in the past.

2. Observational constraints

Using atomic absorption edge filters with a narrow bandpass, spectral information in the ultrasoft X-ray components, i.e. Be-, B- and C-band, was obtained in the WISCONSIN Survey (cf. McCammon & Sanders, 1990). For energies between $0.5 - 1.5\,keV$ (M-, I- and J-bands), one had to rely on pulse height distribution. A somewhat surprising result was the approximate constancy of the Be/B band ratio with increasing count rate (Bloch *et al.* , 1986), because a small column density of $N_H = 10^{19}\,cm^{-2}$ represents already unity optical depth for the B-band. Moreover, the effective absorption cross section for the Be-band is a factor of 6 larger than for the B-band. Observations show that most of the nearby Galactic HI is extended and diffuse, with significant clumping being very unlikely (Lockman *et al.* , 1986). Therefore, an obvious interpretation is that our LISM could be a local cavity, filled with hot plasma and devoid of neutral hydrogen. The observed anticorrelation between ultrasoft X-rays and HI is the basis of the so-called *displacement model* (Sanders *et al.* , 1977; Tanaka & Bleeker, 1977; Snowden *et al.* , 1990). Assuming that *all* of the observed background below 0.3 keV originates in the cavity, it was thought that the measured X-ray intensity along a given line of sight was directly proportional to its extension. Since there is an increase of flux by a factor of 2 to 3 with galactic latitude, a Local Hot Bubble (LHB) was conceived, extending about 200 pc perpendicular and 30 pc into the galactic plane. In such a geometrical model, a complicated 3-dimensional shaping of the LHB was obtained from the varying X-ray intensities along different lines of sight (Snowden *et al.* , 1990). A temperature of $10^6\,K$ was assigned by fitting a Raymond and Smith (1977) equilibrium plasma model to the broad band spectrum, i.e. reproducing the observed C/B/Be band ratios in the WISCONSIN Survey. The essential free parameter in this model is the X-ray emissivity or the plasma thermal pressure, respectively; the best-fit model yields $n_e = 4.7 \times 10^{-3}\,cm^{-3}$ for the electron density, and hence $p/k \approx 9000\,cm^{-3}\,K$. For the Local Cloud, $p/k \approx 2600\,cm^{-3}\,K$ (Bertaux *et al.* , 1985) and thus additional components, such as magnetic pressure, are needed for support. On the other hand the existence of a regular magnetic field reduces the mean free path of thermal electrons perpendicular to the field, thereby impeding conduction efficiently.

Now turning to the M-bands (0.5 - 1.0 keV) emission, it was found in the WISCONSIN Survey, and later confirmed by the ROSAT All Sky Survey (RASS), that the emission is fairly isotropic, if some prominent sources (Loop I, Cygnus SB, Eridanus cavity etc.) are subtracted. This is not easily explained, because both discrete disk and extragalactic sources would certainly exhibit latitudinal intensity variations, the latter due to photoelectric absorption, the former because of their Galactic scale height distribution.

It is certainly fair to state that recent satellite missions like EUVE, DXS and, in particular, ROSAT have fundamentally changed our concept of the LISM and the LB.

One of the first deep pointed ROSAT observations were the so-called shadowing experiments. Due to the fast optics of the XRT (Trümper, 1983) and the sensitivity of the PSPC instrument, it was found (Snowden *et al.* , 1991) that the X-ray intensity, I_x, of a line of sight passing through the Draco nebula was substantially attenuated. Specifically, a satisfactory fit for the C-band count rate was obtained by a simple extinction law, $I_x = I_f + I_b \exp[-\sigma(N_H)N_H]$, with I_f and I_b denoting the foreground and the background intensity and $\sigma(N_H)$ the HI absorption cross section, respectively. Accordingly, roughly 50% of the emission is from beyond the Draco cloud; with distance limits between 300 - 1500 pc, this was clearly *outside* the LHB, and thus in contradiction with the standard assumption of the displacement model. Furthermore, this was direct evidence for the *diffuse* nature of the emission. In order to obtain a limit for I_f, a line of sight that passes through a heavy absorber, like the molecular cloud MBM 12 (distance $\sim 65\,pc$), was chosen. Since almost no C-band shadow was observed ($I_f \approx 0.8\,I_x$), this places a lower limit for C-band emission from the LB. On the other hand, an M-bands shadow was detected, giving a 2-σ upper limit of 30% of the emission in these energy bands (Snowden *et al.* , 1993) originating *inside* the LB. This is interesting, because a $10^6\,K$-LHB-plasma in equilibrium would only produce typically a few percent of the M-bands emission. Thus one may conclude, that a spatial separation of low and high energy SXR bands is dubious. If however some fraction of the M-bands emission originates *inside* the LB, the problem of isotropy becomes less severe.

The Local Cloud is not the only "cool" ($T \approx 7000\,K$) diffuse HI region, embedded in the LB. Kerp *et al.* (1993) found a C-band shadow cast by an HI filament located at a known distance of 60±20 pc. An upper limit for the temperature from the 21-cm line width gives $T \approx 150\,K$. Assuming a similar extension along the l.o.s than perpendicular to it, a density of $n_{HI} \approx 75\,cm^{-3}$ is obtained, giving a pressure in agreement with the LHB model. However, due to the large temperature gradient, conduction should play a rôle, since the pressure contribution of the magnetic field should be negligibly small, and hence observable EUV line emission should be detected.

Due to the short mean free path of EUV photons, shadows are much deeper than for SXRs; they are therefore an excellent probe of the VLISM. Recent observations by the EUVE satellite have given upper limits on the diffuse EUV flux (Jelinsky *et al.* , 1995). Moreover, the high spectral resolution of the instruments in the 70 – 760 Å range allows the detection of emission lines, which should be vastly abundant in a plasma in CIE around $10^6\,K$ (cf. Fig. 2). However, the only lines detected were HeI and HeII, with intensities consistent with local geocoronal and/or interplanetary scattering of solar radiation. Also the limits for the plasma emission measure are a factor of 5 to 10 below of what is expected from the B- and C-band emission from a plasma between $10^{5.7} - 10^{6.4}\,K$, using a Landini and Monsignori-Fossi (1990) plasma code with normal abundances. In addition, the emission spectrum predicted by a conductive interface model (Slavin, 1989) can apparently be rejected at the 99.7% confidence level.

Preliminary results from the DXS Bragg crystal spectrometers (Sanders *et al.* , 1993) ($0.15 \leq E \leq 0.284\,keV$), with high spectral resolution show the existence of emission lines, thus confirming the thermal origin of the diffuse emission. They also indicate that CIE model fits, cannot reproduce the observed spectrum satisfactorily.

There are two peculiar lines of sight (l.o.s.) that cut right through the LB. They may cause severe problems for any hot plasma model, if they were typical for the LB: (i) The l.o.s towards β CMa (distance roughly 200 pc), which is largely filled with HII and shows $\langle n_e \rangle \sim 2 \times 10^{-2}\,cm^{-3}$ and a very low temperature $T \leq 5 \times 10^4\,K$ (Gry *et al.* , 1985) and (ii) the dispersion measure of the nearby pulsar P0950+08 ($D \approx 130\,pc$ from parallax), giving $\langle n_e \rangle \sim 2.3 \times 10^{-2}\,cm^{-3}$ (Reynolds, 1990). However, it may be possible that (i) is not typical and (ii) may refer to a pulsar that is already beyond the LB.

Also considerable effort has been put into modelling the LB as the result of a single massive blast wave. This is certainly the most physical way of understanding the origin of the LB, and the results are valuable as a zero order approach. Basically, the models come in two flavours: a young ($\leq 10^5\,yrs$ active remnant, which produces SXR emitting plasma right behind a strong shock (e.g. Cox & Anderson, 1982), or a very old remnant, in which the hot interior radiates in SXRs (e.g. Innes & Hartquist, 1984; Edgar & Cox, 1993). The problems arise, because reproducing both the intensity and the observed spectrum puts tight constraints on the evolution of the remnant and the boundary conditions. For example, the models of Edgar & Cox (1993) require a rather large ambient magnetic field ($\sim 11\,\mu$G), in order to reproduce a sufficiently large intensity.

3. Modelling

In the last Section, I have discussed the most important LB observations and the major problems associated with the interpretation by the standard displacement model. Here I will present a new class of models, in which the assumption of CIE has been dropped. It is well-known, that radiative recom-

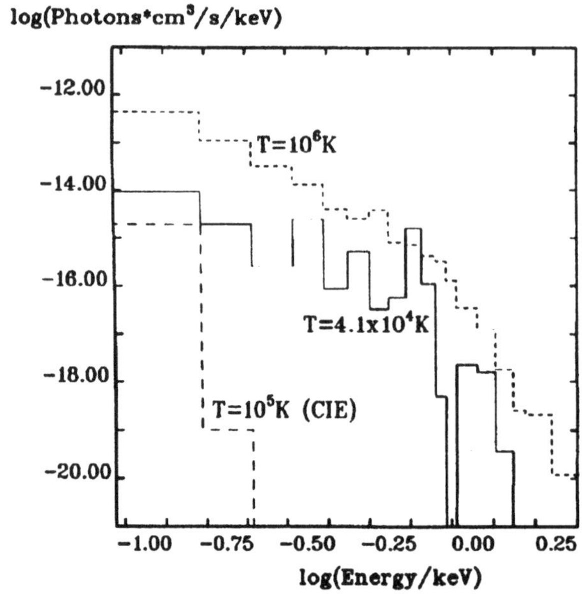

Figure 1. Non-equilibrium emission spectrum (normalized to n_e^2) of a fast adiabatically cooling (initial temperature $T_0 = 2.5 \times 10^6\, K$) plasma (Breitschwerdt and Schmutzler, 1994). The short dashed line shows the spectrum at $T = 10^6\, K$, the solid line at $T = 4.1 \times 10^4\, K$. The long dashed line is a $10^5\, K$ CIE spectrum for comparison.

bination, which is the dominant cooling process in optically thin plasmas, and collisional ionization are not inverse processes, and therefore an external radiation source is needed, in order to maintain equilibrium. Otherwise, if an initially hot plasma ($T \geq 10^6\, K$) in CIE is allowed to cool isochorically or isobarically, substantial deviations from equilibrium occur, increasing with time (Kafatos, 1973; Shapiro & Moore, 1976; Schmutzler & Tscharnuter, 1993). This is even more severe if the gas can expand, and therefore adiabatic cooling can introduce the shortest time scale.

Under these circumstances, the dynamics of the gas cannot be separated from its thermal evolution, and self-consistent calculations have to be performed (Breitschwerdt, 1994; Breitschwerdt & Schmutzler, 1994). As a result, high ionization stages will be "frozen" into the plasma, thus preserving a memory of its origin. The emission spectrum is a superposition of line emission and recombination continuum.

Note the similarity in Fig. 1 between the non-equilibrium X-ray spectrum at $10^6\, K$ and at $4.1 \times 10^4\, K$. The corresponding CIE spectrum would not show any noticeable SXRs. Such a result is of fundamental importance, because it shows that in general the form of the spectrum and the existence of well-known lines do not allow to derive a *unique* temperature of the plasma. Instead its dynamical and thermal history has to be known.

In the following, a model is described that can account for the observations presented in the previous section. It is clear that a successful model of the SXRB requires at least two components: a *local* one (LB), to reproduce the existence of the ultrasoft X-rays and a *distant* one (halo and possibly nearby SBs) to explain the findings of the shadowing experiments. There is observational evidence for a disk-halo connection in spiral galaxies via "chimneys" (Norman & Ikeuchi, 1989), allowing mass, momentum and energy exchange between disk and halo. Such a picture is supported by recent ROSAT observations of the nearby edge-on galaxies NGC 891 (Bregman & Pildis, 1994) and NGC 4631 (Wang *et al.* , 1995), which show the existence of extended X-ray halos in normal spiral galaxies. If the FIR luminosity is taken to be proportional to the star formation rate, these galaxies are comparable to our own (Dettmar, 1992; Table III); in particular NGC 891 is often referred to as a twin of the Galaxy. It has been shown (Breitschwerdt & Schmutzler, 1994) that, with respect to the distant component, a dynamically and thermally self-consistent calculation of a halo outflow towards the North Galactic Pole can reproduce the correct band ratios of count rates from 0.08 - 1.5 keV in the WISCONSIN Survey. Also absorption by an extended HI halo component (Lockman *et al.* , 1986) was included, showing the need for a more local contribution to the ultrasoft component. It is important to note, that a qualitatively new feature of non-equilibrium models is the emission in all of the SXR bands.

As a possible scenario for the origin of the LB, it has been suggested that it is the relic of an old SB, that has been re-energized some 10^6 yrs ago ((Breitschwerdt & Schmutzler, 1994); for details cf. rapporteur paper of the author for Working Group 2). If its further evolution has been dominated by fast adiabatic expansion, the emission spectrum in the SXR bands is characterized by recombination continuum and similar to Fig. 1. Since the plasma kinetic temperature may be as low as $4.2 \times 10^4\, K$, and the density as high as $2.4 \times 10^{-2}\, cm^{-3}$ in order to accommodate to the pulsar dispersion measure, the resulting $p/k \approx 2000$, a value close to the one of the Local Cloud. Temperature gradients between its boundary and the LB are not significant, and therefore conduction should only play a minor rôle. In general, co-existence with HI filaments does not pose a real problem in such a picture.

Since in a non-equilibrium LB model there is simultaneous emission in the B-, C- and M-bands, there seems to be no problem in reproducing the 3/4 keV upper limit intensity towards MBM12. Most interestingly, due to the much lower kinetic temperature, collisional excitation of EUV lines is largely suppressed. Thus the calculations predict a deficiency of EUV lines (s. Fig. 2), in agreement with the EUVE observations, mentioned in the previous Section.

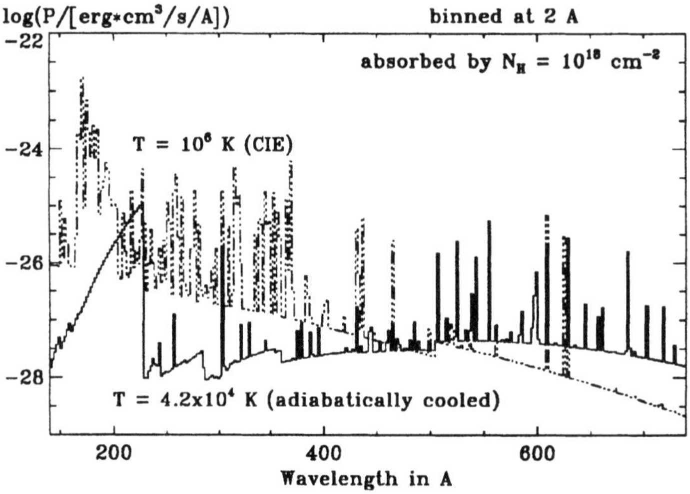

Figure 2. Non-equilibrium EUV emission spectrum (normalized to n_e^2) of the LB in comparison to the CIE spectrum of the LHB; N_H denotes the HI column density of the Local Cloud.

4. Conclusions

Although there are lots of high quality observations, there is still uncertainty about the basic properties of the LB, i.e. the density, temperature (and hence pressure), its extension, amount and distribution of mixed in (partially) neutral material (clouds and HI filaments). Ironically enough, it is just the wealth of information that puts too many constraints on a simple model, from which the LB properties have to be inferred, since none of them can be observed *directly*. These are not only important physical input parameters for modelling the heliosphere, but are also intimately related to the origin of the LB, which still remains a mystery. But the case is not hopeless, because the more detailed information we will obtain in the future by high spatial and spectral resolution in the UV, EUV and SXRs, the more sophisticated modelling we will need. The observational constraints will successively narrow down the number of free parameters. In particular, the observation of individual spectral lines will give us unequivocal information on the excitation conditions, specifically on whether there is still room for CIE models or not.

The model discussed in the previous Section is merely a first step in this direction. No claim is made that it solves the problem of the origin of the LB. Its attractiveness lies in the fact that it connects the fitting of the observations with the origin of the plasma, i.e. its dynamical and thermal history, and that it makes predictions that can be falsified.

If one is willing to accept a "Copernican" view of the LISM, one might get new insight into the physics of the ISM in general. As it appears at present, the LB is just one component of a fairly complex SXRB (including SBs, and a disk-halo outflow component), but it does not seem to be distinctively unusual. The Loop I bubble is a nearby example of a SB, and there is evidence that it is interacting with the LB (cf. rapporteur paper of Working Group 2). A more detailed study of spectral types of the stellar content within the LB, might provide information on whether our LB is also the result of an ancient SB event.

Acknowledgements

I thank my colleagues Drs. R. Egger and M. Freyberg for helpful discussions. The author acknowledges support from the *Deutsche Forschungsgemeinschaft* (DFG) by a Heisenberg Fellowship. Part of this work was carried out while I was a guest at the Max-Planck-Institut für Kernphysik in Heidelberg. I am grateful to Prof. J. Geiss and the ISS Institute for their hospitality and financial support to attend the workshop.

References

Bertaux, J.L., Lallement, R., Kurt, V.G., Mironova, E.N., 1985, Astronomy and Astrophysics, **150**, 1.
Bloch, J.J., Jahoda, K., Juda, M., McCammon, D., Sanders, W.T., Snowden, S.L., Zhang, J., 1986, Astrophysical Journal, Letters to the Editor, **308**, 59.
Bowyer, C.S., Field, G.B., Mack, J.E., 1968, NATURE, **217**, 32.
Bregman, J.N., Pildis, R.A., 1994, Astrophysical Journal, **420**, 570.
Breitschwerdt, D., 1994, Habilitationsschrift, Universität Heidelberg, 158p.
Breitschwerdt, D., Schmutzler, T., 1994, NATURE, **371**, 774.
Chassefière, E., Bertaux, J.L., Lallement, R., Sandel, B.R., Broadfoot, L., 1988, Astronomy and Astrophysics, **199**, 304.
Cox, D.P., Anderson, P.R., 1982, Astrophysical Journal, **253**, 268.
Cox, D.P., Reynolds, R.J., 1987, Annual Review of Astronomy and Astrophysics, **25**, 303.
Cox, D.P., Smith, B.W., 1974, Astrophysical Journal, Letters to the Editor, **189**, 105.
Dettmar, R.-J., 1992, *Fundamentals of Cosmic Physics,* **15**, 143.
Edgar, R.J., Cox, D.P., 1993, Astrophysical Journal, **413**, 190.
Giacconi, R., Gurski, H. Paolini, F., Rossi, B.B., 1962, Physical Review Letters, **9**, 439.
Gry, C., York, D.G., Vidal-Madjar, A., 1985, Astrophysical Journal, **296**, 593.
Henry, R.C., Fritz, G., Meekens, J.F., Friedman, H., Bryam, E.T., 1968, Astrophysical Journal, Letters to the Editor, **163**, L11.
Innes, D.E., Hartquist, T.W., 1984, Monthly Notices of the RAS, **209**, 7.
Jelinsky, P., Vallerga, J.V., Edelstein, J., 1995, Astrophysical Journal, **442**, 653.
Jenkins, E.B., Meloy, D.A., 1974, Astrophysical Journal, Letters to the Editor, **193**, 121.
Kafatos, M., 1973, Astrophysical Journal, **182**, 433.
Kerp, J., Herbstmeier, U., Mebold, U., 1993, Astronomy and Astrophysics, **268**, L21.

Landini, M., Monsignori-Fossi, B.C., 1990, Astronomy and Astrophysics, Supplement Series, **82**, 229.
Lockman, F.J., Hobbs, L.M., Shull, J.M., 1986, Astrophysical Journal, **301**, 380.
McCammon, D., Sanders, W.T., 1990, Annual Review of Astronomy and Astrophysics, **28**, 657.
McKee, C.F., Ostriker, J.P., 1977, Astrophysical Journal, **218**, 148.
Norman, C.A., Ikeuchi, S., 1989, Astrophysical Journal, **345**, 372.
Raymond, J.C., Smith, B.W., 1977, Astrophysical Journal, Supplement Series, **35**, 419.
Reynolds, R.J., 1990, Astrophysical Journal, **348**, 153.
Sanders, W.T., et al., 1993, *Proc. SPIE*, **2006**, 221.
Sanders, W.T., Kraushaar, W.L., Nousek, J.A., Fried, P.M., 1977, Astrophysical Journal, Letters to the Editor, **217**, 87.
Schmutzler, T., Tscharnuter, W.M., 1993, Astronomy and Astrophysics, **273**, 318.
Shapiro, P.R., Moore, R.T., 1976, Astrophysical Journal, **207**, 460.
Slavin, J.D., 1989, Astrophysical Journal, **346**, 718.
Snowden, S.L., Cox, D.P., McCammon, D., Sanders, W.T., 1990, Astrophysical Journal, **354**, 211.
Snowden, S.L., McCammon, D., Verter, F., 1993, Astrophysical Journal, Letters to the Editor, **409**, 21.
Snowden, S.L., Mebold, U., Herbstmeier, U., Hirth, W., Schmitt, J.H.M.M., 1991, *Science*, **252**, 1529.
Tanaka, Y., Bleeker, J.A.M., 1977, Space Science Reviews, **20**, 815.
Trümper, J., 1983, *Adv. Space Res.*, **2(4)**, 241.
Wang, Q.D., Walterbos, R.A.M., Steakley, M.F., Norman, C.A., Braun, R., 1995, Astrophysical Journal, **439**, 176.

Address for correspondence: Max-Planck-Institut für extraterrestrische Physik, Postfach 1603, D-85740 Garching, Germany

THE LOCAL BUBBLE

Origin and Evolution

D. BREITSCHWERDT,* R. EGGER and M.J. FREYBERG
Max-Planck-Institut für extraterrestrische Physik, Postfach 1603, D-85740 Garching, Germany

P.C. FRISCH
Univ. Chicago, Dept. Astronomy Astrophysics, 5640 S. Ellis Ave, Chicago, IL 60637, USA

J.V. VALLERGA
Center for EUV Astrophysics, Kittredge Street, University of California, Berkeley, CA 94720, USA

Abstract. A summary of the lively discussions in working group 2 (WG2) on the origin and evolution of the LB is given below. The debate focussed mainly on the problem of how to pin down the physical properties of the LB. As a first step, we had to critically assess the observational constraints that are put on any model, predominantly from the SXR and EUV data. Next we were discussing models on the origin and evolution of the LB, which are able to explain the observations and which would allow to infer basic LB properties. A simple model, emphasizing the self-consistent dynamical and thermal evolution of a non-equilibrium plasma is presented. We also found, that picturing the LB as an isolated phenomenon is not supported by the data. Instead, the LB environment and its influence on the evolution of the LB have to be taken into consideration. Two different views are presented here. Either the LB and the Loop I superbubble are two physically separate phenomena, which are currently interacting, or star formation epochs in the Scorpius-Centaurus Association interacting with the molecular gas in the Aquila Rift and the interarm region around the Sun may have sculpted the configuration of interstellar matter within 200 parsecs, including the Local Bubble. However, so far observational constraints are insufficient to establish a canonical model of the evolution and the origin of the LB.

Key words: Local Interstellar Medium, EUV emission, Soft X-ray Background, Local Bubble models

Abbreviations: CIE – collisional ionization equilibrium; ISM – Interstellar Medium; LHB – Local Hot Bubble; LB – Local Bubble; LISM – Local ISM; SB – superbubble; SCA – Scorpius-Centaurus OB Association; SN – supernova; SNR – SN remnant, SXR – soft X-ray; SXRB – SXR Background; VLISM – Very Local ISM

1. Introduction

There is no unambiguous and direct way of measuring the physical properties, such as density, temperature, pressure and extension of the LB (cf. D. Breitschwerdt, this volume). Instead, we have to resort to modelling and try to explain all the available data. For example, the most straightforward method of measuring the density inside the LB would be the sampling of pulsar dispersion measures of pulsars with known parallaxes within 100 pc. So far, only one nearby pulsar (distance of 130 pc) is known, giving an average electron density of $2.3 \times 10^{-2}\,\mathrm{cm}^{-3}$ (Reynolds, 1990). Also un upper limit of the temperature could be obtained from absorption line widths. Again, from the COPERNICUS observations of interstellar OVI absorption lines (Jenkins & Meloy, 1974), along lines of sight towards stars with known distances less than 100 pc, we obtain: $5.54 \leq \log(T[\mathrm{K}]) \leq 5.94$. This

* Heisenberg Fellow

Space Science Reviews **78**: 183–198, 1996.

is however not conclusive, in order to decide between hot equilibrium and cooler non-equilibrium LB models, in particular if uncertainties in the atomic physics data, entering the various plasma codes, are taken into account.

Although there is some agreement that the average radius of the LB should be of order 100 pc, its latitudinal extensions are somewhat uncertain. According to the displacement model (e.g. Snowden *et al.* 1990a), it should vary between 30 pc in the plane to 200 pc at high Galactic latitudes. However, taking into account that there is a substantial contribution in the 1/4 keV band from beyond the LB, a factor of 3 in the major/minor axis ratio seems more plausible. This is also supported by recent absorption line studies towards nearby stars (Fruscione *et al.* 1994; cf. R. Egger, below).

In the following contributions, the observational basis for the more local ISM (Section 2: EUVE data discussed by J. Vallerga) and the more distant parts of both the LB and its environment, including the Galactic halo (Section 3: ROSAT data discussed by M. Freyberg) is outlined. In the following, models for the origin (Section 4: D. Breitschwerdt) and for the interaction with the neighbouring Loop I bubble (Section 5: R. Egger) are presented. An alternative scenario whereby the Local Bubble cavity is the result of superbubble shells from star formation epochs in the Scorpius-Centaurus Association interacting with the Aquila Rift molecular cloud in the direction of galactic rotation, and the interarm region surrounding the Sun in the direction perpendicular to galactic rotation, is given (Section 6: P. Frisch).

2. EUV Observations of the Local Bubble

2.1. Introduction

Thermal radiation from optically thin plasmas at temperatures of 10^5–10^7 K occurs preferentially in narrow emission lines, and a significant fraction of the energy is emitted in the extreme ultraviolet (EUV) bandpass (70–912Å). Recent observations with the Extreme Ultraviolet Explorer EUVE satellite (Bowyer & Malina, 1991) failed to detect any non-local diffuse emission in the EUV band (170–730Å) to which it was sensitive (Jelinsky *et al.*, 1995). Because it was not designed or optimized for diffuse observations, EUVE's spectral resolution to diffuse emission is rather poor and similar to the broad band X-ray surveys done by rocket in the C (160–284 eV) and B bands (120–188 eV), limits to expected thermal emission are somewhat model dependent. By assuming thermal equilibrium of the emitting plasma and solar abundances (no depletion), the derived plasma emission measure upper limits were a factor of 5 to 10 below the rocket measured C- and B-band emission measures over the temperature range $10^{5.7}$–$10^{6.4}$ K.

Jelinsky et al. (1995) explored possible scenarios that could reconcile the discrepancy of their results with those of the broad band X-ray surveys, in particular, the B band which does not sample the distant soft x-ray emission beyond the Local Bubble as does the C band. Scenarios in which the hot gas responsible for the soft X-ray background was heated recently (less than 10^5–10^6 years ago) by an active blast wave, most likely caused by a supernova (Cox & Reynolds, 1987), result in a gas not in thermal equilibrium and without cosmic abundances. Refractory elements such as Fe and Si are depleted onto dust grains in cold dense clouds, and evaporative mechanisms that return them to the gas phase (e.g., thermal sputtering in shocks) and then raise the elements to high ionization states require timescales longer than 10^5 years (Cox & Reynolds, 1987). The dominant line features at $T \sim 10^6$ K are high ionization lines of iron and silicon, two elements normally depleted from cosmic abundances in the cold ISM. The emitting gas responsible for the soft X-ray diffuse background may have depleted abundances, which would change the spectral distribution of the emission and explain the non-detection of Jelinsky et al. (1995). Bloch (1988) used this argument to explain why the Be band (77 - 111 eV) rocket data also did not show evidence of the Fe line complex at 172Å. When thermal models are used with depleted elemental abundances the upper limits derived from the EUV data become consistent with the soft X-ray results. Since these models can only be constrained by detections of emission lines, not upper limits, only a limited amount can be said about depletion and non-equilibrium models. Emission lines have been detected by the DXS experiment at shorter wavelengths (Sanders *et al.*, 1993; Sanders *et al.*, 1996), and they concluded from a preliminary analysis that even equilibrium plasma models with depleted abundances could not successfully fit the data.

2.2. Limits on the Conductive Boundary of the Local Cloud

The relatively cool Local Cloud inside the hot Local Bubble will generate a conductive boundary at intermediate temperatures. Such an evaporative boundary could have a strong effect on the ionization state in the Local Cloud as it converts the energy of the hot gas into strongly ionizing radiation in the EUV. Slavin (1989) has developed a detailed model of this interface that includes the dynamic effects of a magnetic field and the non-equilibrium nature of the ionization structure of the interface. Slavin's work includes a prediction of the diffuse spectrum from this interface through a neutral absorbing column of $N_{\rm H\,I} = 1 \times 10^{18}$ cm^{-2}. Slavin's predicted spectrum was inconsistent with the EUVE results at a confidence level of 99.7% (Jelinsky *et al.*, 1995). In particular, the He II emission line at 256Å was predicted by Slavin to be 70 ph cm^{-2} s^{-1} sr^{-1}, which exactly matches the EUVE 3 σ upper limit for monochromatic radiation at this wavelength. Slavin does suggest that decreasing the evaporative mass-loss rate corresponding to cases of lower heat conductivity and more tangential magnetic fields will reduce the predicted flux. This places a strong empirical constraint on the Local Cloud conductive boundary and this constraint will improve as more complete EUVE diffuse spectral survey data become available, including observations of cold clouds within the Local Bubble such as LVC88+36-2 (Kerp et al., 1993).

2.3. Limits on the Local EUV Radiation Field

Cheng & Bruhweiler (1990) pointed out that the diffuse flux would dominate the EUV radiation field over that of hot stellar sources at the surface of the Local Cloud as well as near the Sun. Since much of this flux is in bright lines at wavelengths short enough to penetrate deeply into the Local Cloud, the local ionization of helium strongly depends on the magnitude of this flux; and there is much debate on the local levels of hydrogen and helium ionization (Cox & Reynolds 1987; Vallerga, Frisch and Slavin, these proceedings).

By taking a normal abundance, thermal spectrum at $T = 10^6$ K with an emission measure consistent with EUVE upper limits, the upper limit to the helium ionizing flux between 170 and 504Å is 73 photons cm^{-2} s^{-1} sr^{-1} (assuming a column density of 2×10^{18} cm^{-2} to the hot gas). If the upper limits for the portion of the sky surveyed is extrapolated to the whole sky, then the total expected helium ionizing flux is 927 photons cm^{-2} s^{-1}. The corresponding helium photoionization parameter, $\Gamma_{\rm He}$, is less than 1.4×10^{-15} s^{-1}, where $\Gamma_{\rm He}$ is the integral over wavelength of the flux weighted by the neutral helium cross section. This number is a factor of $\sim$12 below that calculated by Cheng & Bruhweiler for the same emission source. These upper limits now place the diffuse contribution of helium ionization by the Local Bubble gas on the same order as that of stellar sources. For a description of

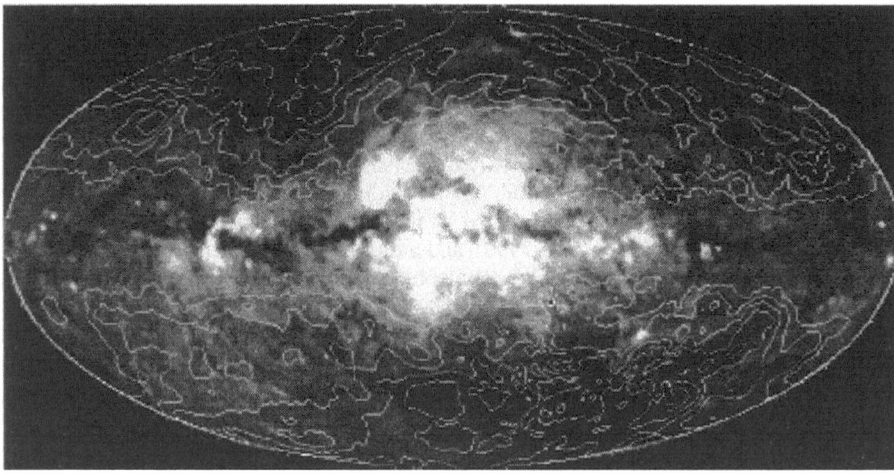

Figure 1. Map of the PSPC *hardness ratio* $HR1(l,b)$ = (hard-soft)/(hard+soft) where "soft" = $0.1 - 0.4$ keV and "hard" = $0.5 - 2.0$ keV in galactic coordinates with $l = 0°$ centered [taken from Freyberg 1994]. Contours represent column densities of HI and range from $\log(N_H) = 19.8$ to 21.2 with 0.2 spacing (Dickey & Lockman, 1990). Darker shading indicates larger values of HR1.

the EUV radiation field and its effects on the Local Cloud see the contribution to these proceedings by Vallerga.

3. The Local Bubble and Beyond

3.1. INTRODUCTION

Talking about the *local* interstellar medium in the context of soft X-rays means distances of *at least* 100 pc. In the 1/4 keV band the optical depth $\tau = 1$ is reached at about a column density of $N_H = 1.2 \times 10^{20}\,\mathrm{cm}^{-2}$ (approximately at $d \sim 100$ pc). Looking for the relation of distance d and $N_H(d)$ for various sight lines could help to determine the 3-dimensional structure of the neutral gas. Using a recent compilation of Lyα absorption line measurements (Fruscione *et al.*, 1994) the shape of the Local Bubble (LB) can be constrained as follows: towards the galactic poles the bubble has an extent of about $120 - 180$ pc while in the galactic plane it is only $60 - 100$ pc. This variation is similar to the variations of the soft X-ray background (Snowden *et al.*, 1995) at energies below 0.3 keV. It can be shown that the observed anti-correlation between the 3/4 keV emission and the galactic N_H suggests a fairly high temperature ($\geq 10^{6.3}$ K) for the background component assuming a CIE plasma. Figure 1 shows an X-ray color image (HR1, cf. caption) of the ROSAT All-Sky Survey. The most prominent characteristics are the strong asymmetry between the northern and southern hemisphere as well as the deep absorption feature in the plane. At higher latitudes (low N_H) an anti-correlation of HR1 with N_H can be found while at low latitudes the picture is more complicated.

3.2. X-RAY SHADOWING

The discovery of (certain) X-ray shadows in the 1/4 keV band with the ROSAT PSPC had undoubtedly established the existence of soft X-ray emission from beyond the LB. The band intensity ratio R2/R1 $[(0.2-0.3)\,\mathrm{keV}/(0.1-0.2)\,\mathrm{keV}]$, similar to the Wisconsin C/B ratio, is sensitive to the absorption of and thus to the distance to the emitting region. In the Wisconsin Survey a large-scale gradient in this ratio has been found (Snowden *et al.*, 1990b) which may be explained by a possibly higher temperature in the direction towards the galactic center. ROSAT survey data show that this ratio is far from uniform: there is a correlation of decreasing R2/R1 with decreasing N_H; this could be interpreted as absorption of an emitter beyond the bulk of the neutral gas. However, significant deviations are also found at high latitudes; a detailed analysis of the ratio R2/R1 will be presented (Snowden *et al.*, 1996), (Freyberg *et al.*, 1996). The 3/4 keV band distribution is still not adequately explained. The band intensity ratio R4/R1 $[(0.4-0.9)\,\mathrm{keV}/(0.1-0.28)\,\mathrm{keV}]$ shows a constant component with R4/R1 $\sim$ 0.14 in the galactic anti-center direction (Freyberg *et al.*, 1996): This relation can be expected, if a reasonable fraction of the emission, responsible for the observed R4 intensity, originates in front of the absorbing material ($\sim$ 100 pc, comparable to the displacement model (Snowden *et al.*, 1990a)), which could not easily be produced in terms of a $10^{6.0}$ K thermal plasma in CIE. This could either mean (1) the LB temperature may be higher ($T \sim 10^{6.3}$ K) or (2) a hotter component present or (3) the assumption of CIE may be invalid.

3.3. GALACTIC X-RAY HALO(S)

Around several galaxies X-ray halos have been found, e.g. around NGC 891 (which is very similar to our own Galaxy). It appears therefore natural to assume that our Galaxy has an X-ray halo as well. Herbstmeier *et al.* 1995 have detected galactic soft X-ray emission shadowed by clouds at distances greater than 1 kpc. To disentangle the isotropic extragalactic X-ray background from a Galactic halo, ROSAT survey data were searched for spatial variations of the absorbed component (Freyberg 1994, 1996). The simple models used in the analysis indicate the existence of an extended X-ray halo; the statistical significance for derived parameters is however poor. Shadowing observations in the regions *Draco* (Snowden *et al.*, 1991) and *Ursa Major* (Snowden *et al.*, 1994) point towards a patchy halo, because the distant component showed large variations over an angular scale of 40°. Major uncertainties in the determination of Galactic halo intensities are introduced by (1) the unknown spectral shape of the extragalactic X-ray background at low energies, by (2) the unknown spatial extent of the hot local component in different directions, and (3) unknown distribution of absorbing and emitting material along the sight line.

3.4. CONCLUSIONS

The ROSAT mission has established the existence of soft X-ray emission from beyond the LB. Analysis of X-ray shadows and of the large-scale distribution of absorbed X-ray emission suggests a pervasive galactic halo with complicated spatial structure. Progress can be expected in the modeling of the (spectral as well as spatial) X-ray structure of the LB. This will put additional constraints on the distribution of any distant (absorbed) soft X-ray emission. We do not think that too many observational constraints spoil the broth but that reality rather looks like a four-course meal (that may demand further observations with higher spectral resolution): a LB, other bubbles like our own, a galactic halo, and an isotropic extragalactic background.

4. A simple model for the origin of the Local Bubble

4.1. INTRODUCTION

Evaluating the different observations associated with the LB, in particular in the EUV and SXR's recently obtained by EUVE, ROSAT and DXS (cf. contributions by J. Vallerga, M. Freyberg and the review paper by the D. Breitschwerdt, these proceedings), it appears that there exist obvious discrepancies between the data and the predicted spectra in the framework of a LHB equilibrium model, that cannot be easily explained away. For example there is evidence for the existence of cold, neutral gas within the much hotter LB gas (Kerp *et al.*, 1993). Moreover, the results of a high resolution spectral study of the local SXRB with the DXS mission (Sanders *et al.*, 1996) cannot be consistently interpreted by CIE models.

Here we sketch a completely different approach, in which the dynamics is treated in a rather simple way, but the assumption of collisional ionization equilibrium (CIE) has been dropped. Nonetheless, as has been discussed before, the dynamical and thermal evolution of the plasma have to be treated self-consistently (Breitschwerdt & Schmutzler, 1994). The reason is, because in an optically thin radiating plasma, there exist largely different time scales, associated both with the atomic physics, and the dynamics of the plasma. Specifically, in the case of fast adiabatic expansion, collisional ionization and radiative recombination can be drastically out of equilibrium.

There have been hydrodynamical simulations of the evolution of supernova remnants, in which the time-dependence of the ionization structure has been followed up (e.g. Cox & Anderson 1982). However, it is not clear to which extent the thermal feedback into the dynamics has been included, i.e. a clear distinction between the kinetic and potential part of the internal energy has been made (Breitschwerdt & Schmutzler, 1994). Moreover, since the blast wave was assumed to propagate into the warm or hot ISM, the spectra showed delayed *ionization*. In the following a scenario, as it is called these days, is outlined in which an old and *hot* bubble expands rapidly, thus emphasizing the effect of delayed *recombination*.

4.2. A SIMPLE NON-EQUILIBRIUM MODEL

The origin of the LB is somewhat ill constrained. We make the assumption, that a succession of stellar winds and SN events created a cavity, filled with hot plasma (SN ejecta and shocked ISM), bounded by a dense shell. This is a fairly common phenomenon in the ISM, and the best example is just around the corner, i.e. the Loop I bubble, with still on-going SN events (cf. R. Egger, below).

The present mass of the LB is largely uncertain due to the unknown density and volume. Taking a radius of $R \approx 120$ pc, it should be somewhere in between 1000 and 5100 $M_\odot$, depending on the density. The lower value refers to the density of the displacement model, the higher value to an average density given by the dispersion measure of pulsar PSR 0950+08 (Reynolds, 1990). In either case this value is larger than the ejecta mass from successive SNe with a total energy of $\sim 10^{52}$ ergs. Thus mass transfer from the swept-up shell is needed. Since heat conduction is efficiently suppressed over a tangential (contact) discontinuity, turbulent mixing will probably dominate.

In order to contribute to the total SN energy of an OB association, a star should be earlier than B3; the typical number of such stars in a modest association is $N_* = 20$, within a diameter of much less than 100 pc. This number is certainly smaller than that for Loop I, in which already 40 SNe have gone off (cf. R. Egger, below). Assuming an initial mass function with exponent $\alpha = d\log(N_*)/d\log(M_*) = -1.6$, and that the most massive star be of type O7V (McCray & Kafatos, 1987), i.e. $N_*(M_* = 35\,M_\odot) = 1$, we find $N_*(M_* = 7\,M_\odot) = 13$. The total SN energy is then roughly $E_{\rm tot} \sim N_*(M_* > 7\,M_\odot)\,\langle E_{SN}\rangle \approx 20 \times 5 \times 10^{50}$ ergs = 10^{52} ergs. The corresponding mechanical luminosity is thus $L_{SB} \approx 6.3 \times 10^{36}$ ergs/s, (McCray & Kafatos, 1987). As a fingerprint, we should expect about 50 stars of spectral type B9 within 100 pc. According to the Yale Bright Star Catalog, this is confirmed by a number of 83 known B9 stars (T. Berghöfer, private communication).

If we insist, that the present average density inside the LB should be around $n_{LB} = 2.4 \times 10^{-2}\,{\rm cm}^{-3}$, there has to be considerable mixing during the evolution of the SB, and consequently the ambient medium has to be rather dense, i.e. a molecular cloud with $n_0 = n(H_2) \sim 10^4\,{\rm cm}^{-3}$. This can only hold on average, since the medium is in reality quite clumpy, and therefore deviations from spherical symmetry are inevitable. After a time $t_{SB} \sim 10^7$ yrs, the radius of the bubble is roughly $R_{SB} = 85\,[(N_*/n_0)\,(E_{SN}/5 \times 10^{50}\,{\rm ergs})]^{1/5}\,(t/10^7\,{\rm yrs})^{3/5} \approx 24\,{\rm pc}$. The ejecta mass maybe around $M_{ej} \sim 200\,M_\odot$, and therefore about 4900 $M_\odot$ have to be mixed in. The mixing process based on an acoustic instability of the surrounding shell (Kahn & Breitschwerdt, 1989), can do the job even in the presence of a tangential field, which may be as high as 100 μG. About 45% of the SN hydrodynamical energy is contained in the hot interior, and thus the ejecta temperature is of order $(2/3)\,0.45\,(N_*\,E_{SN})\,\bar{m}/(M_{ej}k_B) \approx 2.2 \times 10^8$ K. Thus the resulting mixing temperature is about 8.6×10^6 K, where $\bar{m} = 2 \times 10^{-24}$ g and k_B are the average ion mass and Boltzmann's constant, respectively. We note that at temperatures above 10^6 K the assumption of collisional ionization equilibrium is not too bad. The cooling time scale is then given by $t_c \sim \kappa^{3/2}/q \approx 6 \times 10^5$ yrs, where the constant $q = 4 \times 10^{32}\,{\rm cm}^6/({\rm s}^4\,{\rm g})$ and $\kappa = P/\varrho^{5/3}$ is the adiabatic parameter. Suppose that the last SN went off at $t_d = 1.5 \times 10^6$ yrs ago, which is comparable to the cooling time. Therefore, during further evolution the temperature will decrease both by adiabatic and radiative cooling.

For simplicity, we assume that at this stage the SB encountered the edge of the cloud, where the ambient density jumps presumably by a factor of 10 - 100 and then falls off as $\varrho_1 = \tilde{\varrho}(r/R_c)^{-\beta}$, with r, R_c and $\tilde{\varrho}$ denoting the radial variable, the distance from the centre of the SB to the edge of the cloud and the density at this position, respectively. Since $R_{SB} \ll R_{LB}$, we can describe the further evolution of the SB, which from now on we will call the LB, by a similarity solution. During this stage, which is characterized by *fast adiabatic* expansion, the radius will be given by $R_{\rm LB} = A\,t^{2/(5-\beta)}$, with $A = [3(5-\beta)^2\,N_*\,E_{\rm SN}/(4\pi(3-\beta)\tilde{\varrho}R_c^\beta)]^{1/(5-\beta)}$ during the Sedov-Taylor phase, and at later stages by $R_{\rm LB} = R_1[1 + (4-\beta)(\dot{R}_1/R_1)\,(t - t_1)]^{1/(4-\beta)}$, when the shell is momentum driven; $R_1 = R_{LB}(t_1)$, with $t = t_1$ being the time of transition. Note that for large density gradients ($\beta > 3$) the outer shell can fragment due to Rayleigh-Taylor instability. If $\beta = 2$, then $\varrho_{\rm LB} \propto R_{\rm LB}^{-3} \propto t^{-2}$ or $\propto t^{-3/2}$, respectively.

If $\tilde{\varrho}(R_c) \sim 2 \times 10^{-22}\,{\rm g/cm}^3$ and taking $\beta = 2$, we obtain at the present time, i.e. after $t = 1.5 \times 10^6$ yrs, $R_{LB} = 87 \times (t/10^6\,{\rm yrs})^{2/3}\,{\rm pc} \approx 114\,{\rm pc}$, according to the adiabatic expansion law. The temperature decrease due to adiabatic expansion for a gas with ratio of specific heats $\gamma = 5/3$ will be $T_{LB} = T_{\rm SB}\,(R_c/R_{\rm LB})^{3(\gamma-1)}$, i.e. a factor of 25 lower and hence $T_{LB} \approx 3 \times 10^5$ K, which will be further reduced by line cooling. The present average density is $2.4 \times 10^{-2}\,{\rm cm}^{-3}$, as it should be in order to be consistent with the pulsar dispersion measure. Moreover, the thermal pressure is less than $P/k_B \approx 7200\,{\rm K\,cm}^{-3}$, and therefore the co-existence with HI material inside the LB is not a genuine problem. As has been mentioned before, the deficiency of EUV lines is also a natural consequence of such a model. A characteristic spectrum is shown and discussed elsewhere (review paper by D. Breitschwerdt, this volume).

4.3. CONCLUSIONS

The model discussed here emphasizes a quite different aspect in modelling the origin of the LB than previous ones, and should be understood just as one member of a new class. Due to the fact that basic properties of the LB are not known from direct observations, the boundary conditions of any model are not well constrained. Consequently many, sometimes arbitrary, assumptions have to be made on the way. Therefore the results can at best be taken *cum grano salis*, rather than being a realistic description of the evolution of the LB. Future observations, in particular those with higher spectral resolution, will improve the situation considerably.

5. Interaction between Local Bubble and Loop I

5.1. INTRODUCTION

Loop I is a prominent feature of the soft X-ray sky. It covers a solid angle of $7/6\,\pi$ centered roughly around the Galactic centre direction. For long it has been supposed to be the product of activity within the nearby ($\approx$ 170 pc) young stellar association in Scorpio and Centaurus (Weaver, 1979). It has been shown that Loop I is well described as a superbubble produced by the collective stellar winds and by about 40 consecutive supernova explosions from the Sco-Cen OB association (Egger, 1995). This stellar activity is responsible for an expanding dense neutral shell of $\sim$160 pc radius surrounding the hot Loop I bubble as well as for the soft X-rays emerging from its interior. From X-ray spectral analysis, assuming CIE, one can derive a mean temperature of 4×10^6 K and an average density of $2.5 \times 10^{-3}\,\mathrm{cm}^{-3}$ for the hot gas in the bubble.

The close proximity of this neighbouring bubble suggests some kind of interaction with our own Local Bubble which has an average radius of $\sim$100 pc (see the contributions of D. Breitschwerdt and M. Freyberg in this article). Here we present some observational facts that support the scenario of a collision between those two interstellar bubbles.

5.2. OBSERVATIONS AND DISCUSSION

Figure 2 (*left*) shows a ROSAT Survey map centered on Loop I in the energy range 0.1–2.0 keV. The solid white circle of 58° radius represents a circle fit to the radio continuum Loop I. The dashed lines outline an annular X-ray shadow in front of Loop I which is indicated by a deficiency of X-rays form the corresponding region (Egger & Aschenbach, 1995). A corresponding feature is apparent in an H I map of the same field (data from Dickey & Lockman 1990) (*right*). Here, the H I column densities have been divided by cosec$|b_{II}|$ in order to compensate for the main galactic contribution and to enhance the more local structures.

An annular volume of dense neutral gas is predicted to form at the interaction zone between two colliding interstellar bubbles. Hydrodynamical computations (e.g. Yoshioka & Ikeuchi 1990) reveal that the density within the cool interaction ring is by a factor of 20 to 30 higher than that of the ambient medium. The observed neutral gas ring appears to be somewhat distorted and incomplete which may be due to deviations from spherical symmetry of either the Local Bubble or the Loop I superbubble or both. Besides this, it resembles quite well the simulated scenario of the two colliding shells as seen from inside one of them (figure 3).

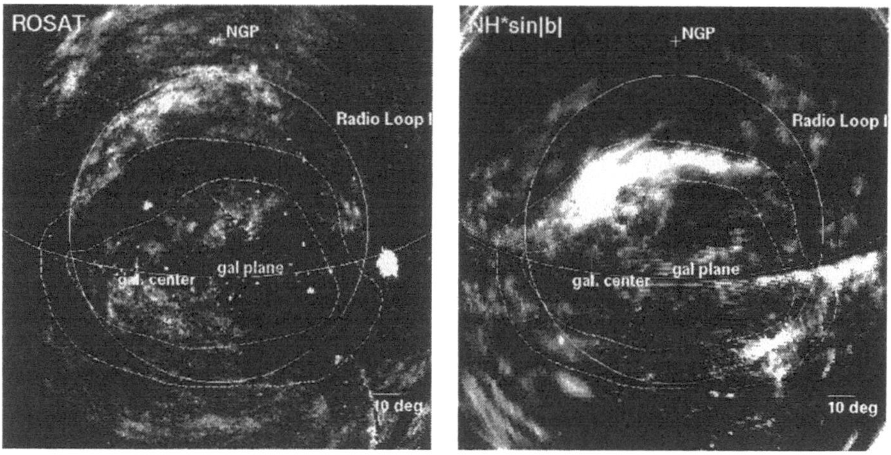

Figure 2. *Left* : ROSAT Survey map centred on Loop I in the energy range 0.1–2.0 keV. The circle of 58° radius follows the radio continuum loop. The dashed lines outline the contours of the annular shadow. *Right* : H I map divided by cosec$|b_{II}|$.

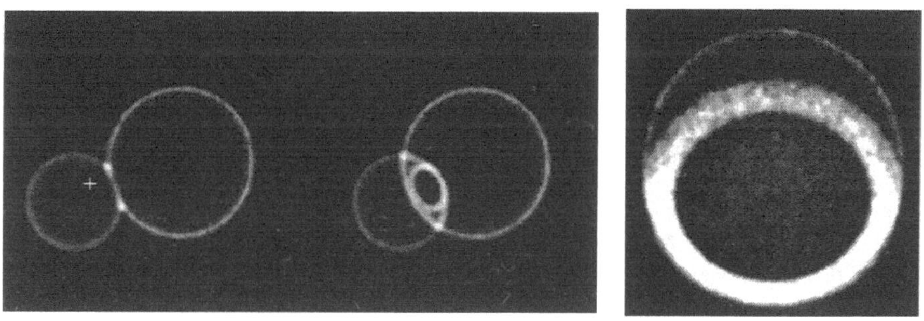

Figure 3. *Left* : Projected column density of a simulation of two colliding bubbles seen from aside. *Middle* : The same, rotated by 30°. *Right* : The same as seen from inside. The viewing point is marked by the cross in the left image. The viewing direction is towards the larger bubble.

Star counts using the ROSAT Wide field camera (Warwick *et al.*, 1993) confirm the presence of an annular structure in the wall of the Local Bubble in the direction of Loop I. The distance of the interaction feature can be estimated by absorption line studies of nearby stars which are projected on the ring (figure 4, data compiled by Fruscione *et al.* 1994). It is found that the N_H jumps from $< 10^{20}\mathrm{cm}^{-2}$ to $> 7 \times 10^{20}\mathrm{cm}^{-2}$ near the distance of about 70 pc, which places the dense gas right between the Local Bubble and Loop I. From the column density a particle density within the ring of $n \sim 15\,\mathrm{cm}^{-3}$ is estimated, which exceeds the ambient density ($n_0 \sim 0.6\,\mathrm{cm}^{-3}$) by a factor of 25, in good agreement with the theoretical prediction. Optical and UV spectral analysis of stars near the centre of Loop I (Centurion & Vladilo, 1991) reveal evidence for the presence of a neutral gas wall of $N_H \sim 10^{20}\mathrm{cm}^{-2}$ at a distance of 40 ± 25 pc. Such a neutral wall between the two colliding bubbles is also predicted from the abovementioned theoretical work. Furthermore, it is in agreement with the fact that the sightlines towards the center of Loop I, i.e. through the center of the ring feature, are almost opaque in the lowest ROSAT energy band (0.1 – 0.2

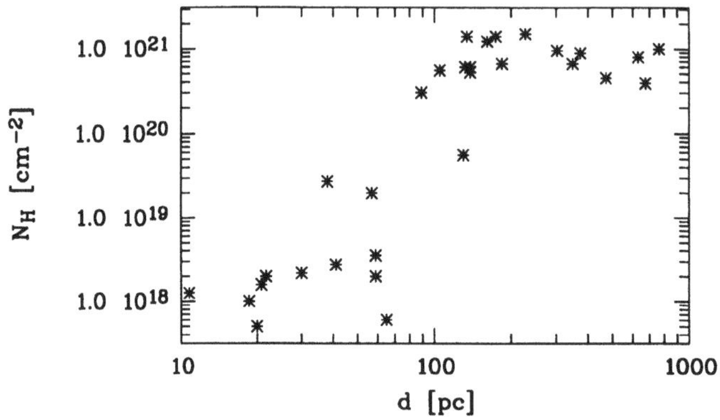

Figure 4. H I column densities towards stars projected on the interaction ring. N_H rises to $\sim 7 \times 10^{20}\text{cm}^{-2}$ near the distance of about 70 pc.

keV). This means, that the soft X-rays are attenuated by an intervening neutral gas wall with a column density of $N_H \sim 10^{20}\text{cm}^{-2}$. This strongly supports the picture of two separate interstellar bubbles, which have collided but have not merged. In other words, the 4×10^6 K hot gas interior to the Loop I supershell has probably not mixed with the cooler gas in the Local Bubble.

On the other hand, the pressure in the Loop I bubble is approximately twice as high as in the Local Bubble, if we adopt the canonical "Hot Bubble model" ($T = 10^6$ K, $n = 5 \times 10^{-3}\,\text{cm}^{-3}$). From this one would expect, that the neutral wall may be bent towards the interior of the Local Bubble, and thus, towards the Solar system. It may well be, that the outer parts of this bent neutral gas sheet are already quite close to us. Single clouds at the surface of this wall, formed by turbulent instabilities, may even have already reached the solar system, making up the local interstellar cloud.

5.3. CONCLUSIONS

We conclude that the observed annular H I feature is a product of the interaction between the Local Bubble and the Loop I supershell. Influences on our nearby interstellar environment may have arisen from the collision of these two bubbles.

6. The Local Bubble as an Asymmetric Superbubble

6.1. INTRODUCTION

This section presents the view that the 'Local Bubble' is part of an asymmetrically shaped superbubble created by stellar activity in the Scorpius-Centaurus Association (SCA) interacting with the interarm region around the Sun. This model is in contrast to a model where the Loop I (or North Polar Spur) and Local Bubble are due to two independent supernova events (e. g. Davelaar *et al.* 1980, section 4 this paper). In the picture in this section, the Local Bubble was sculpted by stellar winds and supernova explosions associated with epochs of star formation in the SCA over the past 15 million years. The superbubble formed thereby would have expanded freely into, and evacuated, the interarm region around the

Sun where average interstellar densities were low. The evacuated interarm region is what is known as the Local Bubble; it also corresponds to the interior of Gould's Belt (Stothers & Frogel, 1974), which itself may be an artifact of spiral arm structure.

In the asymmetric superbubble expansion model (Frisch, 1981, 1995) the Sun is currently located at the edge of, or in, a superbubble shell fragment about 4,000,000 * years old. This age, derived from the kinematics of the interstellar gas surrounding the Sun combined with superbubble shell models (which give shell radius and velocity as $\sim t^{0.6}$ and $\sim t^{-0.4}$, t is the shell age), corresponds to the age of the Upper Scorpius subgroup of the SCA. The Loop I supernova remnant (also known as the "North Polar Spur", after the radio continuum feature associated with it) may have been formed by the supernova event which released ζ Oph as a runaway star $\sim$ 1.1 10^6 years ago (Blaauw, 1961). The low density region surrounding the Sun is referred to as an interarm region even though the exact configuration of the Milky Way spiral arm structure at the solar location is poorly defined. This low density interarm region is well documented in reddening studies (Lucke, 1978), and ultraviolet data (Frisch & York, 1986), and connects to the "Puppis window" at l$\sim$ 240°, which is shown by 21-cm data to correspond to one of the lowest column density sightlines in the galactic disk (Stacy & Jackson, 1982). The volume of very low density gas around the Sun may have been partially reheated by emission from the Vela supernova explosion or its predecessor $\geq$10,500 years ago (see Frisch and Slavin, this volume).

The Local Bubble concept is based on multiwavelength data. Observations of the soft X-ray background (energy < 0.2 keV) suggest the Sun is located inside a region containing hot X-ray emitting plasma such as expected within a supernova remnant (McCammon *et al.*, 1983; Snowden *et al.*, 1990a). The distribution of interstellar absorption lines (Frisch & York, 1983; Paresce, 1984) and extreme ultraviolet point sources (Diamond *et al.*, 1995), reveal an absence of widespread cool gas within 60–80 parsecs of the Sun, although subparsec sized diffuse (or dense) interstellar clouds are seen nearby (Munch & Unsold, 1962; Welty, Morton, & Hobbs, 1996; Kerp *et al.*, 1993; Frisch, this volume). The complex of interstellar clouds surrounding the solar system reveal abundance and kinematic patterns indicating the Sun is located inside either shocked interstellar gas or a fragment of a superbubble shell where interstellar grain destruction has occurred, with the Scorpius-Centaurus Association driving this flow (Frisch, 1981). These combined data led to the vaguely defined concept of the Local Bubble.

6.2. Scenario

Over the past 15 million years, three epochs of star formation have occurred in the Scorpius-Centaurus Association – the formation of the Upper Scorpius (4-5 Myr ago), Upper Centaurus-Lupus (14-15 Myr ago), and Lower Centaurus-Crux (11-12 Myr ago) subgroups centered about 150 pc from the Sun between galactic longitudes 270° to 0°. The flow of interstellar gas past the solar system gives a gas velocity of -18 to -21 km s^{-1} from galactic coordinates l$\sim$335° and b$\sim$-2°. ** This upwind direction indicates gas expanding from the SCA. In addition, iron and calcium are an order of magnitude more abundant in the cloud surrounding the solar system than more distant cooler diffuse clouds, suggesting the interstellar gas flowing past the Sun has been shocked, destroying interstellar

* There is a misprint in Frisch, 1995, where this age is mistakenly listed as 400,000 years.

** This is the flow vector that is obtained after converting from the observed velocity and upwind direction (in the rest frame of the Sun) into the local standard of rest frame (the velocity frame of reference in which the velocity vectors of nearby stars average to zero). To convert from observed (i. e. heliocentric) velocities to the LSR velocity frame, it is assumed that the Sun is moving with a velocity of 16.5 km s^{-1} towards the direction l=53° and b=25°.

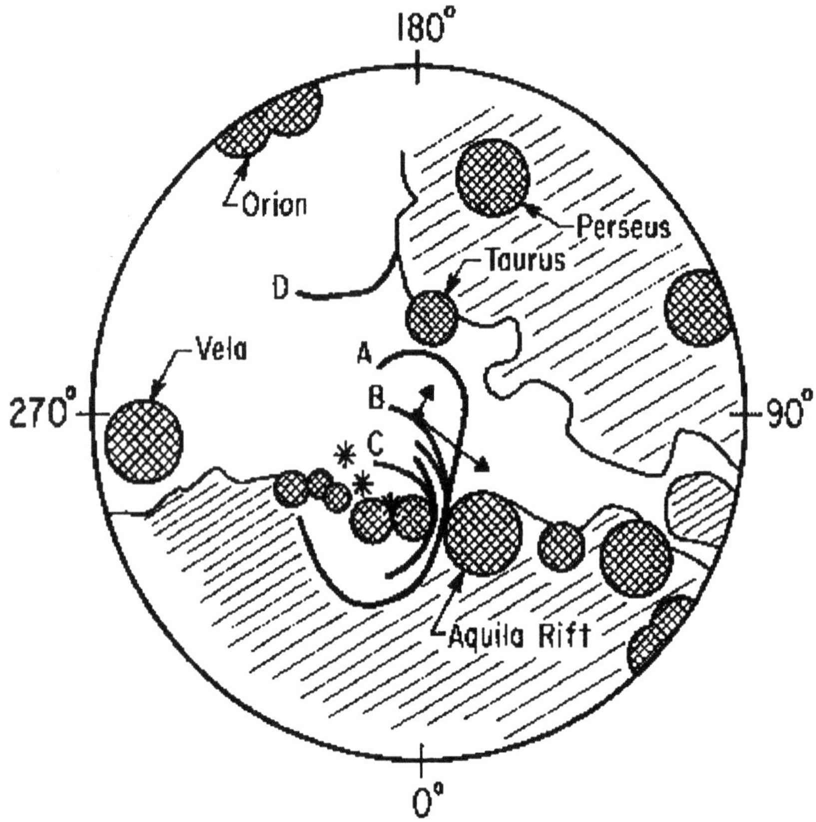

Figure 5. Cartoon illustrating bubble boundaries projected onto the plane of the galaxy, as sculpted by successive epochs of star formation in the Scorpius-Centaurus Association (SCA) expanding preferentially into the low density interarm region around the Sun (from Frisch, 1996). The radius of the figure is 500 pc, and the Sun is located at the intersection of the two arrows. The long arrow gives the direction of solar motion, while the short arrow gives the direction of flow of the interstellar cloud surrounding the solar system in the local standard of rest. The higher densities of diffuse gas seen within 50 pc of the Sun between galactic longitudes 20° – 90° represent diffuse gas ablated from the parent molecular cloud complex, including the Aquila Rift, when impacted by the combined stellar winds and supernova shock fronts from star creation and destruction in the SCA. Spiral arm structure is shown by the CO molecular clouds (cross-hatched) (Dame *et al.* 1991). Superbubble shell fragments due to epochs of star formation in SCA within the last 15 million years ago are also shown. A is the outermost boundary of the SCA superbubble shell, with N(H)> 7×10^{19} cm^{-2}. B is the ~4,000,000 year shell fragment around the Sun, and C is the ~250,000 year old Loop I supernova remnant shell. D is the superbubble shell corresponding to Orion's Cloak.

dust (Frisch, 1981). These facts led to the conjecture that the Sun was located near the edge of a supernova remnant associated with the SCA which has expanded to the solar vicinity (Frisch, 1981).

This scenario was made quantitative using models for superbubble shell formation and the fact the superbubble must have expanded asymmetrically into the low density interarm region around the Sun. De Geus (1991) has studied superbubble shell formation in SCA under the assumption that the shells expanded symmetrically into a homogeneous moderately dense ISM, driven by SCA winds and supernova. His scenario was modified by Frisch (1995), to include the much lower densities that would be encountered as the superbubble expanded in the direction perpendicular to galactic rotation towards the interarm region. Shell expansion models (MacLow & McCray, 1988) were combined with estimates of the kinetic energy input by subgroup star formation and destruction (De Geus, 1991), and data on the ISM surrounding the Sun.

Expansion parallel to the direction of galactic rotation would have been inhibited by high ambient interstellar densities (n$\sim$ 10^4 cm^{-3}) in the parent molecular clouds and in the Aquila Rift molecular gas, $\sim$100-150 pc from the Sun between l$\sim$ 20° and l$\sim$ 40°. In this direction where high preshock densities were encountered, the superbubble shell expanded 110 pc. Bubble expansion stopped when the shock front reached the dense gas of the Aquila Rift, as is seen easily when the narrow (0.3° wide) low latitude extension of the North Polar Spur (the radio continuum associated with Loop I), extending from (l, b)=(24°, 6°) to (22°, 3°) (Sofue & Reich, 1979), is compared with the boundary of the Aquila Rift gas, as traced out by CO (Dame &Thaddeus, 1985). The gas ablated from the Aquila Rift molecular cloud provides a simple explanation as to why the Loop I 21 cm filaments have higher column densities between l=0° – 60° than between l= 270° – 330°.

Expansion perpendicular to galactic rotation would have encountered the interarm region. The superbubble formed by the initial epoch of star formation 14-15 million years ago swept past the solar location, through preshock interarm gas with average density $\sim$ 0.3 cm $^{-3}$, forming the cloud seen today about 70 pc away towards Orion and giving a total shell radius of 210 pc (including the 140 pc distance to the Upper Scorpius Lupus subgroup). The interarm preshock density is constrained by the observed densities in the swept up shell from the expansion. In this way, the SCA superbubble has burst out into the adjacent interarm region.

The second epoch of star formation 11-12 million years ago would have expanded through, and reheated, the evacuated cavity of the original superbubble, and merged with the shell gas seen in the anti-center hemisphere. Between the second and third epochs of star formation, there was a 5-7 million year lull in stellar activity allowing residual molecular cloud material in the star forming regions to evaporate and fill one end of the evacuated cavity with diffuse gas. For instance, if 5% of an initial 50 pc radius volume were filled with 2 pc diameter clouds with initial density 10^4 cm^{-3}, within 4 million years enough gas would have evaporated from these embedded clouds to fill one end of the cavity with a gas of average density 2 cm^{-3} (Frisch, 1995). The superbubble associated with the Upper Scorpius subgroup would have initially evolved in this denser evaporated material, rather than an evacuated bubble from earlier subgroup formation.

Figure 5 shows the configuration of the superbubble shells as sculpted by star formation activity in SCA combined with the interarm location of the Sun. In the figure, the higher densities of diffuse gas seen within 50 pc of the Sun between galactic longitudes 20° and 90° represent diffuse gas ablated from the parent molecular cloud complex, including the Aquila Rift, when impacted by the combined stellar winds and supernova shock fronts from star creation and destruction in the SCA. This higher density gas is in contrast with the low density gas, 0.1 cm^{-3}, seen between galactic longitudes 180° and 270° (e. g. Frisch, this volume). Also, the expansion of the Loop I superbubble has been 'stopped' by the dense molecular cloud of the Aquila Rift, at 100–200 pc, from which it is ablating material (Frisch, 1995). The boundaries of the original superbubble event (A), the shell at the solar location (B), and the boundary of Loop I (C) are shown. The region between A and C is generally referred to as the Local Bubble. D shows the superbubble associated with Orion and known as "Orion's Cloak".

6.3. Discussion

The scenario outlined in this section is in contrast with other views of the Local Bubble origin, which attribute the Local Bubble origin to an event not connected with the Scorpius-Centaurus Association. The N(H)$\sim 10^{20}$ cm^{-2} boundary between the Local Bubble and Loop I bubble discussed by Egger (cf. section 5) corresponds to the shell fragment labeled C in the figure. In the scenario in this section, the Loop I boundary represents interstellar gas evaporated from molecular clouds embedded in the Scorpius-Centaurus Association which has been swept up by the explosive event forming Loop I (which is perhaps the same event which released the runaway star ζ Oph), rather than the collision of two superbubble shells. The total column density of the shell fragment around the Sun, labeled B in the figure, is less than 5 x 10^{19} cm^{-2} (Frisch, 1995), and would shadow only X-ray emission between distances 30 pc – 80 pc (where the shell of Loop I appears). The outermost boundary of the SCA superbubble shell, labeled A on the figure, is detected as cool clouds in several regions and has N(H)> 7 x 10^{19} cm^{-2}. The magnetic field embedded in the upwind interstellar gas, and parallel to the surface of the cloud surrounding the solar system (Frisch, this volume), is consistent with the magnetic field seen threading the Loop I superbubble shell fragments and traced out by both optical polarization and radio polarization measurements. However, inferring from field strengths derived elsewhere in the bubble using Zeeman splitting of the H I 21 cm line, suggests B=5.2 μG captured in the bubble fragment at the solar location (Frisch & York, 1991). This is larger than the value expected for the cloud surrounding the solar system (Frisch, this volume), and would support the view that the advancing superbubble shell is still upwind of the Sun, but within a few parsecs. *

In this context, it is important to note that this magnetic field would inhibit the evaporative mass loss at the conductive boundary of the local cloud surrounding the solar system, bringing the Slavin cloud model (Slavin, 1989) into agreement with EUVE data.

7. Conclusions

Understanding our local interstellar environment is not only a challenge per se, but bears important consequences for our picture of the ISM in general. Moreover, the interpretation of heliospheric observations critically depends on the physical conditions that ions, dust grains and cosmic ray particles encounter on their way to the solar system. We are still far from knowing the three dimensional distribution of gas (neutral and ionized) and its typical (if there is any) temperature and density. But we expect that the situation will improve in the near future thanks to more detailed observations, accompanied hopefully by more sophisticated modelling.

As a summary of our discussion in WG2 we can offer bad and good news. The bad news is, that there was not even a rough agreement among the participants on the origin of the LB, although it was conceded that the LB should be no longer regarded as an isolated phenomenon; but we are still at the stage of collecting ideas. Hence the good news is, at least in our view, that the Max-Planck-Institut für Extraterrestrische Physik will host a major conference on the subject of the Local Bubble in spring 1997, to which everyone, who is interested in this most exciting topic, is invited to participate.

* In Frisch (1995) this upwind advancing superbubble is called the 'squall line' gas, and the cloud surrounding the solar system, which is the preshock gas for the superbubble shell, is referred to as the 'local fluff'.

Acknowledgements

DB acknowledges support from the *Deutsche Forschungsgemeinschaft* (DFG) by a Heisenberg Fellowship. PCF would like to thank the National Aeronautics and Space Administration for its support of her research. The ROSAT project is supported by the German Bundesministerium für Bildung, Wissenschaft, Forschung und Technologie (BMBF/DARA) and the Max-Planck-Gesellschaft (MPG).

References

Blaauw, A.: 1961, Bulletin Astronomical Institutes of the Netherlands **15,** 265.
Bloch, J.J.: 1988, PhD Thesis, University of Madison, Wisconsin.
Bowyer, S. and Malina, R.F.: 1991, in *Extreme Ultraviolet Astronomy,* R.F. Malina and S. Bowyer (eds.), (New York: Pergamon), pp. 397.
Breitschwerdt, D. and Schmutzler, T.: 1994, *Nature* **371**, 774.
Centurion, M. and Vladilo, G.: 1991, *Astrophysical Journal* **372**, 494.
Cheng, K.P. and Bruhweiler, F.C.: 1990, *Astrophysical Journal* **364**, 573.
Cox, D.P. and Anderson, P.R.: 1982, *Astrophysical Journal* **253**, 268.
Cox, D.P. and Reynolds, R.J.: 1987, *Annual Review of Astronomy and Astrophysics* **25**, 303.
Dame, T. M. and Thaddeus, P.: 1985, *Astrophysical Journal* **297**, 751.
Dame, T. M., Ungerechts, H., Cohen, R. S., de Geus, E. J., Grenier, I. A., May, J., Murphy, D. C., Nyman, L. A., and Thaddeus, P.: 1987, *Astrophysical Journal* **322**, 706.
Davelaar, J., Bleeker, J.A.M., and Deerenberg, A.J.M.: 1980, *Astronomy and Astrophysics* **92**, 231.
De Geus, E. J.: 1991, *Astronomy and Astrophysics* **262**, 258.
Diamond, C. J., Jewell, S. J., and Ponman, T. J.: 1995, *Monthly Notices of the RAS* **274**, 589.
Dickey, J.M. and Lockman, F.J.: 1990, *Annual Review of Astronomy and Astrophysics* **28**, 215.
Egger, R.J.: 1995, in *The Physics of the Interstellar and the Intergalactic Medium*, A. Ferrara, C. Heiles, C.F. McKee, and P. Shapiro (eds.), Pub. Astron. Soc. of the Pacific Nr. 80.
Egger, R.J. and Aschenbach, B.: 1995, *Astronomy and Astrophysics* **294**, L25.
Freyberg, M.J.: 1994, PhD Thesis, LMU München (*in German*).
Freyberg, M.J.: 1996, in *The Physics of Galactic Halos*, H. Lesch, R.-J. Dettmar, U. Mebold, R. Schlickeiser (eds.), *in press.*
Freyberg, M.J. *et al.*: 1996, in preparation.
Frisch, P. C.: 1981, *Nature* **293**, 377.
Frisch, P. C.: 1995, *Space Science Reviews* **72**, 499.
Frisch, P. C.: 1996, in preparation.
Frisch, P. C. and York, D. G.: 1983, *Astrophysical Journal* **271**, L59.
Frisch, P. C. and York, D. G.: 1986, in R. Smoluchowski, J. N. Bahcall, M. Matthews (eds.), *The Galaxy and the Solar System*, (Tucson: Univ. Arizona Press) pp. 83.
Frisch, P.C. and York, D.G.: 1991, in *Extreme Ultraviolet Astronomy,* R.F. Malina and S. Bowyer (eds.), (New York: Pergamon), p. 322.
Fruscione, A., Hawkins, I., Jelinsky, P., and Wiecigroch, A.: 1994, *Astrophysical Journal, Supplement Series* **94**, 127.
Graham, J. R., Levenson, N. A., Hester, J. J., Raymond, J. C., and Petre, R.: 1995, *Astrophysical Journal* **444**, 787.
Herbstmeier, U., Mebold, U., Snowden, S.L., Hartmann, D., Burton, W.B., Moritz, P., Kalberla, P., and Egger, R.J.: 1995, *Astronomy and Astrophysics* **298**, 606.
Jelinsky, P., Vallerga, J., and Edelstein, J.: 1995, *Astrophysical Journal* **442**, 653.
Jenkins, E.B. and Meloy, D.A.: 1974, *Astrophysical Journal, Letters to the Editor* **193**, 121.
Kahn, F.D. and Breitschwerdt, D.: 1989, *Monthly Notices of the RAS* **242**, 209.
Kerp, J., Herbstmeier, U., and Mebold, U.: 1993, *Astronomy and Astrophysics* **268**, L21.
Lucke, P. B.: 1978, *Astronomy and Astrophysics* **64**, 367.
MacLow, M.-M. and McCray, R.: 1988, *Astrophysical Journal* **324**, 776.

McCammon, D., Burrows, D.N., Sanders, W.T., and Kraushaar, W.L.: 1983, *Astrophysical Journal* **269**, 107.
McCray, R. and Kafatos, M.: 1987 *Astrophysical Journal* **317**, 190.
Munch, G. and Unsold, A.: 1962, *Astrophysical Journal* **135**, 711.
Paresce, F.: 1984, *Astronomical Journal* **89**, 1022.
Reynolds, R.J.: 1990, *Astrophysical Journal* **348**, 153.
Sanders, W.T. *et al.*: 1993, in *Proc. SPIE*, **2006**, 221.
Sanders, W.T., Edgar, R.J., and Liedahl, D.A.: 1996, in *Röntgenstrahlung from the Universe*, H.U. Zimmermann, J.E. Trümper, H. Yorke (eds.), MPE-Report 263, p. 339.
Slavin, J.D.: 1989, *Astrophysical Journal* **346**, 718.
Snowden, S.L., Cox, D.P., McCammon, D., and Sanders, W.T.: 1990, *Astrophysical Journal* **354**, 211.
Snowden, S.L., Schmitt, J.H.M.M., and Edwards, B.C.: 1990, *Astrophysical Journal* **364**, 118.
Snowden, S.L., Mebold, U., Hirth, W., Herbstmeier, U., and Schmitt, J.H.M.M.: 1991, *Science* **252**, 1529.
Snowden, S.L., Hasinger, G., Jahoda, K., Lockman, F.J., McCammon, D., and Sanders, W.T.: 1994, *Astrophysical Journal* **430**, 601.
Snowden, S.L., Freyberg, M.J., Plucinsky, P.P., Schmitt, J.H.M.M., Trümper, J., Voges, W., Edgar, R.J., McCammon, D., and Sanders, W.T.: 1995, *Astrophysical Journal* **454**, 643.
Snowden, S.L. *et al.*: 1996, in preparation.
Sofue, Y. and Reich, W.: 1979, *Astronomy and Astrophysics, Supplement Series* **38**, 251.
Stacy, J. G. and Jackson, P. D.: 1982, *Astronomy and Astrophysics, Supplement Series* **50**, 377.
Stothers, R. and Frogel, J. A.: 1974, *Astronomical Journal* **79**, 456.
Vallerga, J. and Welsh, B.: 1995, *Astrophysical Journal* **444**, 202.
Warwick *et al.*: 1993, *Monthly Notices of the RAS* **262**, 28.
Weaver, H.: 1979, in *Proc. of IAU Symp. No. 84*, p. 295.
Welty, D. E., Morton, D. C., and Hobbs, L. M.: 1996, *Astrophysical Journal*, in press.
Yoshioka, S. and Ikeuchi, S.: 1990, *Astrophysical Journal* **360**, 352.

THE INTERSTELLAR GAS FLOW THROUGH THE HELIOSPHERIC INTERFACE REGION

HANS J. FAHR
Institut für Astrophysik und Extraterrestrische Forschung der Universität Bonn, Auf dem Hügel 71, D–53121 BONN (F.R.Germany)

Abstract. The relative motion of the solar system with respect to the ambient interstellar medium forms a plasma interface region where the eventually subsonic, interstellar and solar wind plasma flows adapt to each other. In this region ahead of the solar system magnetohydrodynamically perturbed plasma flows are formed which, however, can be penetrated by interstellar neutral atoms at their approach towards the inner heliosphere. Thereby the distribution function of neutral interstellar gas species by means of charge exchange processes in the heliosphere attain gas-specific imprints from the perturbed moments in this plasma region. In recent years one has become interested in the influence of this interface plasma on the helium-to-oxygen-to-hydrogen ratios since observational facts from pick-up ion and anomalous cosmic ray data have meanwhile become available shrinking down the inaccuracy in these ratios to fairly low numbers. The aim thus is to study these ratios as they result from alternative forms of interface structures under debate at present and then to identify the best-fitting interface model. The fact is stressed in this article here that a more correct description of the interface needs a careful consideration of the magnetic fields and the plasma temperature anisotropies which are involved in the actually prevailing counterflow configuration.

1. Introductory Sketch of the solar environment

The thermodynamic state of the VLISM (Very Local Inter–Stellar Medium) unfortunately even at present is only poorly known (see e.g. Cox and Reynolds, 1987, Frisch, 1990,1995, Reynolds, 1990, Lallement, 1993). It is even unclear whether or not a quasistationary near–equilibrium state of the VLISM can be expected which would allow to be described by the Saha equation. In case of a supernova shock passing over the local interstellar region within the past 10^5 to 10^6 years the VLISM could be rather expected to represent a non–stationary, non–equilibrium state which still now undergoes continuing temporal changes in temperatures and in the fractional degrees of ionization of the different atomic species. Then of course without exact knowledge of the time and of the nature of this shock event no theoretical expectations concerning fractional ionizations in the present LISM could be derived from any theoretical basis at all.

The thermodynamical condition of a cloudy interstellar medium being subject to periodically repeating supernova shock events has been studied by McKee and Ostriker (1977) and by McCray and Snow (1979). Different from the earlier result by Field et al.(1969), predicting a two-phase interstellar medium on the pure basis of pressure equilibrium at the phase boundaries, they instead propagated a multi-phase interstellar medium with essentially four co-existing ISM phases, i.e. the cold phase, the warm neutral phase, the warm ionized phase, and the hot ionized phase. The latter phase with temperatures of between 10^5 K $\leq T \leq 10^6$ K, densities of $\cong 10^{-2}$ cm^{-3}, and ionization degrees of $\xi \cong 1.0$ should have a spatial filling factor of $f \cong 0.8$, whereas the other phases if taken together should only fill less than 20 percent of the interstellar space (i.e. $f \leq 0.2$).

A few years ago it was found that the X-ray source Geminga is in fact an X-ray pulsar with a rotation period of 0.23 sec representing the nucleus of an earlier O-type star which exploded by a supernova event. Taking the deceleration rate in the rotation period of this

Space Science Reviews **78**: 199–212, 1996.

pulsar and extrapolating back to the relativistic minimum pulsar period it turns out that the Geminga-associated supernova event took place about 340000 years ago (Teske, 1993), when Geminga, retraced on the basis of its present peculiar motion over this period, was only about 300 lightyears away from the solar system. The associated supernova shock passage carved a huge region of complete ionization into the nearby interstellar medium and may be responsible for the so-called "local bubble" which can nowadays clearly be identified by interstellar absorption line data from IUE (see e.g. Paresce, 1984). Taking this as indication that a rapid non-equilibrium ionization of the local medium only has taken place about 300000 years ago will then clearly point to a high probability that at present no stationary time–independent conditions can be expected since interstellar recombination periods are of the order of $\geq 10^7$ years. Unless recombination to a local photo-ionization equilibrium had taken place till now, we had to face the fact that the LISM conditions do temporally change even today.

In view of this puzzling situation the easiest would be to use the Saha equation for a determination of the VLISM ionization state. This equation determines a degree of ionization ξ connected with the minimum of the thermodynamical LISM Gibbs potential, i.e. $\partial G/\partial\xi = 0$, and only requires the knowledge of two interstellar parameters, namely temperature and total density $N = n_H + n_H^+$ of the LISM. Assuming that at least the ionization of the VLISM hydrogen is close to such an equilibrium state, one would gain the ratio $(n_{\mathrm{H},\infty}^+/n_{\mathrm{H},\infty}) = \xi_{\mathrm{H},\infty}$ simply by use of Saha's equation. Since electrons can be assumed to mainly recombine with H^+–ions and to have about the same density as H^+–ions, i.e. $n_{\mathrm{H}}^+ = n_e; N_\infty = n_{e,\infty} + n_{\mathrm{H},\infty}$, the adequate Saha equation could then be written in the form (see Rybicki and Lightman, 1979):

$$\frac{n_e}{N} = \frac{\xi_{\mathrm{H}}}{1+\xi_{\mathrm{H}}} = \frac{1}{2D_e^3 N}\left[\left[1 + 4ND_e^3 \exp\left(\frac{E_{\mathrm{H}}}{KT_{\mathrm{p},\infty}}\right)\right]^{\frac{1}{2}} - 1\right]\exp\left(\frac{E_{\mathrm{H}}}{KT_{\mathrm{p},\infty}}\right) \qquad (1)$$

where E_{H} = 13,6 eV is the ionization energy of hydrogen and where D_e is the thermal DeBroglie wavelength of the electrons given by (Pathria, 1973):

$$\mathrm{D_e} = \frac{\mathrm{h}}{\sqrt{2\pi \mathrm{m_e} \mathrm{K} \mathrm{T}_{\mathrm{p},\infty}}}. \qquad (2)$$

Here h and m_e are the Planck constant and the electron mass, respectively.

For specific values of N_∞ and $T_{\mathrm{p},\infty}$ one can then calculate with Eq.(1) the quantity: $n_{\mathrm{H}}^+ \cong n_e \cong \xi_{\mathrm{H}} N/(1+\xi_{\mathrm{H}})$. As a clear hint it thereby comes out that, based on the most believed LISM parameters, i.e. $N_\infty \cong 0.1\ \mathrm{cm}^{-3}$; T_∞ = 8000 K, a value for the fractional ionization of hydrogen of less than $\xi_{\mathrm{H}} = 10^{-2}$ would be obtained yielding LISM electron densities of $10^{-3}\ \mathrm{cm}^{-3}$. This contrasts with many good indications for much higher local electron densities as mentioned e.g. by Frisch (1990) or Reynolds (1990). For gases which are strongly coupled to hydrogen by means of charge exchange processes, like oxygen or nitrogen, a local charge exchange equilibrium with protons , rather than an ionization equilibrium, may then be required leading to a fractional ionization, for instance of oxygen, given by $\xi_{\mathrm{O},\infty} = \xi_{\mathrm{H},\infty} < \sigma_{\mathrm{ex}}(\mathrm{H}^+,\mathrm{O})) > / < \sigma_{\mathrm{ex}}(\mathrm{O}^+,\mathrm{H}) >$, where the quantities σ_{ex} are charge exchange cross sections of the respective collision partners, and the <>–brackets denote averaging over collision partner velocities. Based on cosmic abundances α_j (see Cameron, 1973) one then obtains with the total density N_∞, and with $\xi_{\mathrm{O},\infty}$ and α_{O}, the absolute density values of neutral oxygen atoms and ions in the VLISM.

One should, however, envisage the probability that the state of the nearby ISM may be characterized as a transition phase extending between two thermodynamically very different interstellar gas phases of the types discussed already by McKee and Ostriker (1977). Just now the solar system appears to move through a local interstellar cloud of a cross dimension of about 30 pc (Frisch and York, 1983). As a consequence of the mediated, thermal contact of this cloud to the surrounding tenuous and 10^6 K -hot, interstellar HII-plasma one must expect the presence of strong temperature gradients at the periphery of this cloud and consequently of non-equilibrium ionization conditions (Slavin, 1989, Böhringer and Hartquist, 1987). It can therefore be envisaged that the degree of ionization in this medium systematically decreases from the periphery towards the center of the Local Interstellar Cloud (LIC). This expresses the fact that consequently the ionization state of the "VLISM" very much depends on where the solar system at present is located with respect to the center of the LIC. Balbus (1986) and Böhringer and Hartquist (1987) have shown quantitative calculations of stationary profiles for such interstellar transition regions between hot and cool ISM phases. They even include in their calculations magnetically controlled heat conduction, UV– and X–ray radiation transport, and evaporative gas flows emanating from the surface of the central dense cloud. Looking at their results it nevertheless becomes obvious that for the specific situation of the VLISM nothing reliable can be concluded. It may solely be revealed in these calculations that the actual ionization state, at whatever position in such a transition region, is mainly determined by a local photoionization equilibrium governed by the local radiation field.

Based on this confidence estimates of the local photoionization state have been carried out e.g. by Frisch (1990) and by Reynolds (1990). They have attempted to calculate average photoionization equilibria considering the Lyman–continuum emission of nearby White Dwarf stars and to determine electron densities for equilibria at which recombination would just be canceled by Lyman–photoionizations. Hereby the results obtained by Reynolds (1990) who in addition to the Lyman–continuum radiation takes into account a diffuse X–ray background for the calculation of actual photoionization rates are fairly different from those of Frisch (1990). These discrepancies become even more pronounced if less conservative assumptions for the local X–ray background are taken(Bloch et al., 1986). In addition the results of the above authors are very sensitive to the actual LISM temperature T_∞ used for their estimates (12000 K $\geq T_{p,\infty} \geq$ 7000 K). Hence also on this basis no clear decision can be made which LISM ionization state can be expected. Especially no information on the fractional ionization of the other elements, besides hydrogen, can be gained. Nevertheless, here we shall describe the fractional ionization $\xi_H = n_H^+/n_H$ of hydrogen by use of the formula deduced by Reynolds (1990):

$$\xi_H = \left(\frac{\Gamma}{A(T_\infty)n_H}\right)^{\frac{1}{2}} \tag{3}$$

where A denotes the temperature–dependent recombination coefficient which for a temperature of T_∞ = 8000 K is given by $A(8000) = 3.1\ 10^{-13}$ cm^3/s. The quantity Γ denotes the photoionization frequency determined by the actual, local radiation field. Taking the radiation from known White Dwarf stars Reynolds determines $\Gamma = \Gamma_{min} = 2.5\ 10^{-16}$ s^{-1} which may be considered as a minimum value since no X-ray background is included. Speculating on contributions to the photoionization from the soft X–ray background Reynolds discusses values for Γ which are higher by factors of between 10 to 50 (see also Cheng and Bruhweiler, 1990, or Davidsen, 1993). One may decide to use $\Gamma_{max} = 10\Gamma_{min}$ as a maximum value for the photoionization frequency in this paper. From Eq. 3 one then obtains with A = A(8000):

$$\xi_H = \frac{n_H^+}{n_H} = \xi^* \sqrt{\frac{0.1}{n_H}}, \text{ with} : n_H = \frac{N_H}{(1+\xi_H)}\ ; n_H^+ = \frac{N_H \xi_H}{(1+\xi_H)} \tag{4}$$

where N_H denotes the total H–density, and n_H^+ and n_H are the densities of ionized and neutral hydrogen atoms. The value for the resulting ξ^* is then either given by: $\xi^* = \xi_{min} = 0.09$, or by: $\xi^* = \xi_{max} = 0.3$.

As already stated earlier, since the charge exchange rates for VLISM oxygen and nitrogen atoms are higher roughly by factors of 10^4 compared to corresponding photoionization rates, in these cases rather a charge exchange equilibrium than a photoionization equilibrium must be established. Since O–atoms mainly exchange their charge with H^+–ions the following charge exchange equilibrium thus is required at the unperturbed VLISM (for details see Fahr et al., 1995):

$$\left(\frac{n_0^+}{n_0}\right) = \xi_0 = \xi_H \frac{< \sigma_{ex}^{OH^+} >}{< \sigma_{ex}^{HO^+} >}, \tag{5}$$

where the quantities in <>–brackets give the velocity–averaged, relevant O–H charge exchange cross sections which according to Stebbings et al. (1977) then finally lead to the result: $\xi_0 \cong (8/9)\xi_H$.

2. Interface properties and imprints to the neutral LISM gases

The general phenomenon of the interface action on neutral LISM gas atoms consists in a coupling of the locally perturbed plasma properties in the interface region to the neutral gas medium by means of charge exchange induced changes of the velocity moments of the "neutral" distribution functions. Expressed in brief, this means that the deviation of the LISM plasma flow around the heliopause due to the magnetohydrodynamical interaction with the solar system is communicated to the neutral gas atoms, even though the latter do not interact hydrodynamically. At large solar distances one may expect to find a symbiotic state of coexistence between neutral and ionized species. This means that bulk velocities and temperatures of both species due to rapid charge exchanges ("rapid" with respect to all periods of competing temporal changes) can be expected to be equal, i.e. $T_{i\infty} = T_{n\infty} = T_\infty$. Surely for an unmagnetized LISM, but even in the presence of sufficiently homogeneous and stationary LISM magnetic fields, one can also safely adopt that these temperatures are isotropic, i.e. $T_{\infty\|} = T_{\infty\perp}$. The question is more complicate what concerns the LISM electron temperatures since the problem arises whether these electrons originating from photoionization by EUV photons may have the time to assimilate to the ion temperature. Such an assimilation can occur via Coulomb collisions of these photoelectrons with protons within a period of:

$$\tau_e \cong \left[n_{i\infty} < \sigma_e c_e > \frac{2 m_e m_p}{(m_e + m_p)^2} \right]^{-1} \cong 10^{11} \text{sec}. \tag{6}$$

where σ_e and c_e are the Coulomb cross section and the velocity of electrons relative to protons, respectively, and the LISM ion density has been adopted here with $n_i = 10^{-2}$ cm^{-3}. It thus turns out that the Coulomb assimilation time is comparable with the charge exchange time τ_{ex}, that means it is, however, about four orders of magnitude smaller than the photoionization times calculated by Reynolds (1990). This means that in the unperturbed LISM photoelectrons should have ample time to assimilate their temperatures to that of the protons, i.e. $T_{i\infty} = T_{e\infty} = T_\infty$. All what was said in the upper paragraph is invalidated when the LISM approaches the solar system to within a few 10^2 AU. Here the LISM protons, passing with a bulk velocity of about 20 km/s over 100 AU within a transit time $\tau_i = 10^9$ sec, experience substantial changes of their velocity moments within times small compared to both τ_e and τ_{ex}. Thus neither the temperature of the electrons nor that of the neutral atoms will strictly follow the proton temperature in this perturbed region.

Arguing within the frame of a Parker-type interface model (Parker, 1963) one obtains the proton bulk velocity from a stream function potential which on the symmetry axis (LISM wind axis = z–axis) describes the deceleration of the LISM plasma advancing towards the heliopause. Due to Bernoulli's law this deceleration is connected with a temperature increase given by (see Axford, 1972, or Fahr and Neutsch, 1982):

$$T_i(r,z) = \frac{m_p}{K}\left[\frac{1}{4}V_\infty^2 + \frac{KT_\infty}{m_p} - \frac{1}{4}V^2(z,r)\right], \tag{7}$$

where $T_i = T_e = T$ has been assumed, and where r and z are the radial and axial distance of a space point in the LISM, and where V is the local flow velocity given by (see Fahr et al., 1990):

$$V^2(r,z) = V_\infty^2\left[1 + \frac{2\bar{z}}{\bar{r}^3} + \frac{1}{\bar{r}^4}\right], \tag{8}$$

where the bars on top of the space coordinates mean normalizations by the axial stand-off distance L of the heliopause and where axial distances in the upwind hemisphere are counted negative. As one can see at the stagnation point r = -z = -L one obtains $V^2 = 0$ and accordingly with Equ. 7 the highest LISM proton temperature $T_{is} = T_{i\infty} + (1/4)m_pV_\infty^2/K$.

The electron temperature in fact may not have the chance to follow this temperature increase within the time τ_i and thus will only adapt with a substantial phase lag, i.e. $T_e < T_i$. The same happens to the neutral atoms which are not reflecting the proton temperature increase in phase. In addition the LISM plasma at the advance towards the heliopause may be forced to develop a temperature anisotropy if magnetic forces are operating. The adaptation of the neutral atoms or electrons to this proton temperature anisotropy will be incomplete. Electrons may not at all assimilate since their internal relaxation periods, dominated by electron-electron collisions, are smaller than τ_i by about one order of magnitude. Neutral atoms, on the other hand, may only gradually adapt to the proton anisotropy.

Assuming an essentially collision-free magnetized LISM plasma with dominating magnetic field energies, one may be advised to apply the CGL-theory here (Chew et al., 1954). With the assumption of a conservation of the proton magnetic moment, this CGL-theory derives at least for heat-conduction-free plasmas the so-called "double adiabatic equations of state" which are given by:

$$\frac{\partial}{\partial s}\left(\frac{P_{i\perp}}{n_i B}\right) = 0 \text{ and} : \frac{\partial}{\partial s}\left(\frac{P_{i\parallel}B^2}{n_i^3}\right) = 0. \tag{9}$$

where $P_{i\perp}$ and $P_{i\parallel}$ are proton pressures with respect to the direction perpendicular and parallel to the LISM field **B**, respectively, and where s is the streamline–, or fieldline–, coordinate, being identical in the CGL-approach since here $\mathbf{B} \circ \mathbf{V} = BV$ (i.e. strong magnetic LISM field; low β–value of the plasma: $\beta = 8\pi n_i k T_i / B^2 \leq 10^{-1}$).

For a crude orientation one can attempt to estimate the temperature anisotropies $A_i = T_{i\parallel}/T_{i\perp}$ occurring in a magnetized interface described on the basis of Parker's interface flow or a magnetized twin-shock interface as discussed by Baranov and Zaitsev (1995). In the submagnetosonic case for example, at the presence of LISM magnetic fields the Parker-type flow configuration could only be consistent if MHD-forces not taken into account in the original Parker flow concept would not cause additional forces deviating the flow, i.e. if $\mathbf{V} \times \mathbf{B} = 0$. One evident solution for Parker's flow, as well as for the twin–shock flow, then would be suggested in the form (see Fahr et al., 1988):

$$\mathbf{B} = \mathbf{B}_\infty(\mathbf{V}/V_\infty) = \sqrt{4\pi\rho_{i\infty}}(\mathbf{V}/M_{A\infty}), \tag{10}$$

where $M_{A\infty}$ is the LISM Alfvénic Mach number. Assuming the validity of Equ. (9) with Equs. (10) and (8) one could then derive the following behaviour of the proton anisotropy $A_i(z,r)$:

$$A_i(r,z) = A_{i\infty}(B_\infty/B)^3 = A_{i\infty}(V_\infty/V)^3 = A_{i\infty}\left[1 + \frac{2\bar{z}}{\bar{r}^3} + \frac{1}{\bar{r}^4}\right]^{-\frac{3}{2}}, \tag{11}$$

clearly revealing that close to the stagnation point the proton anisotropy shall degenerate into $A_i \Rightarrow \infty$, i.e. $T_i = T_{i\|}$; $T_{i\perp} \cong 0$, even if at large distances a value of $A_{i\infty} = 1$ can be adopted. This means that under the above conditions of a magnetized LISM plasma flow one has to face the fact that, especially for low β–values, the ion distribution function is anisotropic. The same can be concluded for the stagnation region in the magnetized twin–shock interface discussed by Baranov and Zaitsev (1995). One therefore has to face the fact that in the frame of a magnetized plasma interface high degrees of temperature anisotropies may occur in the regions of stagnating plasmas.

As a next step of clarification we shall try to answer the question how the distribution function f_n of neutral hydrogen atoms adapts to this proton anisotropy due to resonant charge exchange interaction with the LISM plasma. The answer to this question, at least in the form of how fast an adaptation to A_i occurs, can be given by a solution of the following Boltzmann equation:

$$\frac{\partial f_n(\mathbf{v},t)}{\partial t} = n_i f_i(\mathbf{v}) \int \sigma_{ex}(W')W' f_n(\mathbf{v}',t)d^3v' - n_i f_n(\mathbf{v},t) \int \sigma_{ex}(W')W' f_i(\mathbf{v}')d^3v' - \frac{[f_n(\mathbf{v},t) - f_{n0}(v)]}{\tau_n} \tag{12}$$

where a homogeneous plasma condition described by a time-independent, anisotropic bi-Maxwellian ion distribution is assumed given by:

$$f_i(\mathbf{v}) = \left(\frac{m_p}{2\pi K T_{i\|}}\right)^{\frac{3}{2}} A_i \exp\left[-\frac{m_p v^2}{2KT_{i\|}}(\cos^2\theta + A_i \sin^2\theta)\right], \tag{13}$$

and where a relaxation term according to concepts given by Bhatnager et al. (1954) has been used to describe the relaxation of the actual distribution function f_n to the associated mono-Maxwellian f_{n0} within a relaxation time τ_n due to atom–atom collisions. The charge exchange cross section $\sigma_{ex}(W')$ has been taken from Maher and Tinsley (1977), and $W' = |\mathbf{v} - \mathbf{v}'|$ is the relative velocity between ion and atom. Furthermore with the Van der Waals cross section for H-H–collisions given by Brinkmann (1970) the relaxation period evaluates to:

$$\tau_n = \frac{6.62\ 10^{10}}{n_n\sqrt{T_n}}\text{sec.}$$

Representing now also the function f_n as a bi-Maxwellian analogous to Equ. 13, however, with $f_n(T_{n\|}(t), A_n(t))$ starting with $A_n(t=0) = 1$, one can then calculate with the Boltzmann equation (12) the change in time of $T_{n\|}(t)$ and $A_n(t)$, if the proton anisotropy is kept constant, e.g. at $A_i = 2$. The result is shown in Figure 1 for different values of n_i and n_n. It is obvious that the adaptation to the proton anisotropy takes place over periods of between $1 \cdot 10^9$ to $8 \cdot 10^9$ sec and never within this period is complete. It is in fact the less complete the higher is the density n_n at constant n_i, or the higher is the ratio (n_n/n_i).

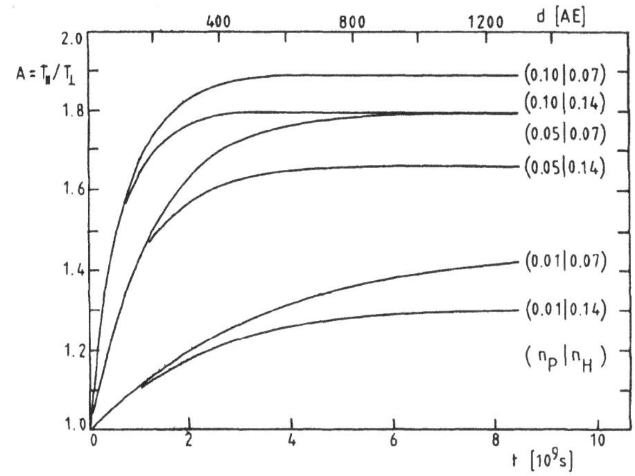

Figure 1. Shown for various combinations of LISM proton and atom densities, $n_p = n_i$ and $n_H = n_n$, is the temperature anisotropy A_n of neutral LISM hydrogen atoms adapting to (or moving with 23 km/s through) an LISM proton plasma with a constant anisotropy $A_i = 2.0$. The initial atom anisotropy is $A_n(t = 0) = 1.0$.

We shall now address the problem of the penetration of neutral LISM atoms through the perturbed plasma interface region. We shall treat this problem as a stationary one (i.e only convective, no local time derivatives are considered) assuming that full knowledge on the local plasma properties is available on the basis of one of the hydrodynamic plasma counterflow models which are in the literature at present (Parker, 1963, Baranov et al., 1979, Fahr and Neutsch, 1983, Fahr, 1986, Fahr et al., 1988, Matsuda et al., 1989, Baranov, 1990, Suess and Nerney, 1990, Baranov and Malama, 1993, Steinolfson et al., 1994, Scherer et al., 1994, Pauls et al., 1995, Fahr and Scherer, 1995, Khabibrakhmanov and Summers, 1996). Though this knowledge is of a hydrodynamic form we shall convert it here into a kinetic form by assuming that the hydrodynamic plasma properties, like density n_i, bulk velocity $\mathbf{V}_i$, temperature T_i, are representable as velocity moments of an underlying associated, shifted Tri–Maxwellian distribution function: $f_i = f_{i0}(\mathbf{v}, \mathbf{r}, n_i(\mathbf{r}), \mathbf{V}_i(\mathbf{r}), T_i(\mathbf{r}))$. The change of the distribution function $f_n(\mathbf{r}, \mathbf{v})$ is then adequately described by the following Boltzmann equation:

$$v\frac{\partial f_n(\mathbf{v},t)}{\partial s} = n_i f_i(\mathbf{v}) \int \sigma_{ex}(W')W' f_n(\mathbf{v}',\mathbf{r})d^3v' - n_i f_n(\mathbf{v},t) \int \sigma_{ex}(W')W' f_i(\mathbf{v}',\mathbf{r})d^3v' - \left[\frac{f_n(\mathbf{v},\mathbf{r}) - f_{n0}(v)}{\tau_n}\right] \tag{14}$$

For explanations of the details of this equation the reader is referred to Equ.(12). The differential operator on the LHS of Equ.(14) is the convective time derivative of the function f_n. The quantity "s" hereby denotes the line element along the dynamical trajectory of atoms with a velocity $\mathbf{v}$ which can be considered as constant in the interface since no explicit forces are influencing the neutral atoms.

Explicit solutions of the above integro-differential equation (14), however, with the neglect of the relaxation term, have been generated by publications in the past (see Ripken and Fahr, 1983, Fahr and Ripken, 1984, Fahr, 1986, 1991, Osterbart and Fahr, 1992, Fahr et al., 1995). For the sake of a broader generality, we here want to introduce a new method of solving the above equation by only requiring that throughout the interface region the function f_n can sufficiently well be represented by a generalized Maxwellian given in the following form (see Fahr and Himmes, 1990):

$$f_n(\mathbf{v},\mathbf{r}) = C_0 \exp\left[-(\mathbf{v}-\mathbf{V})\circ \overleftrightarrow{\mathrm{A}} \circ(\mathbf{v}-\mathbf{V})^t\right], \tag{15}$$

Here in Cartesian coordinates the local atom bulk velocity is denoted by $\mathbf{V} = \{U, V, W\}$. Three temperature components T_x, T_y, T_z are admitted with an arbitrary, local orientation of the Maxwellian velocity ellipsoid with respect to the coordinate axis described by the non-diagonal elements of the (3 × 3)-matrix $\overleftrightarrow{\mathrm{A}}$ which otherwise is symmetric and positively definite. The constant C_0 is determined such that the f_n is normalized to yield the atom density n_n. In our above notation we take the vector $\mathbf{g}^t$ as the transposed vector to $\mathbf{g}$.

For a matrix $A = (A_{ik})$ with i,k = x,y,z one then evolves the above distribution function f_n in the following, more explicit form:

$$f_n(v, r) = \exp[\, - \, (A_{xx}u^2 + A_{yy}v^2 + A_{zz}w^2 + 2A_{xy}uv + 2A_{xz}uw + 2A_{yz}vw + 2B_x u + 2B_y v + 2B_z w + C)], \tag{16}$$

where the following abbreviations were used:

$$B_x = A_{xx}U + A_{yy}V + A_{zz}W \tag{17}$$
$$B_y = A_{yx}U + A_{yy}V + A_{yz}W \tag{18}$$
$$B_z = A_{zx}U + A_{zy}V + A_{zz}W \tag{19}$$

and:

$$C = V_i A_{ik} V_k^t - \ln(C_0). \tag{20}$$

On the symmetry axis the non-diagonal elements of the matrix A_{ik} are vanishing, i.e. the velocity ellipsoid has its main axes coincident with the Cartesian axes, and hence the distribution function then is retained in the simplified form ($\overleftrightarrow{\mathrm{A}}$= main axis matrix):

$$f_n(u, v, w, z) = \exp[-(\; A_{xx}u^2 + A_{yy}v^2 + A_{zz}w^2 + 2B_x u + 2B_y v + 2B_z w + C)], \tag{21}$$

When the Boltzmann equation (14) now is applied to the above form of the distribution function f_n for points on the symmetry axis, i.e. x = y = 0, then one obtains the following differential equation as a result:

$$\begin{aligned} - \quad & f_n(u, v, w, z) \left[u\frac{\mathrm{d}}{\mathrm{dx}} + v\frac{\mathrm{d}}{\mathrm{dy}} + w\frac{\mathrm{d}}{\mathrm{dz}} \right] (A_{xx}u^2 + A_{yy}v^2 + A_{zz}w^2 \\ & +2B_x u + 2B_y v + 2B_z w + C) = n_i f_i(\mathbf{v}) \int \sigma_{ex}(W')W' f_n(\mathbf{v}', \mathbf{r}) \mathrm{d}^3 v' - \\ & n_i f_n(\mathbf{v}, t) \int \sigma_{ex}(W')W' f_i(\mathbf{v}', \mathbf{r}) \mathrm{d}^3 v' - \frac{[f_n - f_{n0}]}{\tau_n} \end{aligned} \tag{22}$$

Since on the symmetry axis one can expect to have vanishing gradients in the x– and y–directions, here the above partial differential equation thus reduces to:

$$\begin{aligned} - \quad & f_n(u, v, w, z)[A^*_{xx}u^3 + A^*_{yy}uv^2 + A^*_{zz}uw^2 + 2B^*_x u^2 + 2B^*_y uv + \\ & 2B^*_z uw + C^*] = n_i f_i(\mathbf{v}) \int \sigma_{ex}(W')W' f_n(\mathbf{v}', \mathbf{r}) \mathrm{d}^3 v' - \\ & n_i f_n(\mathbf{v}, t) \int \sigma_{ex}(W')W' f_i(\mathbf{v}', \mathbf{r}) \mathrm{d}^3 v' - \frac{[f_n - f_{n0}]}{\tau_n} \end{aligned} \tag{23}$$

where the quantities with an asterisk as superscript are derivatives with respect to z. In this equation we have kept 7 unknown quantities, including the density n_n hidden in the quantity

C which parametrizes the distribution function f_n. These quantities which vary with the axial coordinate z can be sufficiently well determined by solving a critical or minimal set of differential equations of the above type, e.g. by taking at least 7 individual, different velocity values $\{u, v, w\} = \{u_e, v_e, w_e\}$ which should be reasonably well distributed over the probability ranges defined by the distribution function f_n itself. These velocities thus are selected by 7 triples of normal random numbers $\{\epsilon_{1e}, \epsilon_{2e}, \epsilon_{3e}\}$ according to the following algebraic relations:

$$\begin{aligned} A_{xx}u_e^2 + 2B_x u_e &= -\ln(\epsilon_{1e}) \\ A_{yy}v_e^2 + 2B_y v_e &= -\ln(\epsilon_{2e}) \\ A_{zz}w_e^2 + 2B_z w_e &= -\ln(\epsilon_{3e}) \end{aligned} \tag{24}$$

With these individual velocity triples one obtains a system of 7 linear equations serving for the determination of the derivatives of all quantities which parametrize the distribution function.

$$\begin{pmatrix} <u^3> & <uv^2> & <uw^2> & <u^2> & <uv> & <uw> & <C> \\ <u^3> & <uv^2> & <uw^2> & <u^2> & <uv> & <uw> & <C> \\ <u^3> & <uv^2> & <uw^2> & <u^2> & <uv> & <uw> & <C> \\ <u^3> & <uv^2> & <uw^2> & <u^2> & <uv> & <uw> & <C> \\ <u^3> & <uv^2> & <uw^2> & <u^2> & <uv> & <uw> & <C> \\ <u^3> & <uv^2> & <uw^2> & <u^2> & <uv> & <uw> & <C> \\ <u^3> & <uv^2> & <uw^2> & <u^2> & <uv> & <uw> & <C> \end{pmatrix}_{1,\dots,7} \times \begin{pmatrix} A^*_{xx} \\ A^*_{yy} \\ A^*_{zz} \\ 2B^*_x \\ 2B^*_y \\ 2B^*_z \\ C^* \end{pmatrix}$$

$$= \begin{pmatrix} \mathrm{P}^+ - \mathrm{P}^- - \mathrm{P}^\mathrm{R} \\ \\ \\ \\ \\ \\ \\ \end{pmatrix}_{1,\dots,7}$$

where the terms in <>– brackets evidently mean expressions like, e.g.:

$$< u^3 >_e = u_e^3 f_n(u_e, v_e, w_e) \tag{25}$$

etc., and where the terms $|\mathrm{P}^+ - \mathrm{P}^- - \mathrm{P}^\mathrm{R}|_e$ represent the collision operators for charge exchange productions, charge exchange losses, and for collisional atom-atom relaxation processes, respectively, evaluated for the individual velocities $\{u_e, v_e, w_e\}$. The explicit form of these integrals can easily be obtained from expressions given e.g. by Burgers(1969), or Barakat and Schunk (1982a/b), if the slowly varying charge exchange cross section (see Maher and Tinsley, 1977) for simplicity reasons is taken out of the integrals P^+ and P^- and is evaluated at the mean relative velocity $< W' >$ yielding:

$$\mathrm{P}^+ \cong n_i f_i \sigma_{ex}(< W_n' >) < W_n' > \text{ and : } \mathrm{P}^- \cong n_i f_n \sigma_{ex}(< W_i' >) < W_i' >$$

The system of linear equations (25) can now be solved for each new coordinate step z on the symmetry axis and thus helps to find the solutions of all parametrizing quantities according to the following relations:

$$A_{zz}(z_i + \Delta z) = A_{zz}(z_i) + A^*_{zz}(z_i)\Delta z \text{ ; etc } \ldots \tag{26}$$

An example for solutions derived with this mathematical procedure is shown in Figure 2.

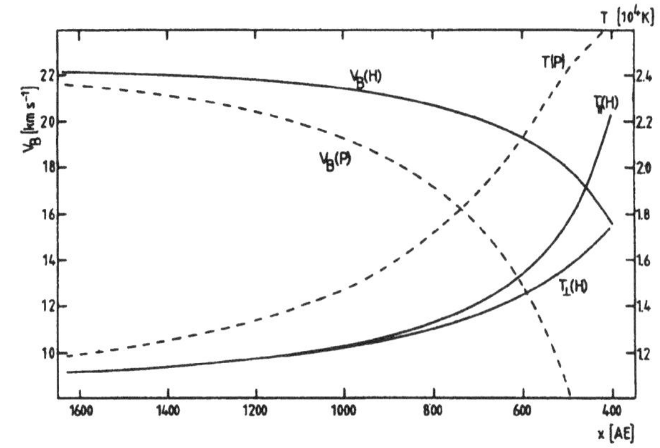

Figure 2. Shown by the full curves are the bulk velocity and the temperature components of LISM hydrogen advancing on the stagnation line towards the solar system through the perturbed LISM plasma region upstream of the heliopause (L = 380 AU). The dashed curves give the LISM proton properties derived from the Parker model (V_{∞} =23 km/s; T_{∞} = 10^4 K; n_{∞}^{+} = 0.015 cm^{-3}).

3. Interface transmissions of neutral LISM gases

No matter which of the existing models describing the above mentioned plasma interface is taken as a basis (Parker, 1963, Baranov et al., 1974, Fahr et al., 1988, Suess and Nerney, 1990, Baranov and Malama, 1993, Scherer et al. 1994, Steinolfson et al., 1994, Pauls et al., 1995, Baranov and Zaitsev, 1995, Khabibrakhmanov and Summers, 1996) in general a qualitatively identical result is achieved concerning the predicted changes of the first velocity moments of neutral interstellar species at the passage over this interface. This briefly means: Density and bulk velocity are decreased, temperatures are increased. Here, for the purpose of a more general comparison, we shall only study two alternative interface types, namely:

a) the one-shock interface already given by Parker (1963)

and

b) the twin-shock interface by Baranov and Malama (1993).

To study these alternative cases we follow the procedure developed by Fahr et al. (1995) to calculate the transmission function by solving the Boltzmann integro-differential equation expanding the distribution function into collision hierarchies. While this approach earlier was only applied to the problem of the VLISM hydrogen filtration by Osterbart and Fahr (1992), here we shall extend this method to the study of both the VLISM hydrogen and VLISM oxygen filtration. Hydrogen and oxygen are the most interesting candidates to study because of their high relative abundance and because of the fact that they are coupled the strongest to the interface due to resonant or quasi-resonant charge exchange with protons. The principles of this method are described by Fahr et al. (1995) and are not repeated here in detail.

As a consequence of relatively poor knowledges on the VLISM state it is especially hard to exclude the existence of an outer interstellar shock ahead of the heliopause. This outer shock which in case of its presence the neutrals VLISM atoms had to pass may have relevance for the selective transmission functions though this was not taken into account in earlier calculations by Osterbart and Fahr (1992). This makes it very interesting to investigate whether earlier results concerning hydrogen and oxygen penetration into the heliosphere could be substantially changed by the existence of an outer shock. Since only recently about 3 papers exist in the literature computing the penetration of VLISM hydrogen through a hydrodynamical twin-shock interface (Baranov and Malama, 1993, Steinolfson et al., 1994, Pauls et al., 1995). All calculations are based on numerical 2–d interface determinations which are not available in analytical forms. Since here mainly the

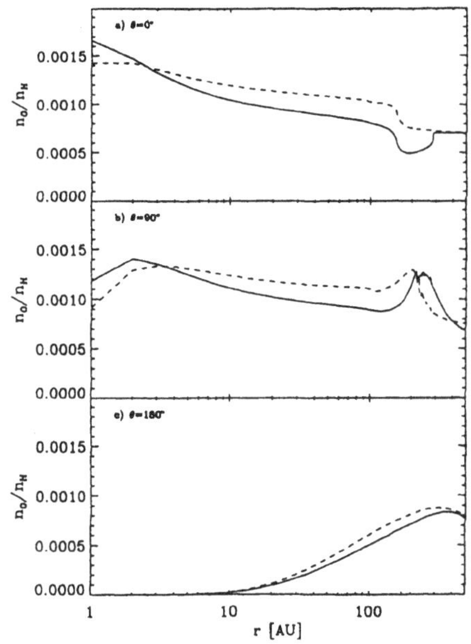

Figure 3. Oxygen to hydrogen ratios versus solar distance at different offset angles with respect to the VLISM flow, 1. (dashed curve) for the one-shock interface model, and 2. (full curve) for the twin-shock interface model.

innerheliospheric oxygen–to–hydrogen ratios are of interest it turns out that there is a lack of knowledge on LISM oxygen filtration in such a twin-shock interface. Here we have attempted to obtain this knowledge by means of a simulated twin-shock interface on the basis of a quasi-Parker interface model. Starting with identical LISM parameters as used by Baranov and Malama (1993) (i.e. $V_\infty = 25$ km/s, $M_\infty = V_\infty / C_\infty = 2.0$) we first apply the classical Rankine-Hugoniot shock relations for a normal shock to this preshock LISM flow and obtain the post-shock LISM conditions. Based on these latter post-shock LISM plasma conditions we then make use of the analytical form of the Parker interface treating a subsonic plasma flow, however, making sure that the heliopause stand-off distance becomes identical with that found in the twin-shock interface by Baranov and Malama (1993).

In Figure 3 we have shown the resulting oxygen-to-hydrogen ratios both for the normal Parker interface and for the twin-shock quasi-Parker interface described above, both calculated for identical LISM plasma parameters. The LISM hydrogen density hereby is adopted with the value n_H = 0.14 cm^{-3} used by Baranov and Malama (1993). Together with their value for the LISM proton density, n_H^+ = 0.07 cm^{-3}, one then arrives at a total LISM density of $N_H = n_H^+ + n_H = 0.21$ cm^{-3}. With the cosmic abundance of oxygen and with the assumption of charge exchange equilibrium between oxygen and hydrogen in the VLISM one then arrives at a consistent value for the LISM oxygen density $n_O = 1 \circ 10^{-4}$ cm^{-3}.

As shown in Figure 3 the twin–shock interface causes a strong pile–up of hydrogen in the region of the stagnating LISM plasma near the heliopause and also leads to slightly higher hydrogen densities all over the heliosphere. This interface–dependence is not revealed in the corresponding oxygen densities, except perhaps for the oxygen density pile–up at the heliopause flanks. As a consequence there appears a general decrease in the oxygen–to–hydrogen ratios in the twin–shock interface all over the heliosphere. The upcoming differences in the filtrations of VLISM hydrogen in the twin–shock quasi–Parker interface and in the Baranov–Malama interface can separately be studied in Figure 4. The results there show that the hydrogen density pile–up in the region of the stagnating VLISM plasma

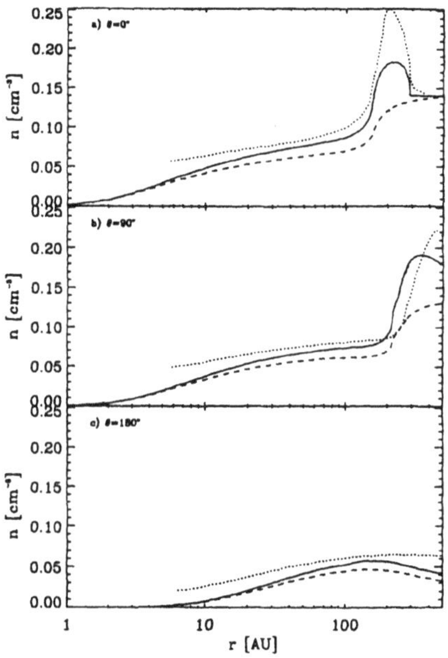

Figure 4. Hydrogen densities obtained by: 1.(dashed curve) = one shock model, 2.(full curves) = simulated twin–shock model, and 3.(dotted curve) = results by Baranov and Malama (1993).

near heliopause is more pronounced in the latter interface which also produces slightly higher hydrogen densities all over in the heliosphere. However, here the differences are fairly small qualifying the calculated results on filtration effects in the simulated twin–shock interface as reasonable approaches.

There is one evident reason why the density pile–up in the Baranov–Malama interface is more strongly pronounced as in our simulated interface. This is the fact that in the latter one an incompressible post-shock LISM plasma is treated whereas in the first a compressible plasma with somewhat higher densities (15 percent at the stagnation point) is acting.

4. Brief conclusions

Inspecting finally the presently available data on neutral LISM gas atoms (Witte et al., 1993) or on associated pick-up ions (Geiss et al., 1993, Gloeckler et al., 1993) in view of the above mentioned theoretical studies one may come to the following conclusions, also drawn in similar forms by Fahr et al.(1995):

I) The location of the solar wind shock is not very sensitive to the pick–up flux data. One can conclude that a necessary decrease of the theoretical H^+–pick–up fluxes to yield a better fit to the data could be in principle achieved with reduction of the shock distance from 80 AU to 50 AU. One should have in mind, however, that a similar reduction effect would as well result from a cosmic ray modified, high–entropy termination shock as described by Barnes (1994, 1995)) or by Chalov and Fahr (1993, 1994) which leads to fairly high post–shock compression ratios, fairly high post–shock densities, and hence in line with that enhanced filtrations.

II) Pick–up ion fluxes are essentially insensitive to the fact whether or not there is an outer LISM shock forming the outer LISM sheath plasma. As can be seen the twin–shock interface has the tendency to favour the LISM hydrogen penetration into the inner heliosphere, but thereby decreasing the oxygen–to–hydrogen density ratios only by small numbers. In view of the fact that the resulting oxygen pick–up ion fluxes, for the purpose of a better fit to the data, anyway needed to be increased relative to the hydrogen pick–up ion fluxes it is perhaps suggested that the one–shock interface has the virtue to allow the better theoretical representation of this fact.

III) It is quite evident that an ionized LISM component is definitely needed for a satisfactorily good representation of the pick–up ion data. As it seems, higher fractional ionizations of $\xi \geq 0.3$ for LISM hydrogen can be favoured in this respect. They would also suitably influence the oxygen–to–hydrogen ratios.

IV) Higher fractional ionizations compared to those resulting from Reynolds (1990) with his conservative photoionization rate would in fact be favourable for the pick–up ion data interpretation as was already evident from the previous point.

V) It should also be clear that an adequate description of the interface can only be expected if the effects of interstellar magnetic fields are taken into account. LISM magnetic fields shall tilt the interface configuration with respect to the inflow of neutral LISM gases (see Fahr et al., 1988) and shall induce temperature anisotropies of the plasma species which are partly then transferred to neutral atoms (see this article).

References

Axford, W. I. (1972): In "SOLAR WIND II", NASA-Special Report SP-308, pp 609-653

Balbus, S.A. (1986) Astrophys.J., 304, 787

Barakat, A.R., Schunk, R.W. (1982a): J.Phys.D:Applied Phys.,15,1195

Barakat, A.R., Schunk, R.W. (1982b): J.Phys.D:Applied Phys.,15, 2189

Baranov, V. B. (1990): Space Sci. Rev., 52, 89-120

Baranov, V. B., Lebedev, M. G., and Ruderman, M. S.: (1979), Astrophys. Space Sci., 66, 441-451

Baranov, V.B., and Malama, Y.: (1993), J.Geophys.Res.,98, 157

Baranov, V.B. and Zaitsev, N.A. (1995): Astron.Astrophys., 304, 631

Barnes, A.B. (1994): J.Geophys.Res., 99, 6553

Barnes, A.B. (1995): Space Science Rev., 72, 233

Bhatnager, P.L., Gross, E.P., Krook, M. (1954): Phys.Rev., 94, 511

Bloch, J.J., Jahoda, K. Juda, M., McCammon, D., Sanders, W.T. (1986): Astrophs.J., 308, L59

Böhringer, H., and Hartquist, T.W. (1987): Mon.Not.Roy.Astron.Soc.,228, 915 - 931

Brinkmann, R.T. (1970): Planet.Space Sci., 18, 449

Burgers, J.M. (1969): Flow Equations for Composite Gases, New York, Academic Press

Cameron, A.G.W. (1973): Space Science Rev., 14, 392 - 412

Chalov, S.V., Fahr, H.J. (1994): Astron.Astrophys., 288, 973

Chalov, S.V., Fahr, H.J. (1995): Planet.Space Sci., 43, 1035

Chew, G.F., Goldberger, M.L., Low, F.E. (1956): Proc.Roy.Soc.London, 236,112

Cheng, K.P., and Bruhweiler, F.C. (1990): Astrophys.J., 364, 573 - 585

Cox, D. P. and Reynolds, R. J. (1987): In "Ann. Rev. Astron. Astrophys.", 25, 303-344

Davidsen, A.F. (1993): SCIENCE, 259, 327 - 334

Fahr, H. J. (1986) Advances Space Res., 6(2), 13-25

Fahr, H. J. (1990): In: Proc. of COSPAR Symposium on "Physics of the outer heliosphere", Warsaw (Poland), Sept. 1989, Pergamon Press, pp. 327 - 346

Fahr, H.J. (1991) Astron Astrophys.241, 251-259

Fahr, H. J. and Neutsch, W.:(1983), Astron. Astrophys., 118, 57-65

Fahr, H. J., Grzedzielski, S., and Ratkiewicz, R.,(1988): Annales Geophys., 6(4), 337-354

Fahr, H.J. and Himmes, A. (1990): DFG-Progress Report (SFB-131):"Plasma- Gas Interactions in Space with special Emphasis to the Heliospheric Scenario".

Fahr, H.J., Ripken, H.W. (1984): Astron.Astrophys., 139, 551

Fahr, H.J., Scherer, K. (1995): Astrophys.Space Sci., 225, 21

Fahr, H.J., Fichtner, H., Neutsch.W. (1990): Planet.Space Sci., 38, 383

Fahr, H.J., Osterbart, R., Rucinski, D. (1995): Astron.Astrophys., 294, 587
Field, G.B., Goldsmith, D.W., Habing, H.J. (1969): Astrophys.J.,155, L149
Frisch, P.C. (1990): 1 st COSPAR Colloquium on "The outer heliosphere", Warsaw 1989, pages 19-28, Pergamon Press
Frisch, P.C. (1995): Space Science Rev., 72, 499
Frisch, P.C. and York, D.G. (1983): Astrophys.J., 271, L59
Gloeckler, G., Geiss, J., Balsiger, H., Fisk, L.A., Galvin, A.B., Ipavich, F.M., Ogilvie, K.W., von Steiger, R., and Wilken, B. (1993): Science, 261, 70 - 73
Geiss, J., Gloeckler, G., Mall, U., von Steiger, R., Galvin, A.B., Ogilvie, K.W. (1994): Astron.Astrophys., 282, 924
Khabibrakhmanov, I.K., and Summers, D. (1996): J.Geophys.Res., 101, 7609
Lallement, R. (1993): Advances Space Res., 13(6), 113
McKee, C.F., and Ostriker, J.P. (1977): Astrophys.J., 218, 148
McCray, R. and Snow, T.P. (1979): Ann.Rev.Astron.Astrophys.,17, 213
Maher, L.J., Tinsley, B.A. (1977): J.Geophys.Res., 82, 689
Matsuda, T., Fujimoto, Y., Shima, E., Sawada, K. , and Inaguchi, T. (1989): Progress of Theoret. Physics, 81, 810-822
Neutsch, W. and Fahr, H. J. (1983): Mon. Not. Roy. Astr. Soc., 202, 735-752
Osterbart, R., Fahr, H.J. (1992): Astron.Astrophys., 264, 260
Paresce, F. (1984):Astron Journal, 89, 1022
Parker, E. N. (1963): In "Interplanetary dynamical processes", Interscience Publishers New York
Pathria, R.K. (1973): Statistical Mechanics, Pergamon Press, New York
Pauls, H.L., Zank, G.P., Williams, L.L. (1995): J.Geophys.Res., 100, 21595
Rybicki,G.B., and Lightman, A.P. (1979): Radiative Processes in Astrophysics, John Wiley & Sons Inc., New York
Reynolds, R.J. (1990): 1.st COSPAR Colloquium "The Outer Heliosphere", Warsaw 1989, pages 101-104, Pergamon Press
Ripken, H. W. and Fahr, H. J. (1983) Astron. Astrophys., 122, 181-192
Scherer, K., Fahr, H.J., Ratkiewicz, R. (1994): Astron.Astrophys., 287, 219
Slavin, J.D. (1989):Astrophys.J., 346, 718
Stebbings, R. F., Smith, A. C. H., Ehrhardt, H. (1964): J.Geophys.Res., 69, 2349
Steinolfson, R.S., Pizzo,V.J.,and Holzer.T. (1994): Geophys.Res.Letters, 21(4), 245 - 248
Suess, S. T., and Nerney, S. (1990): J. Geophys. Res., 95, 6403-6412
Teske, R.G. (1993): ASTRONOMY, Dec.1993, 30
Witte, M., Rosenbauer, H., Banaszkiewicz, M., Fahr. H. J. (1993): Advances in Space Research, 13(6), 121 - 126

LISM STRUCTURE – FRAGMENTED SUPERBUBBLE SHELL?

P. C. FRISCH
Univ. Chicago, Dept. Astronomy Astrophysics, 5640 S. Ellis Ave, Chicago, IL 60637, USA

Abstract. Small scale structure in local interstellar matter (LISM) is considered. Overall morphology of the local cloud complex is inferred from Ca II absorption lines and observations of H I in white dwarf stars. Clouds with column densities ranging from 2–100 $\times$ 10^{17} cm^{-2} are found within 20 pc of the Sun. Cold (50 K) dense ($\sim 10^5$ cm^{-3}) small (5–10 au) clouds could be embedded and currently undetected in the upwind gas. The Sun appears to be embedded in a filament of gas with thickness $\leq$0.7 pc, and cross-wise column density $\leq$ 2 $\times$ 10^{17} cm^{-2}. The local magnetic field direction is parallel to the filament, suggesting that the physical process causing the filamentation is MHD related. Enhanced abundances of refractory elements and LISM kinematics indicate outflowing gas from the Scorpius-Centaurus Association. The local flow vector and λ Sco data are consistent with a 4,000,000 year old superbubble shell at $\sim$–22 km s^{-1}, which is a shock front passing through preshock gas at $\sim$–12 km s^{-1}, and yielding cooled postshock gas at $\sim$–26 km s^{-1}in the upwind direction. A preshock magnetic field strength of 1.6 μG, and postshock field strength of 5.2 μG embedded in the superbubble shell, are consistent with the data.

Key words: Local Interstellar Medium, Superbubbles, Local Bubble models

Abbreviations: LISM – Local ISM; SIC – Surrounding Interstellar Cloud; LIC – Local Interstellar Cloud

1. Introduction

Small scale spatial structure in local interstellar matter ("LISM") increases the fraction of interstellar gas in cloud surface regions, and indicates inhomogeneous physical properties on subparsec scale sizes. Unresolved velocity differences between adjacent structures will mimic turbulence, and reduce the accuracy with which the velocity of the interstellar cloud surrounding the solar system can be inferred from observations of absorption lines in stars. The physical properties of the surrounding interstellar cloud ("SIC") are inferred from observations of stars distanced 1.3–200 pc from the Sun, whereas the heliopause is only $\sim$0.0005 pc distant. Therefore, sightline averaged cloud properties do not necessarily correspond to the physical properties of the interstellar cloud which surrounds the solar system.

Interstellar matter establishes the galactic environment of the Sun, and the solar wind regulates the interaction between the solar system and the SIC. Over the past 5-10 million years the solar system has traveled through a region of space relatively empty of interstellar clouds, with average densities less than 0.0002 cm^{-3} (Frisch & York, 1986). Within the past few hundred thousand years, the Sun encountered the interstellar cloud complex local to the Sun. At the SIC low densities, the solar wind effectively excludes interstellar gas from the 1 AU region containing the Earth. However, 10%–15% of diffuse ISM is contained in cold dense structures (T$\sim$50 K, n$\sim$ 6000 – 10^5 cm^{-3}) with scale sizes of 5 AU to 100 AU (Frail *et al.*, 1991). The

Space Science Reviews **78**: 213–222, 1996.

lowest column density seen for these structures is 9.3×10^{18} cm^{-2}. Evidently, if a density structure such as these were embedded in the upwind cloud, the heliopause radius would collapse to 1–2 au, strongly altering the 1 AU environment of the Earth.

These possibilities motivate a study of the small scale structure in local interstellar matter.

2. LISM Morphology

There are some salient asymmetries in the distribution of interstellar gas within 35 pc of the Sun. Figs. 1, 2 summarize the overall distribution of this gas, based on Ca II data towards 42 stars and N(H) data towards an additional 16 white dwarf stars. The average space density of H I in the LISM, n(H I)$\sim$ 0.1 cm^{-3}, is used to scale the overall morphology of the LISM. Since the ionization potential of Ca II is 6.1 eV, which is less than the 13.6 eV H I ionization potential, $\sim$10% of the LISM calcium is in the form Ca II ($\sim$90 % is Ca III). Observations towards η UMa give a LISM conversion factor N(Ca II)/N(H I+H II)$\sim$ 10^{-8}, since N(H I+H II)=10^{18} cm^{-2} and N(Ca II)=10^{-8} cm^{-2} (Frisch & Welty, 1996). This same factor is found for α Oph, with log N(Ca II)=11.53 cm^{-2} (Crawford & Dunkin, 1995) and log N(H I)=19.48 cm^{-2} (Frisch, York, & Fowler, 1987). When all available data sampling the LISM are considered, N(Ca II)/N(H I+H II) varies by a factor of 10 between sightlines, as does N(Mg II)/N(H I+H II). N(Fe II)/N(H I+H II) varies locally by a factor of 5. Although Ca II is a trace element and sensitive to both density ($\sim n^2$) and abundance variations, the greater amount of data available on Ca II than either Mg II or Fe II make it a useful tracer of nearby LISM.

Using this scaling relation, the local cloud complex extends 3 parsecs towards η UMa, a star in the North Galactic Pole region, but 30 pc towards α Gru, a star near the South Galactic Pole region (Frisch & York, 1991). In the anti-center region the local cloud complex extends 3-5 pc, while in the galactic center region it extends $\sim$30 pc. Thus, the Sun is located closest to the "top" and anti-center portion of local cloud complex. This configuration is consistent with the structure determined from Mg II lines (Genova *et al.*, 1990). Figs. 1, 2 summarize the overall LISM morphology, based on Ca II absorption lines and the Ca II/H I=10^{-8} conversion factor, and white dwarf H I data.

3. Small Scale Structure

The LISM shows individual interstellar "clouds" with column densities N(H I)= $2 - 100 \times 10^{17}$ cm^{-2}. Clouds at the low end of this column density range will be masked in sightlines with larger column densities, unless the clouds are well separated in velocity. Structure in interstellar matter is seen as variations in interstellar

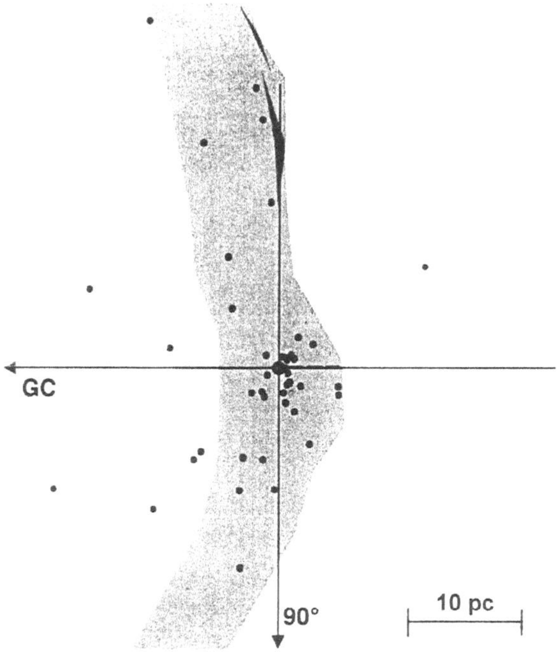

Figure 1. Overall distribution of interstellar gas within 30 pc of the Sun, based on observations of interstellar Ca II lines in 42 stars and observations of 16 white dwarf stars. The conversion factor N(Ca II)/N(H I+H II)$\sim$ 10^{-8} is used. The SIC was assumed to have an average density of 0.1 cm^{-3} for this plot. Data for stars with known circumstellar gas or peculiar emission lines are omitted. Data points with high excursions from the surface represent sightlines with denser clouds. The viewpoint is the north galactic pole.

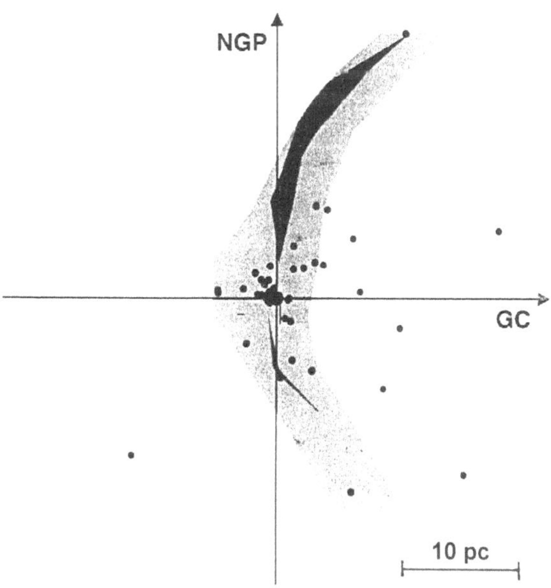

Figure 2. Same as Figure 1, but viewed from a longitude of 270°.

column densities over small angular scales (Munch & Unsold, 1962; Diamond *et al.*, 1995; Frail *et al.*, 1991; Meyer & Blades, 1996), and as spectrally resolved Doppler broadened interstellar absorption components. In the LISM, the galactic interval around the tangential sightline towards the Loop I supernova remnant bubble, l$\sim$0°–30°, shows pronounced angular variation in Ca II column densities for stars within 20 pc: towards α Oph (l=36°, b=+23°, d=20 pc), log N(Ca II)=11.53 cm^{-2} (Crawford & Dunkin, 1995); whereas towards γ Oph (l=28°, b=+15°, d=29 pc), log N(Ca II)=10.86 cm^{-2}; and towards λ Aql (l=31°, b=–6°, d=30 pc), log N(Ca II)<9.59 cm^{-2} (Vallerga *et al.*, 1993). The interstellar gas in this region also shows a complex velocity structure: three Ca II absorption line components are seen towards α Aql (l=48°, b=–9°, d=5 pc), and log N(Ca II)=10.26 cm^{-2} (Bertin *et al.*, 1993). Typical minimum observed Ca II line strengths of 1 mÅ are seen in the LISM, corresponding to log N(H I)$\sim$18.0 cm^{-2} clouds. Towards Sirius (2.7 pc), two absorption features are seen with column densities log N(H I)=17.23 cm^{-2} (Bertin *et al.*, 1995). At the 3.6 km sec^{-1} resolution of the Hubble Space Telescope GHRS, it is estimated that only 10% of the interstellar velocity components are resolved for unsaturated lines (Welty, Morton, & Hobbs, 1996). The range of column densities for cloud structures near the Sun is shown in Figure 3, where N(Ca II) is plotted versus galactic longitude for individual absorption components sampling ISM within 50 pc. Ultimately, high spectral resolution observations ($\sim$0.4 km s^{-1}) in the ultraviolet region (1000-3000 Å) will be required to uniquely define the cloud structure in the LISM.

The presence of clouds with column densities ranging from 2-100 $\times$ 10^{17} cm^{-2} in the LISM makes it difficult to uniquely determine the SIC velocity, since clouds with 3 $\times$ 10^{17} cm^{-2} may be blended with higher column density features. This is certainly the case in the upwind direction where column densities exceed 10^{19} cm^{-2}. These uncertainties are shown by the fact that absorption components at the LIC * velocity are not found in the Fe II and Mg II lines towards α Cen (1.3 pc) (Lallement *et al.*, 1995). They are also not seen in the Ca II line towards η UMa, which has a very low column density (10^{18} cm^{-2}) (Frisch & Welty, 1996). Either the LIC velocity is not the velocity of the SIC, or in both cases the SIC cloud extent is very short in the sightline ($\leq$0.6 pc). However, the heliocentric upwind direction is fixed by observations of SIC H I and He I in the solar system, so the SIC velocity would deviate from the LIC velocity vector in magnitude, rather than direction.

One interesting question is whether dense, low column density, cold clouds (n$\sim$10^5 cm^{-3}, log N(H I)$\sim$10^{19} cm^{-2}, T$\sim$50 K, thickness $\sim$5–10 au) embedded in the upwind gas might have gone undetected? Typically, Ca II in cold diffuse interstellar clouds is depleted by a factor of 20 more than in LISM-type gas; thus the predicted Ca II column density associated with such a cloud would be log N(Ca II)=9.7 cm^{-2}. At 50 K, the Ca II Doppler constant is b= 0.14 km s^{-1}

* The "Local Interstellar Cloud" (LIC) is defined by the velocity vector of 25.7 km s^{-1} flowing from the direction l=5.9°, b=+16.7° (Lallement & Bertin, 1992). If the LIC velocity describes the velocity of the surrounding cloud, then the LIC is the SIC.

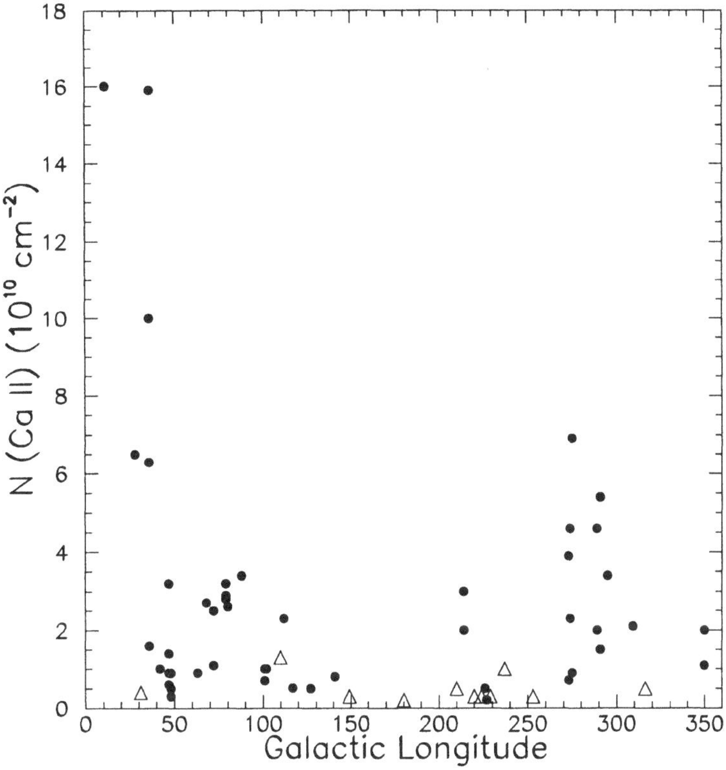

Figure 3. N(Ca II) is plotted versus galactic longitude for individual absorption components for stars sampling ISM within 50 pc. Stars in the "high Mg II column density" region of Genova *et al.* (1990) (viz. low galactic latitudes and the galactic center hemisphere, where log N(Mg II)> 12.1 cm^{-2}) are plotted as filled circles, while stars in low Mg II regions are plotted as open circles. Upper limits are plotted as open triangles. Data for stars with known circumstellar gas or peculiar emission lines are omitted.

(corresponding to a full-width-half-max of 0.2 km s^{-1}). Absorption from the cold gas would be weak, and contribute only 0.4 mÅ absorption to the Ca II 3933 Å line, and if blended in velocity with adjacent warm gas, it would *not* be resolved by existing observations. Therefore, one must conclude that dense small structures could be embedded in nearby upwind interstellar gas. If the ultraviolet Zn II line were available for a nearby star in the upwind direction, for instance, then it would be possible to determine if dense structures are present since depletion would not be a factor.

4. Structure of SIC

Given the evidence for clouds with column densities 2×10^{17} cm^{-2} in the LISM, the overall cloud morphology in Figs 1, 2 becomes the envelope of the distribution of nearby interstellar gas. In order to construct the surrounding cloud structure, one may assume that the cloud is moving in a direction parallel to the cloud surface normal, and then use the component column density towards a nearby star to constrain the distance to the cloud surface in that direction. This assumption rests on the result that the kinematics of nearby interstellar matter (d<20 pc) is consistent with an expanding superbubble origin (Frisch, 1996). In this case, the LIC cloud column density towards the star Sirius led to the conclusion that the Sun has entered the LIC cloud within the last 2,000-8,000 years, and that the Sun is about $\sim$0.1 pc from the surface in the downwind direction (Frisch, 1994). When the upper limits on the Ca II column density towards λ Aql (Vallerga *et al.*, 1993) and the LIC Fe II column density towards α Cen (Lallement *et al.*, 1995) are included, then we are forced to conclude that the Sun is located in a filament of diffuse interstellar gas with total thickness < 0.7 pc. This filament is illustrated in Figure 4, along with additional cloud structures representing cloud multiplicity seen towards nearby stars. If this picture is correct, however, then most of the interstellar gas observed towards α Aur, in the downwind direction, must be in separate, but co-moving, filaments (or clouds) from the surrounding cloud. The thickness of the SIC filament corresponds to a column density of $\sim 2 \times 10^{17}$ cm^{-2}. Filamentary structures are not unusual in interstellar gas; for instance they are seen where the Cygnus superbubble interacts with interstellar clouds (Graham *et al.*, 1995).

The LISM interstellar magnetic field direction traced by observations of polarization in the nearest stars (d=2–30 pc) is directed towards l$\sim$70°, and therefore is parallel to the SIC cloud surface. The polarization is within 3 pc of the Sun in the upwind direction, and there is little change with polarization strength as a function of star distance. Therefore, this magnetic field may be captured in the SIC filament. Alternately, the polarizing magnetic field could be captured in the higher column density, higher velocity gas within $\sim$1 pc of the Sun in the upwind direction. Additional discussion of the Tinbergen polarization data can be found in Frisch (1991, 1996). The facts that the SIC properties are consistent with a filamentary structure, and that the local magnetic field is parallel to this filament, suggest to the author that the physical process causing the filamentation is linked to a cloud fragmenting due to instabilities. For SIC values n=0.15 cm^{-3}, T=7,000 K, estimated B=1.6 μG, the ratio of gas pressure to magnetic pressure, $\beta \sim 1$.

5. Superbubble

The kinematics and abundance patterns in local interstellar matter (LISM), combined with the large angle subtended by the Loop I supernova remnant, led to the view that the Sun is immersed in the shell of the Loop I remnant, associated with the Scorpius-Ophiuchus Association superbubble (Frisch, 1981, and see Breitschwerdt

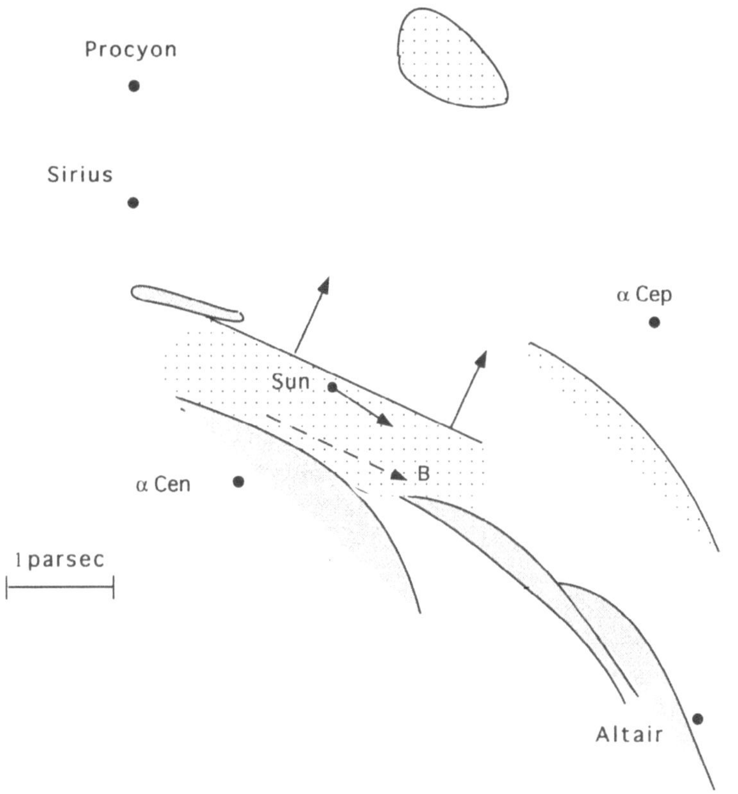

Figure 4. Morphology of the interstellar cloud surrounding the Sun, inferred by assuming that the cloud velocity vector is parallel to the surface normal (as would be expected for an expanding superbubble shell), and constrained by LIC column densities (or limits) observed towards Sirius, α Cen, and λ Aql (see text). Cloud multiplicity towards Sirius, α Cen, and Altair is also illustrated.

et al., this volume). The cloud surrounding the solar system has been modeled as part of a superbubble shell formed ~4,000,000 years ago around this association,

Looking first at the kinematics of the LISM gas: this gas flows past the Sun from an upwind direction (in the local standard of rest, LSR) directed at the Scorpius-Ophiuchus Association, but the flow is not uniform. The upwind velocity vector, after transforming into the LSR velocity frame, corresponds to V~–19, l~ 335°, b~ 0° (with the exact value dependent on the assumed flow vector and local standard of rest transformation). For the velocities of 24 interstellar components observed optically for stars within 50 pc of the Sun, the RMS variation from the local flow velocity vector is 2.4 km s^{-1}. In contrast, the RMS variation from the LSR for 23 component velocities is 6.5 km s^{-1}. Thus, it is possible to select a set of LISM clouds from a sample of nearby stars which give a relatively coherent flow at the local flow velocity, but it is not possible to select a set of components (from this same sample of stars) which are at rest in the LSR. * In conclusion, the fit of the local flow velocity vector to the velocities of interstellar absorption components in nearby stars is significantly better than the fit of the LSR velocity. The dispersion shows, however, that the flow is not a uniform and linear. The velocity vector of the LISM establishes this gas as outflowing from the Scorpius-Centaurus Association.

The second characteristic indicating that the LISM gas has been processed through a shock front is the anomalously high gas abundance pattern seen in this material. For the LISM clouds, typically N(Ca II)/N(H I+H II)=10^{-8}. The enhanced abundance of refractories appears to be a global property of interstellar matter within 20-30 pc of the Sun, since it is found towards both α Oph, with log N(H I)=19.48 cm^{-2} (Frisch, York, & Fowler, 1987), and towards η UMa, with log N(H I+H II)=18.00 cm^{-2} (Frisch & York, 1991)). In contrast, the ratio 4.8×10^{-10} is found towards the cold cloud complex in ζ Oph. The enhanced abundance pattern of refractory elements is also seen for the elements Ti II, Fe II and Mg II in the LISM. IUE satellite data show that Fe II and Mg II are depleted towards η UMa by factors of 9 and 12, respectively. Better quality Hubble data show average depletions for the LIC components in α CMa, α CMi, α Aur, ϵ CMa of a factor of 6 for Fe II, and a factor of 11 for Mg II. These depletions can be contrasted by the ζ Oph cold cloud depletion of a factor of 275 for Fe II, for example, and about 25 for Mg II (after correcting the λ1240 line pair oscillator strengths upwards by a factor of 3, Welty, private communication).

In the superbubble scenario, the LISM is gas evaporated from the surfaces of molecular clouds embedded in the Scorpius-Ophiuchus Association during earlier and more active periods (much as the clouds embedded in the Cygnus Loop superbubble are currently evaporating due to interactions with blast waves, (Graham *et al.*, 1995)). The characteristics of the upstream gas are consistent with a superbubble shell formed 4,000,000 years ago, with radius 160 pc and local standard of rest velocity 22.4 km s^{-1}(corresponding to the upwind gas velocity), expanding into a cloud of density n$\sim$0.1 cm^{-3} (Frisch, 1995), but this simple shell model neglects magnetic fields which would be significant. The shell is a "fossil superbubble shell", and unless grain-grain collisions are effective, grain destruction would not be taking place currently at this relatively low cloud velocity. Superbubble shell velocities are given by V$\sim$age$^{-0.4}$, so the grain destruction would have taken place either during the cloud evaporation or within about 10,000 years after the shell formed initially. The enhanced LISM abundances could be due to dust grain destruction at earlier epochs when the shell velocity exceeded $\sim$100 km $s^{-1}$and grain mantle destruction via sputtering would have been effective (e. g. see Draine, 1995).

Building on the view that the nearest interstellar gas has been processed through a shock front, cloud components seen by the Copernicus satellite towards the low column density star λ Sco (l=372°, b=–2°, d=85 pc) have been interpreted in

* For each star, where more than one component is present, the component used for the calculation was selected so as to minimize the chosen variation. The measurement uncertainties typically are 2-3 km s^{-1}. See Frisch (1996) for additional information.

terms of a shock front moving through the gas (Frisch & York, 1991). The Ca II component heliocentric velocities seen towards this star are at –26.6 km s^{-1} and -19.1 km s^{-1} (Welty, Morton, & Hobbs, 1996), which convert to LSR velocities -19.6 km s^{-1} and -12.1 km s^{-1} The rest of the velocities quoted in this paragraph are in the LSR velocity frame. For an adiabatic shock in a perfect gas, with no magnetic field, the two components observed optically in λ Sco correspond to the preshock and postshock gas. In this model, the preshock gas was identified as the cloud at –12.1 km s^{-1} (York, 1993, found this cloud to be mainly ionized), and the postshock gas was identified as the mainly neutral thin sheet of gas at –19.6 km s^{-1} cloud. The shock front was identified as the gas at –22 km s^{-1} These clouds are upwind of the Sun. A more detailed discussion of this model can be found in Frisch and York 1991).

An adiabatic shock model can also be fit to the data towards α Oph, using Ca II velocities (Crawford & Dunkin, 1995). In this direction, the shock front would be the -14 km s^{-1} (LSR velocities) component, the preshock and postshock gas would correspond to the –5.6 km $s^{-1}$and –8.2 km s^{-1} components.

For both sightlines, magnetic fields are present since the superbubble shell appears to be threaded by a magnetic field. Based on Zeeman splitting observations in other sightlines through the Loop I superbubble, the magnetic field in the postshock gas is estimated at $<B>=-5.2$ μG, which would give a preshock value of $<B>=-1.3$ μG (Frisch & York, 1991). These fields would modify the simple shock approximations used for this shock model, and this question needs to be considered in more detail.

6. Conclusions

In this review, the overall morphology of the LISM gas is inferred from interstellar Ca II absorption lines and observations of H I in white dwarf stars. Clouds with column densities ranging from 2–300 $\times$ 10^{17} cm^{-2} are found within 20 pc of the Sun. Because of the increased depletion in cold clouds, cold (50 K) dense ($\sim 10^5$ cm^{-3}) small (5–10 au) clouds could be embedded undetected in the upwind gas. The factor of 10 variation found in the LISM Ca II and Mg II abundances, and factor of 5 seen in the LISM Fe II abundance, indicate density variations must be present. These structures require ultraviolet data of lines such as Zn II in order to be detected. With simple assumptions, one concludes that the Sun is embedded in a filament of gas with thickness $\leq$0.7 pc, and cross-wise column density $\leq 2 \times 10^{17}$ cm^{-2}; the local magnetic field appears to be parallel to the filament suggesting that the physical process causing the filamentation may be MHD related. Enhanced abundances of refractory elements, combined with the flow direction of local gas, indicate an origin associated with the Scorpius-Centaurus Association. A simple picture consistent with the data towards λ Sco is found if a 4,000,000 year old superbubble shell at $\sim$–22 km s^{-2} is a shock front passing through preshock gas at $\sim$–12 km s^{-2} which is ionized (or partially ionized), giving cooler and denser postshock gas at $\sim$–26 km s^{-2} in the upwind direction (LSR velocities). The preshock magnetic field would be 1.6 μG, and the postshock magnetic field, embedded in the superbubble shell, would be –5.2 μG. However, since a simple adiabatic shock model was considered, which should not apply to an old superbubble shell, this problem needs to be looked at in more detail.

Acknowledgements

I would like to thank the National Aeronautics and Space Administration for research support.

References

Bertin, P., Lallement, R., Ferlet, R., and Vidal-Madjar, A. 1993, Astronomy and Astrophysics, **278**, 549

Bertin, P., Vidal-Madjar, A., Lallement, R., Ferlet, R., and Lemoine, M., 1995, Astronomy and Astrophysics, **302**, 889

Crawford, I. A., Dunkin, S. K., 1995, Monthly Notices of the RAS, **273**, 219.

Diamond, C. J., Jewell, S. J., Ponman, T. J.: 1995, Monthly Notices of the RAS, **274** 589.

Draine, B. T., 1991, Astroph. Sp. Sci., **233**, 111

Frisch, P. C.: 1981, Nature, **293**, 377

Frisch, P. C.: 1991, in *COSPAR Colloquia Series, No. 1, Physics of the Outer Heliosphere*, S. Grzedzielski and E. Page (eds), (London: Pergamon), p. 19.

Frisch, P. C.: 1994, Science, **265**, 1423

Frisch, P. C.: 1995, Space Science Reviews, **72**, 499

Frisch, P. C., and Welty, D. E., 1996, in preparation.

Frisch, P. C., York, D. G.: 1986, in R. Smoluchowski, J. N. Bahcall, M. Matthews (eds.), *The Galaxy and the Solar System*, (Tucson: Univ. Arizona Press) pp. 83

Frisch, P. C.: 1996, *Cosmic Winds and the Heliosphere,* Jokipii, J. R., Sonett, C. P., Giampapa, M. S. (eds.), (Tucson: U. Arizona).

Frisch, P. C., York, D. G., and Fowler, J. R., 1987, Astrophysical Journal, **320**, 842

Frisch, P. C., York, D. G., 1991, in *Extreme Ultraviolet Astronomy,* R.F. Malina and S. Bowyer (eds.), (New York: Pergamon), p. 322

Genova, R., Molaro, P., Vladilo, G., and Beckman, J. E. 1990, Astrophysical Journal, **355,** 150.

Graham, J. R., Levenson, N. A., Hester, J. J., Raymond, J. C., and Petre, R., 1995, Astrophysical Journal, **444**, 787

Lallement, R., Bertin, P.: 1992, Astronomy and Astrophysics, **266**, 479

Frail, D. A., Weisberg, J. M., Cordes, J. M., Mathers, C.: 1994, Astrophysical Journal, **436,** 144.

Lallement, R., Ferlet, R., Lagrange, A. M., Lemoine, M., and Vidal-Madjar, A. 1995, Astronomy and Astrophysics, **304**, 461

Meyer, D. M., and Blades, J. C., 1996, Astrophysical Journal, **464,** 179L.

Munch, G. and Unsold, A., 1962, Astrophysical Journal, **135,** 711. 1994, Astrophysical Journal, **436,** 144.

Slavin, J.D., 1989, Astrophysical Journal, **346,** 718.

Welty, D. E., Morton, D. C., and Hobbs, L. M., 1996, Astrophysical Journal, in press.

Vallerga, J. V., Vedder, P. W., Craig, N., and Welsh, B. Y., 1993, Astrophysical Journal, **411,** 729

York, D. G., 1983, Astrophysical Journal, **264,** 172

RELATIVE IONIZATIONS IN THE NEAREST INTERSTELLAR GAS

P. C. FRISCH
Univ. Chicago, Dept. Astronomy Astrophysics, 5640 S. Ellis Ave, Chicago, IL 60637

J. D. SLAVIN
NASA Ames Research Center, MS 245-3, Moffett Field, CA 94035-1000

Abstract. We compare CLOUDY predictions for the equilibrium ionization in the interstellar cloud surrounding the solar system with pick-up ion data. The incident radiation field includes contributions from hot stars, the emission from the conductive cloud boundary and the diffuse FUV background. To within the observational uncertainties, CLOUDY predictions for the ratios n(He°)/n(O°), n(N°)/n(O°), n(Ne°)/n(O°), and n(He°)/n(Ne°) are consistent with pick-up ion data, provided that O°and N°are filtered by $\sim$ 50% in the heliopause region and the outer heliosphere as predicted by others. Thus, the steady-state ionization model and assumed radiation field appear approximately valid. However, the youth and low intervening column density towards the Vela pulsar leave open the possibility that the parent supernova explosion ~10,500 years ago, and 200 pc distant, may also have affected LISM ionization, although the mechanism is uncertain. Support for this last possibility is provided by the apparent signature of the Vela explosion in the terrestrial geological record.

Key words: Local Interstellar Medium, EUV emission, Soft X-ray Background, Local Bubble models

Abbreviations: ISM – Interstellar Medium; FUV – Far Ultraviolet; EUV – Extreme Ultraviolet; SNR – SN remnant; SXRB – SXR Background; LISM – Local Interstellar Matter

1. Introduction

The level of ionization in the interstellar cloud surrounding the solar system regulates the relative abundances of interstellar neutrals penetrating the heliosphere. In this paper, we look at two aspects of ionization in the surrounding cloud in order to evaluate the fractional ionizations of the interstellar parent populations of the pick-up ions and anomalous cosmic rays. First, we examine whether the surrounding cloud is in steady-state ionization by considering the time dependent ionization due to the initial UV/X-ray flash of the most recent identifiable nearby supernova, i.e. the event which created the Vela pulsar 10,500 years ago. The effect on local cloud ionization of the UV/X-ray flash from the supernova shock breakout is found to be significant only for flash luminosities of $> 10^{49}$ ergs, but other supernova related processes may be relevant. Next, we use the ionization equilibrium code CLOUDY (Ferland, 1993) to make initial estimates of the fractional ionizations of H, He, C, N, O, and Ne in the surrounding cloud using an input radiation field due to hot stars, white dwarfs and EUV emission from the local cloud boundary. The cloud ionization and N(H°)/N(He°) ratios are highly sensitive to the relative contributions of the possible radiation sources. These predictions agree, to within observational uncertainties, with observed column density ratios N(HeI)/N(HI) seen towards nearby white dwarfs, as well as relative abundances found in the pick-up ion population, providing oxygen and nitrogen are filtered by charge exchange in the heliopause region and heliosphere.

Space Science Reviews **78**: 223–228, 1996.

2. Vela Flash

The ratio N(H I)/N(He I) in the cloud surrounding the solar system depends critically on the $500 < \lambda < 900$Å radiation field, and the diffuse component of this field is only partially known. Given the proximity of the solar system to active OB associations, however, there is no *a priori* basis for assuming a constant radiation field over the past million years. The Vela pulsar has a spin-down age of 9,000–27,000 years, and a distance of 200–600 pc citeaschenbach,strom. Vela, located at l=264°, b=–3°, borders the region of very low average interstellar densities in the third galactic quadrant. The proton recombination time in a temperature 8,000 K, density 0.1 cm^{-3} gas is ~650,000 years, and is longer in hotter gas, so insignificant recombination of hydrogen would have occurred since any ionization event associated with the Vela supernova explosion.

Vela's location and relatively young age indicate that it may have contributed to the ionization of the cloud surrounding the solar system. One possible mechanism for this ionization is the initial UV/X-ray flash which occurred as the supernova shock broke out of the stellar photosphere. Another possible source of ionizing radiation may be γ-rays decayed from ^{56}Co and other radioactive elements, which are thermalized in the expanding envelope. This radiation source would be effective over longer timescales (>100 days) (Arnett *et al.*, 1989). Although the details of the Vela explosion are unknown, the presence of a pulsar indicates core collapse. Other Vela related ionizing mechanisms that could be significant include a UV-bright progenitor, collisional ionization by galactic cosmic rays accelerated in the supernova shock, X-rays produced by synchrotron emission from the rotating neutron star, or emission from the shock front propagating into the surrounding ISM. We consider only the flash from initial shock breakout and (briefly) comptonized γ-rays here.

To estimate the photoionization locally due to the Vela flash, we adopt an age for Vela of 10,500 years, based on the presence in the terrestrial geological record of organic residue attributed to modification of the mesosphere chemistry by supernova produced gamma rays and hard X-rays ($\lambda < 100$Å) citebrakenridge. This age is consistent with the pulsar spin-down age (Taylor *et al.*, 1993). With this age, Vela's distance is ~ 200 pc, based on the angular distance from the pulsar of fragments of matter ejected by the blast (Aschenbach, Egger, & Trumper, 1995). The near side of the Gum Nebula has a distance 200 pc citefranco, and the nebular radius is 125 pc citeleahy, indicating Vela is located in the near side of the Gum Nebula and therefore in front of most of obscuring interstellar matter associated with the Gum.

The photoionization rate for hydrogen from the UV/X-ray flash of a supernova is less than $\Gamma = \sigma_o L/(4\pi d^2 E_o)$ s^{-1}. Here σ_o is the hydrogen photoionization cross section at threshold, E_o=13.6 eV, and L is the average luminosity of the supernova during the shock breakout flash. Based on SN 1987A data, the total amount of energy emitted in radiation during the flash is ~ 10^{47} ergs over one day (Woosley, 1988), giving L=1 × 10^{42} ergs s^{-1}. For $\sigma_o = 6.3 \times 10^{-18}$ cm^{-2} and d~200 pc, a maximum hydrogen ionization rate of $\Gamma \sim 6 \times 10^{-8}$ s^{-1} is obtained. In order for hydrogen to be significantly ionized by the flash, $\Gamma \times \tau \sim 1$ is needed, whereas for a flash duration, τ, of one day $\Gamma \times \tau \sim 5 \times 10^{-3}$ is found. Only if the UV/X-ray flash energy of Vela were as high as 10^{49} ergs would it be a significant source of ionization at the local cloud surface.

Another way to evaluate the ability of the flash to ionize the LIC is by looking at the energetics. The energy needed to ionize the hydrogen in the cloud is $N_{\rm H} A_{\rm cl} E_0$, where $N_{\rm H}$ is the column density of (initially) neutral hydrogen along a line of sight towards Vela, $E_0 = 13.6$ eV, and $A_{\rm cl}$ is the area of the cloud face towards Vela. Only a fraction, $\Omega_{\rm cl}/4\pi = A_{\rm cl}/4\pi d^2$, of the N_{ph} ionizing photons emitted during the flash will be absorbed by the LIC, where d is the distance to Vela. Thus for the flash to ionize the cloud requires that $N_{ph}\bar{E}A_{cl}/4\pi d^2 \geq N_H A_{cl} E_0$ ($\bar{E}$ is the average energy of an ionizing photon) or that the total amount of energy emitted in ionizing photons exceed $N_H E_0 4\pi d^2$. For $N_H = 10^{18}$ cm^{-2} and $d = 200$ pc this requires $\sim 10^{50}$ ergs. Again this is much more energy than has been calculated to be contained in the shock breakout UV/X-ray flash. On the other hand, this flash may help explain the high degree of ionization seen in the ϵ CMa direction (Gry, this volume), which is 100 pc from Vela in a low interstellar density region.

Gamma rays from the decay of radioactive ^{56}Co power radiative emission from the supernova envelope over time scales of $\sim$ 100 days. The total energy emitted as radiation from SN 1987A over the first year was $\sim 10^{49}$ ergs s^{-1}, so the energetics may limit effective ionization to outer cloud regions. The primary difficulty with this mechanism is that the opacity of the SN envelope probably prevents the lower energy X-rays, which can efficiently ionize hydrogen and helium, from escaping. This possibility as well as the others mentioned above need to be explored in more detail.

3. CLOUDY Equilibrium Models

CLOUDY is a fortran photo-ionization/thermal equilibrium code designed to predict the steady-state physical conditions in astrophysical plasmas as a function of the incident radiation field, density and cloud composition (Ferland, 1993). The code calculates temperature at each point in the cloud by finding the temperature at which heating and cooling balance. Both the attenuated incident and diffuse radiation fields are calculated. The ionization for each point in the cloud is determined by the ionizing radiation field and the temperature and density. The geometry option which we used is plane parallel which also implies a distant source with parallel rays. This is not entirely consistent with the geometry of the cloud boundary emission which we use as a source, but is as close as the code allows.

Cloud ionization depends both on the spectrum of the illuminating radiation and on the radiative transfer of the radiation to the cloud interior. For a flat extreme ultraviolet spectrum, helium ionizing radiation penetrates twice as far into the cloud as hydrogen ionizing radiation because $1/e$ attenuation is achieved for N(H)= 2×10^{17} cm^{-2} for H, and N(H)= 5×10^{17} cm^{-2} for He. However, the ionization edges of helium and hydrogen are at 504 Å and 912 Å, respectively, so the shape of the EUV spectrum becomes the main factor dominating the relative ionizations of ions and elements with ionization potentials 13–40 eV.

The radiation field input to the code is the sum of the far ultraviolet field in the 1000 Å–2900 Å interval (Gondhalekar, Phillips, & Wilson, 1980), the radiation field of the brightest EUV source in the sky, ϵ CMa (Vallerga & Welsh, 1995), EUV radiation from bright white dwarfs (Dupuis et al., 1996), and the hard EUV field predicted for the conductive interface between the local cloud (modeled as spherically symmetric) and surrounding plasma (Slavin, 1989). For calculating the boundary emission (a similar calculation to (Slavin, 1989)), the cloud is assumed to have a radius of 3 pc, density 0.2 cm^{-2}, magnetic field 5 μG, and a conductive reduction factor of $\eta = 0.1$ ($\eta = \cos^2\theta = 0.1$ for $\theta = 72°$, where θ is the angle between the magnetic field and the radial directions). The surrounding hot gas of the Local Bubble is assumed to have a temperature of 10^6 K. When cloud boundary emission is omitted, a cloud temperature of $\sim$400 K is found, in disagreement with the observed $\sim$7000 K temperature. Also, H ionization is $\sim$94% while He ionization is $\sim$1%, in disagreement with the average ratio N(H°)/N(He°)$\sim$14 white dwarf value (Dupuis et al., 1996; Frisch, 1995). Therefore, either the boundary emission or some other moderately hard photoionization source is needed to provide the relatively high heating rate per ionization inferred from the data. As mentioned above, the geometrical assumptions of the boundary emission calculation and the radiative transfer calculation of CLOUDY are not entirely consistent. However, given the complex LISM structure (Frisch, this volume) we do not expect to model the Local Cloud geometry accurately, and therefore utilize this approximation.

It was found that even with the inclusion of the emission from the evaporative boundary, insufficient heating was generated in the cloud for the assumption of solar abundances and no dust. It has long been known that in warm neutral and warm partially ionized regions of the ISM, the observed temperatures require grain photoelectric heating to be significant. We have used the "grains" command along with "ism" standard depleted abundances in CLOUDY to determine if the heating problem could be solved by the presence of dust. (The "ism" abundances relative to hydrogen for the elements of interest here are: He: 9.8×10^{-2}, C: 2.51×10^{-4}, N: 7.94×10^{-5}, O: 5.01×10^{-4}, Ne: 1.23×10^{-4}.) We have found that indeed the lowered abundances and added heating result in temperatures near the observed 7000 K value. The results we present below employ these assumptions. The actual abundances and dust content in the LIC are not well specified, although reduced depletions of Fe and Ca, as typical of warm or shocked interstellar gas, are found (Frisch, 1981), and Ulysses dust data show dust is present in the cloud (Gruen et al., 1994). We plan to examine the evidence more carefully for future calculations.

In order to use the observed spectrum from the bright EUV sources as input to CLOUDY (i.e. incident upon the outer surface of the Local Cloud) we needed to de-absorb them. An absorbing column density of N(H°)=9×10^{17} cm^{-2} was assumed for this and the CLOUDY calculation is stopped when that column density is reached. This occurs 2.6 pc from the cloud surface and so this is the assumed solar location. With these assumptions, CLOUDY predicts the following values for interstellar gas at the solar location: n(C°)/n(O°) = 0.0024, n(C°)/n(N°) = 0.018, n(He°)/n(O°) = 166, n(N°)/n(O°) = 0.14, n(Ne°)/n(O°) = 0.16.

The Ulysses SWICS pick-up ion data have been used to determine abundances in the interstellar gas penetrating the heliosphere (Geiss et al., 1994). From the pick-up ions, n(He°)/n(O°)= 290(+190, −100), n(N°)/n(O°)= 0.13(+0.06, −0.06), n(Ne°)/n(O°)= 0.20(+0.12, −0.09), n(C°)/n(O°)$\leq$0.15. Since 50% filtration of O°is predicted due to charge exchange with protons between the plasma sheath and 5 AU (Fahr, Osterbart & Rucinsky, 1995; Fahr & Osterbart, 1995), and since $N°$ will charge exchange similarly with protons, the ratios n(C°)/n(O°), n(He°)/n(O°), n(C°)/n(N°), and n(Ne°)/n(O°) predicted by CLOUDY should be raised by a factor of two for comparison with the pick-up ion data. After corrections for filtration, the pick-up ion ratios compare remarkably well with the predictions of CLOUDY, given the uncertainties in the spectrum of the incident radiation and filtration. It should also be noted that in the model the ratios vary considerably with distance into the cloud, and thus cloud geometry could play a part in any differences between the model and observations.

Figure 1 shows the ratio of n(He°)/n(H°) as a function of distance to cloud surface. The main effective ionizing radiation source for helium is the cloud boundary emission, while radiation from ϵ CMa dominates the hydrogen ionization. At the solar location n(He°)/n(H°)$\sim$1/12, but integrating to the cloud surface yields somewhat larger values since hydrogen becomes relatively more ionized towards the cloud surface. The observed ionizations of hydrogen versus helium are very sensitive to the relative column lengths between the Sun and cloud surface over which the neutrals form, and to the spectrum of the radiation field, so that if different surfaces of the cloud are illuminated by different radiation fields the N(H°)/N(He°) ratio will not be constant in different look directions through the cloud. This is the case since white dwarf observations give ratios varying between 12 and 19 (Dupuis et al., 1996), although some of these stars may sample more distant interstellar gas.

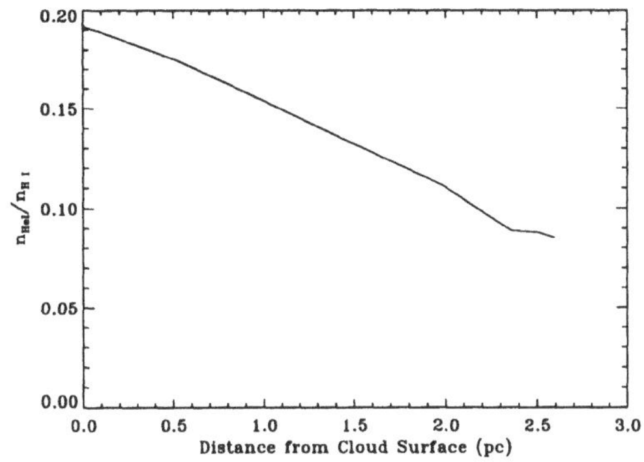

Figure 1. Ratio of n(He°)/n(H°) as a function to the LISM cloud surface, as calculated by CLOUDY. An absorbing column density of N(H°)=9 $\times$ 10^{17} cm^{-3} is reached at 2.6 pc.

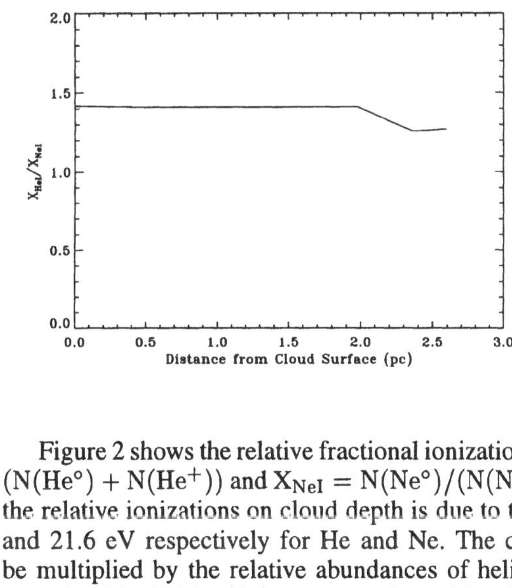

Figure 2. CLOUDY calculation of the relative helium and neon ion fractions, as a function of distance to the cloud surface. X_{HeI}=N(He°)/(N(He°)+N(He^+)), and X_{NeI}=N(Ne°)/(N(Ne°)+N(Ne^+)).

Figure 2 shows the relative fractional ionizations of helium and neon: $X_{HeI} = N(He°)/(N(He°) + N(He^+))$ and $X_{NeI} = N(Ne°)/(N(Ne°) + N(Ne^+))$. The weak dependence of the relative ionizations on cloud depth is due to the similar ionization potentials, 24.6 eV and 21.6 eV respectively for He and Ne. The calculated ratio, $X_{HeI}/X_{NeI} \sim 1.27$ can be multiplied by the relative abundances of helium and neon (A_{He}/A_{Ne} = 800) to give n(He)/n(Ne)=1016, which compares favorably with the pick-up ion value of 1450 (+1290, –820) (Geiss et al., 1994).

4. Conclusions

If filtration of O°and N°occur in the outer solar system and plasma sheath, the above CLOUDY steady-state predictions give remarkably good agreement with the pick-up ion data. After correcting the CLOUDY predictions for 50% filtration of O°and N°, the predicted ratios for n(He°)/n(O°), n(N°)/n(O°), n(Ne°)/n(O°), and n(He°)/n(Ne°) are 332, 0.14, 0.32, 1016 respectively. These can be compared with the pick-up ion data ratios of 290 (+190, –100), 0.13 (+0.06, –0.06), 0.20 (+0.12, –0.09), 1450 (+1290, –820), respectively.

These results, based on the assumption of steady state ionization and an incident radiation field which includes ϵ CMa radiation, EUV white dwarf radiation, and the hard EUV radiation field from the conductive interface of the local cloud boundary, are consistent with the pick-up ion data. However, non-equilibrium ionization cannot be ruled out since the youth, distance, and low intervening column density of the Vela pulsar indicate that the parent supernova explosion may also have affected LISM ionization. Support for this idea is also provided by the apparent signature of the Vela explosion in the terrestrial geological record. An improved understanding of both the LISM morphology and EUV radiation sources can be expected to improve the accuracy of these models, but we are encouraged by the agreement between this simple model and pick-up ion data.

Acknowledgements

PCF would like to thank the National Aeronautics and Space Administration for its support of her research. JDS acknowledges support by NASA's Space Physics Supporting Research and Technology, Suborbital, Theory and Educational Programs.

References

Arnett, W. D., Bahcall, J. N., Kirshner, R. P., and Woosley, S. E., 1989, Annual Review of Astronomy and Astrophysics, **27,** 629.
Aschenbach, B., Egger, R., & Trumper., 1995, NATURE, **373,** 587.
Bloch, J.J., 1988, PhD Thesis, University of Madison, Wisconsin.
Bowyer, S., Malina, R.F., 1991, in *Extreme Ultraviolet Astronomy,* R.F. Malina and S. Bowyer (eds.), (New York: Pergamon), p. 397.
Brakenridge, G. R.: 1981, Icarus, **46**, 81
Cheng, K.P., Bruhweiler, F.C., 1990, Astrophysical Journal, **364,** 573.
Dupuis, J., Vennes, S., Bowyer, S., Pradhan, A. K., and Thejll, P. 1996, Astrophysical Journal, **455,** 574.
Fahr, H.J., Osterbart, R., and Rucinsky, D., 1995, Astronomy and Astrophysics, **294**, 587.
Fahr, H.J., and Osterbart, R., 1995, Adv. Space Res., **16**, 125.
Ferland, G. J., 1993, University of Kentucky Department of Physics and Astronomy Internal Report.
Franco, G. A., 1990, Astronomy and Astrophysics, **227**, 499
Frisch, P. C.: 1981, Nature, **293**, 377
Frisch, P. C.: 1995, Space Science Reviews, **72**, 499
Frisch, P. C.: 1996, in preparation.
Frisch, P. C., York, D. G., 1986, in R. Smoluchowski, J. N. Bahcall, M. Matthews (eds.), *The Galaxy and the Solar System*, (Tucson: Univ. Arizona Press) pp. 83
Geiss, J., Gloeckler, G., Mall, U., Von Steiger, R., Galvin, A. B., and Ogilvie, K. W., 1994, Astronomy and Astrophysics, **282**, 924
Gondhalekar, P. M., Phillips, A. P., & Wilson, R., 1980, Astronomy and Astrophysics, **85**, 272
Gruen, E., Gustafson, B., Mann, I., Gaguhl, M., Morfill, G. E., Staubach, P., Taylor, A., and Zook, H. A. 1994, Astronomy and Astrophysics, **286**, 915
Jenkins, E.B., Meloy, D.A., 1974, Astrophysical Journal, Letters to the Editor, **193,** 121.
Kahn, F.D., Breitschwerdt, D., 1989, Monthly Notices of the RAS, **242,** 209.
Leahy, D. A., Nousek, J., & Garmire, G.: 1992, Astrophysical Journal,**385**, 561
MacLow, M.-M., McCray, R.: 1988, Astrophysical Journal, **324**, 776
McCray, R., Kafatos, M., 1987 Astrophysical Journal, **317,** 190.
Reynolds, R.J., 1990, Astrophysical Journal, **348,** 153.
Slavin, J.D., 1989, Astrophysical Journal, **346,** 718.
Strom, R., Johnston, H. M., Verbunt, F., & Aschenbach, B., 1995, NATURE, **373,** 590
Taylor, J. H., Manchester, R. N., and Lyne, A. G., 1993, Astrophysical Journal, Supplement Series, **88**, 529.
Vallerga, J., Welsh, B., 1995, Astrophysical Journal, **444**, 202.
Woosley, S.E., 1988, Astrophysical Journal, **330**, 218.

PROPERTIES OF THE INTERSTELLAR GAS INSIDE THE HELIOSPHERE

JOHANNES GEISS
International Space Science Institute, Hallerstraße 6, CH-3012 Bern, Switzerland.

MANFRED WITTE
Max-Planck-Institut für Aeronomie, D-37191 Katlenburg-Lindau, Germany.

Abstract. Methods and results of investigations of the interstellar gas inside the heliosphere are summarized and discussed. Flow parameters of H and He and the relative abundances of H, He, N, O, and Ne in the distant heliosphere are given. Charge exchange processes in front of the heliosphere affect the flow of hydrogen and oxygen through the heliopause. The speed of hydrogen is reduced by ~6 km/s, and screening leads to a reduction of the O/He and H/He ratios in the neutral gas entering the heliosphere. When the screening effect and the acceleration processes leading to the anomalous cosmic rays (ACR) are sufficiently understood, abundances in the LIC can be derived from measurements inside the heliosphere. Since isotopic ratios are virtually not changed by screening or by EUV and solar wind ionisation, relative abundances of isotopes in the gaseous phase of the LIC can be determined with no or minor correction from investigations of the neutral gas, pickup ions and ACR particles.

1. Introduction

Measurements of physical parameters and of the composition of the interstellar gas that has penetrated into the heliosphere find two kinds of application: (1) The nature of the interaction of the expanding solar atmosphere with the local interstellar cloud can be inferred from a comparison of velocities and temperatures of individual gaseous species and from differences in chemical composition inside and outside the heliopause. (2) When this interaction is sufficiently well understood, its influence on physical parameters and gas composition can be assessed and, where necessary, corrected for, in order to derive physical properties as well as chemical and isotopic abundances in the LIC.

In this paper the results and highlights are summarized, as they were presented and discussed in WG3. This working group had the mandate to compile and compare results and interpretations concerning "The Interstellar Gas in the Heliosphere: Density, Temperature, Flow Vector, and Chemical Composition". In the following three sections, methods of observation and experimental techniques are briefly mentioned and tables with results are given, with references to the original publications in this volume or elsewhere. Implications concerning the LIC-heliosphere interaction and the composition in the LIC are discussed in this paper as well as in the reports of other working groups.

Space Science Reviews **78**: 229–238, 1996.

2. Recent Results on the Physical Parameters of Hydrogen and Helium

In this section we compile and discuss recent results on physical parameters of the interstellar gas in the heliosphere, i.e. the flow vectors, temperatures and densities of hydrogen and helium. New data became available due to new and/or improved, sophisticated experimental methods, covering different observational ranges:

Local, direct observations: Close to the Sun (within $\sim$ 5 AU) the local distribution of the neutral helium has been measured, using the impact-ionization method (Witte *et al.*, 1992, 1993, 1996) with the GAS-instrument on Ulysses (ULS), and densities and velocity distribution functions of the pick-up ions H^+, He^+ and He^{++} were determined with the ULS/SWICS instrument (Gloeckler *et al.*, 1993; Gloeckler, 1996; see also Möbius *et al.*, 1996; Möbius, 1996).

Remote observations inside the heliosphere: The hydrogen distribution in the vicinity of the Sun and in the inner heliosphere has been further studied with the GHRS spectrometer on board the Hubble Space Telescope (HST). With the high spectral resolution of this instrument Doppler shifts in the Ly-α glow can be resolved providing unprecedented information on the heliospheric hydrogen parameters (Lallement, 1996).

Observations in the Local Interstellar Cloud: With new observations from the HST of the Doppler-shift in the absorption lines of nearby stars the Doppler Triangulation Method (DTM) provided further evidence on the structure of the local interstellar medium and the motion of the solar system in the Local Interstellar Cloud (cf. Lallement, 1996).

The results obtained from the various observations – as presented at this workshop – are compiled in Table I. The data in the first and third part of the table refer to the distant heliosphere. The velocity vector of hydrogen determined in the LIC is given in the second part. For details concerning the methods of determination, the reader is referred to the original papers. Some highlights as well as some problems will be further discussed here.

Most noticeable is the considerable agreement in the helium parameters obtained recently by different methods of determination. Based now on a much larger data set obtained with the ULS/GAS instrument, the uncertainties in the velocity and temperature measurements of earlier results (Witte *et al.*, 1993) have been considerably reduced (Witte *et al.*, 1996). Using a multiparameter fitting process, a comprehensive study of the influence of potential systematic and statistical errors has been performed (Banaszkiewicz *et al.*, 1996) and the errors quoted in Table I are the present best estimates for the confidence levels. In addition, with the determination of the actual loss rates from the simultaneous observation of particles on "direct" and "indirect" orbits it became possible to derive a helium density range of $\sim 1.4 - 1.7 \times 10^{-2}$ cm^{-3}, thus reducing the uncertainty due to the poor knowledge of the actual loss rate or its estimates using proxy data like the 10.7 cm solar radio flux (cf. Rucinski *et al.*, 1996).

Table I
Hydrogen and Helium Parameters.

METHOD	VELOCITY (km/s)	DIRECTION $\beta(°)$	$\lambda(°)$	TEMPE-RATURE (K)	DENSITY 10^{-2} cm^{-3}	@ IONIZA-TION RATE 10^{-7} s^{-1}	OBSER-VATION	REFERENCES
INTERSTELLAR HELIUM								
pick-up He^+	23-30			4800 - 7200	0.9–1.2		AMPTE	Möbius, 1996
pick-up He^+					**1.5**	**0.75**	ULS/SWICS	Gloeckler, 1996
pick-up He^{++}					1.5	0.85	ULS/SWICS	Gloeckler, 1996
direct	**25.3±0.4**	**74±0.8**	**−5.6±0.4**	**7000±600**	**1.4–1.7**	**0.6–1.4**	ULS/GAS	Witte *et al.*, 1996
UV	19–24			8000	0.5–1.4		Prognoz V1, V2	Chassefière *et al.*, 1988
VELOCITY VECTOR OF THE LOCAL INTERSTELLAR CLOUD								
Doppler-shifted absorption lines	**25.7**	74.5	−7.8				HST	Lallement, 1996
HELIOSPHERIC HYDROGEN								
pick-up H^+					11.5±2.5	**5.5**	ULS/SWICS	Gloeckler, 1996
UV	**18 – 20**			**8000**		5.5	HST	Lallement, 1996
UV		72	−7		17		GLL	Ajello *et al.*, 1994
UV	19 – 21	74	−7	8000	6.5		Prognoz	Bertaux *et al.*, 1985
					14			Quémarais *et al.*, 1994

Improvements in the results from the pickup ion measurements were also obtained. Gloeckler *et al.* (1995) and Möbius *et al.* (1996) pointed out that the scattering mean free path, isotropizing the spatial distribution of the pick-up ions plays an important role in the interpretation of the pick-up ion measurements. Using the ULS/SWICS data, Gloeckler *et al.* (1995) and Gloeckler (1996) found large ($\sim$ 1 AU) scattering mean free paths for hydrogen in the fast solar wind regime over the solar poles which are significantly larger than assumed in earlier investigations. In addition, from these data the loss rates of the neutral particles as well as the production rates of the pick-up ions can be determined. As a result a density of $n_{\mathrm{He}} \sim 0.015$ cm^{-3} was derived, in agreement with the result of the ULS/GAS experiment, but significantly larger than the values quoted in earlier pickup ion investigations.

The agreement in the neutral helium densities determined from two different but complementary methods as well as the agreement of the flow vectors of the helium inside the heliosphere (25.3 km s^{-1}, 74°, -5°) obtained from the in-situ measurements and outside the heliosphere from the Doppler Triangulation Method (25.7 km s^{-1}, 74.5°, -7.8°) makes the helium parameters presently the most precisely determined ones, as compared to other constituents of the interstellar medium.

The agreement of the measured velocity vectors is also confirming the general assumption that the interstellar helium is not significantly affected when passing through the heliospheric interface.

The determination of the hydrogen parameters, based until recently mainly on the measurements of the Ly-α glow with hydrogen cells, has been substantially improved by the observations with the GHRS-spectrometer on HST (cf. Lallement, 1996). The observed Doppler shifts in the Ly-α line is best fit by a bulk velocity of 19–20 km s^{-1} and a temperature of 8000 K. An excess in the line width in the up-wind direction can be explained by a velocity dispersion due to a flow of two components, i.e. a minor component having retained the original velocity (at $\sim$ 26 km s^{-1}) and the dominant component with a bulk velocity of 19–20 km s^{-1} resulting from charge exchange reactions at the heliosphere interface (Lallement, 1996).

Whereas the helium density and the difference of about 5–6 km s^{-1} between heliospheric hydrogen (19–20 km s^{-1}) and helium (25–26 km s^{-1}) is now well established, the density of the heliospheric hydrogen is presently the most uncertain parameter. During the last decade, n_{H} values in the range of 0.065 to 0.17 cm^{-3} were obtained by different techniques (cf. Table I). The situation has improved recently. The latest Ly-α glow investigations give $n_{\mathrm{H}} = 0.14 - 0.17$ cm^{-3} (Ajello *et al.*, 1994; Quémarais *et al.*, 1994), while the most recent pick-up ion measurements give $n_{\mathrm{H}} = 0.115$ cm^{-3} (Gloeckler, 1996).

Table II
Neutral interstellar gas abundances in the distant heliosphere.

	Interstellar Gas Distant Heliosphere	Solar System [g]
H/He	7.7 ± 1.3 [a]	10
O/He (10^{-3})	$3.5^{+1.8}_{-1.4}$ [b]	8.7
N/O	0.13 ± 0.06 [b]	0.13
Ne/O	$0.20^{+0.12}_{-0.09}$ [b]	0.14
N/He (10^{-3})	$0.5^{+0.5}_{-0.3}$ [c]	1.2
Ne/He (10^{-3})	$0.7^{+1.0}_{-0.5}$ [c]	1.3
C/O	0.0053 ± 0.0036 [d]	0.42
O/H (10^{-3})	$0.46^{+0.64}_{-0.22}$ [e]	0.85
$[\mathrm{O/H}]_{\mathrm{total}}$ (10^{-3})	$0.61^{+0.56}_{-0.29}$ [f]	0.85

[a] Gloeckler (1996)
[b] Geiss *et al.* (1994)
[c] Calculated from the data given by Geiss *et al.* (1994)
[d] Cummings and Stone (1996)
[e] Combining the data of Gloeckler (1996) and Geiss *et al.* (1994)
[f] Assuming 25% of oxygen in grains
[g] Anders and Grevesse (1989)

3. Elemental Abundances

Several methods have been applied or are being developed for studying abundances of LIC material inside the heliosphere: Resonant photon backscattering (Bertaux *et al.*, 1985; Lallement *et al.*, 1993; Quémerais *et al.*1994; Lallement, 1996), neutral gas analyses, either in situ (Witte *et al.*, 1993, 1996) or by capture in foils (Bühler *et al.*, 1993), pickup ion mass spectrometry (cf. Gloeckler, 1996), and anomalous cosmic ray measurements (Cummings and Stone, 1996; Leske *et al.*, 1996). From the data obtained by these methods, the relative abundances of the neutral component of the interstellar gas in the "distant heliosphere" are derived, after taking into account effects directly caused by the sun in its vicinity, i.e. gravitational focussing, photon pressure and ionisation by solar UV and solar wind. The results for elemental abundances in the distant heliosphere are given in Table II. In cases where several measurements are available, the most recent result or the result with the smallest error is listed. The limits of error are those given in the original publications.

In deriving the neutral gas abundances in the distant heliosphere, possible contaminations from other sources have to be considered. For hydrogen and the noble gases, this problem is easily avoided by keeping away from obvious sources such as planetary magnetospheres. Recent measurements (Geiss *et al.*, 1996) indicate that significant fractions of the O^+ and N^+ pickup ions are due to evaporation from grains for solar distances < 3 AU. Away from the ecliptic plane, the majority of these grains are of interstellar origin (Baguhl *et al.*, 1996). Since the O and N data given in Table II are derived from pickup ion fluxes measured near 5 AU and from speed distributions near the cutoff speed of twice the solar wind speed (corresponding to ions produced near 4 to 5 AU), these results are unaffected by contamination from dust.

The situation with carbon is different. In the inner heliosphere, the dominant source of C^+ are dust grains (Geiss *et al.*, 1995, 1996). Even at the aphelion of Ulysses ($\sim$ 5.4 AU), C^+ pickup ions that are created from the interstellar gas could not be distinguished from C^+ ions derived from grains.

In the ACR which derive their ions from the whole volume of the supersonic heliosphere, the interstellar gas may still be a major source of C^+ (Cummings and Stone, 1996). However, grains could make a significant contribution, as a comparison of different source strengths in the heliosphere suggests (Geiss *et al.*, 1996, Table I). The CLOUDY model of Frisch and Slavin (1996) gives C/O = 0.0017 for the neutral gas in the LIC. If this model correctly reflects the degrees of ionisation of C and O, we have again an indication that a major part of ACR carbon is not related to the interstellar gas.

The interstellar gas is quite well established as the principal source of the other identified ACR components, i.e. H, He, N, O, Ne (cf. Fisk *et al.*, 1974; Cummings and Stone, 1996; Geiss *et al.*, 1996).

In column (3) of Table II we give the solar system abundances which represent the elemental abundances in the solar nebula as a whole (atoms, ions, molecules, grains). Since the formation of the solar system 4.6 Gy ago, the chemical composition of the galaxy has further evolved, with generally increasing abundances of heavier elements relative to hydrogen. These increases are probably only of the order of 10–20%, and therefore, are not too important in view of the present uncertainties in the measured abundances.

When comparing the abundances in the distant heliosphere (column 2) with total elemental abundances estimated for the LIC, three factors must be taken into account):

(1) Some of the material in the LIC is contained in grains. the relative proportion depends on the chemical properties of the element in question. Virtually all H, He and Ne is in the gas phase.

(2) A fraction of the material in the gas phase of the LIC is ionized and does not enter the heliosphere. The degree of ionisation depends on the EUV flux and spectrum, estimates for several elements are given by Frisch and Slavin (1996), and for H and He by Vallerga (1996).

(3) Charge transfer between ions and atoms in the upstream region of the heliosphere leads to a depletion of certain species in the neutral gas crossing the heliopause (Ripken and Fahr, 1983). Most affected by this "screening effect" are hydrogen and oxygen, because the atoms and ions of these two elements undergo

resonant charge exchange with H^+ and H, respectively. Helium atoms, on the other hand, ought to be least affected, because of its small charge transfer cross section with protons. We therefore consider neutral helium as a standard and define the Screening Effect S_E for an element Z as

$$S_E(Z) = 1 - \frac{[n(Z)/n(\mathrm{He})]_{\mathrm{DH}}}{[n(Z)/n(\mathrm{He})]_{\mathrm{LIC}}} \tag{1}$$

The number densities n both in the distant heliosphere (DH) and in the LIC refer to the neutral component of the gas. S_E will depend on the angle from the apex direction, which would be reflected in a latitudinal dependence of ACR fluxes. Since pickup ions in the inner solar system (R < 10 AU) are produced from interstellar atoms which entered the heliosphere not far from the apex, screening should produce no significant latitudinal effect for them.

Helium and neon should be least affected by screening. These light noble gases will not be contained in grains, and, since He and Ne have high and similar ionisation potentials, we expect the Ne/He ratio in the distant heliosphere to be close to $(\mathrm{Ne/He})_{\mathrm{LIC}}$. So far, the error limits for Ne/He are high, due to limited counting statistics of Ne^+.

An average HI/HeI ratio of 14 is observed in the LIC (Dupuis *et al.*, 1996), in agreement with the values estimated from EUV fluxes by Vallerga (1996) and Frisch and Slavin (1996). Combining this with the H/He ratio of 7.7 in the distant heliosphere (Gloeckler, 1996) gives $S_E(\mathrm{H}) = 0.5$. We note that the screening effect would be stronger, if fluxes instead of densities would be used in the definition of S_E (Eq. 1), because the speed of H is reduced by $\sim$ 6 km/s in the interaction region.

Earlier model calculations suggested that S_E(O) could be much higher than S_E(H). However, by comparing the relevant exchange processes, Geiss *et al.* (1994a) concluded $S_E(\mathrm{O}) \leq S_E(\mathrm{H})$, which was confirmed by model calculations of Fahr *et al.*(1995). Assuming that the fractional ionisation of H and O in the LIC is the same and assuming that 25% of the total oxygen in the LIC is contained in grains, we have calculated the $(\mathrm{O/H})_{\mathrm{total}}$ ratio in the LIC from the H/He and O/He ratios in the distant heliosphere. The result is included in Table II. Comparison with the O/H ratio in the solar system shows that the data are still compatible with $S_E(\mathrm{O}) \leq S_E(\mathrm{H})$. Since this condition is derived from atomic data (Geiss *et al.*, 1994a) and therefore ought to be quite independent of model assumptions, it would be worthwhile to reduce the uncertainties of O/He both in the LIC and in the distant heliosphere, in order to have a more precise check on the relative magnitudes of S_E(H) and S_E(O).

The abundance ratios in the distant heliosphere will be improved by use of additional data being accumulated, both from Ulysses and from spacecraft operating near 1 AU. We note, however, that the possibilities of investigating interstellar pickup ions at 1 AU are limited, because here grains are the main contributor of C^+, N^+ and O^+ (Geiss *et al.*, 1996). Also interstellar H^+ is difficult to investigate at 1 AU, because its abundance relative to the solar wind protons is several orders of magnitude lower than it is near 5 AU, where interstellar and solar protons can be very well separated (cf. Gloeckler, 1996). It might still be possible to identify interstellar H^+, N^+ or O^+ at 1 AU, especially in the anti-apex region. Since interstellar grains and heavier interstellar gas components are presumably focussed in a similar way into the anti-apex direction, it will be difficult at 1 AU to distinguish between the interstellar grain and gas sources for elements such as N or O. Thus among all interstellar gas components, He and Ne are by far the most suited for quantitative investigations of neutrals or pickup ions by spacecraft operating near 1 AU. ACR particle analyses will be much less subject to these limitations, because they are produced from neutrals throughout the supersonic heliosphere.

4. Isotopic Abundances

Isotopic abundances in the distant heliosphere and in the LIC can be much more directly related than chemical abundances. Ionisation rates by interstellar and solar EUV and by solar wind charge exchange are virtually the same for isotopes, and also the screening effect S_E of isotopes should be very similar. Thus isotopic analyses of pickup ions and of the ACR is a very promising way to obtain interstellar nuclear abundances. In the future, these two methods will be complemented by isotopic analyses of neutral interstellar particles and interstellar grains.

In Table III we list the isotopic ratios obtained so far for the LIC. Data from three methods are included, namely EUV absorption spectroscopy against nearby stars, pickup ion mass spectrometry, and ACR isotopic analyses.

The D/H and $^3He/^4He$ ratios have been determined in various parts of the galaxy. For the first time, however, we have in the LIC a determination of these two isotopic ratios in the same reservoir of the present-day galaxy. This is significant, because the D and ^3He abundances are strongly interrelated by low-temperature nuclear processes in stars. From a comparison of their abundances in the LIC and in the solar nebula important conclusions can be drawn regarding the primordial production of the light nuclei and the evolution of their abundances in the galaxy (Linsky, 1996; Gloeckler and Geiss, 1996).

Except for hydrogen and helium, the isotopic abundances of the elements listed in Table III are still too limited in precision to become important for astrophysical application. However, Table III gives an indication of the potential of such measurements. As explained above, isotopic data obtained from ACR particles, pickup ions and the neutral interstellar gas can be directly interpreted in terms of isotopic abundances in the LIC. No or only minor corrections are needed to account for separation effects at the heliopause or in the inner heliosphere. If the progress of recent years continues, we may foresee meaningful isotopic analyses for a set of important elements in the interstellar gas and in interstellar grains. The LIC would then represent a reservoir in the present galaxy with a defined nucleosynthetic

Table III
Isotopic abundances in the local interstellar gas.

Isotopic Ratio	Abundance Ratio LIC (Gas Phase)	Abundance Ratio Protosolar Nebula
D/H	$(1.6 \pm 0.2) \times 10^{-5}$ [a)]	*2.6×10^{-5}* [e)]
^{3}He/^{4}He	$(2.2 \pm 0.8) \times 10^{-4}$ [b)]	*1.5×10^{-5}* [e)]
(D+^{3}He)/H	$(3.8 \pm 1.0) \times 10^{-5}$ [c)]	$(4.1 \pm 1.0) \times 10^{-5}$ [f)]
^{15}N/^{14}N	< 0.023 [d)]	0.0037 [g)]
^{18}O/^{16}O	< 0.0034 [d)]	0.0020 [h)]
^{22}Ne/^{20}Ne	$0.077^{+0.085}_{-0.023}$ [d)]	0.073 ± 0.002 [i)]

[a)] Linsky *et al.* (1993); Linsky (1996); from Lyman-α absorption in the spectra of nearby stars.
[b)] Gloeckler and Geiss (1996); from interstellar pickup ions.
[c)] The sum of [a)] and [b)]
[d)] Leske *et al.* (1996); from ACR particles.
[e)] Solar wind data give (D+^{3}He)/H, cf. [f)]. The ^{3}He/^{4}He ratio in the planetary component of meteoritic gases (Black, 1972; Eberhardt, 1974) has been used to estimate the fraction of ^{3}He contributing to the sum D+^{3}He (cf. Black, 1972; Geiss and Reeves, 1972; Geiss, 1993). The resultant D/H is compatible with the D/H ratios of $\sim (2-4) \times 10^{-5}$ obtained for Jupiter (cf. Gautier and Owen, 1983; Carlson *et al.*, 1993). The values for D/H and ^{3}He/^{4}He in the solar nebula are given in italics and without error limits, in order to indicate that they are less directly determined than the corresponding data for the LIC. Results by the Galileo probe (Niemann *et al.*, 1992) on the hydrogen and helium isotopes in Jupiter's atmosphere could contribute to arriving at firmer conclusions regarding D/H and ^{3}He/^{4}He in the solar nebula.
[f)] Since the D+^{3}He in the protosolar nebula can be directly derived (cf. Geiss and Reeves, 1972) from solar wind ^{3}He measurements (Geiss *et al.*, 1972; Coplan *et al.*, 1984; Bodmer *et al.*, 1995), this sum is included here.
[g)] Given here is the ^{15}N/^{14}N in the Earth's atmosphere, although it is probably not exactly representative of the protosolar value (cf. Kerridge *et al.*, 1992).
[h)] Anders and Grevesse, 1989.
[i)] Measured in the solar wind (Geiss *et al.*, 1972). The same value was found in the separated solar wind component of the trapped gases in lunar surface material (Benkert *et al.*, 1993).

status, which can be compared to the older solar nebula. Even though they are not of common descent, comparing two isotopically well defined clouds will give us direct information on the evolution of the galaxy during the last 5 billion years.

References

Ajello, J.M., Pryor, W.R., Barth, C.A., Hord, C.W., Steward, A.I.F., Simmons, K.E., and Hall, D.T.: 1994, *Astron. and Astrophys.* **289**, 283.
Anders, E. and Grevesse, N.: 1989, *Geochim. Cosmochim. Acta* **53**, 197–214.
Baguhl, M, Grün, E., and Landgraf, M.: 1996, *Space Sci. Rev.*, this issue.
Banaszkiewicz, M., Witte, M., and Rosenbauer, H.: 1996, *Astron. and Astrophys. Suppl. Ser.*, in press.
Benkert, J.P., Baur, H., Signer, P., and Wieler, R.: 1993, *J. Geophys. Res.* **98**, 13147–13162.
Bertaux, J.L., Lallement, J.L., Kurt, V.G., Mironova, E.N.: 1985, *Astron. and Astrophys.* **150**, 1.
Black, D.C.: 1972, *Geochim. Cosmochim. Acta* **36**, 347–375.
Bodmer, R., Bochsler, P., Geiss, J., von Steiger, R., and Gloeckler, G.: 1995, *Space Sci. Rev.* **72**, 61–64.

Bühler, F., Lind, D.L., Geiss, J., and Eugster, O.: 1993, in *A.S. Levine, NASA Conf. Publ.* **3194**, Part 2, 705–722.
Carlson, B.E., Lacic, A.A., and Rossow, W.B.: 1993, *J. Geophys. Res.* **98**, 5251-5290.
Chassefière E., Dalaudier F., Bertaux J.L.: 1988, *Astron. Astrophys.* **201**, 113.
Coplan, M. A., Ogilvie, K.W., Bochsler, P. and Geiss, J.: 1984, *Solar Phys.* **93**, 415–434.
Cummings, A.C. and Stone, E.C.: 1996, *Space Sci. Rev.*, this issue.
Dupuis, J., Vennes, S., Bowyer, S., Pradhan, A.K. and Thejll, P.: 1996, *Astrophys. J.* **455**, 574–589.
Eberhardt, P.: 1974, *Earth and Planet. Sci. Lett.* **24**, 182–187.
Fahr, H.J., Osterbart, R. and Rucinski, D.: 1995, *Astron. and Astrophys.* **294**, 587.
Fisk, L.A., Koslovsky, B. and Ramati, R.: 1974, *Astrophys. J. Letters* **190**, L35.
Frisch, P. and Slavin, J.: 1996, *Space Sci. Rev.*, this issue.
Gautier, D. and Owen, T.: 1983, *Nature* **304**, 691–694.
Geiss, J.: 1993, in Prantzos, N., Vangioni-Flam, E., Cassé, M. (eds.), *Origin and Evolution of the Elements*, Cambridge University Press, pp. 112–131.
Geiss, J., Bühler, F., Cerutti, H., Eberhardt, P. and Fillieux, C.: 1972, in *Apollo 16 Preliminary Science Report*, **NASA SP-315**, Washington, D.C., Chapter 14.
Geiss, J. and Reeves, H.: 1972, *Astron. and Astrophys.* **18**, 126–132.
Geiss, J., Gloeckler, G., Mall, U., von Steiger, R., Galvin A.B. and Ogilvie, K.W.: 1994, *Astron. and Astrophys.* **282**, 924–933.
Geiss, J., Gloeckler, G., and von Steiger, R.: 1994a, *Phil. Trans. Royal. Soc. A* **349** 213–226.
Geiss, J., Gloeckler, G., Fisk, L. A., and von Steiger, R.: 1995, *J. Geophys. Res.* **100**, 23,373–23,377.
Geiss, J., Gloeckler, G., and von Steiger, R.: 1996, *Space Sci. Rev.*, this issue.
Gloeckler, G.: 1996, *Space Sci. Rev.*, this issue.
Gloeckler, G., Geiss, J., Balsiger, H., Fisk, L. A., Galvin, A. B., Ipavich, F. M., Ogilvie, K. W., von Steiger, R., and Wilken, B.: 1993, *Science* **261**, 70–73.
Gloeckler, G., Schwadron, N.A., Fisk, L.A. and Geiss, J.: 1995, *Geophys. Res. Letters* **22**, 2665–2668.
Gloeckler, G. and Geiss, J.: 1996, *Nature* **381**, 210–212.
Kerridge, J.F., Bochsler, P. Eugster, O. and Geiss, J.: 1992, *Lunar and Planetary Science XXII*, Lunar and Planetary Science Institute, Houston, 239–248.
Lallement, R.: 1996, *Space Sci. Rev.*, this issue.
Lallement R., Bertaux J.L., and Clarke J.: 1993, *Science* **260**, 1095.
Leske, R.A., Mewaldt, R.A., Cummings, A.C., Cummings, J.R., and Stone, E.C.: 1996, *Space Sci. Rev.*, this issue.
Linsky, J.L.: 1996, *Space Sci. Rev.*, this issue.
Linsky, J.L. *et al.*: 1993, *Astrophys. J.* **402**, 694–709.
Möbius, E.: 1996, *Space Sci. Rev.*, this issue.
Möbius, E., Rucinski, *et al.*: 1996, *Ann. Geophysicae* **14**, 492–496.
Niemann, H.B. *et al.*: 1992, *Space Sci. Rev.* **60**, 111.
Quémerais, E.J., Bertaux, J.L., Sandel, B.R., and Lallement, R.: 1994, *Astron. and Astrophys.* **290**, 961.
Ripken, H.W. and Fahr, H.J.: 1983, *Astron. and Astrophys.* **122**, 181–192.
Rucinski, D., Cummings, A.C., Gloeckler, G., Lazarus, A.J., Möbius, E., and Witte, M.: 1996, *Space Sci. Rev.*, this issue.
Vallerga, J.: 1996, *Space Sci. Rev.*, this issue.
Witte, M., Rosenbauer, H., Keppler, E., Fahr, H., Hemmerich, P., Lauche, H., Loidl, A. and Zwick, R.: 1992, *Astron. Astrophys. Suppl. Ser.* **92**, 333.
Witte, M., Rosenbauer, H., Banaszkiewicz, M. and Fahr, H.: 1993, *Adv. Space Res.* **13**, (6)121.
Witte, M., Banaszkiewicz, M. and Rosenbauer, H.: 1996, *Space Sci. Rev.*, this issue.

LOCAL CLOUDS : DISTRIBUTION, DENSITY AND KINEMATICS THROUGH GROUND-BASED AND HST SPECTROSCOPY

CECILE GRY
Laboratoire d'Astronomie Spatiale (CNRS), BP8, F-13376 Marseille cedex 12

Abstract. The distribution, kinematics and physical properties of the interstellar matter surrounding the Sun can be inferred from ground-based and UV spectroscopic observations. On a 200 pc scale the local interstellar matter appears inhomogeneous and asymmetric. Although it generally flows towards the lower density region, it is composed of numerous small components a few parsecs in size with slightly different velocities. On a smaller scale the extent and the nature of the Local Cloud which flows over the Sun are discussed based on HST-GHRS observations of nearby stars.

Key words: ISM : general, abundances, clouds, density - solar neighborhood

1. Introductory remark

What is implicitly understood as the "local" interstellar medium can vary from one author to another depending on the scope of their study and interest. Therefore the subject of "local clouds" will be addressed here in two different scales : the large scale will designate here the first 200 parsecs, involving large samples of target stars, while the small scale will imply describing the contours of an individual cloud.

2. The Local Interstellar Medium at Large Scale (~200 pc)

2.1. DISTRIBUTION

An obvious characteristic of the distribution of matter in the local interstellar medium is its inhomogeneity which has been known since the beginning of its study by the absorption lines. Indeed two close lines of sight can have largely differing column densities, and if one looks at a plot of column density as a function of distance, it is clear that column density is not a strictly monotonic function of distance (see for example the plots in Frisch and York 1991).
The asymmetry in the distribution is another important characteristic that has been known for more than 10 years. If one looks at maps giving the contours of equal column density (e.g. Frisch and York 1983, Paresce 1984) one can notice a rapid increase of column density in the Sco-Oph direction (more or less the Galactic center direction), where N(HI) reaches 10^{20} cm^{-2} at a few tens of parsec, and 2×10^{21} at cm^{-2} before 200 pc. In the opposite direction, however, in the IIIrd quadrant, there is a lack of neutral gas in particular in the Canis Majoris direction

Space Science Reviews **78**: 239–246, 1996.

(l~ 225°) where N(HI) does not exceed 2×10^{18} cm^{-2} at 200 pc (Gry et al. 1985, Vallerga et al. 1993, Cassinelli et al. 1996).

2.2. KINEMATICS

All studies of interstellar absorption lines show a complex velocity structure in almost all directions, on all scales. Whatever the spectral resolution achieved - 10 km/s for the ultraviolet Copernicus observations, 3.5 km/s for the actual ultraviolet observations with HST and for many ground-based observations, and even less than 1 km/s for some of them (and not always the most recent ones, e.g. Hobbs 1969 !) - the interstellar profiles show velocity structures down to the very resolution.
Vallerga et al. (1993) has studied the spatial correlation of the velocities, with the CaII lines in a sample of stars closer than 50 pc. They looked for clouds that intersect multiple lines of sight and show that in fact most components do **not** have a counterpart (with a precision of 2 km/s) in a nearby sight-line. They concluded that most CaII clouds subtend angles less than 15°. At the mean distance of these stars, 30 pc, 15° correspond to a distance of 7 pc, which is thus a typical upper limit for the size of the CaII clouds.
Several authors have tried to identify among all velocity components a velocity vector defined by the direction (l,b) and the amplitude V that would show a coherent motion of the local interstellar medium.
First Crutcher (1982) used the TiII absorption line in 66 sightlines among which only 7 are shorter than 100 pc. He found a 77% match with the following vector for the downwind flow (heliocentric) : [l = 205°, b = −10°, V=28km/s]. Frisch and York (1986) performed a similar study with stars closer than 100 pc, and found a somewhat different vector : [l = 214°, b = −15°, V = 27 km/s]. Welsh et al. (1991) used NaI lines for stars between 50 and 200 pc and found yet another vector : [l = 232°, b = −29°, V = 22 km/s]. All these studies infer a similar velocity vector, indicating that there is some coherency in the general flow of material in the first 200 pc. The material flows more or less from the higher density regions towards the empty region. However there are significative discrepancies between the different methods, which come from their sampling slightly different media. This definitely shows that on the scales considered the LISM does not have a coherent motion, but that every small volume of gas is moving slightly differently from the mean of the LISM.
Only when considering a very reduced number of very short sightlines, did Lallement and Bertin (1992) succeed in identifying a common velocity vector which also coincides with the interstellar wind felt inside the solar system (see Lallement, this volume). Their vector thus traces the smallest volume of gas in which the Sun is embedded, that presents a coherent motion. The volume responding to this definition is now called the LIC (Local Interstellar Cloud).

Table I

Density indicators derived from measurements of MgII and FeII with the HST-GHRS.

(1)	(2)	(3)	(4)	(5)	(6)	(7)	(8)	(9)	(10)	(11)
star (ref)	l,b	Dist (pc)	N_{LIC}(H) 10^{18} at cm^{-2}		n_{LIC}(H) at cm^{-3}		N_{total}(H) 10^{18}at cm^{-2}		n_{total}(H) at cm^{-3}	
N(H) from			MgII	FeII	MgII	FeII	MgII	FeII	MgII	FeII
α Psa (1)	20.5,-65	6.7	-	2.6	-	.13	-	3.9	-	.19
α Aql (1)	48,-9	5	1.15	1.7-3.4	.08	.11-.23	3.6	5.3	.24	.35
α Lyr (1)	67.5,19	7.5	-	4.8-9.0	-	.21-.4	-	17	-	.8
δ Cas (1)	127,-2	27	1.5	-	.018	-	1.5	-	.02	-
G191B2B (2)	156,7	48	.99	2	.007	.014	5.4	8.4	.04	.06
α Aur (3)	162.5,4.5	12.5	1.6	2.1	.043	.056	1.6	2.1	.04	.06
α CMi (3)	214,13	3.5	.53	.77	.05	.073	.82	1.4	.08	.13
α CMa (4)	227,-9	2.7	.39	.59	.048	.073	.60	.97	.07	.12
ϵ CMa (5)	240,-11	187	.69	.93	.0012	.0016	1.3	1.5	.002	.003
α Cen (6)	316,-1	1.3	-	-	-	-	1.1	1.9	.28	.41

References : (1) Lallement et al. 1995 ; (2) Lemoine et al. 1996 ; (3) Linsky et al. 1995 ; (4) Lallement et al. 1994 ; (5) Gry et al. 1995 ; (6) Linsky and Wood 1996

3. HST-GHRS Study of the LISM at Small Scale.

Ultraviolet lines are considerably more sensitive to low column densities than optical lines partly because they more often trace the dominant ionization stage of the elements. They are therefore more useful for studying the interstellar gas on very small scales than the ground-based surveys. However, only recently with the GHRS on board the HST, have we been able to observe the ultraviolet lines with the required resolution to distinguish the absorption due to the LIC from the other velocity components. For these reasons the GHRS is the most suitable instrument to study the local interstellar medium at very small scale.
The GHRS sight-lines suitable for this study are listed in Table I. They are also displayed in Figures 1 and 2 as a function of their galactic longitude and distances from the Sun. Note that the figures do not represent projections onto the galactic plane, however all stars except α PsA are relatively close to the plane. Therefore in the following, the distribution of matter with galactic latitude will not be discussed.

3.1. Density Indicators

N(HI) is difficult to estimate directly, mainly because of the heavy saturation of the Lyman α line as well as of the uncertainty in the stellar line in the case of cool stars. It is therefore more convenient to compare column densities and thereby mean densities, inferred from other elements measured with a better accuracy. Usually only MgII, FeII and MgI measurements are available at the highest resolution for two reasons : a momentaneous failure of the GHRS Echelle Spectrometer (Ech-A) working at short wavelengths prior to the HST repair in December 1993 and the low fluxes of cool stars at these wavelengths (only the interstellar Lyman α line is observable for those stars at short wavelengths thanks to the photospheric emission line).

Both MgII and FeII present the advantage of being the dominant ionization stage in HI regions, although they can also account for HII regions as their ionization potential is slightly higher than that of HI. However they have the drawback of being largely depleted on grains, therefore only with the assumption that the depletion is constant in the local medium can they be considered as tracers of the total quantity of matter in the sight-lines. With this assumption we derive two independent mean density indicators by the following :

$n_{FeII}(H) = N(FeII)/A_w(Fe)\ 1/D = 6.9 \times 10^5\ N(FeII)\ 1/D$

$n_{MgII}(H) = N(MgII)/A_w(Mg)\ 10^{.67}\ 1/D = 2.3 \times 10^5\ N(FeII)\ 1/D$

where A_w(Fe) and A_w(Mg) are the abundances in the warm diffuse medium as given by Jenkins et al. (1986), and $10^{.67}$ is the correcting factor found by Sofia et al. (1994) in the case of ξ Per to account for a suspected error in the MgII 1239A line f-value.

The results are gathered in Table I.

A component is detected at the predicted LIC velocity for all sight-lines but α Cen. Extra components are also detected except for two sight-lines ; therefore, for each sight-line we show both the LIC mean density which represents the density if the LIC extends out to the star (columns 6 and 7), and the total mean density with all velocity components together (columns 10 and 11).

Depletion

By comparing column 7 to column 6 and column 11 to column 10, one can note that n(H) as derived from FeII is a factor of 1.5 $\pm$ 0.2 larger than that derived from MgII. This shows that one of the depletion values might be wrong by 50 % and as there seems to be a greater uncertainty on the Mg depletion because the f-value is not well known, $n_{FeII}(H)$ will be used for the further discussions. In any case for our purpose, the important fact to note is that the abundance ratio $n_{FeII}(H)/n_{MgII}(H)$ is almost constant. In effect if there were important variations in the depletion among or inside the local clouds, this ratio would be likely to show variations also. In consequence, although $n_{FeII}(H)$ and $n_{MgII}(H)$ with the above definitions cannot formally be taken as density estimates because they depend on the choice made for the depletion values, they however do serve as indicators to discuss the relative densities in the sight-lines.

Ionization

Our density indicators do not measure only the neutral HI density but also account for some ionized gas, as the ionization potentials of both MgII and FeII are slightly above that of HI. For some of the lines of sight, we have a measure of N(HI), given in Table II :

Table II
Comparison of our FeII-derived densities with the direct measurements of HI.

	N(HI)	n_{FeII}(H)/n(HI)	ref
G191-B2B	1.85×10^{18}	1.08	(2)
α Aur	1.74×10^{18}	1.21	(3)
α CMi	7.5×10^{17}	1.03	(3)
α CMa	$0.9 - 2.5 \times 10^{17}$	2.3–6.5	(7)
ϵ CMa	$\leq 5 \times 10^{17}$	≥ 1.9	(5)
α Cen (cloud G)	$3 - 12 \times 10^{17}$	1.3–5	(6)

References : (7) Bertin et al. 1996, others : as in Table I

For the first three lines of sight listed above, our LIC column density derived from FeII is close to the measured HI column density. This could mean that the LIC is essentially neutral, unless the depletion of Fe II has been underestimated (this is not what is suggested by the ratio n_{FeII}(H)/n_{MgII}(H)). However for the two CMa sight-lines as well as for the G cloud in the α Cen sight-line, the comparison could show that in this region the LIC (or G) is somewhat ionized. This could be explained by the fact that these sight-lines are the most exposed to the radiation from ϵ CMa that was shown by Vallerga and Welsh (1995) to largely dominate the ionizing radiation field near the Sun.

3.2. The Local Cloud (LIC) Contours

Hypothesis 1 : the density is constant in the LIC With the adoption of a density value n_{LIC}, we can draw the cloud contours by deriving the extension in each direction with the help of the column density listed in Table I column (5). As the real density in the LIC is by definition higher than the mean densities listed in column 7, the highest density sight-lines provide a lower limit for the density n_{LIC}. Although the highest density sight-lines (first quadrant) are also those known with the less precision, they show that n_{LIC} is at least equal to 0.2 cm^{-3} (cases of α Lyr, α Aql). With this lower value, the LIC would extend as far as the star α Lyr and possibly α Aql, however in both sight-lines two more components are seen in addition to the LIC. Therefore, if it is believed that all components are physically separated clouds, then n_{LIC} has to be considerably higher than 0.2 cm^{-3}. Figure 1 represents the contour of the LIC with the assumption of a constant density $n_{LIC} = 0.4$ cm^{-3}. In this case, the cloud surface is situated at less than 1 pc from the sun in the direction of Canis Majoris (in particular Sirius), which is also the direction of the void at larger scale.

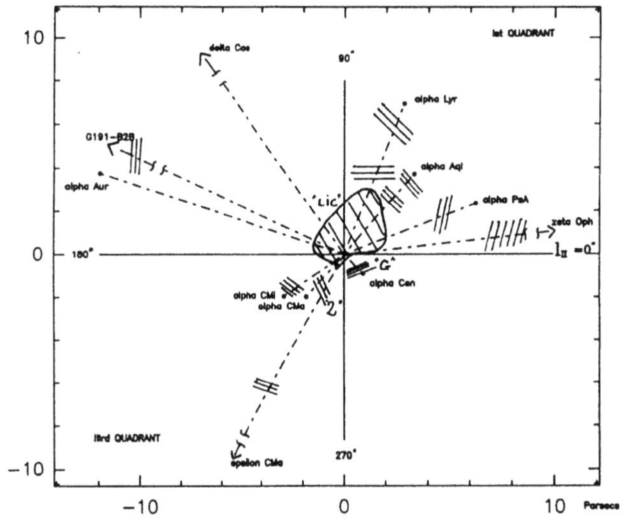

Figure 1. LIC extension derived from the FeII column densities in the hypothesis 1, i.e. the LIC is a small component like any other one and its density is homogeneous. Here the contours are drawn for $n_{LIC} = 0.4\ cm^{-3}$. The hachures indicate schematically the presence of other components in the sight-lines.

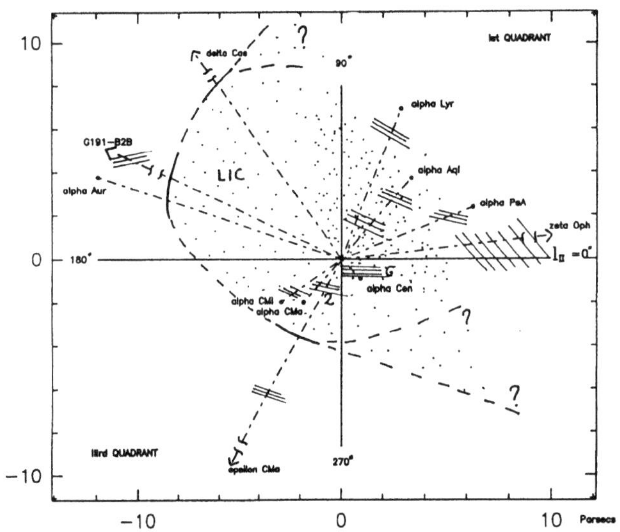

Figure 2. As in figure 1 for hypothesis 2, i.e. the LIC is a more diffuse medium in which some of the other components are embedded and there is a density gradient from the IIIrd towards the Ist quadrant. The contours in the low density side are derived from the mean density in both α CMi and α CMa sight-lines ($\sim 0.07\ cm^{-3}$).

Hypothesis 2 : the LIC is a more diffuse medium in which some of the extra components are embedded and it presents a density gradient. In this case the density would be lower close to the surface of the cloud, i.e. more or less in the direction of the IIIrd quadrant, than in the direction of what could be the "core" of the cloud, i.e. towards the highest column densities. In the IInd and IIIrd quadrants, the LIC density could be as low as .073 cm^{-3}, the mean density in both α CMa and α CMi sight-lines. Such a low value for the density is possible only if we consider that the extra components observed in both sight-lines are embedded inside the LIC, thus representing velocity -and density- inhomogeneities. In this case, the

LIC extension in this general direction is derived from the column density in the ϵ CMa sight-line : around 4 pc. In the IInd quadrant, it is given by the column density towards G191-B2B and α Aur : about 9 pc.
Note that this description could explain why the LIC is not detected in the sight-line of α Cen, which is the closest star in the sample. In effect, if the density is also close to .07 cm^{-3} in the direction to α Cen, then the LIC column density would be less than about 3×10^{17} cm^{-2} and it could be hidden by the 6 times more important absorption of the principal component in that sight-line.
In the Ist quadrant, the LIC density is higher by at least a factor 2 and in the present hypothesis the LIC could extend passed the stars, therefore the data set does not allow to constrain the extension and/or the density gradient. A detailed analysis of longer sight-lines might be useful in this respect. For example in the line of sight to ζ Oph (140 pc) studied by Sembach et al. (1994) a component is seen at the LIC velocity with a column density of 6×10^{19} ; however in this component the LIC could be blended with a more distant cloud.

3.3. CLOUD INTERFACE WITH THE HOT BUBBLE

The particularity of the IIIrd quadrant is the very low column density of gas up to distances of at least 200 pc. The line of sight to ϵ CMa (at 187 pc) intercepts mostly local gas. Its two main clouds have been detected in the much shorter line of sight to α CMa and so, apart from a third weaker component, it is essentially empty of neutral gas after the 3 first parsecs (Gry et al. 1995).
However this space is most often believed to be filled with hot coronal gas of the so-called Local Bubble, and thereby this line of sight provides a good opportunity of a long pathlength in the bubble gas. Indeed, we have detected at the LIC and at the third component velocities highly ionized species (SiIII and CIV) that at least for CIV are interpreted as collisional ionization in a high temperature gas. The interpretation of this could be naturally that the LIC presents an interface with the hot bubble gas. But no detectable absorption from these species is seen for the second component, also present in the α CMa sight-line. The reason for this non-detection could be that the column density of the second component interface is lower than that of the LIC but it could also be that the second component do not have an interface simply because it is not in contact with the hot bubble gas. This tends to support hypothesis 2, where the LIC extends out to Sirius and other stars and is embedding some of the extra components.

4. Conclusion - Summary

The local ISM is composed by many diffuse cloud components, having sizes of at most a few parsecs. Towards the general direction of the Galactic center and the Ist galactic plane quadrant its column density increases rapidly and it does

not seem to be bounded. Its extension is on the contrary very limited -to a few parsecs- in the IIIrd quadrant. The mean direction of motion points towards the low column densities, towards the IIIrd quadrant. However the small components all have velocities differing by a few km/s.
On a small scale, the gas component in which the Sun is embedded (the LIC) has been identified by Lallement and Bertin (1992), and it corresponds to the interstellar wind in the solar system. Its absorption has been detected in all lines of sight towards nearby stars observed with HST-GHRS except the shortest one. From a discussion of its density based on these observations, two possible descriptions of the LIC are proposed here. In the first one the LIC is a small cloud of the kind of the other components with a homogeneous density larger than 0.2 at cm^{-3} and an extension of less than 1 parsec in the IIIrd quadrant and a few parsecs in the upwind direction. In the second it is a more diffuse medium in which some of the other components are embedded, it extends out to about 4 parsecs with a progressive decrease in density in the CMa direction, and it shows a higher density and no yet known limits in the opposite direction. The latter hypothesis is supported by the detection of the interface between the LIC and the hot bubble gas in the ϵ CMa sight-line.

References

Bertin P., Vidal-Madjar A., Lallement R., Ferlet R., Lemoine M.: 1995, A&A **302**, 889.
Cassinelli J.P.,Cohen D.H., Macfarlane J.J., Drew J.E., Lynas-Gray A.E., Hubeny I., Vallerga J.V., Welsh B.Y., Hoare M.G.: 1996, ApJ, in press.
Crutcher R.M.: 1982, ApJ **254**, 82.
Frisch P.C., York D.G.: 1983, ApJ **271**, L59.
Frisch P.C., York D.G.: 1991, EUV Astronomy, Malina and Bowyer, eds., Pergamon Press.
Gry C., York D.G., Vidal-Madjar A.: 1985, ApJ **296**, 593.
Gry C., Lemonon L., Vidal-Madjar A., Lemoine M., Ferlet R.: 1995, A&A **302**, 497.
Hobbs L.M.: 1969, ApJ **157**, 135.
Jenkins E.B., Savage B.D., Spitzer L.: 1986, ApJ **301**, 355.
Lallement R., Bertin P.: 1992, A&A **266**, 479.
Lallement R., Bertin P., Ferlet R., Vidal-Madjar A., Bertaux J.L.: 1994, A&A **286**, 898.
Lallement R., Ferlet R., Lagrange A.M., Lemoine M., Vidal-Madjar A.: 1995, **A&A** 304, 461.
Lemoine M., Vidal-Madjar A., Bertin P., Ferlet R., Gry C., Lallement R.: 1996, A&A **306**, 601.
Linsky J.L., Diplas A., Wood B.E., Brown A., Ayres T.R., Savage B.D.: 1995, ApJ **451**, 335.
Linsky J.L., Wood B.E.: 1996, ApJ, in press.
Paresce F.: 1984, AJ **89**, 1022.
Sembach K.R., Savage B.D., Jenkins E.B.: 1994, ApJ **421**, 585.
Sofia U.J., Cardelli J.A., Savage B.D.: 1994, ApJ **430**, 650.
Vallerga J.V., Welsh B.Y.: 1995, ApJ **444**, 702.
Vallerga J.V., Vedder P.W., Welsh B.Y.: 1993, ApJ **414**, L65.
Vallerga J.V., Vedder P.W., Craig N., Welsh B.Y.: 1993, ApJ **411**, 729.

Address for correspondence: ISO Science Operation Center, PO Box 50727, 28080 Madrid, Spain

POSSIBLE SHOCK WAVE IN THE LOCAL INTERSTELLAR PLASMA, VERY CLOSE TO THE HELIOSPHERE

S. GRZEDZIELSKI
Service d' Aéronomie du CNRS, B.P. No. 3, F-91371 Verrieres-le-Buisson, France, and Space Research Centre, Polish Academy of Sciences, Bartycka 18A, 00-716 Warsaw, Poland

R. LALLEMENT
Service d' Aéronomie du CNRS, B.P. No. 3, F-91371 Verrieres-le-Buisson, France

Abstract. We show, using the HST - GHRS data on velocity and temperature in the nearby interstellar medium, that the observed 3 - 4 km s^{-1} relative velocity between the Local Interstellar Cloud (LIC) and the so-called G-cloud located in the Galactic Center hemisphere can be quite naturally explained assuming that the two clouds do interact with each other. In the proposed interpretation the two media are separated by a (quasi-perpendicular) MHD shock front propagating from the LIC into the G-cloud. The LIC plasma is then nothing else but the shocked (compression 1.3 -1.4) gas of the G-cloud. A 1-D single-fluid solution of the Rankine - Hugoniot equations can fit the most probable observed values of the relative velocity (3.75 km/s), LIC (6700 K) and G-cloud (5400 K) kinetic temperatures, if the plasma-beta of the LIC plasma is in the range 1.3 - 1.5 (Table 1). This corresponds to a super - fast magnetosonic motion of the heliosphere through the LIC, independently of LIC density. The LIC magnetic field strength is 1.9 (3.1) μG for the LIC electron density n_e = 0.04 (0.10) cm^{-3}. In this case the shock is less than 30 000 AU away and moves at about 10 km s^{-1} relative to the LIC plasma. The Sun is chasing the shock and should catch up with it in about 10^4 years. If the heliospheric VLF emissions cutoff at 1.8 kHz is indicative of n_e (LIC) = 0.04 cm^{-3} (Gurnett et al., 1993), the (pure plasma) bowshock ahead of the heliopause could be the source of quasi-continuous heliospheric 2-kHz emission band. We believe that with the expected increase in the performance of modern spectroscopic instrumentation the proposed method of magnetic field evaluation may in the future find wider application in the studies of the interstellar medium.

1. Introduction

Studies of the interaction of the heliosphere with the external interstellar surrounding involve spatial scales of the order of 10^2 to 10^3 AU. They require knowledge of the neutral and ionized species of the very local interstellar plasma as well as information on the magnetic field and cosmic ray populations at these scales. One of the major difficulties is that, in contrast to the case of neutral gases, there is no direct local information on the ionized component and on the direction and strength of the magnetic field. All data on the ion and electron densities in the Very Local Interstellar Medium (VLISM) analyzed so far are based on measurements on the lines of sight to the nearest stars and therefore represent averages over distances of, at least, a few hundreds of thousands of AU. The situation with the magnetic field is even more difficult as meaningful averages in this case refer to distances on the order of tens of parsecs or more.

We aim in this paper at narrowing the gap between the scales normal to astronomy and those that are proper to heliospheric physics. We believe this now becomes

Space Science Reviews **78**: 247–258, 1996.

possible due to the excellent observational possibilities offered by instruments such as the HST UV spectrograph (GHRS). The ability to discern spectroscopically very subtle features in the line profiles of such common interstellar species as Mg^+, Fe^+, D, Ca^+, Na, H calls for a more refined analysis of the velocity structures in the solar neighborhood. In particular, as in some cases one observes both velocities and temperatures of adjacent plasma regions in the VLISM one may try to find out whether such differences could not be indicative of the existence of discontinuity surfaces (like contact/tangential discontinuities or shock waves) that are so common to space plasmas. Since such structures, if properly identified, imply conformity of parameters characterizing the medium with the well known constraints imposed by the conservation laws used in hydrodynamics (or, more generally in MHD) one may hope to learn, by analysis of these conditions, something more about such an "unobserved" parameter of the medium as magnetic energy density in the vicinity of the heliosphere.

As we discuss it in the present paper such a possibility does now exist in practice. We consider the possible existence of a MHD shock wave in the interstellar medium at a distance on the order of one or two tenths of a parsec. Consistency with the Rankine-Hugoniot conditions requires shock compression to be rather moderate (1.3 - 1.4) and the plasma-beta of the VLISM of the order of 1. As small velocity differences seem to be visible in several directions on the sky, this shock may be just one of a multitude of such discontinuities criss-crossing the local interstellar medium (Lallement at al., 1996).

2. Velocity and temperature differences in the local gas

A peculiar velocity structure seems to be observed in the very near interstellar medium in the direction of α Centauri (l_{II} = 316 deg, b_{II} = -1 deg ; d = 1.3 pc). Towards this star the observed, line-of-sight projected velocity (- 17.5 km/s) of the intervening gas is larger (i.e. more negative) by about 2 km/s (Lallement et al., 1995, Linsky and Wood, 1996) than the expected projected velocity (-15.7 km/s) of the LIC that was previously identified (Lallement and Bertin, 1992; Witte et al., 1992, Bertin et al., 1993) as the coherent interstellar structure in which our Sun is at present embedded. The difference is too large to be easily dismissed as simply due to observation errors. Rather, the data suggest that in a solid angle of about 1 sr around α Centauri the nearby gas is moving at a somewhat higher speed (in solar frame) than the LIC. This could indicate that in this area one may see another very local interstellar cloud, not identical with the LIC. As the area in question lies in the general direction of the galactic center, the other cloud is often referred to as "the G-cloud". Detailed analysis of the interstellar velocity vectors (Lallement et al., 1990; 1995) yields the following velocity vectors for the LIC (index L) and the G-cloud (index G):

$v_L = 26\pm1$ km/s, $l_L = 6\pm3$ deg , $b_L = 16\pm3$ deg ; (1)

$v_G = 29\pm1$ km/s, $l_G = 4.5\pm4$ deg , $b_G = 20.5\pm4$ deg (2)

The galactic coordinates l,b above denote,respective, l_{II} ,b_{II} of the apex the corresponding cloud is moving from in the solar frame.

The important physical quantity is the relative velocity vector v_{rel} of the two clouds. v_{rel} will be defined in such a way as to describe the velocity of the G-cloud in the frame of LIC :

$v_{rel} = v_G - v_L$. (3)

The accuracy of determination of this vector can be inferred from Fig. 1. This figure shows in galactic coordinates the 3-D beam of v_{rel} vectors resulting from the individual velocities v_L and v_G sampling all combinations of extremities (plus the center) of the observational error boxes (1) and (2). The observational spread of directions and lengths of v_{rel} is caused by the fact that v_{rel} is a small vector obtained by taking a difference of two larger ones. As seen in Fig. 1 the dominant part of the v_{rel}-beam so defined is directed towards the galactic anti-center hemisphere and southwards. The magnitude of v_{rel} seems to be contained within the interval

2 km/s $< v_{rel} <$ 6 km/s, (4)

with the most probable value (called vnom, the nominal value of v_{rel}) taken equal to vnom = 3.75 km/s . (5)

The interpretation that the G-cloud velocity values refer to an entity physically distinct from the LIC is supported by recent HST-GHRS velocity determinations of the local gas on the α Cen line of sight by Linsky and Wood (1996). Similar conclusion can be drawn from temperature determinations.

Profile analysis (fitting) of the interstellar absorption lines allows to determine the temperature of the parcel of gas responsible for absorption. Though line broadening depends both on the internal mass motions ("turbulence") and on the kinetic temperature of ions (atoms), the two effects can be separated when line profiles caused by species of different atomic mass are compared. This profile fitting technique leads to the LIC kinetic temperature $T_L \sim 6500 - 7500$ K (Lallement et al., 1994, Linsky and Wood, 1996) consistent with the He I temperature of 6700 K inferred from direct sampling of interstellar neutral He by the GAS instrument on the Ulysses mission (Witte et al.,1992).

All above data suggest that the heliosphere is immersed in a gaseous medium with $T_L = 7000 \pm 500$ K . On the other hand the temperature of the interstellar gas towards α Cen (T_G, the temperature in the G-cloud) seems to be definitely lower : $T_G = 5400 \pm 500$ K (Linsky and Wood, 1996). Therefore the temperature determinations point to the same conclusion as the velocity analysis i.e. support the view that the LIC and the G-cloud are two distinct pieces of the local interstellar gas (plasma).

It is important to note that the internal ("turbulent") velocity spread inside each of the clouds amounting to 1.6 km/s inside the LIC on the Capella line of sight (Linsky et al., 1993) and to 1.2 km/s inside the G-cloud on the α Centauri line (Linsky and Wood, 1996) is less than the total velocity difference between the two regions (about 3.5 km/s) and much less than the rms thermal velocity spread of hydrogen atoms in the LIC and in the G-cloud (13.2 and 11.6 km/s, respectively).

Another useful piece of information may be related to the circumstance that the interstellar line profile observed on the line-of-sight to α Cen does not exclude the existence, besides a dominant contribution from the slower and cooler G-cloud gas, of a small absorption by a faster and hotter LIC-type gas (Lallement et al., 1995; Linsky, 1996). While the column density (in units of cm^{-2}) of the G-cloud gas towards α Cen is estimated to correspond to $\log N(H\ I) = 17.8$, only an upper limit for the column density of the LIC-type gas can be inferred : no detectable absorption by deuterium combined with the assumed ratio $D/H = 6.7{*}10^{-6}$ means that the LIC component to α Cen is less than 10% of the total gas content in that direction (Linsky, 1996). Therefore, the extent of the LIC towards α Cen should not, for a uniform density, exceed about 30 000 AU.

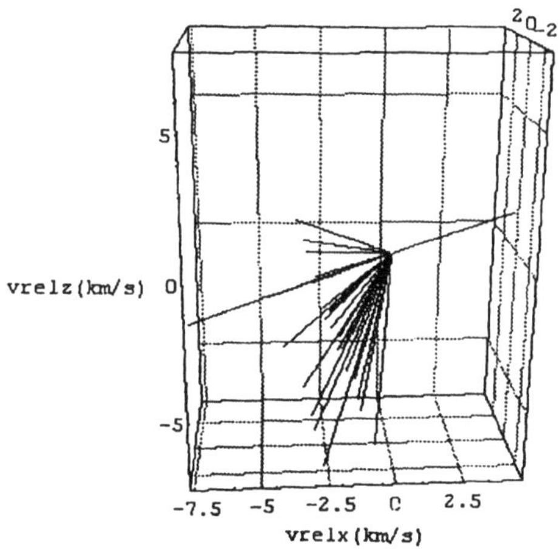

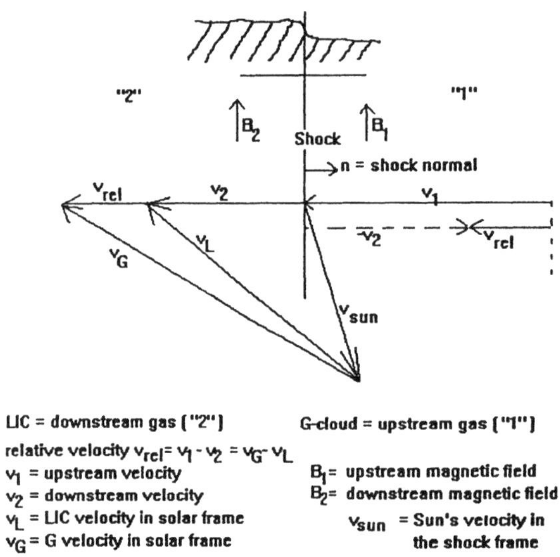

Figure 1. (top): Relative velocity vectors $v_{rel} = v_G - v_L$ corresponding to individual velocities v_L and v_G sampling all combinations of extremities (plus the center) of the corresponding observational error boxes (cf. Sec.2). v_{rel}x and v_{rel}z denote the cartesian coordinates of v_{rel}. The + v_{rel}x and + v_{rel} z axes are directed towards the Galactic Center and the North Galactic Pole, respectively. The long line piercing the focus of the v_{rel} vectors gives the direction of the velocity v_L.

Figure 2. (bottom): Velocity and magnetic field vectors for the assumed model with a perpendicular , fast MHD shock front separating the LIC from the G-cloud. The v_1, v_2 , and v_{rel} vectors are drawn in a shock frame sliding over the front such that they are collinear. The density and magnetic field are higher on the LIC side (identified with the downstream side "2"), as suggested by the step drawn at the top of the picture. The velocities are not to scale and the angles between v_L, v_G and v_{sun} are exaggerated.

3. Interpretation of the observed velocity and temperature differences

3.1. Model of the interaction

We shall try to explain the measured velocity and temperature differences between the LIC and the G-cloud as an effect of 'random' mass motions in the interstellar medium. These differences may be interpreted either (a) in terms of two physically completely separated, non-interacting objects (clouds) or (b) in terms of a gasdynamical interaction of component parts of a larger dynamical entity. In the case (a) there are no physical reasons for any particular relationship to exist between the velocities and the thermodynamical quantities such as densities and temperatures of the two objects. Such a situation, though fully possible, does not lead to any further immediate inferences. In view, however, of the close spatial proximity of the two clouds and of the absence of any evidence for an "empty" space between them, it is instructive to consider the opposite hypothesis (b). As we show in the following, a number of conclusions can be then drawn, at least in the case of the simple model adopted.

In our model we interpret the differences between the LIC and the G-cloud as resulting from a simple 1-D, fast MHD shock front propagating through an initially uniform medium. Furthermore, we assume that the shock is fully perpendicular, i.e. the magnetic field is perpendicular to the shock normal on both sides of the shock. This assumption, though in fact arbitrary, can be also taken as plausible as one may expect the quasi-perpendicular shocks to outlive, because of slower dissipation, the quasi-parallel ones. The interstellar gas is assumed to be an ideal plasma with all transport coefficients vanishing. As the energy related to the shock transition is small (velocity jump of a few km/s) the ionization degrees of H and He are treated as unchanged upon the shock transition.

Another important simplification is that all relevant linear scales that may determine the thickness of the shock are small enough compared with the size of the colliding regions. This is certainly true for the ion Larmor radius scale important for the plasma. This radius is only 1000 km for a 10 km/s proton in a 1 μG magnetic field. As the nearby interstellar medium constitutes a mixture of ionized and neutral gases one has to consider also linear scales corresponding to plasma-neutrals and neutral-neutral interactions. Efficient plasma-neutrals coupling can be mediated by the proton-neutral H atom resonant charge-exchange reaction. For a single temperature mixture of protons and neutral H atoms at T = 7000 K the charge-exchange cross section is $4*10^{-15}$ cm^2 (Maher and Tinsley, 1977) and the proton - H atom charge-exchange time is $1.3*10^8/n$ (H I) which corresponds to a linear scale of 90 AU for a 10 km/s shock speed and nH I = 0.1 atoms/cm^3 . The neutral-neutral coupling is characterized by a larger scale. The elastic H-H collision cross section for 7000 K is $6.4*10^{-16}$ cm^2 (Spitzer, 1968) corresponding for the same conditions as above to a linear scale of 560 AU. Note, however, that even this relatively large shock transition scale in neutral gas is about two orders of magnitudes smaller than the size of the regions in question. These estimates suggest that on scales larger than a few thousand AU the interstellar medium can be treated as a single fluid.

The observed quantities are the temperatures T_L and T_G and the two vectors v_G and v_L. Because $T_G < T_L$, we identify $T_G = T_1$ and $T_L = T_2$, where T_1 and T_2 denote the temperatures upstream and downstream of the shock transition, respectively. Therefore the shock must propagate from the LIC into the G-cloud, which means that the LIC gas should be nothing else but the G-cloud gas after being processed by the shock transition. The vectors v_G and v_L themselves are of no direct interest to the discussion of the shock transition as they depend on the choice of the reference frame. This frame happens to be heliocentric because of observational convenience. The vector of importance is the frame-independent relative velocity of the two media $v_{rel} = v_G - v_L$ which represents the velocity jump across the shock and hence can be written as $v_{rel} = v_1 - v_2$, where v_1 and v_2 denote the upstream and downstream gas velocities in the shock frame. For a perpendicular shock this frame can be chosen such as to have v_{rel}, v_1 and v_2 collinear. What matters then is only the vector's length $v_{rel} = v_1 - v_2 > 0$.

Note that the circumstance that the shock propagates into the G-cloud does not imply that the shock necessarily recedes from the Sun The velocity of the shock front in the solar frame depends on both the heliocentric speed of the LIC and the speed of the shock in the local (upstream or downstream) plasma frame (cf.

discussion in Sec.3.3). The geometrical relations between the various vectors are shown in Fig. 2. in the top of which a sketch of an expected density step is drawn. v_{sun}, v_1 and v_2 denote the velocity of the Sun and of the upstream and downstream plasmas in the frame of the shock. v_L and v_G are the velocities of the LIC and of the G-cloud in the solar frame. Vectors are not to scale and the angles between them are exagerated .

3.2. Rankine-Hugoniot equations for the case discussed

The R-H equations for a perpendicular MHD shock can be reduced (Priest, 1982) to an algebraic equation for the shock compression ratio r

$$2\,(2-\gamma)\,r^2-\gamma\,(1+\gamma)\,\beta_1\,{M_1}^2+\gamma\,r\,\left(2+\beta_1\,\left(2+(-1+\gamma)\,{M_1}^2\right)\right)=0, \quad (6)$$

where r denotes the ratio of mass densities (ρ_i) or magnetic fields (B_i) and is equal to the inverse ratio of velocities (v_i) on both sides of the shock (i = 1 , 2 for upstream , downstream)

$$r=\frac{\rho_2}{\rho_1}=\frac{B_2}{B_1}=\frac{v_1}{v_2}, \quad (7)$$

supplemented by an equation for the ratio of temperatures

$$\frac{T_2}{T_1}=\frac{{c_{s2}}^2}{{c_{s1}}^2}=\frac{1+\frac{1-r^2}{\beta_1}+\gamma\left(1-\frac{1}{r}\right){M_1}^2}{r}\ . \quad (8)$$

In Eqs. (6) and (8) β_1 denotes the upstream plasma-beta of the single fluid (pressure determined by both plasma and the neutral gas)

$$\beta_i=\frac{2{c_{si}}^2}{\gamma\,{c_{ai}}^2}\ , \quad (9)$$

M_1 is the usual (gasdynamic) upstream Mach number

$$M_1=\frac{v_1}{c_{s1}}\ , \quad (10)$$

and γ is the ratio of specific heats taken to be equal to 5/3. The sound (c_{si}) and Alfven (c_{ai}) velocities are given by

$${c_{si}}^2=\frac{\gamma\,k\,T_i\,(1+A+A\,x_H+x_{He})}{(4+A)\,m_H}\ , \quad (11)$$

$${c_{ai}}^2=\frac{{B_i}^2}{4\,\pi\,\rho_i}\ ,$$

where

$$\rho_i=\frac{\left(1+\frac{4}{A}\right)m_H\,n_{HI,i}}{1-x_H}\ , \quad (12)$$

and k, m_H denote the Boltzmann constant and the hydrogen atom mass.

The formulae for (c_{si}) and (ρ_i) are written for a mixture of protons, He^+ ions, electrons, neutral H and neutral He atoms . x_H and x_{He} are the ionization degrees of hydrogen and helium. Helium atoms are supposed to be at most singly ionized. As mentioned before no x_H or x_{He} changes are envisaged at the shock transition.The total helium abundance is described by the ratio H : He = A (by number of atoms).

Now, using Eqs. (7) and (10) one can write

$$v_{rel}=v_1-v_2=\left(1-\frac{1}{r}\right)\,c_{s1}\,M_1\ . \quad (13)$$

Eqs. (6), (8) and (13) constitute a set of three equations for three unknowns r, β_1 and M_1. These equations contain three other quantities: T_2/T_1, c_{s1} and v_{rel}. As it was discussed in Sec. 2 these quantities are all directly observed (though, to obtain c_{s1} from T_1, one needs also assumptions about x_H, x_{He} and A). Therefore one may hope to obtain in principle a well defined answer to the question what kind of a single MHD shock can fit the data.

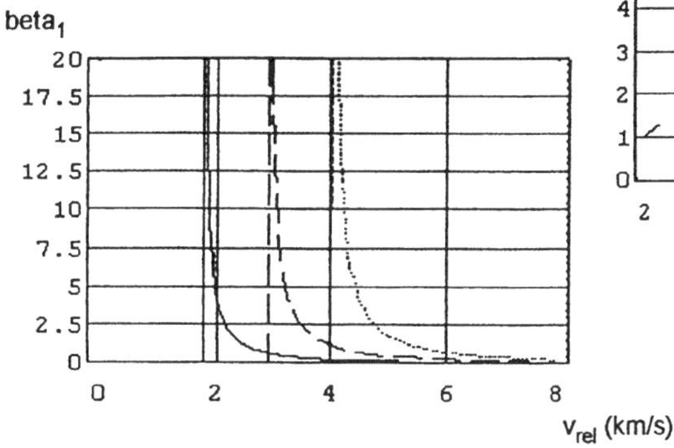

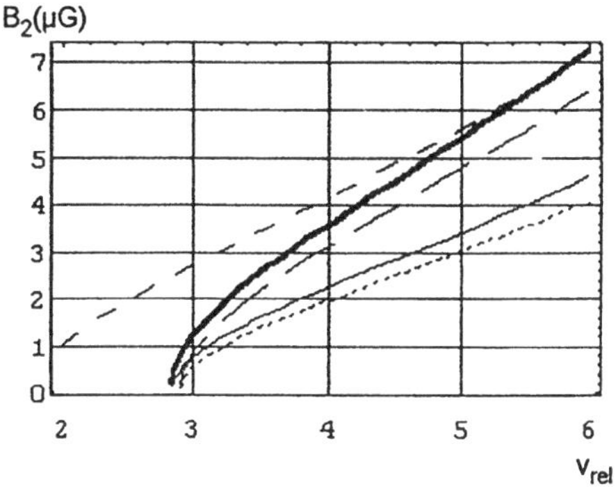

Figure 3. (left): Dependence of beta1 on v_{rel} (in km/s) is shown for three solutions of the Rankine-Hugoniot equations calculated for the perpendicular MHD shock. The solutions differ by the assumed temperature T_1 of the G-cloud. Dotted, dashed and solid curves correspond to $T_1 = 4900$, 5400 and 5900 K, respectively. For each temperature the solution is bounded on the left by a vertical asymptote at some vmin corresponding to vanishing magnetic field (purely gasdynamical shock). There are no shock solutions for $v_{rel} <$ vmin (see text for details). In all cases "soft" ionization and the LIC temperature $T_2 = 6700$ K was assumed. For the most probable value of $T_1 = 5400$ K and the nominal value of the relative velocity (vnom = 3.75 km/s, Eq. (5)) one obtains $\beta_1 = 1.617$, $\beta_2 = 1.481$, which yields the magnetic field in the LIC, $B_2 = 2.67$ G for ne,LIC = 0.10 cm-3 (cf. Se 3.3 for discussion).
Figure 4. (right) The magnetic field in the LIC, B_2, as function of v_{rel} (km/s) for the G -cloud temperature $T_1 = 5400$ K, and the LIC temperature $T_2 = 6700$ K . Solid curves correspond to "hard" ionization conditions and $ne_2 = 0.10$ cm-3 (heavy line) or 0.04 cm-3 (thin line). Non-solid curves correspond to "soft" ionization conditions and ne2 = 0.10 cm-3 (long dashes) or 0.04 cm-3 (dots). For these four solutions the magnetic field tends to zero for v_{rel} below 3 km/s (i.e. each β_1 approaches its vertical asymptote in Fig. 3). For comparison is also shown a solution corresponding to $T_1 = 5900$ K (short dashes), the upper limit suggested by the observations (for n_{e2}=0.04 cm^{-3}).

4. Effective solutions for a shock front between the LIC and the G-cloud

It turns out that the solutions are quite sensitive to v_{rel} which is determined with a relatively low accuracy. Therefore the results will be discussed with v_{rel} treated as a free parameter. Note also that physical solutions require non-negative values of β_i.

In all cases a standard He abundance with H:He = A = 10 is assumed. As regards ionization conditions, two cases are compared, corresponding to "soft" (with $x_H = 0.3$, $x_{He} = 0.1$) and to "hard" (with $x_H = 0.2$, $x_{He} = 0.5$) ionizing radiation fields in the nearby interstellar medium (Cheng and Bruhweiler, 1990; Vallerga and Welsh, 1995). For the temperatures T_1 (of the G-cloud) and T_2 (of the LIC) kno wn from the observations, the set of equations described in Sec.3.2 is equivale nt to a relationship between v_{rel} and one of the plasma-betas (say, β_1 ; then β_2 can be calculated from Eqs. (11)). Fig. 3 presents "soft" ionization solutions for β_1 as function of v_{rel} (km/s) for a single value of T_2 = 6700 K and for three values of T_1 (T_1 = 4900, 5400, 5900 K) corresponding to the low (dotted curve), middle (dashed) and upper (solid curve) value of Linsky and Wood (1995). For each value of T_1 the physical solution corresponding to positive β_1 ends at some minimum value of v_{rel} (called v_{rel} and indicated in Fig. 3 by a vertical line) for which β_1 tends to infinity (vanishing magnetic field). This limit corresponds to a purely gasdynamical shock. There are no shock solutions for $v_{rel} < v_{min}$. This constraint expresses the simple requirement that to sh ock-heat plasma from T_1 to T_2 a certain minimum of kinetic energy has to be dis sipated. Because the solution allowing a purely gasdynamical shock (B = 0) requi res fine tuning of the observed values of T_1, T_2 and v_{rel} , it is obvious that such a situation may only be met in a rather special case. In other words, since T_1, T_2 and v_{rel} come out of the observations quite independently, a realistic approach means looking for a MHD shock solution that could fit the most probable values of these observables. This can be achieved by using Eqs. (6), (8) and (13) to determine β_1 , M_1 and r as functions of T_1, T_2 and v_{rel}. Note however, that neither the magnetic field nor the density can be directly determined. For a particular set of T_1, T_2, v_{rel} the solution gives $\beta_1 = \beta 1(T_1, {}_2, v_{rel})$ and $\rho_2/\rho_1 = B_2/B_1 = r(T_1, {}_2, v_{rel})$. Solving these relationships for B_2 yields $B_2 \;\; n_{e2}^{1/2}$ with the proportionality coefficient now a known function of the observables :

$$B_2 = 2\sqrt{2}\,\sqrt{n_{e2}}\sqrt{\frac{k\,\pi\,r(T_1,T_2,v_{rel})\,T_1\left(1+\frac{1+A}{Ax_H+x_{He}}\right)}{\beta_1(T_1,T_2,v_{rel})}}. \quad (14)$$

The resulting dependence of B_2, the magnetic field in the LIC, on v_{rel} is shown in Fig. 4 for the assumed LIC temperature T_2 = 6700 K. The four curves with B_2 vanishing at v_{rel} somewhat below 3 km/s are all calculated for the most probable value of the G-cloud temperature, T_1 = 5400 K, and correspond to two plausible electron densities $n_{e,LIC}$ = 0.04 and 0.10 cm-3 for both "hard" and "soft" ionization. The fifth, short-dashed curve, extending to lower v_{rel}-values, shows the solution for T_1 = 5900 K, the upper limit of the error box, for $n_{e,LIC}$ = 0.04 cm_{-3}. It corresponds to the leftmost curve in Fig. 3.

It is evident that change in ionization conditions affects the results in a minor way, except in cases when v_{rel} happens to be close to vmin. On the other hand the obtained magnetic field is quite sensitive to changes in the values of T_1 and T_2. A decrease in the temperature jump T_2 - T_1 (for unchanged v_{rel}) increases the magnetic field strength since less shock compression is then required to raise the temperature. For instance, in the case of "hard" ionization, v_{rel} = vnom, and ne2 = 0.04 cm-3 a shift from (T_1,T_2) = (5400, 6700) to (T_1,T_2) = (5900, 6500) increases the value of B_2 from 1.94 to 5.84 μG.

The numerical solution corresponding to the "best" observed values T_1 = 5400, T_2 = 6700, and v_{rel} = vnom = 3.75 km/s is shown in Table 1 for the two ionization cases and for $n_{e,LIC}$ = 0.04 and 0.10 cm-3. The last column in the table shows the velocity v_2 of the downstream plasma in the shock frame (cf. Fig. 2) or the shock speed as seen in the LIC frame (vector "minus v_2" in Fig.2).

The results in Table 1 are sensitive to changes in T_1 and T_2. A +100 K shift in T_1 changes r by -2.5%, B_2 by +18% and v_2 by +11%. A +100 K shift in T_2 results in changes by +2.1%, -15%, and -7.6%, respectively.

The solution given in Table 1 allows one to find the shock motion in the solar frame. The Sun's velocity in the LIC-frame (vector "minus v_L"in Fig. 2) is about 26 km/s and therefore significantly exceeds the speed of the shock in this frame (vector "minus v_2"). In a 1-D configuration the Sun is chasing the shock with a relative velocity (vector "vsun") of 26 km/s - 10 km/s = 16 km/s (in reality the configuration is certainly not purely 1-D and the solar velocity vector will be inclined, though probably much less than shown in Fig. 2). If the present distance to the shock is of the order of 30 000 AU (Sec. 2) the Sun will catch up with the shock after 30 000 AU / 16 km/s = 10^4 years.

5. Conditions upwind of the heliosphere; possibility of 2-kHz emission

The solution for the value of plasma-beta in the LIC (Table 1) provides new input into the discussion of heliosphere interaction with the interstellar plasma. In particular, the question whether the motion of the heliosphere relative to the external medium is super- or sub-fast magnetosonic can be put on a new quantitative basis. This, however, can only be done if the interstellar shock is quasi-perpendicular as previously assumed and the vectors v_L and v_{rel} are not too far from colinearity. How well this last assumption is satisfied can be gauged from Fig. 1. In the following we shall assume that this is indeed the case. The speed of the fast magnetosonic wave in the LIC is then given (cf. Eqs. (9) and (11)) by

$$c_{fL} = \sqrt{c_{aL}^2 + c_{sL}^2} = \sqrt{1 + \frac{2}{\gamma \beta_L}}\, c_{sL} \;. \quad (15)$$

Index L refers here to the LIC and is now equivalent to index 2 employed in the previous sections.

The important point is that for $\beta_L = \beta_2$ determined by the shock solution (Table 1) the fast magnetosonic speed does not anymore depend on the unknown electron density in VLISM, provided A, x_H and x_{He} are fixed. For solutions given in Table 1, one obtains in the "hard" ("soft") case c_{sL} = 9.43 km/s (9.63 km/s) and c_{fL} = 13.00 km/s (12.96 km/s); therefore $c_{fL} < v_L$ = 26 km/s, i.e. the flow is super-fast magnetosonic. Theoretically, if the single-fluid picture for the mixture of plasma and neutral gas could apply, a bowshock should stand ahead of the heliopause. If in addition the nose of the bowshock could be treated as a quasi-perpendicular shock, the compression r would be 2.20 (2.16). The post-shock maximum temperature is then 15260 K (for both cases). This goes part way towards explaining the temperature level of the cloud of hot neutral hydrogen (the so called "hydrogen wall", T = 24 000 - 34 000 K) determined by Linsky and Wood (1996). However, the assumption of a single-fluid mixture of plasma and neutrals is a gross oversimplification. It should break down for linear scales much less than the plasma-neutral and neutral-neutral coupling lengths which are of the order of tens to hundreds of AU (cf. Sec.3.1). Therefore, it may hardly apply to the description of the bowshock. Rather, in the opposite limit, one can assume that on a sufficiently short spatial scale plasma is completely decoupled from the neutral gas . This approach could better describe the more plausible situation of a pure plasma bow shock.

In the case of a single-temperature plasma fully decoupled from the neutral gases the appropriate plasma-beta (called $\beta_{L,pl}$) is related to previously determined β_L by

$$\beta_{L,pl} = \frac{\beta_L n_{pl}}{n_n + n_{pl}} \, , \quad (16)$$

where n_{pl} is the total charged particles density and n_n is the total neutral particle density. For solutions in Table 1 one obtains $\beta_{L,pl} = 0.49185$ (0.651357), $c_{sL} =$ 10.73 (12.96) km/s, and $c_{fL} = 19.90$ (21.85) km/s for the "hard" ("soft") case, independently of the electron density. The flow turns out to be again super-fast magnetosonic with the super-fast Mach number equal to 1.31 (1.19).The shock compression ratio in a quasi-perpendicular situation at the tip of the bowshock "nose" is r = 1.407 (1.259) for the "hard" ("soft") case. The corresponding post-shock temperature is much lower, 8750 (7890) K, far below the values observed by Linsky and Wood (1996).

A bow shock consisting of a sharp, pure plasma transition may, however, produce another observable signature. It is well known that shocks in the solar wind plasma, such as planetary bowshocks (Gurnett et al.,1989) or propagating shocks associated with the solar type II radio bursts (Lengyel-Frey, 1992), are a source of VLF emissions at frequencies close to the local plasma frequency or its harmonic.The 2 and 3 kHz band emissions observed by the plasma wave instrument on Voyager 1 and 2 are currently interpreted as associated with the confines of the heliosphere. The sporadic, relatively bright 3 - kHz band emissions were suggested to originate when strong interplanetary transients excite plasma in the vicinity of the heliopause (Gurnett at al., 1993). These emissions are of no immediate concern to the present discussion. It was, however, also suggested by Gurnett at al. (1993) that the low-frequency cut-off (at 1.8 kHz) of the to date unexplained, weak, quasi-permanent emissions that constitute the 2-kHz band is indicative of a general electron density of 0.04 cm^{-3} in the local interstellar plasma in which the heliosphere is immersed. The plasma frequency for this density is 1.8 kHz, therefore any frequency below that can not propagate into the heliospheric plasma cavity.

Let us suppose following Gurnett at al. (1993) that indeed $n_{e,LIC} = 0.04$/cc. This implies that neutral hydrogen density in LIC is for the "hard" ("soft") case equal to 0.128 (0.090) atoms/cc and neutral helium density equals 0.008 (0.0116) atoms/cc which may be rather on the lower side when compared with the values coming from the UV backscatter observations. The magnetic field in LIC determined by the solution corresponding to $v_{rel} = v_{nom}$ (Table 1) is then equal to 1.9 μG (1.7 μG). Using the (pure plasma) bowshock compression r calculated above for $\beta_{L,pl}$ found from Eq. (16) one obtains the electron density at the "nose" of the bow shock equal to 0.056 (0.050) electrons/cm^3 for the "hard" ("soft") case. The corresponding plasma frequency is 2.14 (2.02) kHz. In consequence one obtains a possibility to explain the emission in the 2-kHz band in terms of a shock induced radiation at the plasma frequency corresponding to the post-shock plasma in the bowshock "nose" region. This radiation can not propagate directly into the heliosphere since such a ray path is prevented by the density pile-up in front of the heliopause, a feature suggested by the theoretical modeling by Baranov and Malama (1995), and Pauls et al. (1995). Before entering, the radiation would have first to travel around this obstacle towards the flanks of the heliopause, being continously only slightly above the local plasma frequency. That could explain the weakness of the observed signal and its variability in function of the solar cycle conditions.

6. Problem of abundance anomalies

One of the issues that are at the moment not understood in the context of the suggested shock - induced differences between the LIC and the G-cloud is the question of the D/Mg^+ ratio. This ratio is four times higher in the downstream plasma (the LIC) than in the upstream medium (the G-cloud)(Linsky and Wood, 1995). This implies, that upon the proposed shock transition either the D abundance increases or the Mg^+ abundance decreases (or both). As most of magnesium, because of low ionization potential, should be in a singly ionized state both upstream and downstream of the shock, the observed change in D/Mg^+ implies rather an increase in deuterium abundance in the post-shock plasma. It is not clear what may cause this enrichment. One of the possibilities could be that dust grains are accelerated at the interstellar shock to velocities of the order of few hundred km/s. Collisions of such grains with the gas atoms may lead to sputtering of grain surface material that is enriched in D (Jura, 1982). We do not attempt to discuss such a mechanism and limit ourselves to pointing out that basing purely on energy grounds such grain acceleration and consequent sputtering is not excluded.

For shock speed v = 10 km/s and B = 2 μG (Table 1) the v x B electric field parallel to the shock surface is E = 0.002 mV/m. A dust grain of mass mg = $A_g m_H$ and charge Z_ge drifting along the shock surface over length L will acquire kinetic energy $1/2{*}A_g m_H u_g^2 = Z_g eEL$. For $m_g = 10^{-14}$ g (say, $A_g = 6{*}\ 10^9$ grain density 2 g/cm^3 and radius = 0.1 μm) one obtains u_g = 200 km/s (1 keV kinetic energy of impinging He atom as seen in the grain frame) if $Z_g L$ = 20 pc, i.e. for instance if Z_g = 20 and L = 1 pc. Assuming sputtering yield 1 per one He atom (hydrogen being not energetic enough to give an effect) one obtains a loss rate per grain of the order of one atom every $1.5{*}10^4$ s (for background helium density of 10^{-2} atoms/cc). For average atom mass of grain material equal to 10mH, the mass loss rate is 10^{-27} g/s per grain, i.e. the grain would release into the surrounding medium a significant fraction of its mass in about 10^5 years. These very crude estimates are quoted here only to point out that it may not be excluded that grains could be significantly affected even by weak interstellar shocks of the type suggested in the present paper. It may be therefore worth to investigate whether grain related processes can not be one of the causes of abundance anomalies observed in the local interstellar medium.

7. Conclusions

We have shown that the observed velocity and temperature differences between the LIC and the G-cloud can be explained as an effect of propagation through the VLISM of a (quasi-perpendicular) fast MHD shock with a moderate compression ratio of 1.3 -1.4. Consistent physical solutions require the shock front to propagate from the LIC into the G-cloud and the VLISM plasma-beta (before shock) to be about 1.5. This allows to determine the local magnetic field strength if the local electron density can be deduced from other measurements. The range of magnetic field intensities in the local interstellar medium obtained in the present model (1.7 - 3.1 μG) is in a good agreement with other current estimates for the local interstellar field (Frisch, 1995). Two new important points concerning magnetic field determination have to be stressed. First, the method proposed in this paper is based on the analysis of differences in plasma temperatures and bulk velocities only, i.e. the

Table I

Shock compression ratio (r), upstream and downstream plasma-beta (β_1 and β_2), LIC (=downstream) magnetic field (B_2) and shock speed in the LIC-frame (v_2) for the best values of observables : T_1 = 5400,T_2 = 6700, v_{rel} = vnom = 3.75 kms^{-1}.

Ionization	r	beta1 beta2	B_2 ne(lic)=0.04	 ne(lic)=0.10	shock speed in LIC v_2 (km s-1)
soft	1.3547	1.617 1.481	1.690	2.672	10.57
hard	1.3535	1.449 1.328	1.945	3.075	10.61

derived values are independent of previous estimates of the magnetic field strength. Second, the spatial scale of the layer of LIC to which these considerations may pertain is quite small by astronomical standard and thus the plasma-beta we found may be indeed representative for the conditions in the immediate surrounding of the heliosphere.

References

V.B. Baranov, Y.G. Malama J.Geophys.Res., 100, 14755, 1995

P. Bertin, R. Lallement, R. Ferlet, A. Vidal-Madjar, J. Geophys. Res., 98, 15193, 1993

K.-P. Cheng, F.C. Bruhweiler, Astrophys. J., 364, 573, 1990

P.C. Frisch, Space Sci. Revs., 72, 499, 1995

D.A. Gurnett, W.S. Kurth, R.L. Poynter, L.J. Granroth, I.H. Cairns, W.M. Macek, S.L. Moses, F.V. Coroniti, C.F. Kennel, D.D. Barbosa, Science, 246, 1494, 1989

D.A. Gurnett, W.S. Kurth, S.C. Allendorf, R.L. Poynter Science, 262, 199, 1993

M. Jura, Advances in Ultraviolet Astronomy, p.54, NASA-CP 2238, 1982

R. Lallement, P. Bertin Astron. Astrophys., 266, 479, 1992

R. Lallement, P. Bertin, R. Ferlet, A., Vidal-Madjar, J.L. Bertaux , Astron.Astrophys., 286, 898, 1994

R. Lallement, R. Ferlet, A. Vidal-Madjar, and C. Gry Physics of the Outer Heliosphere, Grzedzielski S., Page, D.E., (Eds.), Pergamon Press, Oxford 1990, pg 37-42

R. Lallement, R. Ferlet, A.M. Lagrange, M. Lemoine and A. Vidal-Madjar , Astron.Astrophys., 304, 461, 1995

R. Lallement, S. Grzedzielski, R. Ferlet, A. Vidal-Madjar, Science with the Hubble Telescope - II, Space Telescope Science Institute, 1996, P.Benvenuti, F.D. Macchetto, E.J. Schreier, eds.

D. Lengyel-Frey J. Geophys. Res., 97, 1609, 1992

J.F. Linsky, 1996 (this volume)

J.F. Linsky, B.E. Wood Astrophys. J. 463, 254,1996

J.F. Linsky, A. Brown, K. Gayley, A. Diplas, B.D. Savage, T.R. Ayres , W. Landsman, S.T. Shore, S. Heap, Astrophys. J., 402, 694, 1993

L.J. Maher, B.A. Tinsley, J.Geophys.Res. 82, 689, 1977

H.L. Pauls, G.P. Zank, L.L. Williams J. Geophys. Res., 100, 21595, 1995

E.R. Priest D. Reidel Publishing Co., Dordrecht, Holland, 1982

L. Spitzer Diffuse Matter in Space, Chapter 4, John Wiley and Sons, New York, 1968

J.V. Vallerga, B.Y.Welsh, Astrophys. J., 444, 702, 1995

M. Witte, H. Rosenbauer, M. Banaszkiewicz, H.-J. Fahr Adv. Space Res. 13, (6), (6)121-(6)130, 1992

INTERSTELLAR GRAINS IN THE SOLAR SYSTEM: REQUIREMENTS FOR AN ANALYSIS

INGRID MANN
Max-Planck-Institut für Aeronomie, D-37189 Katlenburg–Lindau, Germany

Abstract. We discuss present knowledge about interstellar dust grains in the heliosphere in order to give goals for future investigations. As far as the identification of the interstellar flux from brightness observations is concerned we calculate the influence of interstellar dust entering the solar system on the Zodiacal light and Zodiacal emission brightness. In case of the Zodiacal light produced by the scattering of solar radiation, the brightness from interstellar dust within the solar system is not detectable within the limits of present observations. In the case of the thermal emission a distinction of the brightness from the interstellar dust component may be possible. This would be especially interesting for an analysis of the overall spatial distribution of the interstellar flux in the solar system. As far as the identification of the interstellar flux from impact experiments is concerned, parameters like the impact direction are essential. Since the interstellar dust flux is modified in the outer solar system already, it is helpful to probe its variation with increasing distance from the Sun in interstellar upstream direction.

1. Introduction

It is expected from theory, that a flux of interstellar particles may enter the solar system due to the relative velocity of the Sun with respect to the local interstellar medium (Fahr 1971). The local interstellar environment can be investigated from the absorption of low energy x - ray radiation (see for instance Egger et al. this issue) from the absorption of emission spectra from nearby stars (Lallement 1990 and see Frisch 1995 for a review) and from in-situ measurements of neutral atoms and pickup-ions (cf. Geiss et al. 1994). While charged particles are deflected from entering the heliosphere directly, neutral gas particles can stream into the solar system and are detected at distances as close as 1 AU from the Sun (Witte et al. 1993) before they are ionized. In the case of dust particles, it depends on their electric charge, respectively on the charge to mass ratio, whether they are sufficiently neutral and enter the solar system. This is especially the case for bigger grains. The impact measurements on Ulysses provide the first direct detection of interstellar dust particles within the solar system (Grün et al. 1994) and show, that the flux at 5 AU is in its size distribution already significantly different from what is expected to be the size distribution of interstellar dust. When entering the solar system, the particles are influenced by solar gravity and radiation as well as the solar magnetic field. This causes various dynamical effects and interactions (cf. Morfill and Grün 1979) and may yield a unique opportunity of studying the properties of interstellar grains, however within some restrictions.

A study of this dust component with the goal to understand the nature of interstellar grains has to consider these alterations and besides that the significance of this local dust component on global interstellar scales has to be discussed.

Space Science Reviews **78**: 259–264, 1996.

We describe the present knowledge of the interstellar dust component in the solar system and compare it to the other components of the outer solar system dust cloud. We check whether the dust flux is detectable in the scattered light brightness (Zodiacal Light), respectively in the thermal emission brightness from interplanetary dust (Zodiacal Emission) and we discuss perspectives of in-situ measurements.

2. Interstellar Dust Flux in the Inner Solar System

The flux of the interstellar particles was derived from the Ulysses data at 5 AU to amount to $1.4 \cdot 10^{-4}$ $m^{-2}s^{-1}$ with an average mass of the particles of $8 \cdot 10^{-13}$ g (Grün et al. 1994). The detected interstellar grains cover a very small mass interval in comparison to a total mass interval of detected particles between 10^{-16} to 10^{-8}g. Comparing this to the findings about the interstellar dust, it means, that we see only a limited mass range, i.e. the "bigger" ones among the size spectrum derived from the interstellar extinction. The identification of interstellar particles is based on determination of impact velocity and impact direction (Baguhl et al. 1995). Especially small particles, which are dominant in the interstellar medium are not identified. Two reasons may explain the lack of small grains. First they are preferably deflected from entering the heliosphere due to their presumably high charge to mass ratio and second they are more affected by Lorentz and radiation pressure forces which deplete them from the original interstellar flux direction. Therefore it is expected, that the smaller particles seen in the Ulysses data, which have random impact directions but also relative velocities that are attributed to hyperbolic orbits, may partly be as well of interstellar origin.

Describing a typical absorption of 0.3 mag/kpc for a region between interstellar clouds and an average grain size of 0.12 μm (radius) Giese derived in 1979 a value of $n_{IS} = 1.7 \cdot 10^{-13}$ cm^{-3} for the density of interstellar grains in the outer solar system (Giese 1979) and gave a similar number of $1 \cdot 10^{-13}$ cm^{-3} from the relation between dust and local hydrogen density. Assuming again a grain size of 0.12 μm and a bulk density of 2 g/cm^3 the mass of the grains amounts to $1.5 \cdot 10^{-14}$ g and hence the number density derived from the latter value amounts to $n_{IS} = 1.3 \cdot 10^{-13}$ cm^{-3}. Applying on the other hand the flux and the mass derived from the Ulysses data and a velocity of 26 km/s the value is much smaller $n_{IS} = 5.8 \cdot 10^{-15}$ cm^{-3}. The average mass of detected grains is $3 \cdot 10^{15}$ g, which corresponds to a particle radius of 0.4 μm (Grün et al. 1994). However the total geometric cross sectional area of particles per volume element amounts to $3 \cdot 10^{-23}$ cm^{-1} (i.e. cm^2/cm^3) for the Ulysses flux. The value used in calculations by Giese (1979) amounts to $4.5 \cdot 10^{-23}$ cm^{-1}. This similarity is due to the fact, that the component of small grains, that are not detected in the Ulysses measurements do not contribute to the geometric cross sectional area either.

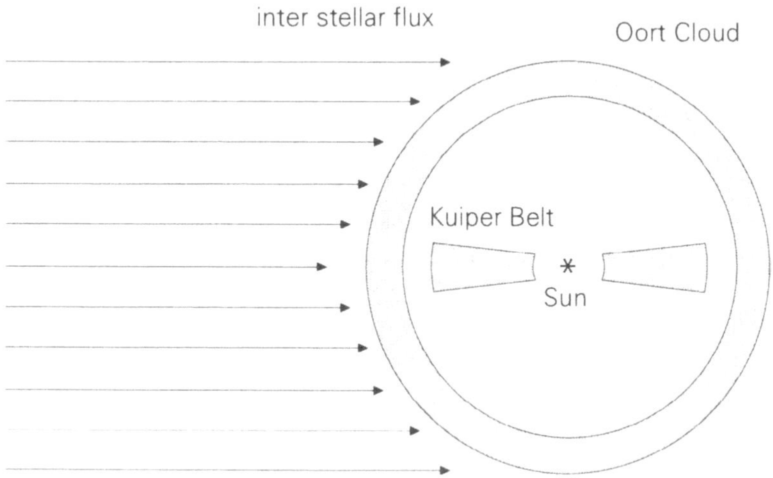

Figure 1. The outer solar system dust cloud: sources of dust are the activity of long period and short period comets, the interstellar flux and possibly effects, related to the Kuiper belt objects.

3. The outer Solar System Dust Cloud

Models derived from brightness observations describe the interplanetary dust cloud at distances of about 0.3 AU to 1.7 AU around the Sun as axially symmetric with respect to an axis through the solar poles (a slight deviation within 3° will not be considered at this point) and in its radial slope proportional to $r^{-\nu}$ with exponent ν close to 1. Measurements in the outer solar system don't prove directly that this slope is continued up to the Jovian orbit. However it seems reasonable to assume, that the density is either continued according to this power law, or taking into account that the asteroids yield a source of dust particles to assume a constant number density from about 3 to 5 AU.

Sources of dust in the outer solar system are the activity of comets and the flux of interstellar bodies entering the solar system (see figure 1). When emitted from a comet particles follow the orbit of its parentbody until gravitational perturbations randomize the orbital elements: argument of the perihelion and ascending node. However, when expelled from the comet the coupling of the particles to the Sun is reduced due to the radiation pressure force. The effect is especially strong for submicron sized grains which in the majority will be in unbound orbits leaving the solar system in hyperbolas (cf. Mann and Grün 1995). Besides that, the collision of the bigger cometary fragments is a source of dust particles. However these effects cannot be assumed to cause a dust cloud, as dense as the Zodiacal cloud in the inner solar system, since for example the activity of a typical comet ceases between the orbits of Jupiter and Saturn. Only sporadic outbursts and splitting of comets may produce large amounts of dust in this region as well as the dust production from Kuiper belt objects is probably small. On the other hand the interstellar flux itself, and its impact on Kuiper belt objects may yield a further source of dust particles. The observation of faint infrared brightness around stars, discussed as the Vega phenomenon, indicates the existence of dust clouds around those stars (see for instance Backman and Paresce 1993), which are in the regions of Kuiper belt objects in scales of our solar system. Effects like evaporation or collision are discussed to explain such dust clouds, effects which similarly may occur the Kuiper belt region of our solar system. Taking into account the long time scales of particles in Keplerian orbits that are far away from the Sun, the flux of interstellar dust may cause significant alteration of objects.

4. The Zodiacal Light Brightness

Model calculations by Giese (1978) have shown that the light scattered by the interstellar dust component in the solar system can be nicely separated from the Zodiacal light (i.e. the scattered component), when observed from a spacecraft like Ulysses in its out of ecliptic orbit with predicted brightness ratios $S_{IS}/S_{IP} > 0.25$. Unfortunately, such type of observations are not available and we are restricted to ground based observations, respectively to satellite data from Earth orbit.

The number density n_{IP} of interplanetary particles derived from Zodiacal light models and extrapolated to 5 AU amounts to between $n_{IP} = 1.4 \cdot 10^{-17}$ cm^{-3} and $n_{IP} = 1.6 \cdot 10^{-12}$ cm^{-3}. These differences between different models can partly be explained in uncertainties in the absolute calibration of the early infrared observations, which some models are based upon, but still there is an uncertainty when comparing different data sets. Extrapolation to 5 AU assuming a radial decrease of the number density with 1/r and an average size of interplanetary particles of 30 μm leads to values of the geometric cross section between $4 \cdot 10^{-22}$ cm^{-1} and $4.5 \cdot 10^{-17}$ cm^{-1} for different Zodiacal cloud models. This shows that the data for the Zodiacal cloud are clearly above the values for the interstellar component. We assume a value of $2.8 \cdot 10^{-17}$ cm^{-3} which can explain the Zodiacal light brightness and is as well in agreement to estimates of the interplanetary flux close to 1 AU. We compared the Zodiacal light brightness from a "pure" interplanetary model to a model that includes a further component of interstellar grains and found that a reasonable distinction is not possible, taking into account the uncertainties of the Zodiacal light models within more than 10%.

The thermal emission brightness is biased to the components at large solar distances, where the contribution of the interstellar flux is increasing. The contribution of the interstellar dust is in the range of 6% at maximum for the case that the Earth is in interstellar upstream direction (see Figure 2). A seasonal variation of the infrared brightness is expected to occur from the influence of the interstel-

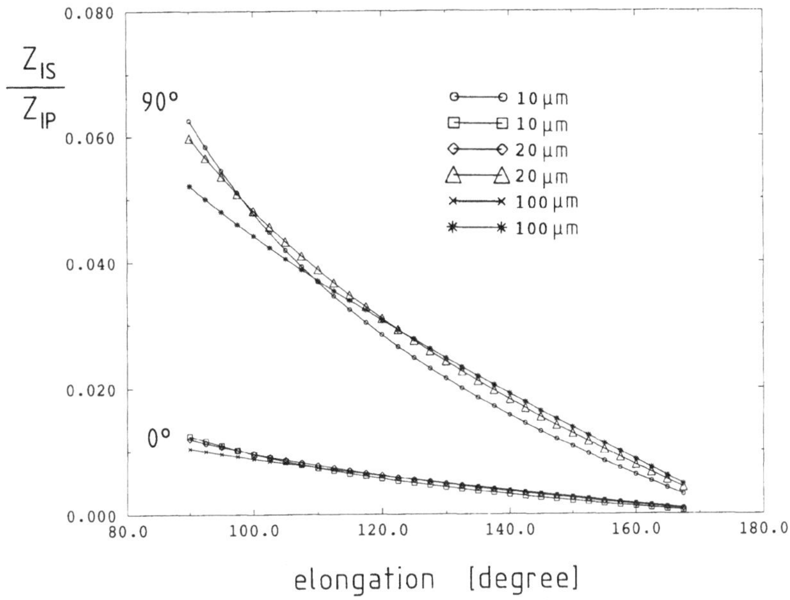

Figure 2. The calculated thermal emission brightness of the interstellar flux at wavelengths 10, 20 and 100 μm for helioecliptic latitudes 0° (line of sight in the ecliptic plane) and 90° (line of sight perpendicular to the ecliptic plane). The figure shows the brightness ratios of the interstellar component Z_{IS} and the interplanetary component Z_{IP} in interstellar upstream direction.

lar dust component. At this point new observations, together with advanced model calculations may yield some insights.

5. Perspectives for Future Investigation

The in-situ detectors that are presently available are capable to determine mass, relative velocity and sometimes even charge and elemental composition of grains. The in situ detection of interstellar dust particles is certainly very interesting in different aspects. The orbital parameters as well as the mass distribution give some insight in the conditions of the dust in the local environment of the solar system as well as the interactions with the heliosphere. The study of the elemental composition would be very helpful for models of the astrophysical processes of grain formation.

The parameters that need to be identified from brightness observation are the overall spatial distribution and the average properties of dust. A knowledge of the spatial distribution is desirable for an understanding of the dust flux and its depletion when entering the solar system. Some clues about the size distribution would be possible from the comparison of mass distributions, derived in-situ and cross-sectional distributions derived from brightness analysis.

Based on our present knowledge the future studies need:

- to identify the flux of interstellar grains from impact directions and relative velocities and to distinguish it from a possible Kuiper belt dust population;
- to determine its mass, respectively size distribution and its variation with increasing distance from the Sun;
- to identify its contribution to the Zodiacal brightness and to determine its spatial distribution in the solar system;
- to determine the cross - sectional distribution, average radiation temperature, albedo and emissivity;
- to determine the elemental composition and its variation in the interstellar upstream direction.

To address these topics, in-situ experiments in the outer solar system are best suited for the analysis of dust fluxes and elemental composition. But there are some limits to in situ experiments: the relation between mass distributions and size of particles are ambiguous and also the impact experiments don't describe the spatial distribution and properties of grains. These are the points, which are best studied with brightness observations.

References

Backman, D.E. and Paresce, F.: 1993, 'The VEGA Phenomenon', in Protostars and Planets III, Eds. E.h. Levy, J.I. Lunine and M. S. Matthews (Tuscon: Univ. Arizona Press), p 1253.

Baguhl, M., Grün, E., Hamilton, D.P. et al.: 1995, 'The Flux of Interstellar Dust observed by Ulysses and Galileo', *Space Sci. Rev.* **72**, 471-476.

Eggert, et al. this issue

Fahr, H.J.: 1971, 'The Interplanetary Hydrogen Cone and its Solar Cycle Variations', *Astronom. Astrophys.* **14**, 263-274.

Frisch, P.C.: 1995,'Characteristics of Nearby Interstellar Matter', *Space Sci. Rev.* **72**, 499-592.

Geiss, J., Gloeckler, G., Mall, U., von Steiger R., Galvin, A.B. and Ogilvie, K.W.: 1994, 'Interstellar Oxygen, Nitrogen and Neon in the Heliosphere',*Astron. Astrophys.* **282**, 924-933.

Giese, R.H.: 1979, 'Zodiacal Light and Local Interstellar Dust: Predictions for an Out-of-Ecliptic Spacecraft',*Astron. Astrophys.* **77**, 223-226.

Grün, E. Gustafson, B., Mann, I. et al.: 1994, 'Interstellar Dust in the Heliosphere',*Astron. Astrophys.* **286**, 915-924.

Lallement, R.: 1990, in S. Grzedzielski an E. Page (eds.), 'Scattering of Solar UV on Local Neutral Gases', COSPAR Colloquia Series, No 1, Physics of the Outer Heliosphere, Pergamon Press, London, pp. 49-60.

Mann, I.: 1995, 'Spatial Distribution and Orbital Properties of Interplanetary Dust at High Latitudes', *Space Sci. Rev.* **72**, 477-482.

Mann, I. and Grün, E.: 1995, 'Dust beyond the Asteroid Belt', *Planet. Space Sci.* **43**, 827-832.

Morfill, G. and Grün, E.: 1979 'The Motion of Changed Dust Particles in Interplanetary Space', *Planet. Space Sci.* **27**, 1269-1282.

Witte, M. Rosenbauer, H., Banaskiewicz, M. and Fahr, H.: 1993, 'Ulysses Neutral Gas Experiment: Determination of the Velocity and the Temperature of the Interstellar Neutral Helium',*Adv. Space Res.* **13(6)**, 121-130.

MODELLING OF THE INTERSTELLAR HYDROGEN DISTRIBUTION IN THE HELIOSPHERE

D. RUCIŃSKI and M. BZOWSKI
Space Research Centre of the Polish Academy of Sciences, Bartycka 18A, 00-716 Warsaw, Poland

Abstract. The detailed knowledge of the distribution of neutral interstellar hydrogen in the interplanetary space is necessary for a reliable interpretation of optical and H^+ pickup ions observations. In the paper, we review the status of the modelling efforts with the emphasis on recent improvements in that field. We discuss in particular the role of the nonstationary, solar cycle-related effects and the consequences of hydrogen filtration through the heliospheric interface region for its distribution in the inner Solar System. We demonstrate also that the use of the simple 'cold' model, neglecting the thermal character of the hydrogen gas ($T \sim 8000$ K), is generally incorrect for the whole region of the inner heliosphere ($R < 5$ AU) since it leads to a substantial underestimation of the local hydrogen density and thus influences the derivation of the H properties in the outer heliosphere/LISM. Referring to recent Ulysses measurements, we point out also the need to consider in the modelling the effects of the latitudinal asymmetry of the ionization rate.

1. Introduction

The permanent presence of the neutral interstellar hydrogen gas in the Solar System, predicted theoretically by Fahr (1968) and Blum and Fahr (1970) was soon confirmed by optical observations of the solar Lyman-α radiation backscattered on neutral H atoms (Thomas and Krassa, 1971; Bertaux and Blamont, 1971). Approaching the Sun, the hydrogen atoms are subjected to the solar gravitation, solar radiation pressure, and various ionization processes, among which the charge-exchange with the solar wind protons plays the dominant role. Whereas all of these processes shape the pattern of hydrogen distribution inside the Solar System, the ionization reactions lead also to the creation of new populations of particles such as the H^+ pickup ions and the neutral component of the solar wind. The traditional local diagnostic method, used in the last two decades to probe the properties of the interstellar H gas, based on observations of the interplanetary Lyman-α glow (see e.g. Lallement, 1990; Quèmerais et al., 1994 and references therein), has been recently complemented by the first direct measurements of H^+ pickup ions in the interplanetary space (Gloeckler et al., 1993). Since the parameters of the interstellar gas are inferred mostly from observations performed in the inner heliosphere (typically at 1 – 5 AU from the Sun), the best possible knowledge of the neutral hydrogen distribution is a key factor for the correct interpretation of the data from these observational methods.

The classical approach, commonly used to describe the interstellar gas distribution in the Solar System, is based on several simplifying assumptions:

Space Science Reviews **78**: 265–276, 1996.

- the neutral, collisionless interstellar gas, flowing with the characteristic bulk velocity V_B, can *freely* penetrate into the heliosphere without being affected by any perturbations at the heliospheric interface region;
- the H atoms approaching the Sun are subject to the ***stationary, spherically symmetric*** effective solar gravitational force (resulting from the combined action of the solar gravity and solar radiation pressure) and to the ***spherically symmetric, stationary ionization***, both decreasing like $1/r^2$ with the heliocentric distance.

Basing on these assumptions, two models describing the local density distributions were worked out. The first one – given in analytical form (see e.g. Fahr, 1968; Axford, 1972) – corresponds to the case of monoenergetic, cold ($T = 0$ K) interstellar gas, whereas the second one, a more realistic 'hot' model, requiring multiple numerical integration over the velocity space (see e.g. Fahr, 1971; Thomas, 1978; Wu and Judge, 1979), describes the distribution of the interplanetary gas with a finite non-zero temperature.

However, it became apparent that due to several simplifying assumptions (stationary character of all interaction processes, neglect of the filtration effect, spherical symmetry of the ionization by the Sun, etc.) these models, though handy and successful in many respects, give only an approximate description of the physical reality. Discussing in the following sections the problem of a more accurate modelling of the physical reality, we focus on the effects that affect the hydrogen distribution in the inner part of the heliosphere ($R < 10$ AU) and thus influence the interpretation of the local optical and pickup ion measurements.

2. Current approach to the modelling

2.1. Thermal character of the hydrogen gas

The thermal character ($T \sim 10^4$ K) of the interstellar gas in the outer heliosphere and its neighbourhood has been confirmed by the analyses of optical observations of backscattered EUV glow (see e.g. Bertaux et al., 1985), direct measurements of neutral He atoms in the interplanetary space (Witte et al., 1993, 1996), and He^+ pickup ion observations (Möbius et al., 1995). Since the 'hot' model of density distribution requires multiple numerical integration, the thermal spread of the gas has been quite often ignored in various estimates and the handy analytical solution for the 'cold' ($T = 0$ K) gas has been applied instead. As it was noted, in certain conditions the temperature effects manifest mainly in a relatively narrow cone around the downwind axis (see e.g. Feldman et al., 1972; Blum et al., 1975). The 'cold' model can still be an adequate approximation for the upwind hemisphere and for sidewind directions, but only for the interstellar species with thermal veloc-

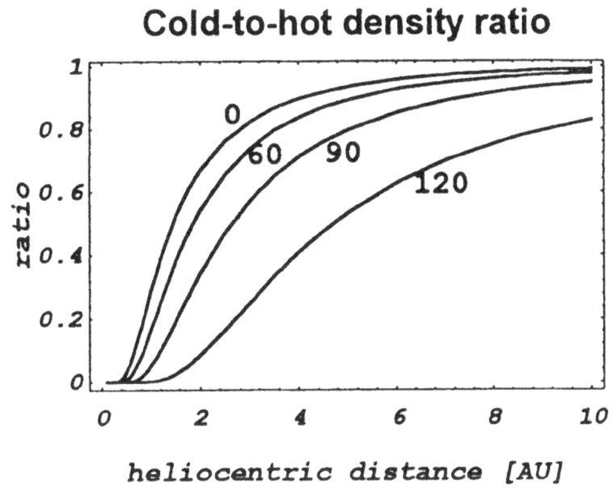

Figure 1. The 'cold-to-hot' ratio of local hydrogen densities calculated on the basis of the 'cold' ($T = 0$ K) and 'hot' ($T = 8000$ K) model for heliocentric distances $R \leq 10$ AU and the following angles θ from the interstellar wind apex: 0 (upwind), 60, 90 and 120 degrees. Other adopted parameters are: $V_B = 20$ km/s; radiation pressure/solar gravity ratio $\mu = 1$; ionization rate at 1 AU $\beta_E = 8.0 \cdot 10^{-7}$ s^{-1}.

ity $V_T = (2kT/m)^{1/2}$ much smaller than their bulk velocity V_B. For typical values of $V_B = 20$ km/s and $T = 8000$ K this condition is fulfilled for helium and heavier species but it is questionable in the case of hydrogen, where $V_T = 11.5$ km/s is not much lower than V_B.

To demonstrate the consequences of neglect of the thermal character of the hydrogen gas in the area of interest ($R < 10$ AU), even for the regions far away from the downwind axis ($0° < \theta < 120°$), we compare the local densities calculated on the basis of the 'cold' and 'hot' model solutions, presenting the relevant cold to hot density ratio in Fig.1. It is evident that the use of the 'cold' model leads to a substantial underestimation of the local hydrogen density in comparison with the thermal case inside at least 5 AU. The discrepancies dramatically increase with the decrease of heliocentric distance and the increase of offset angle θ and smear out only beyond 10 AU from the Sun. Thus, the cold model, while evidently incorrect for the regions near the Sun ($R < 5$ AU), can still be applied for considerations of the global hydrogen distribution over the scale of dozens of AU from the Sun.

Our analysis indicates that the use of the 'cold' approach for modelling of hydrogen distribution near the Sun may adversely affect the correctness of conclusions from local measurements (where a significant contribution to the measured signal comes from the region inside 5 AU), since it may potentially cause a false determination of the H density in the LISM/outer heliosphere. The significance of this fact has been recently taken into account by Gloeckler (1996) where the hot model has been used for the interpretation of the Ulysses H^+ data instead of the cold model applied in the previous analysis (Gloeckler et al., 1993).

2.2. Time-dependent modelling including solar cycle effects

As it comes out from the classical stationary approach (see e.g. Thomas, 1978), the joint action of the 'effective' gravitational force and various ionization processes modifies significantly the hydrogen distribution inside a huge circumsolar region (extending typically to 10 AU upwind and 50 AU downwind) in comparison to its homogeneous distribution in the Local Interstellar Medium (LISM). Since the size of that region is comparable to the distance covered by the inflowing H atoms during the 11-year solar cycle, the postulate to develop a time-dependent model including variations of the radiation pressure and ionization rate, expressed already in the early reviews by Holzer (1977) and Thomas (1978), seems fully legitimate. Due to the complexity of the task only recent studies by Fahr and Scherer (1990), Blum et al. (1993), Kyrölä et al. (1994), Bzowski and Ruciński (1995a) and Ruciński and Bzowski (1995a) brought a progress. Most often, a Monte Carlo scheme involving tracing of the individual H atoms along their non-keplerian trajectories was used. In the present section, we refer to the analysis of the problem by Ruciński and Bzowski (1995a), the most comprehensive and adequate for the solar case. To demonstrate illustratively the scale of the nonstationary effects, we often use as a reference the results obtained on the basis of the 'mean stationary model' (MSM), i.e. calculated for the the radiation pressure and the ionization rate averaged over the solar cycle.

As shown by Ruciński and Bzowski (1995a), the variability of the solar factors exerts a significant imprint on the hydrogen density distribution especially in the inner heliosphere and is the most pronounced in the downwind region. The departures of the density profiles from the MSM presented in Fig.2 at different phases of the solar cycle are clearly visible up to 5 AU on the upwind side (approximately the same occurs in the sidewind direction) and to 10 AU in the downwind region. Further away from the Sun, the differences systematically smear out, practically vanishing beyond 15 – 20 AU upwind and 50 – 60 AU downwind, respectively. The replacement of the MSM model by the stationary model with "instantaneous" values of the radiation pressure and ionization rate, appropriate for the considered phase of the solar cycle, does not reproduce the hydrogen distribution pattern resulting from the variable model, either (Bzowski and Ruciński, 1995b).

While the modulations on the upwind side are predominantly induced by the variations of the ionization rate, in the downwind region they result mainly from the accumulated effect of the radiation pressure on the trajectories of H atoms around their perihelion. Those effects become the dominant factors controlling the amount for neutrals inflowing to that region and are thus responsible for the modulations. The solar cycle-induced variations of the local hydrogen density distribution may influence, as pointed out by Ruciński and Bzowski (1995a), the interpretation of the interplane-

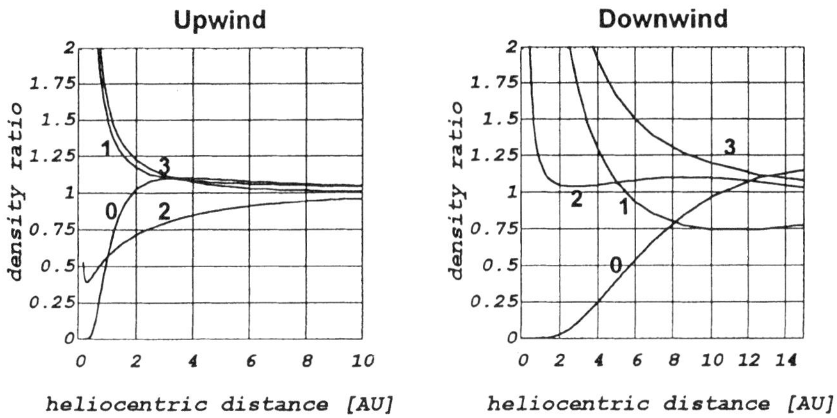

Figure 2. The radial profiles of the variable-to-MSM hydrogen density ratio along upwind (left panel) and downwind directions (right panel) for four phases of the solar cycle, separated in time by 2.75 years. Phase 0 corresponds to the solar maximum; phase 2 to solar minimum, phases 1 and 3 are intermediate.

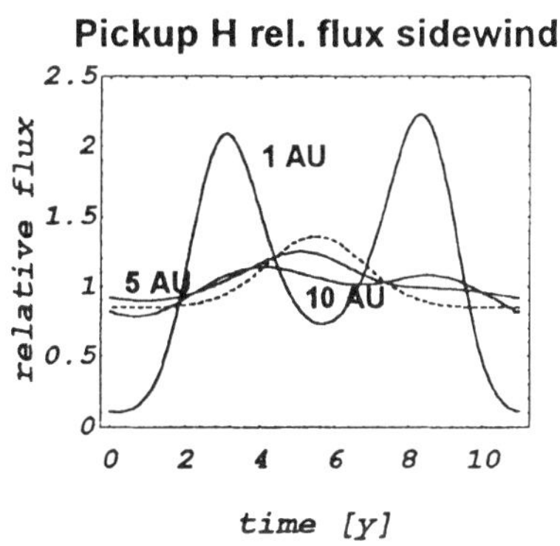

Figure 3. Relative variations of the H^+ pickup ion fluxes over the solar cycle at 1 AU, 3 AU and 5 AU at $\theta = 90°$ with respect to the mean stationary model (MSM) corresponding to the mean values of the ionization rate and radiation pressure. The dotted curve reflects the assumed variability of the ionization rate in the units of its mean value.

tary Lyman-α glow observations carried out from ~ 1 AU. For deep space observations, since the modulations smear out systematically with the distance, the discrepancies of the observed intensities with those predicted by the classical approach cannot be attributed to the solar cycle effects (see also Sec. 2.4). The discussed density modulations near the Sun affect noticeably the local production and resulting fluxes of the H^+ pickup ions (Ruciński and Bzowski, 1995b). To illustrate the effect for the realistic case, we present in Fig.3 a model variation of the flux at the sidewind direction ($\theta = 90°$) for distances 1 AU, 3 AU and 5 AU covering the range of the locations of

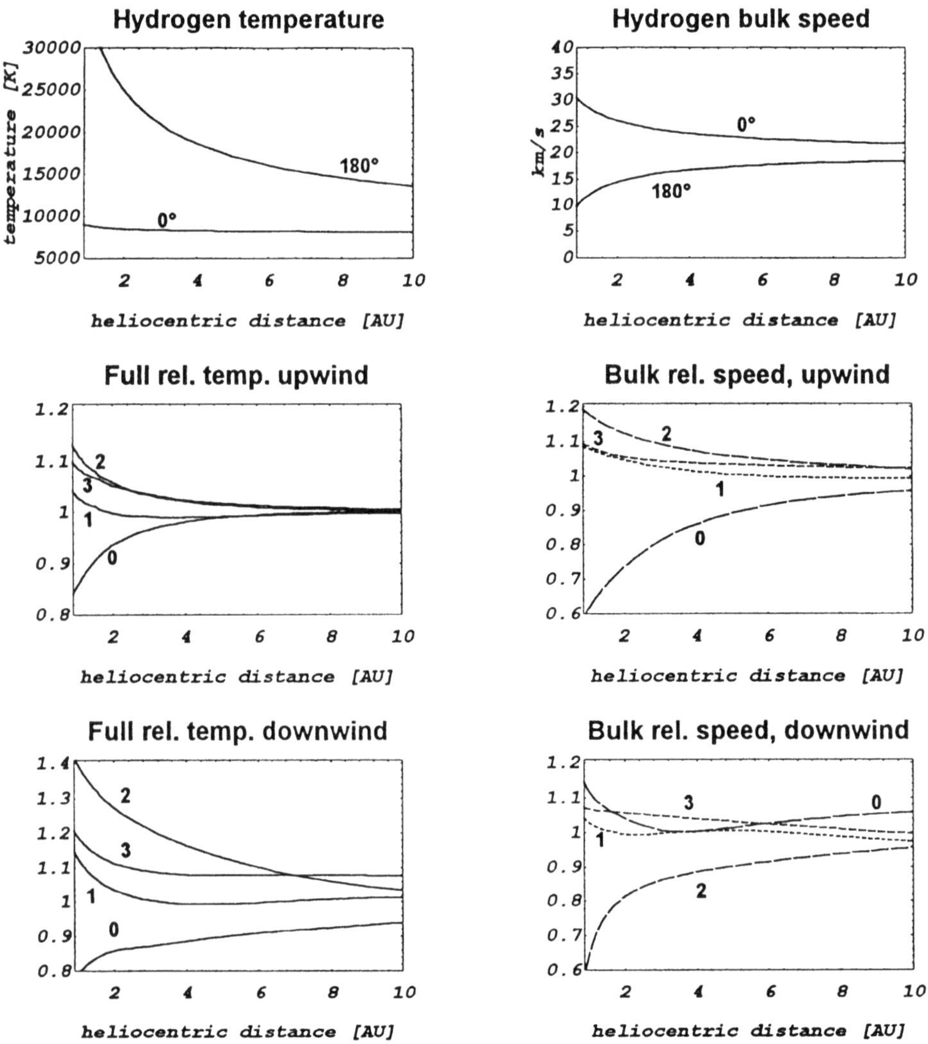

Figure 4. Variations of temperature and bulk velocity along the upwind and downwind axes. The left-hand column contains the profiles of full temperature and the right-hand one of full bulk speed between 1 and 10 AU. The upper row contains the profiles of the absolute values computed using the classical MSM model, the middle and the lower rows contain the profiles of temperature and bulk velocity relative to MSM for four selected phases of the solar cycle.

the Ulysses spacecraft. It is evident that the behaviour of the relative fluxes (in comparison with the MSM model) is significantly different from the behaviour of the ionization rate (equal to the creation rate of the H^+ pickup ions). This means that the variable production of the H^+ pickup ions in the inner Solar System is determined mainly by the modulation of the distribution of the neutral hydrogen background, and only in smaller degree depends on the instantaneous ionization rate.

The solar cycle effects influence also the behaviour of higher moments of hydrogen distribution function, although the variations are significantly less pronounced than the variations of density. As it is shown in Fig.4, the bulk velocity and the temperature variations are significant only within a few AU from the Sun. The bulk velocity reveals practically no departures from the MSM throughout almost the whole heliosphere, exceeding 10% only about 2 – 3 AU from the Sun on the upwind side. However in the downwind region such oscillations can be noticed as far as 25 AU. The full temperature of the hydrogen gas also does not depart considerably from the static case except regions inside 3 AU for the upwind and sidewind directions and 10 AU along downwind axis. What concerns the latter case, one should note, however, that at 5 AU the local temperature of the gas is about 17000 K, so modulations of about 20% expected at that distance translate into the the absolute variation of about ±3500 K.

Concluding our analysis, we point out that although the solar cycle effects influence the hydrogen distribution only within a limited spatial range, the modulations of hydrogen density in the inner Solar System ($R \leq 5$ AU) are quite significant. Since they cannot be generally reproduced with the use of stationary models, a true time-dependent approach is certainly the most reliable way to model the evolution of hydrogen distribution near the Sun.

2.3. Latitudinal asymmetry of the ionization

In the classical scenario a spherically symmetric field of ionization of the inflowing neutral atoms has been assumed. The validity of such an idealistic assumption has already been a subject of examinations. While possible longitudinal asymmetries average out over the 27-day period of solar rotation, the problem of the expected latitudinal asymmetry of hydrogen ionization rate remains still open. In the analyses of the interplanetary Lyman-α glow observations from Mariner 10 (Kumar and Broadfoot, 1979; Witt et al., 1979, 1981), Prognoz 6 (Lallement et al., 1985; Summanen et al., 1993), and Pioneer-Venus (Ajello et al., 1987; Lallement and Stewart, 1990) it was postulated that to fit the observational data (within the framework of the theory applied) a decrease of the ionization rate from equatorial to polar regions by about 25– 40% is required.

Due to the lack of direct out-of-ecliptic measurements the expected departure of the ionization rate from spherical symmetry was supported by indirect estimates of the average solar wind speed increase between the equator and the poles inferred from interplanetary scintillation data (e.g. Kojima and Kakinuma, 1990). Recent direct measurements on Ulysses during its in-ecliptic and out-of-ecliptic phases of the mission allow to revisit the problem and verify earlier results. As shown by Phillips et al. (1995), the average speed of the solar wind increases from about 400 km/s near the solar equator to about 750 km/s at high heliographic latitudes. It has to be noted (see e.g. Ruciński et al., 1996) that for such increase of the solar wind speed, the charge-exchange cross-section decreases by about 25%. This means that even in the case of the constant solar wind mass flux one can expect at least that degree of asymmetry in the charge-exchange ionization rate – the most important loss process for hydrogen. In addition, variations of the solar wind mass flux with heliolatitude, being still a matter of debate, may also contribute to the scale of asymmetry. Moreover, one can expect that the degree of latitudinal asymmetry varies during the solar cycle (e.g. Manoharan, 1993; Lallement and Stewart, 1990). From the Ulysses observations it was recently recognized that the average solar wind flux at polar regions is smaller by 25% than near the equator (Summanen et al., 1996; Marsden, 1996). Combining that with the aforementioned behaviour of the charge-exchange cross-section one may expect that the hydrogen ionization rate at the polar regions are smaller from the equatorial one by 40%. The current estimate of the asymmetry of the ionization rate (25 – 40%) is fully consistent with the earlier predictions from the EUV studies. Since the hydrogen density distribution near the Sun is very sensitive to value of the ionization rate, this effect should be carefully taken into account in the modelling. In particular, it can affect not only the interpretation of the Lyman-α backscattered radiation data, but even stronger the interpretation of the H^+ pickup ion data from different phases of the Ulysses mission, since the contributions to the observed signal come from the regions very close to the Sun ($< 1.5 - 5$ AU).

Another potential latitudinal effect, not considered here in detail, is possible asymmetry of the solar Lyman-α radiation which may cause asymmetry in the solar radiation pressure (see e.g. Pryor et al., 1992). This, in turn, may distort the hydrogen flow in comparison with the spherically symmetric case and thus affect its local distribution in the inner heliosphere.

2.4. Influence of the Filtration on Local Hydrogen Density

Another questionable point of the classical approach is the assumption that interstellar hydrogen can penetrate freely into the Solar System without any perturbations of its flow at the heliospheric interface region, separating the heliosphere from the LISM. Modelling of the hydrogen flow through that region and related phenomena have already been subject of studies by several authors (see e.g. Ripken and Fahr, 1983; Bleszyński, 1987; Osterbart and Fahr, 1992). The most comprehensive analysis of the changes of the H flow parameters during the crossing of the two-shock interface structure with the self-consistent treatment of the gas-plasma interactions (see Baranov et al., 1991) was recently presented by Baranov and Malama (1993). It was demonstrated that the charge-exchange processes in the region between the interstellar bow shock and the heliopause lead to a substantial enhancement of hydrogen density in the central part of that region, creating a structure called the "hydrogen wall". Simultaneously, the flow becomes noticeably decelerated in that area, especially close to the stagnation line. Near the heliopause a very sharp decrease of hydrogen density is expected. In consequence, the hydrogen density at the entrance to the Solar System may be already reduced by about 30 – 60% in comparison with its unperturbed level in LISM. It is also noticeably lower than in the conventional model, where any interface effects are ignored. Also the radial gradient of density in the outer Solar System was postulated to differ substantially from the one resulting from the classical ('no-interface') approximation. The predicted behaviour is not incompatible with the Lyman-α glow observations on Voyager and Pioneer in the outer heliosphere (Hall et al., 1993; Quèmerais et al., 1995). The results similar in general terms to the ones presented by Baranov and Malama (1993), but worked out using somewhat different assumptions on the interface plasma conditions have recently been obtained also by Fahr et al., (1995) and Pauls et al., (1995). In all quoted studies, the modelling was focused mainly on the global picture of hydrogen interaction with the heliospheric interface plasma seen in the macroscale of hundreds or dozens of AU and not on the modifications of the hydrogen distribution close to the Sun caused by filtration. To present a comparison of the 'interface' and 'no-interface' models in the "microscale" ($R \leq 20$ AU), we employ the results of calculations kindly provided by Baranov (1996, private communication).

The comparison is performed for two different cases, corresponding to a relatively modest ionization fraction of hydrogen in LISM (model B1: 75% of neutrals + 25% of ions) and to a very high ionization degree (model B2: 40% of neutrals + 60% of ions). In Fig.5, we compare the results for the upwind ($\theta = 0°$) and sidewind ($\theta = 90°$) directions with those corresponding the classical 'hot' model for the same set of the input parameters ($V_B = 26$ km/s; $T = 6700$ K; $\mu = 0.8$; $\beta_E = 6.3 \cdot 10^{-7}$ s^{-1}). One can easily conclude that the local densities in the whole region of interest are significantly lower when the interface effects are included than in the classical scenario with no interface. The relevant density ratios (classical model-to-Baranov model) within ~ 20 AU from the Sun, shown in Fig.6, vary approximately within the limits of 2.5 to 4, depending on the considered case and the heliocentric distance. The higher discrepancy is expected for the case of high ionization fraction of LISM and it slightly increases with the decrease of the heliocentric distance from 20 AU to 3 AU. Because the electron density and thus the ionization degree of hydrogen in LISM are still poorly known, the comparison presented above should be treated only as an illustrative example and not as a definite answer how the heliospheric interface processes affect the local H distribution near the Sun. Nevertheless, such substantial

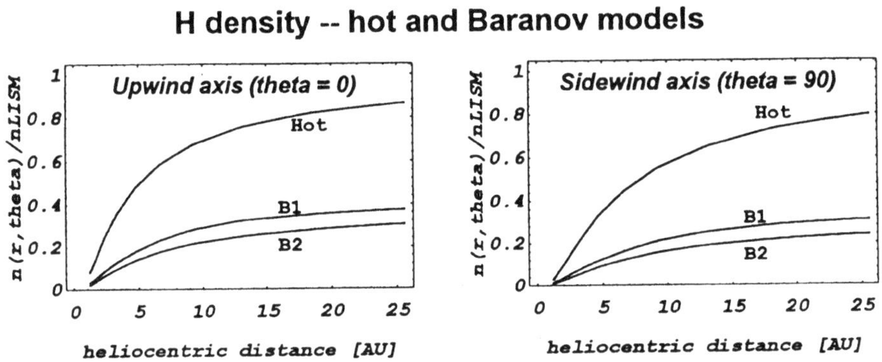

Figure 5. Comparison of radial profiles of hydrogen density (in the units of unperturbed H density in LISM) along the upwind axis (left-hand panel) and sidewind direction (right-hand panel), corresponding to the classical 'hot' model (HOT), and two Baranov models – B1 and B2, reflecting different neutral gas-to-proton density ratios in LISM. For B1 $n_{p,LISM} = 0.1$ cm^{-3} and $n_{H,LISM} = 0.3$ cm^{-3}; for B2 $n_{p,LISM} = 0.3$ cm^{-3} and $n_{H,LISM} = 0.2$ cm^{-3}. The remaining parameters (the same for the 'hot', B1 and B2 models) are: $V_B = 26$ km/s, $T = 6700$ K, $\mu = 0.8$, $\beta_E = 6.3 \cdot 10^{-7}$ s^{-1}.

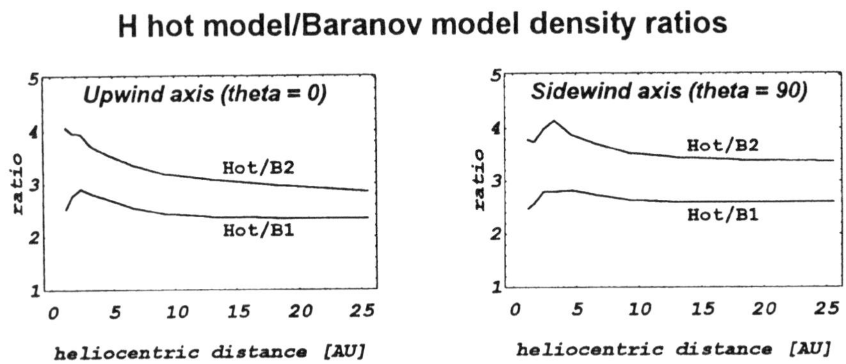

Figure 6. Ratios of the HOT-to-B1, and HOT-to-B2 hydrogen densities along upwind axis (left-hand panel) and sidewind direction (right-hand panel) resulting from Fig.5.

differences occurring in both considered cases suggest that the conventional modelling should be considered with the appropriate caution, since it might not describe appropriately the true local hydrogen density distribution.

3. Outlook

It was shown that several important effects ignored in the classical approach, including nonstationary solar cycle-related effects, asymmetry of the ionization field, and filtration at the heliospheric interface region, can significantly affect the pattern of neutral hydrogen distribution in the inner heliosphere. Since up-to-now all discussed effects have been considered separately, it seems desirable to treat them in a consistent manner in the future modelling efforts. In particular, combining the time-dependent model of gas distribution near the Sun with the model of filtration effects should enable a more reliable interpretation of the past and future in-ecliptic observations. Supplementing such a model with the latitudinal dependence of the solar ionization radiation pressure and illumination would bring a complete, time-dependent, three-dimensional model of the neutral hydrogen distribution in the heliosphere. It should also narrow the broad margin of uncertainty concerning its density in the outer heliosphere.

Acknowledgements

The authors are grateful to Prof. Vladimir Baranov for providing unpublished results from his computational scheme, what enabled us to perform the comparisons discussed in Section 2.4. This work was supported by grant No. 2 P03C 013 11 from the Committee for Scientific Research (Poland).

References

Ajello, J.M., Stewart, A.I., Thomas, G.E., and Graps, A.: 1987, *Astrophys. J.* **317**, 964.
Axford, W.I.: 1972, in 'Solar Wind', NASA SP-308, 609
Baranov, V.B., and Malama, Yu.G.: 1993, *J. Geophys. Res.* **98**, 15157
Baranov, V.B., Lebedev, M.G., and Malama, Yu. G.: 1991, *Astrophys. J.* **375**, 347
Bertaux, J.-L., Lallement, R., Kurt, V.G., and Mironova, E.N.: 1985, *Astron. Astrophys.* **150**, 1
Bertaux, J.-L., and Blamont, J.E.: 1971, *Astron. Astrophys.* **11**, 200
Bleszyński, S.: 1987, *Astron. Astrophys.* **180**, 201
Blum, P.W., Gangopadhyay, P., and Judge, D.L.: 1993, *Astron. Astrophys.* **272**, 549
Blum, P.W., Pfleiderer, J., and Wulf-Mathies, C.: 1975, *Planet. Space Sci.* **23**, 93
Blum, P.W., and Fahr, H.J.: 1970, *Astron. Astrophys.* **4**, 280
Bzowski, M., and Ruciński, D.: 1995a, *Space Sci. Rev.* **72**, 467
Bzowski, M., and Ruciński, D.: 1995b, *Adv. Space Res.* **16(9)**, 131
Fahr, H.J.: 1971, *Astron. Astrophys.* **14**, 263
Fahr, H.J.: 1968, *Astrophys. Space Sci.* **2**, 474
Fahr, H.J., Osterbart, R., and Ruciński, D.: 1995, *Astron. Astrophys.* **294**, 587
Fahr, H.J., and Scherer, K.: 1990, *Astron. Astrophys.* **232**, 556
Feldman, W.C., Lange, J.J., and Scherb, F.: 1972, in Solar Wind, NASA SP-308, 684
Gloeckler, G: 1996, *Space Sci. Rev.*, this issue
Gloeckler, G., Geiss, J., Balsiger, H., Fisk, L.A., Galvin, A.B., Ipavich, F.M., Ogilvie, K.W., von Steiger, R., and Wilken, B.: 1993, *Science* **261**, 70

Hall, D.T., Shemansky, D.E., Judge, D.L., Gangopadhyay, P. and Gruntman, M.A.: 1993, *J. Geophys. Res.* **98**, 15185
Holzer, T.E.: 1977, *Rev. Geophys. Space Phys.* **15**, 467
Kojima, M., and Kakinuma, T.: 1990, *Space Sci. Rev.* **53**, 173
Kumar, S., and Broadfoot, A.L.: 1979, *Astrophys. J.* **228**, 302
Kyrölä, E., Summanen, T., and Råback, P.: 1994, *Astron. Astrophys.* **288**, 299
Lallement, R.: 1990, in 'Physics of the Outer Heliosphere', COSPAR Colloquia Series, Vol.1, eds. S. Grzędzielski, D.E.Page, Pergamon Press, 49
Lallement, R., and Stewart, A.I.: 1990, *Astron. Astrophys.* **227**, 600
Lallement, R., Bertaux, J.-L., and Kurt, V.G.: 1985, *J. Geophys. Res.* **90**, 1413
Manoharan, P.K.: 1993, *Solar Physics* **148**, 153
Marsden, R.G.: 1996, *Space Sci. Rev.*, this issue
Möbius, E., Ruciński, D., Hovestadt, D., and Klecker, B.: 1995, *Astron. Astrophys.* **304**, 505
Osterbart, R., and Fahr, H.J.: 1992, *Astron. Astrophys.* **264**, 260
Pauls, H.L., Zank, G.P., and Williams, L.L.: 1995, *J. Geophys. Res.* **100**, 21595
Phillips, J.L., Bame, S.J., Feldman, W.C., Goldstein, B.E., Gosling, J.T., Hammond, C.M., McComas, D.J., Neugebauer, M., Scime E.E., Suess, S.T.: 1995, *Science* **268**, 1030
Pryor, W.R., Ajello. J.M., Barth, C.A., Hord, C.W., Stewart, A.I.F., Simmons, K.E., McClintock, W.E., Sandel, B.R., and Shemansky, D.E.: 1992, *Astrophys. J.* **394**, 363
Quèmerais , E., Bertaux, J.-L., Sandel, B.R., and Lallement, R.: 1994, *Astron. Astrophys.* **290**, 941
Quèmerais , E., Sandel, B.R,. Lallement, R., and Bertaux, J.-L.: 1995, *Astron. Astrophys.* **299**, 249
Ripken, H.W., and Fahr, H.J.: 1983, *Astron. Astrophys.* **122**, 181
Ruciński, D., Cummings, A.C., Gloeckler, G., Lazarus, A.J., Möbius, E., and Witte, M.: 1996, *Space Sci. Rev.*, this issue
Ruciński, D., and Bzowski, M.: 1995a, *Astron. Astrophys.* **296**, 248
Ruciński, D., and Bzowski, M.: 1995b, *Adv. Space Res.* **16(9)**, 121
Summanen, T., Lallement, R., Bertaux, J.-L., and Kyrölä, E.: 1993, *J. Geophys. Res.* **98**, 13215
Summanen, T., Lallement, R., Quèmerais , E.: 1996, *J. Geophys. Res.*, in press
Thomas, G.E.: 1978, *Ann. Rev. of Earth Planet. Sci.* **6**, 173
Thomas, G.E., and Krassa, R.F.: 1971, *Astron. Astrophys.* **11**, 218
Witt, N., Ajello, J.M., and Blum, P.W.: 1981, *Astron. Astrophys.* **95**, 80
Witt, N., Ajello, J.M., and Blum, P.W.: 1979, *Astron. Astrophys.* **73**, 272
Witte, M., Banaszkiewicz, M., and Rosenbauer, H.: 1996, *Space Sci. Rev.*, this issue
Witte, M., Rosenbauer, H., Banaszkiewicz, M., and Fahr, H.J.: 1993, *Adv. Space Res.* **13(6)**, 121
Wu, F.-M., and Judge, D.L.: 1979, *Astrophys. J.* **231**, 594

Address for correspondence: Dr. Daniel Ruciński, Space Research Centre of the Polish Academy of Sciences, Bartycka 18A, 00-716 Warsaw, Poland

OBSERVATIONS OF THE LOCAL INTERSTELLAR MEDIUM WITH THE EXTREME ULTRAVIOLET EXPLORER

JOHN VALLERGA
Center for EUV Astrophysics, UC Berkeley

Abstract. Because of the strong absorption of extreme ultraviolet radiation by hydrogen and helium, almost every observation with the *Extreme Ultraviolet Explorer* (*EUVE*) satellite is affected by the diffuse clouds of neutral gas in the local interstellar medium (LISM). This paper reviews some of the highlights of the *EUVE* results on the distribution and physical state of the LISM and the implications of these results with respect to the interface of the LISM and the heliosphere. The distribution of sources found with the *EUVE* all-sky surveys shows an enhancement in absorption toward the galactic center. Individual spectra toward nearby continuum sources provide evidence of a greater ionization of helium than hydrogen in the Local Cloud with an mean ratio of H I/He I of 14.7. The spectral distribution of the EUV stellar radiation field has been measured, which provides a lower limit to local H II and He II densities, but this radiation field alone cannot explain the local helium ionization. A combination of *EUVE* measurements of H I, He I, and He II columns plus the measurement of the local He I density with interplanetary probes can place constraints on the local values of the H I density outside the heliosphere to lie between 0.15 and 0.34 cm^{-3} while the H II density ranges between 0.0 and 0.14 cm^{-3}. The thermal pressure ($P/k = nT$) of the Local Cloud is derived to be between 1700 and 2300 cm^{-3} K, a factor of 2 to 3 above previous estimates.

Key words: ISM: general, abundances, clouds—radiation field—solar neighborhood

1. Introduction

NASA's *Extreme Ultraviolet Explorer* satellite (*EUVE*; Bowyer and Malina 1991) was launched on June 7, 1992. It consists of four grazing incidence telescopes and seven detectors. Three of the telescopes ("scanners") were designed to perform an all-sky survey in four photometric bandpasses centered on 100, 200, 400, and 600 Å, respectively. The fourth telescope ("Deep Survey/Spectrometer") performed a deep survey along the ecliptic in two bandpasses and supports three objective grating spectrometers: the Short Wavelength (70–190 Å), the Medium Wavelength (140–380 Å) and the Long Wavelength (280–760 Å). These spectrometers, with a spectral resolution to point sources ($\lambda/\Delta\lambda$) of $\sim$300, are employed in long observations of EUV sources chosen by guest observers.

H and He cross sections are at their maximum in the EUV spectral region which makes *EUVE* the instrument of choice in the study of the local interstellar medium (LISM) since the integrated column density of the neutral gas near the Sun (hereafter referred to as the "Local Cloud") is on the order of $\log N_{\rm H\,I}$ (cm^{-2}) = 18. *EUVE* is sensitive to a range of H I column densities of $\log N_{\rm H\,I} = 17 - 20$. Another advantage of the EUV bandpass in ISM work is the existence of both the He I and He II Lyman edges at 504 Å and 228 Å, respectively. When detected in absorption in continuum spectra, these ISM He edges give the column density directly without the need for a complex stellar source model or ISM H value. *EUVE* has therefore provided, for the first time, direct measurements of the total He column that can be used to determine the H and He ionization fractions of the Local Cloud.

Because *EUVE* measures the Lyman continuum of both H and He, it directly measures the radiation field that photoionizes H and He. The local EUV radiation field is one

Space Science Reviews **78**: 277–288, 1996.

component of the ionization balance equation of the local gas and was only known to a factor of 20 before the launch of *EUVE* (Cox and Reynolds 1987). Previously it was modeled by summing theoretical EUV spectra of nearby hot stars and gas and attenuating the spectra through an assumed distribution of the local neutral gas. *EUVE* can turn this process around by taking the *measured* local EUV radiation field and unattenuating the spectrum through a *measured* distribution of local gas. Models of the photoionization of the Local Cloud can be compared with the measurements of column average H and He ionization fraction discussed above.

This review will attempt to cover the wide range of ISM results collected and published by guest observers using *EUVE*, with an emphasis on the local interstellar medium (LISM) and the Local Cloud. In particular, the distribution, density, and ionization fraction of the gas near the Sun will be emphasized so that limits to these values can be recommended for use in models of the interaction of the heliosphere with the Local Cloud.

2. Distribution of Absorbing Gas

The fact that the EUV spectrum of most individual objects is not known, i.e., standard candles do not exist, complicates the determination of the distribution of local absorbing gas using an all-sky photometric survey. This situation will improve slowly as EUV, UV, and optical spectra of many of these sources are collected, which, when combined with successful identifications and stellar models, will allow a reasonably accurate column measurement. However, much can be done in a statistical sense using $\log N - \log S$ curves (see below) and number densities on the sky.

The Second *EUVE* Source Catalog (Bowyer et al. 1996) lists 734 detected sources. It is an improvement over the first catalog in that better detection algorithms were used, and more of the low-exposure sky areas have been filled in with additional observations. Two types of sources dominate the list of identified objects: hot white dwarfs and active late-type stars. Table I lists the number of identified objects by type. The hot white dwarfs are intrinsically more luminous than the coronally active late-type stars and therefore sample a much larger volume of space. A map of the distribution on the sky of both white dwarfs and late-type stars detected by *EUVE* in the 100 Å bandpass is shown in Figure 1.

A large-scale non-uniformity in the distribution of sources can be seen in the maps with a distinct paucity of sources toward the galactic center region, especially in the map of white dwarf detections. This confirms the result of Warwick et al. 1993 who showed the same results with the Wide Field Camera on the *ROSAT* satellite. In the direction of the galactic center lies the gas whose column density roughly increases with distance from the Sun with an average density of 0.1 cm^{-3} but with density enhancements on the order of 10^3 times the average density (the "squall line," Frisch 1995). Frisch interprets this gas to be a shell of a superbubble event originating in the Sco-Cen Association that expanded in a previously rarefied cavity and is now within a few parsecs of the Sun toward the galactic center. Another obvious white dwarf surface density enhancement is toward the southern third galactic quadrant ($180° < l^{\rm II} < 270°, b^{\rm II} < 0°$): the famous low H I column direction toward β CMa that extends out to 200–300 pc (Gry et al. (1985); Welsh 1991).

Figure 2 shows a plot of the log of the number of sources, N, brighter that a flux, S, versus $\log S(\log N - \log S)$ of all detected white dwarfs and late-type stars. Because the distance to most identified EUV sources is less than 100 pc, the spatial distribution of stellar

Table I
Object Identification by Type of Source Detected in *EUVE* Survey (from Bowyer et al. 1996)

Type	All-Sky Survey	Deep Survey along Ecliptic
Late-type stars (F, G, K, M)	161	27
White dwarfs	98	1
Early-type stars (A, B)	18	0
Cataclysmic variables	13	0
Extragalactic	6	0
Low mass X-ray binaries	2	0
Other	25	0
No identification	191	7
Total	514	35

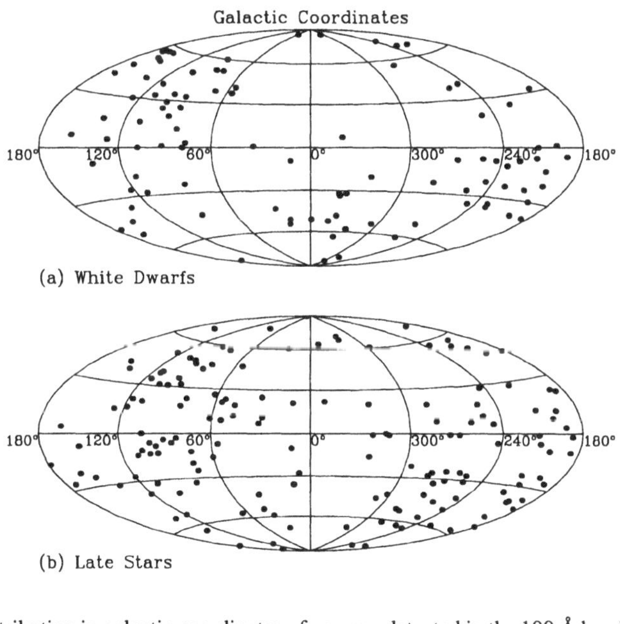

Figure 1. The distribution in galactic coordinates of sources detected in the 100 Å band of the *EUVE* all-sky survey (Bowyer et al. 1996) (a) white dwarfs and (b) late-type (F, G, K, and M) stars. Note the paucity of sources toward the galactic center.

sources is usually assumed to be isotropic. Therefore, $\log N - \log S$ curves of sources of a given type (and luminosity) would have a slope of –3/2 down to the detection limit if not absorbed by an intervening column density of H I. A rollover to a smaller slope would indicate that the source distribution is highly attenuated as the distances are increasing.

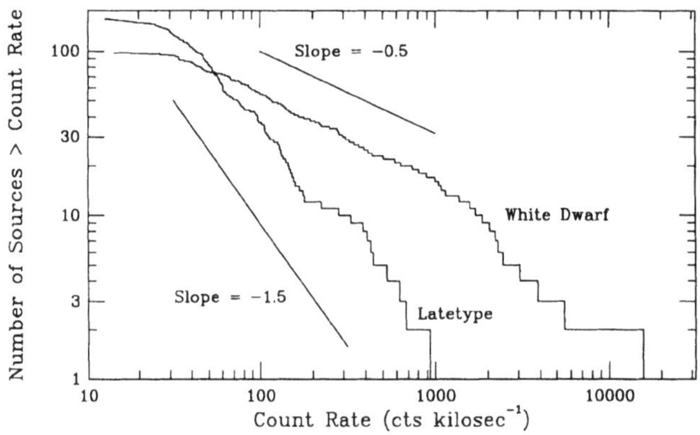

Figure 2. A $\log(N > S)$ vs. $\log(S)$ plot of detected white dwarfs and late-type stars detected by *EUVE* in the 100 Å bandpass of the all-sky survey. Because white dwarfs are inherently brighter in the EUV than late-type stars, the rollover from the –3/2 slope indicates that the white dwarf population samples a larger average column density whereas the nearby and relatively unattenuated late-type stars increase in number until the sensitivity limit of the survey is reached.

Figure 2 shows that the hot white dwarfs are inherently brighter than the late-type stars and that the slope of the distribution quickly falls off from the –3/2 power of an unattenuated sample. The number of late-type stars, in contrast, rise at near the –3/2 slope until the sensitivity of a large fraction of the survey is reached ($\sim$40 cts ks^{-1}). Therefore the nearby late-type stars detected are on average much closer than the $\sim$ 1 optical depth of the 100 Å band (100 pc at a density of 0.1 cm^{-3}).

3. Column Densities Toward Individual Stars

The Lyman continuum absorption for H I (< 912 Å), He I (< 504 Å) and He II (< 228 Å) are all in the *EUVE* spectrometer passbands. This direct determination of the H and He column densities using does not suffer from the uncertainties of other techniques such as UV and optical absorption lines of tracer elements (e.g., N I and Mg II). Using these techniques, errors associated with assumptions about equilibrium, depletion, temperature and atomic physics can all enter into the conversion from measured absorption to H I column. With its limited spectral resolution, *EUVE* cannot, however, resolve individual kinematic components and therefore can only report total column averages to nearby stars.

The absorption edges at 228 Å and 504 Å in a continuum spectrum can also provide stellar model–independent measures of the ISM He columns. These edges are important when separating the contributions of H and He absorption. A limited range of H I columns exist where an accurate He I column can be measured. Given the ten to one ratio for an neutral ISM, when the He I column is large enough to measure the 504 Å edge easily, the H I column absorbs most of the photons above 504 Å. Fortunately, He I has another spectral feature: the auto-ionization feature at 206 Å (Rumph, Bowyer, and Vennes 1994). Because

Table II
ISM H and He Column Densities Derived from *EUVE* Spectroscopy (10^{17} cm^{-2})

Name	N(H I)	N(He I)	N(He II)	Reference
Feige 24	30.0 ± 0.7	1.54± 0.14	<6.6	Dupuis et al. (1995)
G191–B2B	20.7 ± 0.4	1.46± 0.08	<4.5	"
MCT 0455–2812	12.9 ± 0.4	0.87± 0.04	<4.1	"
GD 153	9.8 ± 0.8	0.66± 0.07	<1.0	"
HZ 43	8.7 ± 0.7	0.55± 0.03	<0.7	"
GD 71	6.3 ± 1.6	0.52± 0.08	<1.0	"
GD 246	140.0 ± 20	11.5± 1.0	<3.5–4.0	Vennes et al. (1993)
Adhara (ϵCMa)	9.5 ± 2.5	—	—	Cassinelli et al. (1995)
Mirzam (βCMa)	20.0 ± 2.0	>14	—	Cassinelli et al. (1996)

the feature is at a shorter wavelength, it effectively increases the range of detecting He I columns with *EUVE* to between log N (cm^{-2}) = 16.3 and 18.3.

Most EUV sources that have measured H and He columns have continuum spectra such as hot white dwarfs and the few early-type stars whose photospheric emission has been detected. Coronal sources such as active late-type stars that emit mostly in emission lines could theoretically be used to determine column densities by using carefully chosen line ratios. However, candidate line pairs insensitive to density, temperature, and abundance variations in the corona tend not to be strongly enough separated in wavelength to be very sensitive to a wide range of column densities. Certainly, just the detection of an emission line at wavelengths longer than 504 Å implies a very low H I column density. Yet, of the late-type stars observed spectroscopically by *EUVE* only three have confirmed line detections at these long wavelengths: αCen, α Aur, and α CMi. More accurate column densities to these stars derived from high resolution Lyman alpha measurements from the GHRS instrument aboard *Hubble Space Telescope* now exist (Linsky et al. 1995).

Dupuis et al. (1995) reported on the observations of six DA white dwarfs with the *EUVE* spectrometers (Table II). Though all are considered to have H rich atmospheres, three of these sources have enough metallicity ($Z \sim 10^{-5}$) to show a complex structure in the spectra, especially below 260 Å, while the other three are consistent with a metallicity of $Z < 10^{-7}$. All of the spectra exhibit the He I edge at 504 Å to different degrees, correlated to the amount of derived H I column. Table II also gives the ISM parameters derived from their fits to the edges and continuum. For objects with the higher metallicity, the 228 Å edge caused by He II is possibly blended with heavy element opacities, reducing the accuracy of the He II column determination. Dupuis et al. also noted that the edges were not as sharp as a pure step function, and the fit improved when their fitting also included the convergent absorption lines of the ground state series of both He I and He II from the ISM.

The ratios of H I to He I for the white dwarfs in Table II (taken from Dupuis et al. 1995), show that He is more ionized than H in all directions sampled if the canonical cosmic ratio of 10:1 of total H to He is used. The mean value of these measurements is 14.7. The distance to these stars ranges from 42 to 90 pc, and therefore the sightlines could sample other gas than the Local Cloud. But the low H I column values of these objects are consistent with the values measured to nearby stars, and this absorption is most likely local in origin.

Dupuis et al. placed limits on the level of ionization of both H and He in the absorbing gas by using the upper limits to the He II edge at 228 Å. Adding the limit of the He II to the He I gives the total He, assuming the density of He III is very small. Multiplying this total He by 10 gives an upper limit to the total H for comparison to the measured H I. The tightest limits were toward HZ 43 because it is the brightest white dwarf at the 228 Å edge and it does not have the spectral structure of high metallicity. The column ionization fraction toward HZ 43 (and therefore of the Local Cloud) is less than 0.4 for H and less than 0.5 for He. These limits are inconsistent with the high range of electron density derived from the Mg II to Mg I ratio toward Sirius measured in the UV (Lallement et al. 1994). The implications of these results on the Local Cloud densities near the Sun are discussed in § 5 below.

An actual *EUVE* detection of He II in the LISM was made toward the white dwarf GD 246 (Vennes et al. 1993). The He ionization fraction was ~25% with a total He column of $1.40 - 1.65 \times 10^{18}$ cm^{-2}. The H I column density was less accurately determined to lie between $1.2 - 1.8 \times 10^{19}$ cm^{-2}. The upper limit to the H ionization fraction was 27%. Because the H column is roughly a factor of 10 higher than the typical Local Cloud value, this line of sight must sample other diffuse cloud(s), but the ionization fractions reported are consistent with the values determined locally. This then is a major discovery of *EUVE*, that He is partially ionized in the warm diffuse clouds in the LISM.

One of the surprising results of the *EUVE* all-sky survey was the detection of two early-type stars in the 600 Å bandpass, ϵ CMa and β CMa (Adhara and Mirzam). Adhara turned out to be the brightest EUV source in the sky (Vallerga, Vedder, and Welsh 1993). This result was surprising for two reasons. First, the H I column density toward Adhara was a factor of 2 less than toward its neighbor Mirzam even though they were at similar distances (187 and 210 pc, respectively). Second, the photospheric flux in the Lyman continuum was a factor of 30 greater than predicted using standard local thermodynamic equilibrium models (Cassinelli et al. 1995; Cassinelli et al. 1996).

Models of the photospheric emission for these sources in the Lyman continuum cannot yet predict the absolute flux. But the shape of the intrinsic continuum can be predicted well enough above 504 Å such that the strong spectral signature of H I absorption can be used to place strong limits on the neutral column to Adhara from 7 to 12×10^{17} cm^{-2} (Cassinelli et al. 1995). (The uncertainty in the jump at the 504 Å edge in the intrinsic stellar spectrum is such that no useful limits can be placed on the He I ISM column.) The H I absorption is most likely caused by local gas as two of the absorption features of Adhara found at high resolution in the UV match the velocities of the two components found toward Sirius (Gry et al. 1995). The measured H I column range is above the 5×10^{17} cm^{-2} upper limit derived by Gry (1995) using an upper limit of the tracer N I column. However, if the total Mg II column measured by Gry et al. is converted to a H I column using the ratios of Mg II/H I found toward Capella and Procyon (Linsky et al. 1995) then the predicted column would be $\sim 16 \times 10^{17}$ cm^{-2}. This shift in values is a prime example of instances in which indirect measurements of tracer gases in the ISM suffer from uncertainties in abundances compared to the direct measurement of H I itself.

The H I column toward Mirzam is twice that of Adhara, though it is only slightly farther away. But the angular separation of 14° allows clouds of 50 pc radius to exist at that distance without intersecting the line of sight to Adhara. The Mirzam spectrum does not show the 304 Å emission line of He II from the stellar wind that exists in the Adhara spectrum, even though they were expected to be similar based on their spectral type and observed

X-ray intensities. Cassinelli et al. (1996) placed a lower limit to the ISM He I column of 1.4×10^{18} cm^{-2} as an explanation for the unobserved line. This lower limit would imply that the "extra" gas that exists toward Mirzam is composed of highly ionized H and neutral He, similar to an H II region. This composition would be consistent with the postulate by Gry et al. (1985) of a highly ionized region of total H column of 1.8×10^{19} cm^{-2} located in a cloud of 40 – 90 pc in length. The "tunnel" toward Mirzam would therefore be a tunnel of ionization rather than an absence of gas. This possibility is important for the view of the Local Bubble being filled with hot $T = 10^6$ K gas since no enhancement appears in the soft X-ray background toward Canis Majoris (McCammon et al. 1983).

4. EUV Radiation Field

The physical parameters of the gas in the Local Cloud nearest the Sun, such as the density, temperature, ionization fraction, and velocity, establish the boundary conditions into which the solar wind expands and therefore has an important affect on the structure of the heliosphere. Direct measurements of column densities of H I, He I, and He II toward distant stars only give line-of-sight averages that sample different directions of the cloud or even other clouds that lie in the same path. A complementary approach to determine the density and ionization structure of the Local Cloud is to try to measure the causes of ionization. One such cause would be photoionization of H and He by the integrated EUV radiation field. With the launch of *EUVE*, the stellar contribution to this field can now be accurately determined and limits can be placed on the diffuse component of the field. Given the locally measured EUV radiation field, the local ionization fraction of H and He can be determined if ionization equilibrium is assumed or a lower limit to this value can be determined if equilibrium is not assumed. Also, to interpret the line-of-sight averages of H, He, and electron column densities requires the modeling of the photoionization process through the Local Cloud, as these values can change dramatically from the outside of a cloud to the Sun's position.

The all-sky survey in four photometric bands across the EUV spectrum conducted by the *EUVE* satellite identified all of the major contributors to the stellar EUV radiation field (Bowyer et al. 1996). These stars have now been observed with the *EUVE* spectrometers and their spectra have been combined to give the local ionizing field for He as well as H with a typical spectral resolution, $\lambda/\Delta\lambda$, of ~ 300. A previous analysis of just the spectrum of Adhara showed that this one star was the dominant stellar source of H ionization (Vallerga and Welsh 1995). By including all of the bright EUV sources, the total stellar radiation field can be used to estimate the local He ionization fraction as well as the ratio of H I to He I whose measured value differs between the Local Cloud and inside the heliosphere.

The stellar EUV radiation field was created by summing the spectral flux of the 10 brightest white dwarfs in the EUV plus the two B stars, Adhara and Mirzam (Vallerga 1996). Figure 3 shows the resultant spectrum in units of photons cm^{-2} s^{-1} Å^{-1} versus wavelength, and it represents 99% of the total survey count rate in the 600 Å and 400 Å bandpasses, 81% in the 200 Å bandpass, and 61% in the 100 Å bandpass. Weighting this spectrum with the photoionization cross sections of H and He gives the photoionization parameters at the Earth: $\Gamma_{\rm H\,II} = 1.4 \times 10^{-15}$ s^{-1}, $\Gamma_{\rm He\,II} = 7.4 \times 10^{-16}$ s^{-1}, $\Gamma_{\rm He\,III} = 3.3 \times 10^{-17}$ s^{-1}. These photoionization rates would predict a minimum H ionization fraction of 13% near the Sun for an assumed H I density of 0.2 cm^{-3}. But the He ionization fraction would only be 4.3%.

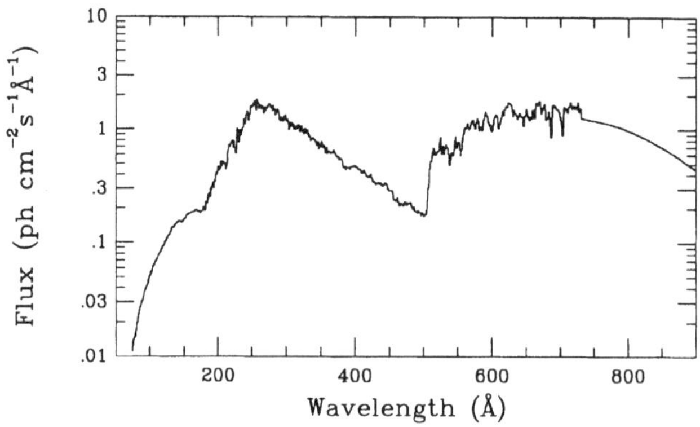

Figure 3. A spectrum of the *stellar* EUV radiation field at Earth consisting of the sum of the EUV flux from the 10 brightest white dwarfs in the all-sky survey plus the B stars Adhara and Mirzam. The strong edge at 504 Å is mostly due to the B stars' intrinsic stellar spectrum as the drop off below 240 Å is due to the metals in the atmospheres of some of the brighter white dwarfs. The *EUVE* spectrometers cannot measure the spectrum above 730 Å, so an extrapolation of the spectrum of Adhara is plotted (Vallerga and Welsh 1995).

Therefore, stellar photoionization by itself cannot explain the enhanced He ionization with respect to H.

In a model of Local Cloud ionization produced before these EUV measurements, Cheng and Bruhweiler (1990) summed model spectra of nearby white dwarfs and Mirzam to create an EUV radiation field, and also included the diffuse flux from the hot ($\log T = 6$) Local Bubble. This coronal-type spectrum of emission lines preferentially ionizes He over H, and a cloud with $\log N_{\rm H} = 18$ column is optically thin to this radiation, resulting in a uniform increase in He ionization. The *EUVE* spectrometers were optimized for measuring the spectrum of point sources but also have residual sensitivity to diffuse EUV radiation between 170 Å and 730 Å. By integrating over 500,000 s of exposure pointed down Earth's shadow, Jelinsky et al. (1995) placed stringent limits on the emission measure of a $\log T = 6$ gas, a factor of $\sim$10 lower than derived from the soft X-ray surveys. They attributed the lack of a detection to either non-equilibrium conditions or depletion of the hot gas. If a $\log T = 6$ coronal spectrum consistent with the emission measure limits of Jelinsky et al. is added to the radiation field of Figure 3, then the He photoionization parameter increases by a factor of 3.3, but He would still be underionized compared to H.

A simple one-dimensional model of the photoionization of the Local Cloud has been developed that calculates the photoionization of H and He from the Sun out to the edge of the cloud (Vallerga 1996). It is similar to the model developed by Cheng and Bruhweiler but it starts with the locally measured radiation field and a local H I density and works its way out through the cloud to a given input column density. Like the Cheng and Bruhweiler model, the one-dimensional model is not meant to solve the three-dimensional radiation transfer and ionization structure of the Local Cloud, but rather to illustrate the ionization structure differences of H and He of a diffuse cloud illuminated by a strong EUV radiation field.

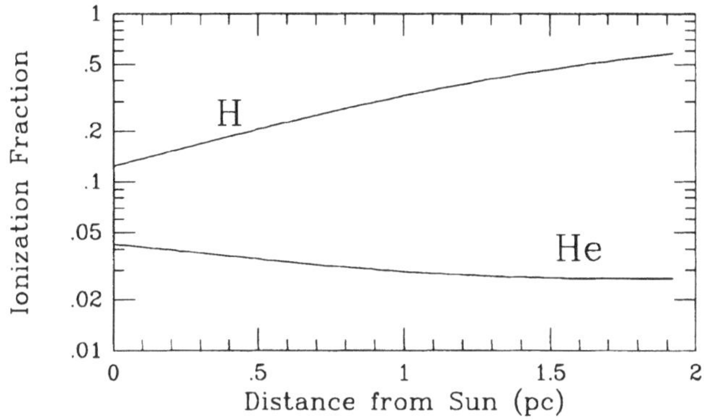

Figure 4. Results of a one-dimensional equilibrium photoionization model of the Local Cloud (Vallerga 1996) given the stellar radiation field of Fig. 3, an assumed density of H I of 0.2 cm^{-3} near the sun, and a total H I column to the cloud edge of 9×10^{17} cm^{-2}. Plotted is the ionization fraction $X(= n(ion)/n(total))$ as a function of distance from the sun to the edge of the cloud. Note the strong ionization of hydrogen at the cloud's edge compared to the shielded region inside. The He ionization fraction is a weak function of distance into the cloud since a cloud of this column is optically thin to radiation that ionizes helium. If the assumption of equilibrium conditions is wrong, these ionization fractions represent lower limits.

Figure 4 is a sample run of this program with the input radiation field given in Figure 3, a local neutral H density of 0.2 cm^{-3}, and a column to the edge of the cloud of $9 \times 10^{17} cm^{-2}$. Note the strong increase in the H ionization ratio as the edge of the cloud is reached, while the He ionization fraction stays roughly constant because the cloud is optically thin to radiation that ionizes He. To explain the ratio of the neutral H to He greater than 10, therefore, requires He to be highly ionized throughout the cloud to make up for the high H ionization at the cloud surface. In this example the $N_{\rm H\,I}/N_{\rm He\,I}$ ratio is only 7. If photoionization alone is invoked to explain the ionization structure of the Local Cloud, then another strong source of He ionization is needed, such as the emission from the conductive interface between the Local Cloud and the Local Bubble as calculated by Slavin (1989). As pointed out by many authors, the recombination timescales at these temperatures and densities are on the order of a million years, so any recent ionizing event such as supernova flashes or shocks would leave unequilibrated fossil ionization (Frisch 1995). Attempts to model the relaxation from such events (Lyu and Bruhweiler 1996) will also have to include the effects of the measured EUV radiation field and explain the enhanced ionization of He.

5. Implications of *EUVE* Results on the Local Cloud: Heliosphere Models

It is widely accepted that the He I density measured within the heliosphere can simply be related to the density of the Local Cloud near the Sun because processes in the heliospheric boundary region or solar radiation pressure leave neutral He relatively unaffected. Previously, the He was measured through the resonance backscattering of solar He lines (Bertaux et al. 1985; Chassefiere et al. 1986), but these techniques depended on complex backscattering models to derive a density. Recently, the density of He I has been measured directly through an impact ionization method (Witte et al. 1993) and indirectly through pickup ion measurements of He II (Mobius et al. 1995).

If we combine this one *locally* measured quantity with the H and He column densities and ratios through the Local Cloud as measured by *EUVE*, then we can derive a strong set of limits to the interstellar local densities and ionization fraction of H and He. The relatively model-independent derivations use only a few straightforward assumptions concerning the conversion of ratios of column averages to ratios of local densities.

The strong limits argument starts with four measurements:

$$\begin{aligned}
\frac{N_{\mathrm{H\,I}}}{N_{\mathrm{He\,I}}} &\approx 14 && \text{Dupuis et al. (1995)} \\
X_{\mathrm{H}} &< 0.4 && \text{"} \\
X_{\mathrm{He}} &< 0.5 && \text{"} \\
n(\mathrm{He\,I}) &= 0.014 \pm .003 && \text{Witte et al. (1993)}
\end{aligned}$$

where N_{ion} represents a column density (cm^{-2}), $n(ion)$ represents a density (cm^{-3}), and $X_{\mathrm{H,He}}$ is the ionization fraction of H and He, respectively, toward HZ 43. These arguments assume that n(He III) is much smaller than n(He II) and may be ignored.

Assuming that H is more strongly ionized than He as the cloud surface is approached, and that the He ionization fraction is roughly constant throughout the cloud (see discussion above on EUV radiation field), then the first equation converts to a locally valid inequality:

$$\frac{n(\mathrm{H\,I})}{n(\mathrm{He\,I})} > 14 \qquad \text{(near the Sun)}$$

First, the maximum values allowed by these measurements for the local density of H and He can be derived:

$$\begin{aligned}
n(\mathrm{He}) &= \frac{n(\mathrm{He\,I})}{(1 - X_{\mathrm{He}})} < 0.034 \\
n(\mathrm{He\,II}) &= X_{\mathrm{He}} * n(\mathrm{He}) < 0.017 \\
n(\mathrm{H}) &= 10 * n(\mathrm{He}) < 0.34 \\
n(\mathrm{H\,II}) &= X_{\mathrm{H}} * n(\mathrm{H}) < 0.14 \\
n_e &= n(\mathrm{H\,II}) + n(\mathrm{He\,II}) < 0.15
\end{aligned}$$

Lower limits to these quantities can also be derived in a similar fashion:

$$\begin{aligned}
n(\mathrm{H\,I}) &> 14 * n(\mathrm{He\,I}) = 0.15 \\
n(\mathrm{H}) &> n(\mathrm{H\,I}) > 0.15 \\
n(\mathrm{He}) &> 0.015 \\
n(\mathrm{He\,II}) &= n(\mathrm{He}) - n(\mathrm{He\,I}) > 0.004
\end{aligned}$$

Table III summarizes these values.

The four initial measurements do not require that H be ionized at all in the Local Cloud near the Sun, and therefore the maximum values of H I would be equal to the total H density. However, a minimum ionization of 15% is required because of the radiation field of the

Table III

Local Limits of the H and He Densities with and without model effects of stellar EUV radiation field (in cm^{-3})

	Excluding EUV RAD Field		Including EUV RAD Field	
Atom/Ion	Lower	Upper	Lower	Upper
He I	0.011	0.017	0.011	0.017
He II	0.004	0.017	0.004	0.017
He	0.015	0.034	0.015	0.014
H I	0.15	0.34	0.15	0.31
H II	0.00	0.14	0.03	0.14
H	0.15	0.34	0.18	0.34

bright EUV source Adhara. Since the amount of ionization caused by the EUV radiation field is somewhat model dependent, values that include this ionization source for H have been separated in Table III.

Recent analyses of H I density inside the heliosphere have increased the estimate to 0.165 ± 0.035 cm^{-3} (Quemerais et al. 1996). The local limits on the densities outside the heliosphere reported in Table III are consistent with these values but imply that the filtration of the H I through the heliosphere interface ranges from 0% to 50%. More importantly, the limit on the value of the proton density constrains the models of the charge exchange interaction at the interface that predict the filtration factor as well as the location of the heliopause (Ripken and Fahr 1983; Lallement et al. 1992). Previous estimates of the H II plasma density used electron densities derived from Mg I and Mg II column densities toward Sirius (Lallement et al. 1994). These estimates suffer from uncertainties in the dielectronic recombination cross section of Mg because of strong temperature dependence, the assumption of equilibrium conditions, and the conversion of a column average to a local value, when it is known that the ionization fraction of H changes dramatically in the direction of Sirius because of the influence of the strong radiation field of Adhara.

The thermal pressure of the Local Cloud ($P/k = nT$) from 1700 to 2300 cm^{-3} K is a factor of $2.5 - 3.0$ higher than the canonical value of 700 cm^{-3} K. However, this new value of the thermal pressure locally is still a factor of 4 to 11 lower than the the Local Bubble pressure ($P \sim 10000 - 20000$; Cox and Reynolds 1987; Bowyer et al. 1995) and this discrepancy has fuelled a controversy over competing models of the galactic ISM (Cox 1995). A higher thermal pressure of the Local Cloud requires less contribution from other pressure sources such as magnetic fields, but these other sources are still required to maintain a pressure confined cloud.

This work has been supported by NASA contract NAS5-29298. The Center for EUV Astrophysics is a division of UC Berkeley's Space Sciences Laboratory.

References

Bertaux, J. L., Lallement, R., Kurt, V. G. and Mironova, E. N.: 1985, *A&A*, **150**, pp. 1
Bowyer, S., Lampton, M., Lewis, J., Wu X., Jelinsky, P. and Malina, R. F.: 1996, *ApJS*, **102**, pp. 129–160
Bowyer, S., Lieu, R., Sidher, S., Lampton, M. and Knude, J.: 1995, *Nature*, **375**, pp. 212–214
Bowyer, S. and Malina, R. F.: 1991, in Extreme Ultraviolet Astronomy, ed. R. F. Malina and S. Bowyer, (New York: Pergamon), pp. 397
Cassinelli, J. P., Cohen, D. H., MacFarlane, J. J., Drew, J. E., Lynas-Gray, T., Hubeny, I., Vallerga, J., Welsh and Hoare, M.: 1996, *ApJ*, April 1.
Cassinelli, J. P., Cohen, D. H., MacFarlane, J. J., Drew, J. E., Hoare, M. G., Lynas-Gray, T., Vallerga, J. V., Vedder, P. W., Welsh, B. Y., Hubeny, I. and Lanz, T.: 1995, *ApJ*, **438**, pp. 932–949
Chassefiere, E., Bertaux, J. L., Lallement, R. and Kurt, V. G.: 1986, *A&A*, **160**, pp. 229
Cheng, K. P. and Bruhweiler, F. C.: 1990, *ApJ*, **364**, pp. 573
Cox, D.: 1995, *Nature*, **375**, pp. 185–186
Cox, D. P. and Reynolds, R. J.: 1987, *An. Rev. A&A*, **25**, pp. 303
Dupuis, J. Vennes, S., Bowyer, S., Pradhan, A. and Thejll, P.: 1995, *ApJ*, **455**, pp. 574–589
Frisch, P.: 1995, *Space Sci. Rev.*, **72**, pp. 499–592
Gry, C., Lemonon, L., Vidal-Madjar, A., Lemoine, M. and Ferlet, R.: 1995, *A&A*, **302**, pp. 497–508
Gry, C., York, D. G. and Vidal-Madjar, A.: 1985, *ApJ*, **296**, pp. 593
Jelinsky, P., Vallerga, J. and Edelstein, J.: 1995, *ApJ*, **442**, pp. 653–661
Lallement, R., Bertin, P., Ferlet, R., Vidal-Madjar, A. and Bertaux, J. L.: 1994, *A&A*, **286**, pp. 898–908
Lallement, R., Malama, Y., Quemerais, E., Bertaux, J. and Zaitzev, N.: 1992, *ApJ*, **396**, pp. 696–703
Linsky, J., Diplas, A., Wood, B., Ayres, T. and Savage, B.: 1995, *ApJ*, **451**, pp. 335
Lyu, C. and Bruhweiler, F.: 1996, *ApJ*, **459**, pp. 216–225
McCammon, D., Burrows, D. N., Sanders, W. T. and Kraushaar, W. L.: 1983, *ApJ*, **269**, pp. 107
Mobius, E., Rucinski, D., Hovestadt, D. and Klecker, B.: 1995, *A&A*, **304**, pp. 505–519
Quemerais, E. Bertaux, J. L., Sandel, B. and Lallement, R.: 1996, *A&A*, **290**, pp. 941
Reynolds, R.: 1986, *AJ*, **92**, pp. 653
Ripken, H. W. and Fahr, H. J.: 1983, *A&A*, **122**, pp. 181
Rumph, T., Bowyer, S. and Vennes, S.: 1994, *AJ*, **107**(6), pp. 2108
Slavin, J. D.: 1989, *ApJ*, **346**, pp. 718
Vallerga, J.: 1996, *ApJ*, submitted
Vallerga, J., Vedder, P. and Welsh, B.: 1993, *ApJ*, **414**, pp. L65
Vallerga, J. and Welsh, B.: 1995, *ApJ*, **444**, pp. 202
Vennes, S., Dupuis, J., Rumph, T., Drake, J. J., Bowyer, S., Chayer, P. and Fontaine, G.: 1993, *ApJ*, **410**, pp. L119
Warwick, R., Barber, C., Hodgkins, S. T. and Pye, J. P.: 1993, *MNRAS*, **262**, pp. 289
Welsh, B. Y.: 1991, *ApJ*, **373**, pp. 556
Witte, M., Rosenbauer, H., Banaskiewicz, M. and Fahr, H.: 1993, *Adv. Space Res.*, **13**(6), pp. 121–1300

Address for correspondence: Center for EUV Astrophysics, 2150 Kittredge Street, University of California, Berkeley, CA 94720-5030, USA

RECENT RESULTS ON THE PARAMETERS OF THE INTERSTELLAR HELIUM FROM THE ULYSSES/GAS EXPERIMENT

M. WITTE, M. BANASZKIEWICZ* and H. ROSENBAUER
Max-Planck-Institut für Aeronomie, D-37191 Katlenburg-Lindau, Germany.

Abstract. Velocity and direction of the flow of the interstellar helium and its temperature and density have been determined from the measurements of the ULYSSES/GAS experiment for two different epochs: during the in-ecliptic path of ULYSSES, representing solar maximum conditions, and during the south to the north pole transition (11/94-6/95), close to the solar minimum conditions. Within the improved error bars the values are consistent with results published earlier.

The determination of the density n_∞ of the interstellar helium at the heliospheric boundary from observations in the inner solar system requires knowledge about the loss processes experienced by the particles on their way to the observer. The simultaneous observation of the helium particles arriving on "direct" and "indirect" orbits at the observer provides a tool to directly determine the effects of the loss processes assumed to be predominantly photoionization and – for particles travelling close to the Sun – electron impact ionization by high-energy solar wind electrons.

Such observations were obtained with the ULYSSES/GAS instrument in February 1995, before the spaceprobe passed its perihelion. From these measurements values for the loss rates and the interstellar density could be derived. Assuming photoionization to be the only loss process reasonable fits to the observations were obtained for an ionization rate $\beta = 1.1 \cdot 10^{-7}$ s^{-1} and a density $n_\infty \approx 1.7 \cdot 10^{-2}$ cm^{-3}. Including, in addition, electron impact ionization, a photoionization $\beta = 0.6 \cdot 10^{-7}$ s^{-1} was sufficient to fit both observations, resulting in a density $n_\infty \approx 1.4 \cdot 10^{-2}$ cm^{-3}.

1. Introduction

Since more than twenty years the interstellar neutral particles as representatives of the local interstellar medium have been of particular interest. Except for high energy galactic cosmic rays these particles are the only ones which are available in the inner solar system for in-situ measurements or short-range remote observations. From the determination of the local parameters (flow velocity and direction, temperature and density) it should be possible to infer the values of these parameters outside the heliosphere ("at infinity").

In the past, first results were obtained from the measurement of the intensity distribution of solar UV-light resonantly backscattered from the local density distribution of interstellar hydrogen and helium atoms (e.g. Bertaux and Blamont, 1971; Thomas and Krassa, 1971; Chassefière et al., 1986, 1988; Dalaudier et al., 1984).

In addition to these ongoing observations (e.g. Bertaux et al., 1995, Pryor et al., 1995) other techniques and methods more recently provided improved results: The observation of pick-up ions (e.g. Möbius et al., 1985, 1995) revealed an independent source for the parameters of, in particular, the interstellar helium distribution. With the ULYSSES/SWICS instrument it was possible for the first time to measure their

* On leave from Space Research Centre, Warsaw, Poland.

Space Science Reviews **78**: 289–296, 1996.

full phase space distribution and derive in a self-consistent way the local loss and production rates of the pick-up ions (Gloeckler et al., 1996).

Also, with new data from the Hubble Space Telescope, the analysis of the dopplershift in the absorption lines in the light of near-by stars provided independent new information about the velocity v_∞ of the motion of the solar system through the local interstellar medium (Lallement et al., 1996).

A completely different approach in the determination of the parameters of the interstellar helium has been performed for the first time with the ULYSSES/GAS-instrument, which detects individual helium atoms. In this paper we shall describe in detail the observations and analysis of the data obtained around the perihelion of ULYSSES, when it was possible for the first time to observe simultaneously particles on "direct" and "indirect" orbits.

In this unique case it was possible to infer the rate of the losses, the helium particles experience on their way to the observer and to determine the density n_∞ of the helium particles outside the heliosphere. Finally, we shall summarize the results obtained so far for the velocity v_∞ and temperature T_∞.

2. Instrument Description

The GAS-instrument on the ULYSSES spaceprobe, which acts like a pinhole camera, is capable to detect individual interstellar helium particles and allows to determine their local flow direction (Witte et al., 1992). From these local angular distributions the parameters of the interstellar particles outside the heliosphere ("at infinity"), the velocity magnitude and direction v_∞ and the temperature T_∞, can be determined in a straight forward manner (Witte et al., 1993) as one can assume that the particles move on hyperbolic orbits around the Sun because radiation pressure and interactions at the heliospheric interface are negligible. The details of mathematical methods involved in the detailed analysis (tomography problem) and their limitations have been extensively described by Banaszkiewicz et al. (1996).

3. Observations

In principle, interstellar helium particles having the same velocity v_∞ at infinity can arrive at a given point in space (observer) only on two hyperbolic orbits named "direct" and indirect", lying in a common plane which is uniquely determined by the focal point (the Sun), the position of the observer and the vector v_∞ (Fig. 1).

As the kinetic energy of these particles is only of the order of a few 10 eV which is close to the detection threshold of the instrument it requires a favorable combination of the velocity vectors of the spaceprobe and the particles to get a sufficiently large relative energy in the instrument's frame of reference (Fig. 2). In the course of the ULYSSES-mission, this is the case for the direct particles during the in-ecliptic

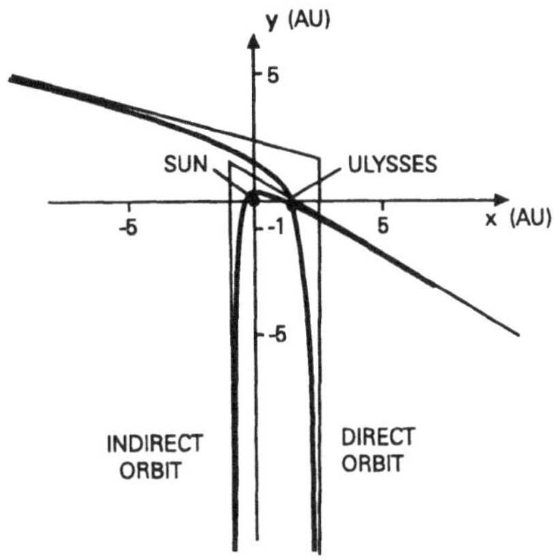

Figure 1. The "direct" and "indirect" hyperbolic orbits of helium particles, that arrive at the position of the spaceprobe at 1.3 AU in February 1995, during the perihelion of ULYSSES. Particles on indirect orbits have had their closest distance to the Sun at 0.2 AU and are ionized much more (~98%)than the particles on the direct orbit. The simultaneous observation of both particle intensities allows the determination of the rate of the photoionization, the pre-dominant loss process.

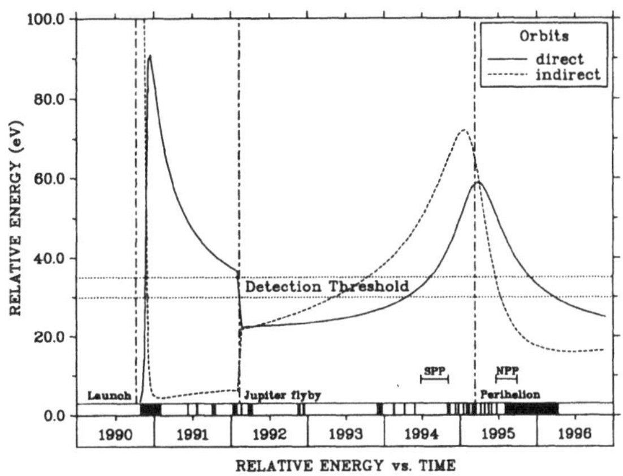

Figure 2. The relative energy of interstellar helium particles on direct and indirect orbits in the instrument's frame of reference. Mainly, this energy is determined by the velocity and flight direction of the spaceprobe with respect to the local flow of the particles (~ 43 km/s). Due to the high velocity after launch (~ 40 km/s) and around perihelion (~ 32 km/s), there are two periods, 1990-1992 and 1994-1996, when the particle energy exceeds the detection threshold of about 35 eV. (Periods, when the instrument was active, are indicated on the abscissa.)

orbit towards Jupiter (October 1990 to February 1992) and around perihelion (end 1994 to early 1996) for particles on both orbits. The angular distributions, which have been measured in February 1995 shortly before the perihelion of ULYSSES, are shown in Fig. 3 (uppermost panels) for particles on direct (left h.s.) and indirect (right h.s.) orbits. They are represented in the instrument's coordinate system of elevation angle (i.e. cone angle with respect to the spacecraft's spin axis) and azimuth angle (the rotation angle) as measured in steps of $2° \times 2.8°$, respectively. The instrument's coordinate system can be directly converted into an inertial coordinate system (e.g. solar ecliptic system) using the available orbit and attitude data of the spacecraft.

While there is a clear signal of the direct particles (~ 8 cts/s), the signal of the indirect particles (~ 0.2 cts/s) hardly exceeds the background, immediately showing the high degree of losses for these particles due to ionization. (To reduce the statistical fluctuations and improve the image a sliding-3×3-box averaging has been applied to the data.)

4. Density Determination

If one wants to infer the density of the interstellar neutral particles at the boundary of the solar system from remote or in-situ measurements in the inner solar system, knowledge about the effectiveness and also about the time dependence of various loss processes acting on the neutral particles on their way into the inner solar system is a prerequisite. In general, it is difficult to obtain quantitative figures because of the various processes involved (e.g. Rucinski et al., 1996). With the observations described above it was possible, however, to estimate the actual, global loss rates valid for the time of the measurement.

In the first step of this analysis, we have assumed that photoionization is the pre-dominant loss process for the interstellar helium particles. In our model (Witte et al., 1993, Banaszkiewicz et al., 1996) used to simulate the observations processes with a $1/r^2$-dependence have been taken into account and are characterized by the ionization rate β_{ion} at 1 AU distance. However, the ionization rate is a free input-parameter to the model and can, in general, not be derived from one measurement of the local density alone.

This problem can be overcome when particles can be observed simultaneously arriving at the position of the observer on "direct" and "indirect" orbits.

In the case of the observation, reported here, particles on "indirect" orbits had their closest distance to the Sun during their perihelion at 0.2 AU while particles on "direct" orbits, still inbound, have their closest distance from the Sun during the arrival at the observer at 1.3 AU (Fig. 1).

Therefore, both groups of particles experience a completely different history with respect to their exposure to the solar UV radiation field, responsible for their photoionization. As the travel time of the indirect particles is only about a month longer as compared to the direct particles, temporal variations of the radiation may be neglected and the same ionization rate can be used for the simulation of both observations eliminating this degree of freedom in the model.

The results of the simulation of the direct particles are shown in the l.h.s. of Fig. 3. While the values of the velocity ($v_\infty, \beta_\infty, \lambda_\infty$) and temperature T_∞ are a result of a fitting process, for the ionization rate reasonable values have been chosen. As a result the model yields densities n_∞ which increase with increasing loss rates to reproduce the observed local densities as one would expect. It is obvious that the simulated distributions match the observed distribution of direct particles equally well, it is not possible to determine a unique pair of ionization rate β_{ion}/density n_∞ from a single measurement. Quantitatively, the extinction factor (ratio of the density at the point of observation to the density at infinity) ranges from 0.73 to 0.54 for $\beta_{ion} = 0.8 \cdot 10^{-7}\ \mathrm{s}^{-1}$ to $0.16 \cdot 10^{-7}\ \mathrm{s}^{-1}$, respectively and is 0.68 for $\beta_{ion} = 1.0 \cdot 10^{-7}\ \mathrm{s}^{-1}$.

Now, in the r.h.s. column, the distributions are shown, which are the predictions of the model for the indirect particles, using the same sets of parameters as for the distributions of direct particles. As one can see, only for ionization rates around

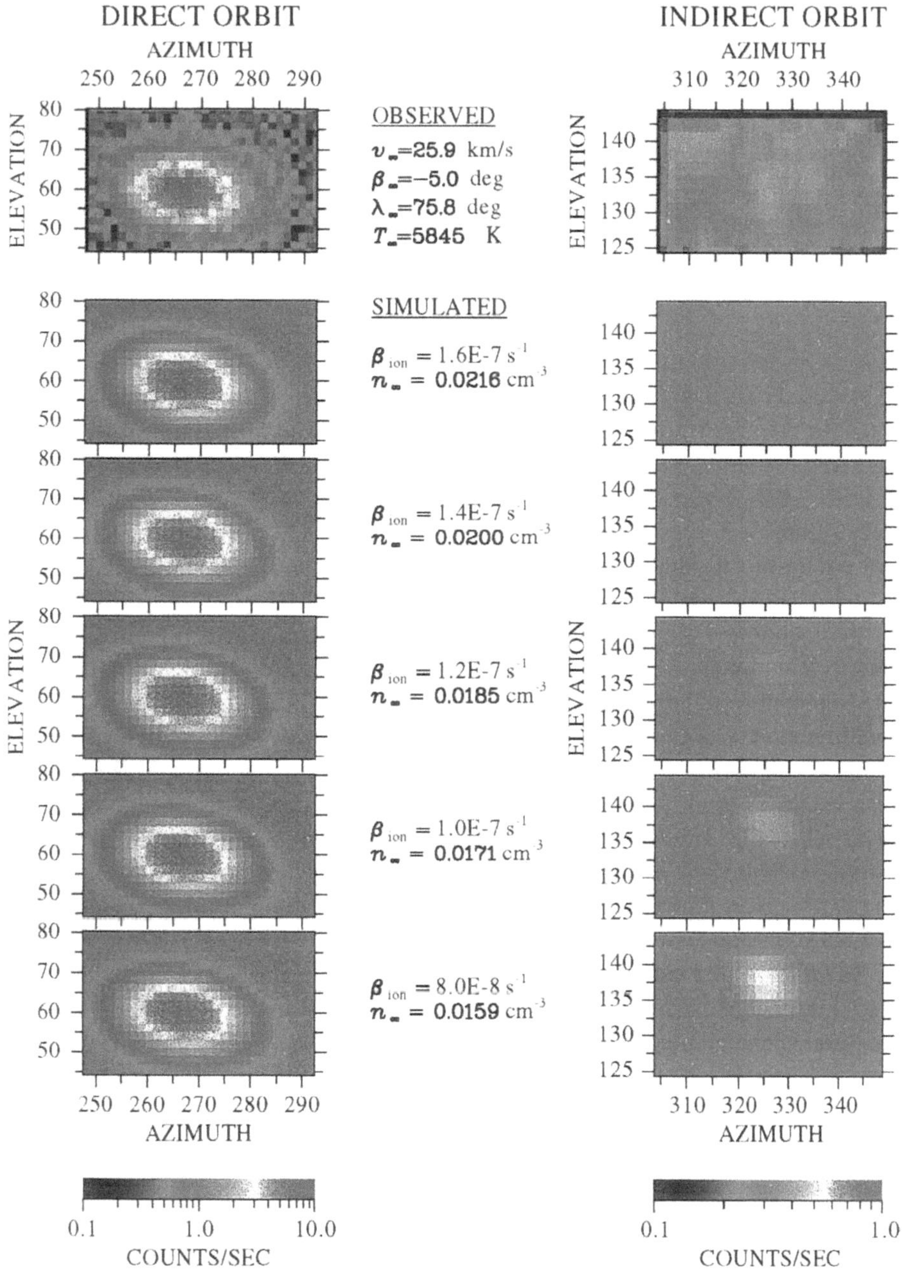

Figure 3. Comparison of observed (uppermost panels) and simulated distributions of particles on direct (left hand column) and indirect orbits. While for the direct particles various sets of density n_∞ and photoionization rate β_{ion} can be found which fit equally well to the observed distribution, the observed distribution of indirect particles can be fitted reasonably only by $\beta_{ion} \approx 0.9 - 1.1 \cdot 10^{-7}$ s^{-1} and a density around $n_\infty \approx 1.7 \cdot 10^{-2}$ cm^{-3}. (Note the change in the countrate scale by a factor of 10.)

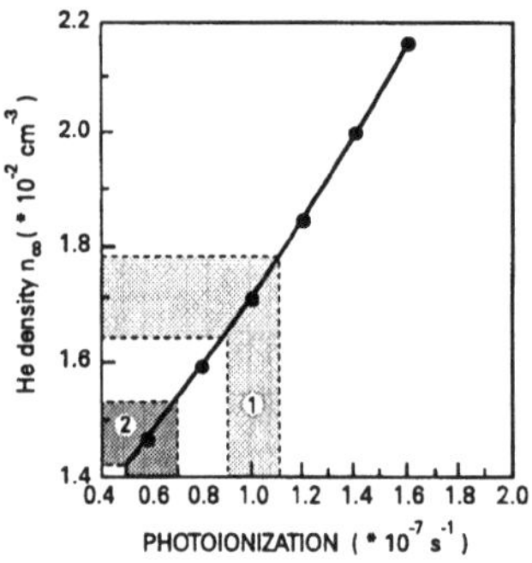

Figure 4. Relation between the photoionization and the helium density (solid line) as derived from the model fits to the observed distributions of particles on direct orbits (Fig. 3, left column) and the ranges of photoionization rate, required to reproduce the observations of indirect particles, assuming photoionization to be the only loss process (1) or including electron-impact ionization (2). (Note, that this relation is valid only for the particular position of ULYSSES close to perihelion ($\sim$ 1 February 1995, R = 1.3 AU).

$\beta_{ion} \approx 1.0 \cdot 10^{-7}$ s^{-1} the count rates from the simulation agree with the observed ones (in the uppermost panel). From a more detailed comparison of the observed data and the simulated ones, we get a likely range for the ionization rate of 0.9 – 1.1 $\cdot$ 10^{-7} s^{-1}, which is partly due to uncertainties in the determination of the background. As a result, a density range $n_\infty \approx 1.64 - 1.78 \cdot 10^{-2}$ cm^{-3} would be required to fit simultaneously the observed local fluxes of direct and indirect particles (Fig. 4, shaded area (1)).

As mentioned already, in this model so far we have assumed photoionization due to a stationary UV-radiation field to be the only noticeable loss process for the interstellar helium particles. However, Rucinski and Fahr (1989) pointed out that in regions close to the Sun an additional ionization process might be important which results from collisions of hot solar wind electrons with the helium atoms (see also Rucinski et al., 1996). As the electrons rapidly cool down in the expanding solar wind, this effect is estimated to have decreased to the order of 20% of the photoionization at 1 AU only. Therefore, this effect is assumed to be negligible for the particles on direct orbits but not for particles on indirect orbits which have a perihelion distance of 0.2 AU only.

In an improved version of the model this loss process has been included, using the same average solar wind conditions as in Rucinski and Fahr (1989). The effect has then been calculated from numerical integration along the particle trajectory.

Applying the same procedure as before, the best fits between the observation and simulation of the direct particles were obtained with the same set of parameters (v_∞, T_∞), as before. Also, the relation between ionization and density remained essentially the same as before (Fig. 4, solid line), verifying that the electron impact ionization has little effect on particles outside 1.3 AU. For the indirect particles the results were significantly different. As part of their losses now are due to the electron impact ionization only a smaller photoionization rate $\beta_{ion} \approx 0.6 \cdot 10^{-7}$ s^{-1} is required to account for the total losses observed. (In this case, the total extinction factor of 0.024 is the product of the extinction factors for photoionization of 0.11 (at $\beta_{ion} = 0.6 \cdot 10^{-7}$ s^{-1}) and 0.22 for electron impact ionization.) As indicated in Fig. 4, region (2), a density in the range $n_\infty \approx 1.4 - 1.5 \cdot 10^{-2}$ cm^{-3} is required to simultaneously fit the observations of the local fluxes on direct and indirect orbits.

Table I

EPOCH			INECLIPTIC 12/90 – 1/92	SOUTH/NORTH 11/94 – 6/95
Velocity	V_∞	(km/s)	25.3 ± 0.4	24.6 ± 1.1
Ecl. longitude	L_∞	(°)	73.9 ± 0.8	74.7 ± 1.3
Ecl. latitude	B_∞	(°)	-5.6 ± 0.4	-4.6 ± 0.7
Temperature	T_∞	(K)	7000 ± 600	5800 ± 700
Density	n_∞	(10^{-2} cm^{-3})	1.7 ± 0.2	1.4 1.7
@ Photoionization	β_{Ion}	(10^{-7} s^{-1})	1.4	0.6* 1.1
No. of distributions			36	10

(* plus electron impact ionization)

5. Discussion and Conclusions

We have extended the analysis of the measurements of the Interstellar Neutral Gas Experiment (Witte et al., 1993) to a period in 1994/1995 when ULYSSES performed its rapid latitudinal transition from the south pole to the north pole of the Sun. The results obtained so far for the velocity vector $\boldsymbol{v}_\infty$ and temperature T_∞ are summarized in Table I.

Within the error bars these results do not significantly deviate from the values obtained in the first epoch in 1990/1992 during the in-ecliptic trajectory of ULYSSES towards Jupiter. Compared to the results, published earlier (Witte et al., 1993) the error bars could be substantially reduced because of the larger data base available and the statistical error analysis applied (Banaszkiewicz et al., 1996). It should be noted that the first epoch was characterized by solar maximum conditions while the measurements in the second epoch took place close to solar minimum.

The determination of the density n_∞ of the interstellar helium outside the heliosphere from measurements in the inner solar system requires knowledge of the loss processes the particles are exposed to during their travel to the observer, e.g. charge exchange, photoionization, electron impact ionization. This is yet regarded the most uncertain aspect in the determination of the interstellar density. (Updates of this topic may be found also in the papers by Cummings et al., 1996; Rucinski et al., 1996; and Geiss and Witte, 1996).

For the analysis of the measurements in the first epoch we had to assume (in accordance to literature) that a photoionization $\beta_{ion} \approx 1.4 \cdot 10^{-7}$ s^{-1} is an appropriate value for the solar maximum conditions, resulting in an interstellar density $n_\infty \approx 1.7 \cdot 10^{-2}$ cm^{-3} (Table I).

Contrary to this, in the second epoch there was direct observational evidence about the acting loss processes, for the first time. The simultaneous observation of particles on direct and indirect orbits, which experience a completely different history with respect to their loss processes, allows to determine the actual loss rates.

Within the limitations of the data due to the weak signal of the indirect particles and problems related with the background determination, a range of most likely photoionization rates could be obtained (Fig. 4). An additional uncertainty comes from the inclusion of the electron impact ionization because of the assumptions which have to be made about this loss process, e.g. the actual electron distributions. However, in this particular case studied, the total effect (with or without electron impact ionization) leads to a variation of the density of the order of 25% only and, as a consequence, attempts to refine the (solar wind-) parameters used to describe this effect would have even smaller effects. It is interesting to note that the results of the density n_∞ and ionization rates obtained from pick-up ion measurements, with the SWICS-instrument on ULYSSES (Gloeckler et al., 1996; Geiss and Witte, 1996) perfectly agree with the results presented here, which are almost a factor of two larger than most previously published values.

Acknowledgements

We appreciate the support of M. Bruns in the data processing. One of us (M.B.) is grateful to the Max-Planck-Society for the stipends to participate in the ULYSSES programme. This work has been supported by the Deutsche Agentur für Raumfahrtangelegenheiten (DARA) GmbH, grant 50 ON 9103 and in part by the Polish Committee for Scientific Research (grant no. 2 P03C 013 11).

References

Banaszkiewicz M., Witte M., Rosenbauer H.: 1996, *Astron. Astrophys., Suppl. Ser.*, in press.
Bertaux J.L. and Blamont J.E.: 1971, *Astron. Astrophys.* **11,** 200.
Bertaux J.L., Kyrölä E.: 1995, Quémerais E., et al., *Solar Physics* **162,** Nos. 1-2, 403.
Chassefière E., Bertaux J.L., Lallement R., Kurt V.G.: 1986, *Astron. Astrophys.* **160,** 229.
Chassefière E., Dalaudier F., Bertaux J.L.: 1988, *Astron. Astrophys.* **201,** 113.
Cummings A.C. and Stone E.: 1996, *Space Sci. Rev.*, this issue.
Dalaudier F., Bertaux J.L., Kurt V.G., Mironova E.N.: 1984 , *Astron. Astrophys.* **134,** 171.
Geiss J. and Witte M.: 1996, *Space Sci. Rev.*, this issue.
Gloeckler G.: 1996, *Space Sci. Rev.*, this issue.
Lallement R.: 1996, *Space Sci. Rev.*, this issue.
Möbius E., Hovestadt D., Klecker B., Scholer M., Gloeckler G. Ipavich M.: 1985, *Nature* **318,** 426.
Möbius E., Rucinski D., Hovestadt D., Klecker B.: 1995, *Astron. Astrophys.* **304,** 505.
Pryor W.R., Barth C.A., Hord C.W., et al.: 1995, *Geophys. Res. Letters,* submitted.
Rucinski D. and Fahr H.J.: 1989, *Astron. Astrophys.* **224,** 280.
Rucinski D., Cummings A.C., Gloeckler G., Lazarus A.J., Möbius E., and Witte M.: 1996, *Space Sci. Rev.*, this issue.
Thomas G.E. and Krassa R.F.: 1971, *Astron. Astrophys.* **11,** 218.
Witte M., Rosenbauer H., Keppler E., Fahr H., Hemmerich P., Lauche H., Loidl A., Zwick R.:1992, *Astron. Astrophys., Suppl. Ser.* **92,** 333.
Witte M., Rosenbauer H., Banaszkiewicz M.: 1993, *Adv. Space Res.* **13,** (6)121.

INTERACTION BETWEEN HS AND LISM

PHYSICAL AND CHEMICAL CHARACTERISTICS OF THE ISM INSIDE AND OUTSIDE THE HELIOSPHERE

R. LALLEMENT[1], J. L. LINSKY[2], J. LEQUEUX[3] and V. B. BARANOV[4]

[1] *Service d' Aéronomie du CNRS, B.P. 3, 91371 Verrières le Buisson, France*

[2] *JILA, Univ. of Colorado and N.I.S.T., Boulder, CO 80309-0440 USA*

[3] *Observatoire de Paris, Avenue de l' Observatoire, 75014 Paris, France*

[4] *Institute for Problems in Mechanics, Russian Academy of Science, Prospect Vernadskogo 101, Moscow, 117526, Russia.*

Abstract. This paper summarizes some of the discussions of working group 8-9 during the ISSI Conference on "The Heliosphere in the Local Interstellar Medium". Because the subject of these working groups has become significantly broader during the last ten years, we have selected three topics for which recent observations have modified and improved our knowledge of the heliosphere and the surrounding interstellar medium. These topics are the number densities and ISM ionization states of hydrogen and helium, the newly discovered hot gas from the "H wall" seen in absorption, and the comparison between ISM and heliospheric minor element abundances. Papers from this volume in which more details on these topics can be found are quoted throughout the report.

1. Helium and Hydrogen Ionization in the Local Interstellar Cloud and the Heliospheric Neutral Number Densities

1.1. A Short Historical Review

A debate about the ionization of H and He and the HI/HeI ratio in the ISM began in the 1970's when the first UV photometers measured the helium 58.4 nm and hydrogen 121.6 nm interstellar glows. Comparisons between these data and models of the interstellar H and He flows predicted a H/He number density ratio of about 6 and definitely smaller than 10 at large distances from the Sun (i.e., before close interaction and destruction by the solar photons and the supersonic solar wind). For a cosmic abundance ratio $H/He = 10$, the observed small HI/HeI ratio was interpreted as a sign of partial H ionization, since interstellar helium was assumed to be essentially neutral. A later interpretation of the small ratio was in terms of HI filtration at the heliospheric interface or both effects simultaneously. If a portion of the HI is prevented from entering the heliosphere, while all of the HeI is allowed to enter, the observed heliospheric HI/HeI ratio will be smaller than the ISM ratio. Because helium ionization requires hard UV photons, and the ISM shows many examples of HII regions in which H is completely ionized, there was then no reason to question the assumption of helium near-neutrality. At the same time photometric measurements of HeI in the heliosphere from backscattered solar resonance line radiation were leading to a larger velocity and a larger temperature compared to hydrogen. The large He temperature contradicted the first models of heliospheric perturbations which predicted heating of H but not of He. (If there were no heating at the heliosphere-ISM interface, the temperatures of He and H should be equal.) Some attempts to explain the apparently different temperatures have been only partially successful.

Subsequently, the direct detection of He with the GAS experiment on Ulysses (Witte *et al.*, this vol.) and the AMPTE (Möbius, this vol.) and SWICS (Gloeckler, this vol.) pick-up ions and other recent measurements, together with improved models for both species, have shown that the helium temperature is about 7,000 K, slightly smaller than the temperature measured for H in the direction perpendicular to the flow (Lallement, this vol.) and in better agreement with the model prediction. Why the photometric measurements of HeI are in disagreement with the direct detections of particles is still unexplained.

At the same time, the H/He number density ratio has been revised upward due to better modeling of the radiative transfer (Quémerais *et al.*, 1994, 1995) and models that go beyond the assumptions that underlie the simple classical model of the neutral H flow (Baranov and Malama, 1993; Lallement *et al.*, 1994, 1995; Clarke *et al.*, 1995). It is unfortunately true, however, that the HI number density is the least precisely known quantity, due in part to solar wind anisotropies (Summanen *et al.*, 1996). After these revisions, the range in the heliospheric HI density (0.08 – 0.17 cm^{-3}) became consistent with the cosmic ratio $H/He = 10$. The lower limit comes from the pick-up ion data analysis (Gloeckler, this vol.), while the upper limit comes from Lyα glow analysis which may be overestimated. The heliospheric density in the flow at about 40 AU (i.e., after the heliospheric interface filtration but before solar UV photons and the supersonic wind can ionize the HI) is very likely smaller than 0.15 cm^{-3} (Lallement, this vol.).

A HI/HeI ratio close to 10 in the heliosphere implies that the circumsolar interstellar medium is globally neutral (both H and He) and that there is negligible filtration of H at the heliospheric interface. At that time HI/HeI was assumed to be no larger than 10. However, this neutrality was contradicted by other arguments which favor significant ionization in the LIC as the deceleration of H with respect to He (Lallement *et al.*, 1993). Another argument against a neutral ISM was the small distance to the heliospheric boundary inferred from the Voyager cosmic ray gradients and temporal evolution (Cummings and Stone, 1990, and this vol.), unless the interstellar magnetic field is entirely responsible for the confinement of the heliosphere.

1.2. Recent EUVE and Ulysses Results

During the last three years, new observations and analyses have provided strong evidence for substantial ionization of the surrounding ISM, and at the same time they have also removed the contradiction discussed above.

Neutral magnesium was observed in the spectrum of Sirius (2.7 pc), and more recently in δ Cas (12.5 pc) (Lallement, this vol.) and ϵ CMa (200 pc) (Gry, this vol.) at the LIC Doppler shift. The MgII/MgI ionization equilibrium suggests a surprisingly high electron density of about 0.1 cm^{-3} or even 0.3 cm^{-3} when one uses the ionization equilibrium calculation for magnesium at 7,000 K including charge exchange reactions between MgI and protons. The NaI/(NaI + NaII) equilibrium also suggests significant, although smaller, electron density (about 0.05 cm^{-3}) than what can be inferred from magnesium (Lallement, this vol.). The reasons for the discrepancies between the Mg and Na determinations are still unclear. The Sirius results are puzzling: local electron densities larger than 0.25 cm^{-3} are precluded by the Voyager measurements since Voyager 1 would have already crossed the termination shock! The Na results seem more reasonable. However, the C/O anomalous cosmic ray ratios analyzed by Frisch (1994, 1995) imply a very large electron density.

New fundamental results have been inferred from the EUVE spectral observations of nearby white dwarfs: the HI/HeI ratio lies between 12 and 18 (Dupuis *et al.*, 1995; Vallerga, this vol.), implying that helium is significantly more ionized than hydrogen! The ionization of He toward GD246 has been measured to be 25%, while the upper limit on H ionization is about 20% (but the amount of material probed shows that there are clouds other than the LIC along this line of sight). This means that a strong ionizing event in the solar neighborhood occurred quite recently (say 10^5 years ago), or that there is a strong EUV background. These results are discussed further by Frisch and Slavin (this vol.).

One of the highlights of this meeting was that the interpretations of Ulysses observations of HeI (Witte *et al.*, this vol.), those of pick-up helium ions (Gloeckler, this vol.), and of

AMPTE pick-up helium (Möbius, this vol.) now appear to converge to a similar result. This convergence is due to new observations of primary and secondary trajectories and to a better modeling of the ionization by electron impact. When combining the HeI number density as measured by these three different techniques (i.e., about 0.013 – 0.017 cm^{-3}) with the EUVE HI/HeI ratio (12 – 18.5), it appears that the LIC HI density should lie within the range 0.18 – 0.31 cm^{-3}. The smallest value is slightly larger than the highest value for the heliospheric number density, which provides evidence for HI losses during the entrance into the heliosphere. We then return to the previous conclusions that a fraction of the HI is prevented from entering the heliosphere, in agreement with models. This result also removes the discrepancies with the observed deceleration and the H wall observation (see Section 3).

1.3. Different Approaches to the ISM Ionization and Ionization Models

Four different calculations of the H and He ionization are presented in this volume using very different approaches. (a) The first approach is a "local" one: it simply uses the heliospheric helium and hydrogen density ranges, the EUVE mean HI/HeI ratio, and the Baranov and Malama relationship between the heliospheric interface filtration factor and the electron density interpolated from a series of models. By combining these parameters, one can estimate the electron density in the surrounding medium and the H and He ionization rates as a function of the actual heliospheric HI density (Lallement, this vol.). This method does not use any model for the interstellar ionization in the LIC. (b) A second calculation provides upper and lower limits to the electron density by using the same "ingredients", except the filtration factor, but includes the EUVE observed upper limits on the H and He ionization toward HZ43. For the lower limit, this model uses the ionizing EUV radiation field (Vallerga, this vol.). (c) A third calculation makes use of a model for the ionization equilibrium of the LIC based on the EUV radiation field from hot stars and the calculated H and He ionization degrees throughout the cloud from the Sun to the cloud surface (Vallerga, this vol.). (d) A fourth calculation uses a model for the LIC thermal and ionization equilibrium controlled by the ionizing radiation of the hot stars, plus an estimated flux from a conductive interface between the LIC and the surrounding hot gas of the "local bubble" (Frisch and Slavin, this vol.).

It is informative to intercompare the results of these four calculations. There are no discrepancies between (a) and (b) if the heliospheric HI density is larger than 0.09 cm^{-3} [see Fig. 7 in (a)], because according to (b) the upper limits on the H and He ionization measured by Dupuis *et al.* (1995) are 40% and 50%, respectively. On the other hand, according to (a) the H and He ionization degrees diverge for a heliospheric HI density larger than about 0.12 cm^{-3}. The He ionization remains large (40%), while the H ionization becomes smaller than 15%. This corresponds to a small electron density (< 0.05 cm^{-3}). None of the Vallerga and the Frisch and Slavin models predicts such small degrees of H ionization when the He ionization is so large. This may mean that the filtration does not follow the assumed Baranov and Malama model prediction, perhaps due to magnetic field effects, or that the heliospheric H density is indeed smaller than 0.12 cm^{-3}.

From (c) the minimum H and He ionization degrees are 13% and 4%, respectively. For an HI density of 0.21 cm^{-3} (the most probably value from the Ulysses/EUVE data described above), the electron density must be larger than about 0.03 cm^{-3}. However, as shown by (c), hot stars alone cannot explain the relatively high He ionization. There must have been another source of ionization, in which case the electron density is likely larger than this lower limit. As shown by Frisch and Slavin (d), the existence of an emissive conductive interface can, on the contrary, explain the ionization degrees of H and He. The HI/HeI ratio from (d) is found to be 12 near the Sun and about 5 at the cloud boundary. For values of the ratio integrated along the line of sight to be larger (between 12 and 18) toward

Table I

Relative abundances of heliospheric vs interstellar minor ions. PUI = pick-up ions (see Gloeckler; Geiss and Witte, this vol.). ACR = anomalous cosmic rays (see Cummings and Stone; and Rucinski *et al.*, this issue). CLOUDY = model results for the ISM near the Sun (see Frisch and Slavin, this vol.). Corrected values assuming 50% filtration for OI and NI are followed by *f*.

Quantity	INSIDE HELIOSPHERE	OUTSIDE HELIOSPHERE
He/O	290 (–100, +190) PUI if filtration factor = 0.8	He/O from He/H (EUVE) and (O/H) (Capella) $(1/14)/4.7 \times 10^{-4} = 150$ 166- 332*f* CLOUDY
N/O	0.13 ± 0.06 PUI	0.14-0.14*f* CLOUDY
Ne/O	0.20 (–0.09,+0.12) PUI	0.16-0.32*f* CLOUDY
C/O	0.006 ACR	0.0024 CLOUDY
C/N	0.03 ACR	0.018 CLOUDY
He/Ne	1450 (+1290, –820) PUI	1016-1016*f* CLOUDY

the nearby stars, these lines of sight must go deeper through the cloud rather than through the cloud surfaces, since He ionization remains constant throughout the cloud while the H ionization decreases strongly with increasing distance from the cloud surface. However, neither the hard EUV background implied by the conductive interface model nor the high ionization species associated with the conductive interface have yet been detected.

2. Oxygen and Carbon Relative Abundances in the ISM and Heliosphere

Spectacular progress has been achieved in the interstellar pick-up ion measurements with the SWICS experiment on board Ulysses (Gloeckler; Geiss and Witte, this vol.). At the same time, there has been a real advance in our understanding of the interstellar pick-up ions, thanks to the new opportunity to compare the relative abundances in the heliosphere directly with the interstellar abundances in the surrounding gas (see Table I). Because the improved spectral resolution allows us to disentangle the local cloudlets and to identify the LIC for the first time, abundance ratios can now be given for the local cloud alone. This is an important improvement with respect to earlier comparisons with the abundances of the solar system or dense interstellar clouds which are not representative of our environment. The new abundances for the LIC remove the uncertainties due to the large variations in the relative abundances from one cloud to the other. These variations are illustrated in Table II. The abundances of iron and magnesium in the LIC toward Capella and Procyon are very similar, but they differ significantly from those in the colder G cloud toward α Cen.

Until now only the OI/HI ratio has been directly measured outside the heliosphere (toward Capella by Linsky *et al.*, 1993) and inside the heliosphere (Geiss *et al.*, 1994; Gloeckler *et al.*, 1993). The OI/HI ratio can be converted into the OI/HeI ratio by using the EUVE mean H/He ratio. This comparison suggests that OI is filtrated by at least 50% at the entrance to the heliosphere (Izmodenov *et al.*, 1996; Lallement, this vol. and Table I).

Table II

Interstellar minor ions relative abundances in the LIC and the G cloud. D(E)= log([N(E)/N(HI)] - log $[E/H]_{Sun}$

	Capella	Procyon	α Cen A,B
N(HI)	1.80×10^{18}	1.15×10^{18}	$(0.42 - 1.0) \times 10^{18}$
N(DI)	2.97×10^{13}	1.83×10^{13}	6.1×10^{12}
N(FeII)	3.01×10^{12}	2.0×10^{12}	2.82×10^{12}
N(MgII)	6.49×10^{12}	3.6×10^{12}	5.25×10^{12}
N(OI)	8.2×10^{14}	...	...
Log(N(FeII)/N(HI))	−5.77	−5.76	−5.05 to −5.65
Log(N(MgII)/N(HI))	−5.44	−5.50	−4.78 to −5.38
Log(N(OI)/N(HI))	−3.32	...	...
D(Fe)	−1.28	−1.27	−0.56 to −1.16
D(Mg)	−1.03	−1.09	−0.37 to −0.97
D(O)	−0.25	...	...
N(DI)/N(MgII)	4.1 ± 0.4	5.1 ± 0.7	1.2 ± 0.2
N(DI)/N(HI) $\times 10^5$	1.60 ± 0.09	(1.6)	0.6 to 1.4

Monte-Carlo simulations of the oxygen flow (Izmodenov *et al.*, 1996; Fahr *et al.*, 1995) imply an electron density > 0.1 cm^{-3}.

For oxygen and other species, comparisons can be made between the pick-up ion data and sophisticated models of the ISM ionization (Frisch and Slavin, this vol.). The first results of such computations and comparisons with observations are encouraging. Table I summarizes some of the comparisons between the pick-up ion data and the interstellar relative abundances computed with the CLOUDY program. However, there is a need for calculations of the filtration factors (see Rucinski *et al.*, this vol.). Minor ion abundances will probably play a major role in the future, because they can be used in conjunction with the He and H ionizations for the interpretation of the observed ionization state of the local ISM. The origin of the high ionization of helium is, however, still not understood.

3. Hydrogen "Wall" and Hot Gas in the ISM

An unexpected and exciting new GHRS observation of the heliospheric interface is the detection of HI Ly-α absorption toward α Cen characterized by a temperature of about 0.3×10^5 K and a smaller velocity with respect to the main flow of interstellar gas. This absorption is identified with the "hydrogen wall" of decelerated and heated gas created through charge-exchange processes near the bow shock (see Linsky, this vol.). In addition to the newly offered possibilities to detect stellar winds from their own "H walls" (Linsky and Wood, 1996), which is beyond the scope of this report, there are important new questions raised by this detection: First, can one constrain the characteristics of the heliospheric boundary from such observations? The answer should be investigated theoretically by parametric studies of the interface structure and by computing the corresponding absorbing column. This has not yet been done, because all studies so far have focused on the characteristics of HI near the Sun for comparison with available data rather than on HI at large distances. In such computations, the use of the exact HI distribution is required, since the H wall cannot be well represented by a Maxwellian gas at a constant velocity. The second question is which are the best target stars which could be used together with α Cen to better constrain the size and shape of the heliosphere? It does not seem easy to find such targets, because there should not be any overlap between the solar H wall absorption, the absorption by interstellar gas at a larger distance, or by the absorption in a possible stellar H wall.

There is another difficulty. The models predict that the interstellar conductive interfaces between warm clouds and the hot gas of the Local Bubble should have temperatures around 10^5 K, which makes it difficult to distinguish such interfaces from H walls. For example, the ϵ Ind H wall has the same 10^5 K temperature. Indeed, there were two recent detections of hot gas at 10^5 K towards Sirius (Bertin *et al.*, 1995) and ϵ CMa (see Gry, this vol.). We are thus facing interesting new problems which are the consequences of the enormous improvements in stellar UV spectroscopy brought about by the HST-GHRS Echelle spectrometer. Hopefully STIS, the future UV spectrograph on board the HST, will continue to improve our understanding of the solar environment.

Acknowledgements

We thank Priscilla Frisch and John Vallerga for their inputs to this report, and all participants and speakers of this working group.

References

Baranov, V.B, & Malama, Y.: 1993, *JGR* **98**, 15157.
Bertin, P., Lallement, R., Ferlet, R., Vidal-Madjar, A., & Bertaux, J.L.: 1993, *JGR* **98**, A9, 15193.
Bertin, P., Vidal-Madjar, A., Lallement, R., Ferlet, R., & Lemoine, M.: 1995, *A&A* **302**, 889.
Clarke, J.T., Lallement, R., Bertaux, J.L., & Quémerais. E.: 1995, *ApJ* **448**, 893.
Cummings, A C., & Stone, E.C.: 1990, 21st Int. Cosmic Ray Conference **6**, 202.
Dupuis, J., Vennes, S., & Bowyer, S.: 1995, *ApJ* **455**, 574.
Fahr, H.J., Osterbart, O. & Rucinski, D.: 1995, *A&A* **294**, 584.
Frisch, P.C.: 1994, *Science* **265**, 1443.
Frisch, P.C.: 1995, *Space Sci. Rev.* **72**, 499.
Geiss, J., *et al.*: 1994, *A&A* **282**, 924.
Gloeckler, G., *et al.*: 1993, *Science* **261**, 70.
Izmodenov, V., Malama, Yu., & Lallement, R.: 1996, *A&A*, in press.
Jelinsky, P., Vallerga, J. & Edelstein, J.: 1994, *ApJ* **442**, 653.
Lallement, R., Ferlet, R., Vidal-Madjar, A., & Gry, C.: 1990, in "Physics of the Outer Heliosphere" (Warsaw, Sept 89), S. Grzedzielski and D.E. Page (Eds.), Pergamon Press.
Lallement, R., & Bertin, P.: 1992, *A&A* **266**, 479.
Lallement, R., Bertaux, J.L., & Clarke, J.T.: 1993, *Science* **260**, 1095.
Lallement, R., Bertin, P., Ferlet, R., Vidal-Madjar, A., & Bertaux, J.L.: 1994, *A&A* **286**, 898.
Lallement, R., *et al.*: 1996, in "Science with the Space Telescope", in press.
Lallement, R. & Ferlet, R.: 1996, *A&A*, in press.
Linsky, J.L., Brown, A., Gayley, K., *et al.*: 1993, *ApJ* **402**, 694.
Linsky, J.L., *et al.*: 1995, *ApJ* **451**, 335.
Linsky, J.L., & Wood, B.E.: 1996, *ApJ* **463**, 254.
Quémerais, E., Bertaux, J.L., Sandel, B.R., & Lallement, R.: 1994, *A&A* **290**, 941.
Quémerais, E., Sandel, B.R., Lallement, R., & Bertaux, J.L.: 1995, *A&A* **299**, 249.
Quémerais, E., *et al.*: 1996, *A&A* **308**, 279.
Rucinski, D., & Bzowski, M.: 1995, *Space Sci. Rev.* **72**, 467.
Summanen, T., Lallement, R., & Quémerais, E.: 1996, *JGR*, in press.
Vallerga, J. & Welsh, B.: 1995, *ApJ* **444**, 202.
Witte M., *et al.*: 1993, *Adv. Space Res.* **6**, (13), 121.

AXISYMMETRIC SELF-CONSISTENT MODEL OF THE SOLAR WIND INTERACTION WITH THE LISM: BASIC RESULTS AND POSSIBLE WAYS OF DEVELOPMENT

V.B. BARANOV AND YU.G. MALAMA
Institute for Problems in Mechanics, Russian Academy of Science, Prospect Vernadskogo 101, Moscow, 117526, Russia.

Abstract. We analyze the main results of the axisymmetric self-consistent model of the solar wind (SW) and supersonic local interstellar medium (LISM) interaction proposed by Baranov and Malama (1993, hereafter BM93, 1995) for an interstellar flow assumed to be composed of protons, electrons and hydrogen atoms. Here, in addition to the resonant charge exchange we also take into account the photoionization and the ionization by electron impact. The characteristics of the plasma in the interface region and inside the heliosphere depend strongly on the ionization degree of the LISM. The distribution function of the H atoms which penetrate the solar system from the LISM is non-Maxwellian, which implies that a pure hydrodynamic description of their motion is not appropriate. The H atom number density is a non-monotonic function of the heliocentric distance and the existence of a "hydrogen wall" in the vicinity of the heliopause is important for the interpretation of solar Lyman-alpha scattering experiments.

The influence of the interface plasma structure on the interstellar oxygen penetration into the solar system is also illustrated. Possible ways of development of the model are analyzed.

1. Introduction

The model of the solar wind interaction with the supersonic flow of the local interstellar medium (LISM) was first suggested by Baranov et al. [1970]. They proceeded from the concept of the Sun's motion relative to the nearest stars (consequently, relative to the interstellar gas at rest) with a velocity of $V_\infty \cong 20$km/s. This velocity is supersonic for a temperature $T_\infty \cong 10^4$K. The first model was constructed in a Newtonian approximation of a thin layer (hypersonic flows) under the assumption that the LISM is a fully ionized gas (two-shock model). The model by Baranov et al. [1970] has become especially interesting after the beginning of the 1970s, when experiments on scattered solar radiation in 1216 and 584Åwavelengths and their interpretation [Bertaux and Blamont, 1971; Thomas and Krassa, 1971; Blum and Fahr, 1970; Fahr, 1974; Weller and Meier, 1974] proved that H and He atoms of the LISM move with supersonic velocity $V_\infty \cong 20 - 25$km/s relative to the Sun. However, the direction of this motion did not coincide with the direction of the solar motion with respect to the nearest stars (apex direction). After the experimental discovery of the penetration of H and He atoms into the solar system it became impossible to construct the model of the solar wind interaction with the LISM on the basis of hydrodynamic equations, because the mean free path of neutral atoms is comparable with the characteristic length of the system, for example, with the size of the heliosphere (the hydrodynamic approximation can be correct for the solar wind interaction with the ionized hydrogen of the LISM only).

At present there is no doubt, that the LISM contains a partially ionized gas. Therefore, it is necessary to develop a correct model of the solar wind interaction

Space Science Reviews **78**: 305–316, 1996.

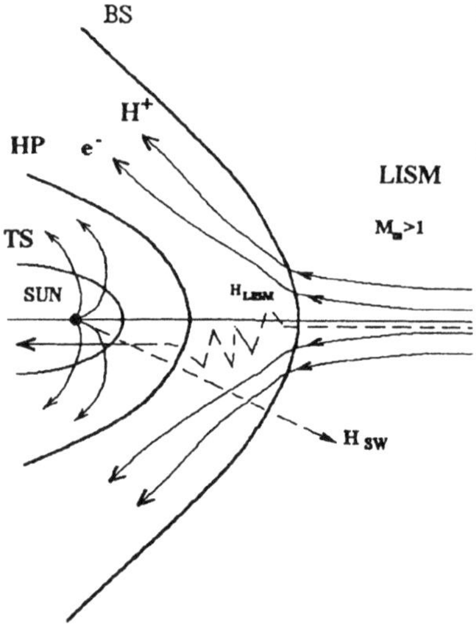

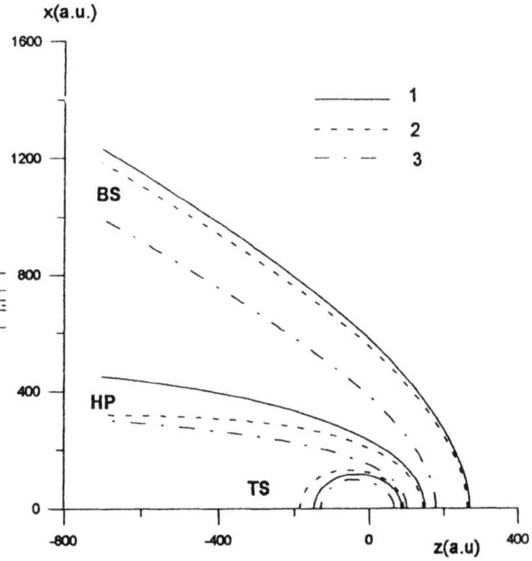

Figure 1. left: Qualitative picture of the solar wind interaction with the supersonic LISM: BS is the bow shock, HP is the heliopause, TS is the termination shock; H_{LISM} are H atoms of the LISM's origin, H_{SW} are energetic H atoms of solar wind origin.
Figure 2. right: Geometrical pattern of the interface for $n_{e\infty} = 0.1\text{cm}^{-3}$, $n_{H\infty} = 0.2\text{cm}^{-3}$ (curves 1) and for $n_{e\infty} = 0.2\text{cm}^{-3}$, $n_{H\infty} = 0.3\text{cm}^{-3}$ (curves 3). The curves 2 correspond to the same conditions as for the curves 1, but when neglecting H atoms ionization due to electron impact.

with the LISM taking into account the mutual influence of plasma (electrons and protons) and neutral (H atoms) components of the flow (helium atoms are negligible in the problem considered because their cosmic abundance is much less than that of H atoms).

Wallis [1975] was the first who showed that plasma and neutral components can influence each other by resonance charge exchange processes. This mutual influence has two aspects: first, the plasma interface between two shocks (the solar wind termination shock TS and the interstellar bow shock BS) in the model of Baranov et al. [1970, 1979] becomes a kind of filter for H atoms penetrating from the LISM to the solar system and, second, the resonance charge exchange processes can change the plasma interface structure and its distance from the Sun. However, it was difficult to obtain quantitative results, because the electron (proton) number density of the LISM ($n_{e\infty} \cong n_{p\infty}$) is a parameter which is measured very poorly. In particular, the observed magnitude of this parameter could range from $0.003cm^{-3}$, deduced from LISM observations of the ionization state of magnesium [Frisch et al.,1990], up to about $0.1cm^{-3}$ according to ionization calculations by integrated celestial UV radiation [Reynolds, 1990]. There is no clear observational upper limit to the local magnitude of $n_{p\infty}$. For example, an interpretation neutral magnesium absorption detected by Goddard High-Resolution Spectrograph (GHRS) on board of the Hubble Space Telescope (HST) in terms of an extremely small Local Interstellar Cloud (LIC), could give rise to a very large proton number density $n_{p\infty} = 0.3cm^{-3}$ [Lallement et al., 1992, 1994], if the LIC temperature is as small as measured toward Capella and for the neutrals in the solar system.

The first papers [Baranov et al., 1979; Baranov and Ruderman, 1979; Ripken and Fahr, 1983; Fahr and Ripken, 1984] did not take into account the influence of the resonance charge exchange on the motion of the plasma component, although they showed that the effect of the filter can be important.

Baranov et al. [1981] were the first who have constructed a self-consistent model taking into account the mutual influence of plasma and neutral components via the processes of the resonance charge exchange to estimate the influence of the LISM's H atoms on the plasma structure of the heliosphere. However, this model had a number of defects (see, for instance, the review by Baranov [1990] and BM93): (1) the motion of the LISM's H atoms was described by hydrodynamical equations, although the mean free path of H atoms and the characteristic length scale of the problem are comparable; (2) the temperature T_H and velocity V_H of H atoms were assumed constants and the change of H atom number density n_H was determined by the continuity equation taking into account the charge exchange loss term only; (3) plasma momentum and energy sources, which are due to resonance charge exchange processes, were used in the hydrodynamical equations in the Maxwellian approximation for the H atom distribution function [Holzer, 1972]. To correct these defects, we decided to use the Monte Carlo method, suggested by Malama [1991], for simulation of the H atom trajectories. This simulation gives us an opportunity to calculate "source" terms of the momentum and energy equations for the plasma component on the basis of the kinetic description of the H-atoms. An iterative method was proposed by Baranov et al. [1991] to solve the problem of the solar wind interaction with the two-component (plasma and H atoms) interstellar gas. A complete numerical solution of the self-consistent axisymmetric problem was obtained by BM93 (below self-consistent two-shock model or, abbreviated, SCTS model).

2. The SCTS model. Mathematical formulation of the problem.

Let us consider the problem of the solar wind interaction with the supersonic flow of the LISM consisting of neutral (H atoms) and plasma (electrons and protons) components. The qualitative structure of the flow formed is given in Figure 1 under the assumption that the interaction between the solar wind and the plasma component of the LISM can be described by hydrodynamical equations. Here the BS is formed in the LISM's plasma due to deceleration of the plasma component. The HP (contact or tangential discontinuity) separates the LISM plasma component, compressed by the BS, and the solar wind, compressed by the TS. Neutral atoms (including H atoms) can penetrate the surface HP into the solar wind, because their mean free path is comparable with the characteristic length of the problem (for example, with the HP heliocentric distance). In so doing, there are two kinds of H atoms: H atoms of the LISM's origin (H_{LISM}) and energetic H atoms of the solar wind origin (H_{SW}) formed due to the charge exchange with solar wind protons. In the following we will distinguish with primary H_{LISM} from secondary H_{LISM} (born following charge exchange processes) and the solar wind H atoms (H_{SW}), which were born in the supersonic solar wind, from thise born in the subsonic region (in the region between TS and HP).

The region between BS and TS we will call the interface. It is interesting to note here that the plasma outside of the interface (pre-shock regions of BS and

TS) is disturbed due to processes of resonance charge exchange and the "pickup" of "new" protons. The solar wind disturbance is mainly caused by H atoms of the LISM (H_{LISM}) whereas the disturbance of the LISM plasma is mainly a result of the interaction with the energetic H atoms of the solar wind (H_{SW}) penetrating the LISM [Gruntman, 1982; Gruntman et al., 1990]. These effects were also taken into account in the model by Baranov and Malama [1993, 1995].

Hydrodynamic equations for the plasma component must take into account the momentum and energy changes due to processes of resonance charge exchange ("source" terms). For a stationary problem, the equations of mass, momentum and energy conservation for ideal gas (without viscosity and thermal conductivity) have the following form (one-fluid approximation for plasma component)

$$\begin{gathered} \nabla \cdot \rho \mathbf{V} = 0, \\ (\mathbf{V}\cdot\nabla)\mathbf{V} + (1/\rho)\nabla p = \mathbf{F}_1[f_H(\mathbf{r},\mathbf{w}_H),\rho,\mathbf{V},p], \\ \nabla\cdot[\rho\mathbf{V}(\varepsilon + p/\rho + V^2/2)] = F_2[f_H(\mathbf{r},\mathbf{w}_H),\rho,\mathbf{V},p], \\ p = (\gamma-1)\rho\varepsilon, \end{gathered} \tag{1}$$

where p, ρ, $\mathbf{V}$ and ε are pressure, mass density, bulk velocity and internal energy of the plasma component, respectively; γ is the ratio of specific heats; $\mathbf{F}_1$ and F_2 are the functionals, describing the change of momentum and energy of the plasma component due to collisions between H atoms and protons which characterize the resonance charge exchange ("source" terms); and $f_{\rm H}(\mathbf{r},\mathbf{w}_{\rm H})$ is the H atom distribution function depending on position-vector $\mathbf{r}$ and the individual velocity $\mathbf{w}_{\rm H}$ of H atoms.

The trajectories of H atoms were calculated by the complicated Monte Carlo scheme with "splitting" of the trajectories [Malama, 1991] in the field of the plasma gasdynamic parameters. Such an approach allows the calculation of $\mathbf{F}_1$ and F_2 in equations (1) within the framework of a kinetic description of H atoms (multiple charge exchange are also taken into account by the Monte Carlo method). The use of this method is identical (Malama, 1991) to the numerical solution of the Boltzmann equation for $f_{\rm H}$

$$\begin{aligned} &\mathbf{w}_{\rm H}\cdot\partial f_{\rm H}(\mathbf{r},\mathbf{w}_{\rm H})/\partial\mathbf{r} + [(\mathbf{F}_r+\mathbf{F}_g)/m_{\rm H}]\cdot\partial f_{\rm H}(\mathbf{r},\mathbf{w}_{\rm H})/\partial\mathbf{w}_{\rm H} = \\ &= f_p(\mathbf{r},\mathbf{w}_{\rm H})\int \mid \mathbf{w}'_{\rm H}-\mathbf{w}_{\rm H}\mid \sigma f_{\rm H}(\mathbf{r},\mathbf{w}'_{\rm H})d\mathbf{w}'_{\rm H} \\ &-f_{\rm H}(\mathbf{r},\mathbf{w}_{\rm H})\int \mid \mathbf{w}_{\rm H}-\mathbf{w}_p\mid \sigma f_p(\mathbf{r},\mathbf{w}_p)d\mathbf{w}_p \end{aligned} \tag{2}$$

In equation (2) f_p is the local Maxwellian distribution function of protons with gasdynamic values $\rho(\mathbf{r}), \mathbf{v}(\mathbf{r})$ and $T(\mathbf{r})$ from equations (1), $\mathbf{w}_p$ is the individual velocity of a proton, $\mathbf{F}_g$ and $\mathbf{F}_r$ are the solar gravitational force and the force of radiation pressure, respectively, and σ is the cross section of the resonance charge exchange, which is determined by the formula $\sigma = (A_1 - A_2\ln u)^2{\rm cm}^{-2}$, where u is the speed of the proton relative to the H atom (in cm/s), $A_1 = 1.64\cdot10^{-7}$, $A_2 = 6.95\cdot10^{-9}$ [Maher and Tinsley, 1977].

If the distribution function $f_{\rm H}$ is known, then the "source" terms $\mathbf{F}_1$ and F_2 in (1) can be calculated exactly according to the Monte Carlo procedures of Malama [1991]. We have in this case

$$\begin{aligned} &\mathbf{F}_1 = 1/n\int d\mathbf{w}\int d\mathbf{w}_p\sigma\mid\mathbf{w}_{\rm H}-\mathbf{w}_p\mid(\mathbf{w}_{\rm H}-\mathbf{w}_p)f_{\rm H}(\mathbf{r},\mathbf{w}_{\rm H})f_p(\mathbf{r},\mathbf{w}_p), \\ &F_2 = m\int d\mathbf{w}\int d\mathbf{w}_p\sigma\mid\mathbf{w}_{\rm H}-\mathbf{w}_p\mid(w_{\rm H}^2/2-w_p^2/2)f_{\rm H}(\mathbf{r},\mathbf{w}_{\rm H})f_p(\mathbf{r},\mathbf{w}_p), \\ &n_{\rm H} = \int d\mathbf{w}_{\rm H}f_{\rm H}(\mathbf{r},\mathbf{w}_{\rm H}), \qquad n_p = \int d\mathbf{w}_p f_p(\mathbf{r},\mathbf{w}_p). \end{aligned} \tag{3}$$

To solve the system of equations (1) - (3) the following boundary conditions for the plasma component were used: the Rankine-Hugoniot relations on the shock waves BS and TS (see Figure 1); the condition of equality of pressures and the no-flow condition (a vanishing normal component of the plasma velocity) through the contact discontinuity (HP); the symmetry conditions (the axisymmetric problem was solved by BM93); the velocities, electron (proton) number densities and Mach numbers are given in the undisturbed LISM (index "∞") and at Earth orbit (index "E"); and the non-reflecting conditions on the right boundary of the computation region. To calculate H atom trajectories and the "source" terms (3) the distribution function f_H was assumed to be Maxwellian in the undisturbed LISM (at infinity) with the temperature T_∞, number density $n_{H\infty}$ and velocity V_∞. In so doing, the motion of H atoms is determined by the solar gravitational force $\mathbf{F}_g$, the force of solar radiation pressure $\mathbf{F}_r$, and the resonance charge exchange. Photoionization of H atoms is also taken into account near the Sun (we have now calculated our model taking also into account the H atom ionization due to electron impact in the region between HP and TS).

An iterative method for the solution of the problem, suggested by Baranov et al.[1991] and completed by Baranov and Malama [1993, 1995] consists of several steps. First the trajectories of H atoms are calculated by the Monte Carlo method in the field of plasma parameters obtained without "source" terms for fully ionized H (see, for instance, the zero iteration made by Baranov et al.[1991]). Then the momentum and energy "sources" $\mathbf{F}_1$ and F_2 in (1) are calculated in this step of the iteration using equations (3). In the first iteration, the hydrodynamical equations (1) with these "sources" are solved using the gasdynamic boundary conditions as formulated above. Then, the new distribution of plasma parameters is used for the next Monte Carlo iteration for H atoms. The gasdynamic problem is solved again with the new "source" terms of this iteration (the second iteration) and so on. This process of iterations is continued until the results of two subsequent iterations practically coincide. To solve the gasdynamic part of the problem numerically BM93 have used the discontinuity-fitting "second order" technique, which is based on the scheme of Godunov's method [Godunov et al.,1976; Falle, 1991]. We have also used the time stabilization method to solve the stationary problem, i.e. we solved the non-stationary version of equations (1).

3. Basic results of the SCTS model and their connection with experimental data.

3.1. Geometrical pattern of the interface

One of the main results of the axisymmetric model, described in Sec.2, is the geometrical pattern of the interface as a function of the H fractional ionization of the LISM. Figure 2 shows the BS, TS and HP shapes and heliocentric distances in the xOz plane for different magnitudes of electron (proton) and H atom number densities in the LISM. Here the Oz axis coincides with axis of symmetry and is antiparallel to the vector of the LISM's velocity V_∞ (the Sun is in the center of the coordinate system). The Ox axis is normal to the Oz axis.

The results presented in Figure 2 take into account the H atom ionization due to electron impact and photoionization and not just the resonance charge exchange

(Baranov and Malama, [1993,1995]). It should be noted here that the effect of ionization due to electron impact can be important in the region between TS and HP (the effect of H atom photoionization can be significant only near the Sun). For comparison the geometrical pattern of the flow without the processes of H atom ionization due to electron impact and photoionization is drawn in Figure 2 by the dashed lines (case 2). All results were obtained for the following values of the specified parameters

$$n_{pE} = 7\text{cm}^{-3}, \qquad V_E = 450\text{km/s}, \qquad M_E = 10$$
$$V_\infty = 26\text{km/s}, \qquad T_\infty = 6700\text{K} \qquad (M_\infty = 1.914), \qquad (4)$$
$$\mu = F_r/F_g = 0.80,$$

where n_p, V_E and M are the proton number density, the solar wind radial velocity at the Earth's orbit and the Mach number, respectively. The values of the LISM parameters in (4) were chosen based on the observations of the Local Interstellar Cloud (LIC) by Lallement and Bertin [1992] and Lallement et al. [1994].

In a series of recent measurements from ground and space [Lallement and Bertin, 1992] the interstellar absorption lines due to the LIC were identified through reconstruction of its velocity vector by Doppler triangulation. It was proved that the solar system is embedded in the LIC moving relative to the Sun with the velocity $V_\infty = 25.7 \pm 0.5$km/s and temperature $T_\infty = 7000$K (supersonic flow with $M_\infty \cong 2$). The derived velocity V_∞ coincides with the vector by Lyman-alpha glow observations [Bertaux et al., 1985]. The motion of the LIC relative to the Sun with the same velocity vector was recently confirmed by the results of the Neutral Gas Experiment [Witte et al., 1993] on board the Ulysses spacecraft, which measured in situ the velocity distribution of the interstellar helium (the effect of the "filter" at the interface is negligible for He atoms).

From Figure 2 we see that in framework of the SCTS model (see Sec.2) the TS and HP positions in the upwind direction range from 65 up to 85 A.U and from 100 up to 148 A.U., respectively, for the range of parameters chosen (a different range of LISM parameters was used by Baranov and Malama [1993, 1995]). The position of the TS in the downwind direction ranges from 130 up to 150 A.U. Therefore, we estimate that the Voyager spacecraft moving in the upwind direction could cross the TS in the near future. The flow in the region between TS and HP is subsonic; and the Mach disc, reflected shock and tangential discontinuity are absent if the resonance charge exchange processes are taken into account. It should be stressed that the use of the discontinuity-fitting technique by BM93 provides the possibility to calculate exactly the TS, HP and BS locations and shapes , and at the same time to satisfy both the Rankine-Hugoniot relations at BS and TS, and the boundary conditions at the HP.

3.2. Predicted Parameters of the Plasma Component in the Distant Solar Wind and Interface Region

Plasma parameters and interplanetary magnetic field of the distant solar wind are measured by direct experiments on board the Voyager 1/2 and Pioneer 10/11 spacecraft. These experiments have shown that the average magnetic field as a function of the distance from the Sun follows Parker's model of the spiral structure (see, for example, Parker, [1963]). In this case the interplanetary magnetic field is negligible for considering the gas dynamics of the supersonic solar wind since $M_A \gg 1$, where M_A is the Alfven Mach number. However, the influence of H atoms penetrating from the LISM into the solar system, on the plasma parameters of the distant solar wind may be strong enough to be observed by spacecraft. This effect increases with decreasing LISM proton number density (at constant H number density in the LISM). The reason is that the role of the plasma "filter" between the BS and the HP decreases with decreasing LISM's H fractional ionization (Baranov and Malama, [1995]).

The following effects in the distant supersonic solar wind due to the process of resonance charge exchange were predicted by the SCTS model: the decrease of the solar wind velocity; the deviation of the proton number density from the law $1/r^2$, where r is the distance from the Sun; and the increase of the plasma temperature. It is interesting to note that there is experimental evidence of such effects obtained by Voyager and Pioneer (see, Richardson et al., [1995]). For example, the deceleration of the solar wind velocity as 30km/s (about 8% of the average velocity 440km/s) was estimated from experiments on Voyager 2 and IMP 8.

Figure 3 shows the electron number density n_e in upwind direction as a function of distance from the Sun. The key features of these distributions, caused by the effect of the LISM H atoms, are: (1) the deviation from the density law, $1/r^2$, before the TS, connected with the decrease of the solar wind velocity; (2) the strong increase in electron number density from TS to HP (Bernoulli's integral and an adiabatic law are not correct here due to the processes of resonant charge exchange); (3) the "pile-up" region of the LISM electrons near the HP, which are important for interpretation of the kHz radiation detected by Voyager (see, for example, Gurnett et al.[1993]; Gurnett and Kurth [1995]).

3.3. Predicted H atom parameters

The distribution of the H atom number density in the whole region of the flow is presented in Figure 4. As mentioned in Sec.2, we distinguish here two kinds of H_{LISM} (primary and secondary) and two kinds of H_{SW}, which were born due to the charge exchange of solar wind protons in the supersonic (upwind of the TS) and subsonic (between TS and HP) regions. From Figure 4 we see that the effect of the accumulation of H atoms moving from the LISM in the region between BS and HP (peaks of the number density in the vicinity of the HP or the effect of the "hydrogen wall") is connected with the secondary H_{LISM} only since the number density of the primary H_{LISM} decreases smoothly as the Sun is approached. This effect was first obtained by Baranov et al. [1991] numerically for the non-self-consistent problem and confirmed by BM93 with the solution of the self-consistent problem. Quemerais et al. [1993, 1995] used the idea of the "hydrogen wall" to explain the results of Lyman α measurements made by the Ultraviolet Spectrometers on board the Voyager 1 and 2 spacecraft since 1993 (see, also, the paper by Hall et al.[1993], where the validity of the distributions presented in Figure 4 is qualitatively confirmed by measurements of the UV interplanetary glow not only on board of Voyager but on Pioneer as well).

It is interesting to note here that the idea of the "H wall" was also used by Linsky and Wood [1995] for the interpretation of observations of interstellar gas absorption along the line-of-sight toward both components of the α Cen visual binary system $-\alpha$ Cen A (G2V) and α Cen B (K1V) - obtained with the GHRS instrument on board of the HST spacecraft. Linsky and Wood [1995] have shown that their observations cannot be explained by the one-component model of the interstellar gas. To explain the experimental results they suggested the "H wall" in

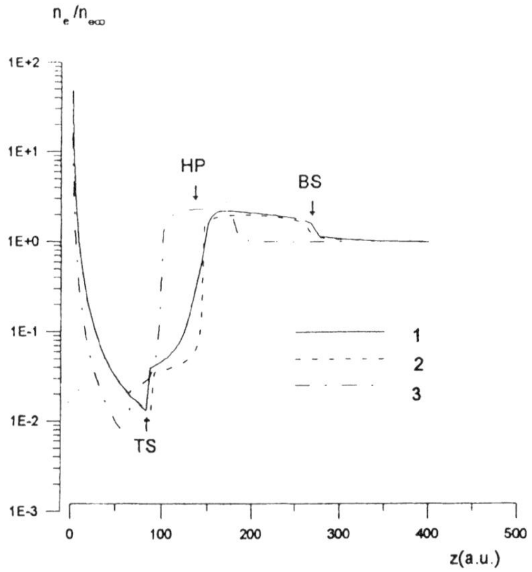

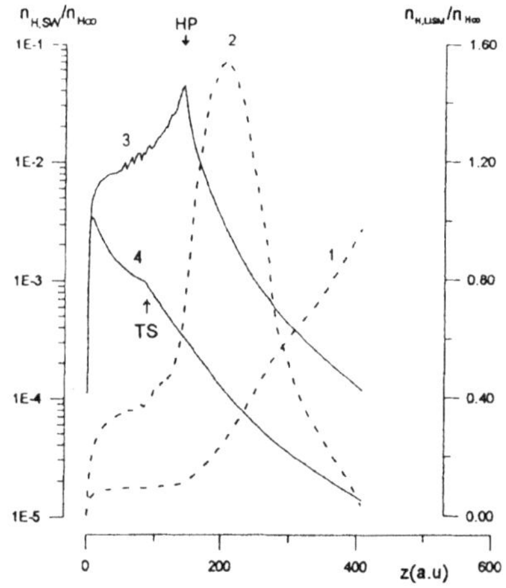

Figure 3. left: Electron number density in the upwind direction as a function of the distance from the Sun for $n_{e\infty} = 0.1\text{cm}^{-3}$, $n_{\text{H}\infty} = 0.2$ (curve 1) and for $n_{e\infty} = 0.2\text{cm}^{-3}$, $n_{\text{H}\infty} = 0.3\text{cm}^{-3}$ (curve 3). Results shown by curve 2 are obtained without taking into account H atom ionization by electron impact and are to be compared with curve 1. *Figure 4. right:* H atom number density in the upwind direction as a function of distance from the Sun for $n_{e\infty} = 0.1\text{cm}^{-3}$, $n_{\text{H}\infty} = 0.2\text{cm}^{-3}$. Dashed lines are for H_{LISM} (curve 1 is for primary H atoms and curve 2 is for secondary H atoms) and solid lines are for H_{SW} (curves 3 and 4 are for H atoms born due to charge exchange in the subsonic and supersonic solar wind, respectively).

the heliosphere as a second component, which contributes significantly to the total H absorption.

Although the effect of H_{LISM} is strong enough to influence the plasma interface structure, the effect of energetic solar wind H atoms (H_{SW}) is also important for the flow considered. This effect, suggested by Gruntman [1982] qualitatively, was first considered by BM93 quantitatively. In particular, the contribution of each H atom component (H_{LISM} and H_{SW}) to the source terms $\mathbf{F}_1$ and F_2 of equations (1) is determined by the following estimates

$$\frac{|\,\mathbf{F}_1\,|_{\text{LISM}}}{|\,\mathbf{F}_1\,|_{\text{SW}}} \simeq \frac{n(\text{H}_{\text{LISM}})V_\infty^2}{n(\text{H}_{SW})V_E^2}, \qquad \frac{|\,F_2\,|_{\text{LISM}}}{|\,F_2\,|_{\text{SW}}} \simeq \frac{n(\text{H}_{LISM})V_\infty^3}{n(\text{H}_{SW})V_E^3}$$

Taking into account these estimates, one can see from Figure 4, where the distribution of the H_{SW} number density is also presented [BM93], that: (1) in the entire LISM region the neutral solar wind H atoms are the main source of heating of the LISM plasma; (2) near the HP we have $|\,\mathbf{F}_1\,|_{\text{LISM}} \cong |\,\mathbf{F}_1\,|_{\text{SW}}$, but $|\,\mathbf{F}_1\,|_{\text{LISM}} > |\,\mathbf{F}_1\,|_{\text{SW}}$ near the BS. Thus, the heating of the LISM plasma by the

H_{SW} atoms leads to the decrease of the Mach number ahead of the BS, which can disappear due to this effect if $n_{H\infty}$ is large enough.

The characteristics of the energetic neutral H atom fluxes in the heliosphere (H_{SW}) have never been explored experimentally. Recent developments in experimental technique (see, for example, Gruntman and Morozov [1982], Gruntman et al.[1990], Gruntman [1993]) suggest that these fluxes can be reliably measured. In particular, techniques for viewing the outer heliosphere in energetic neutral atoms from within were suggested by Hsieh and Gruntman [1993].

3.4. Oxygen atom penetration from the LISM to the solar system

As we have mentioned above there are no direct ways to measure the LISM electron (or proton) number density. Among the various types of diagnostics are the observations of species which can penetrate deeply in the heliosphere such, as the interstellar neutral atoms. Certain of these neutrals suffer significant modifications during the crossing of the plasma interface region (see Figure 1). Models show that the changes in the abundances and velocity distributions due to the interface filtering do vary strongly from one species to another (for example, He atoms penetrate the solar system through the interface practically without changing). As a result, the relative abundances and the velocity distributions of different species inside the heliosphere can be different from the original interstellar abundances and velocity distributions.

The charge exchange cross-section of O and a proton, as well as H and a proton, is large enough for O and H atoms to suffer significant modifications in the plasma interface. We have already considered above the problem of H atom filtering. The problem of O atom penetration from the LISM to the solar system is also intimately related to the problem of the plasma interface structure. The process of O penetration from the LISM to the solar system was investigated theoretically by Izmodenov et al. [1996], who used a Monte Carlo method for calculations of O atom trajectories with plasma parameters from the SCTS model (Baranov and Malama, [1993, 1995]). An example of these calculations is shown in Figure 5. From Figure 5 we see that the effect of the interface filter is important for O atoms penetrating the solar system from the LISM. Previously this fact was discussed by Fahr and Osterbart [1995], who used the parker's model of the interaction between the incompressible interstellar and solar winds.

4. Possible ways of development of the SCTS model.

The present model does not take into account a number of physical phenomena. It ignores the effects of the interplanetary and interstellar magnetic fields. The Alfven Mach number $M_A \gg 1$ in the supersonic solar wind (ahead of TS) and, therefore, the interplanetary magnetic field is negligible in this region. However, the last inequality is not necessarily fulfilled in the region between TS and HP (especially in the vicinity of the stagnation point) and magnetohydrodynamic effects could be important in this region. At present neither the value nor the direction of the LISM magnetic field is known. That is why Baranov and Zaitsev [1995] began to study the effect of the interstellar magnetic field on heliospheric structure in an axisymmetric

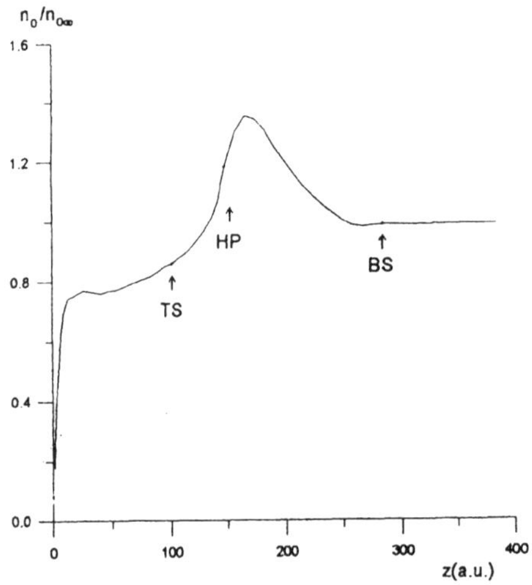

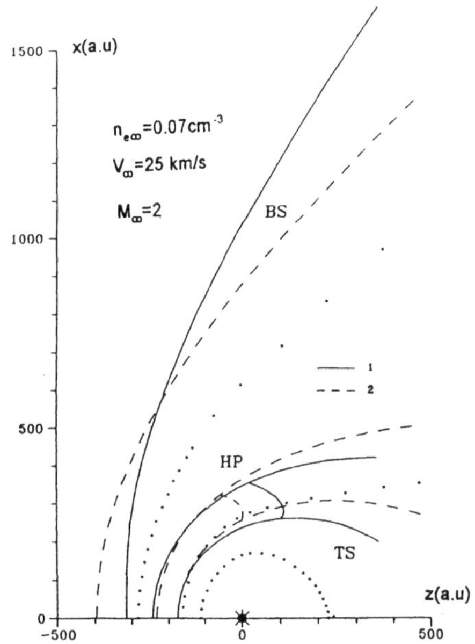

Figure 5. left: Effect of the interface region on oxygen atoms penetrating from the LISM to the solar system. O atoms number density (normalized to the density in the LISM) is presented in the upwind direction for $n_{e\infty} = 0.1\text{cm}^{-3}$, $n_{H\infty} = 0.2\text{cm}^{-3}$ as a function of distance from the Sun (Izmodenov et al.,1996). *Figure 6. right:* Effect of the interstellar magnetic field on the geometrical pattern of the flow (vector of the magnetic field **B** is parallel to the vector of the LISM plasma velocity). Parameter α is equal $B_\infty/V_\infty\rho_\infty^{1/2}$, where ρ_∞ is the LISM mass density. Curves 1 correspond to $\alpha = 2.2$ and curves 2 to $\alpha = 0$ both with $n_{H\infty} = 0$ (Baranov and Zaitsev, 1995); dotted lines correspond to $\alpha = 0$ and $n_{H\infty} = 0.14\text{cm}^{-3}$ (BM93).

approximation (the LISM magnetic field is parallel to the vector of the LISM velocity) and for a one-component (plasma) interstellar gas. A comparison of the effects of the interstellar magnetic field in this case [Baranov and Zaitsev, 1995] and of the H atoms of the LISM [BM93] on the geometrical pattern of the flow is presented in Figure 6 (dashed, solid and dotted lines are used for the non-magnetized and magnetized fully ionized LISM, and for the non-magnetized partially ionized LISM, respectively). We see that the thickness of the region between the HP and BS at the axis of symmetry is reduced by a factor 2.3 and becomes 1.8 times less than when taking into account the resonant charge exchange without magnetic field. On the contrary, the thickness of the region between HP and BS along the line passing through the Sun perpendicular to the z-axis increases by 36%. The thickness of this region affects its transparency to the LISM H atoms penetrating the solar system.

We also did not take into account the effects of galactic and anomalous component cosmic rays (GCR and ACR respectively). We think that these effects do not significantly change the results presented in Sec.3. However, as shown by Chalov and Fahr [1994, 1995], the ACR could decrease the TS strength. A TS of very small strength may not be detected while traversed by a spacecraft.

It should be noted that the BM93 model is a one-fluid model for the plasma component. An attempt to develop a two-fluid model (pick-up and primary

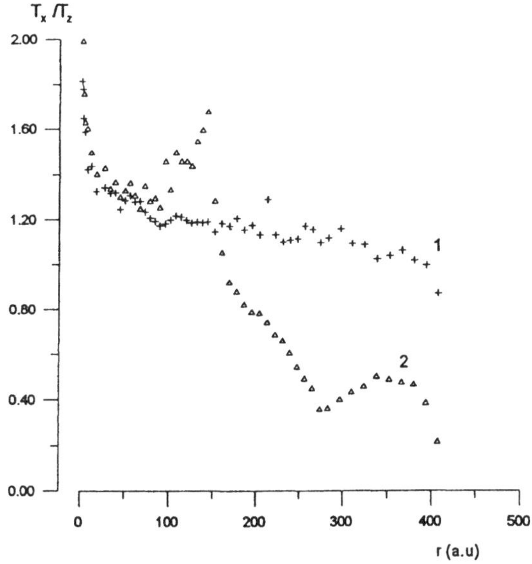

Figure 7. Ratio of *H* atom apparent temperatures (along the Ox and Oz axes) as a function of distance from the Sun for $n_{e\infty} = 0.1 \mathrm{cm}^{-3}$, $n_{H\infty} = 0.2 \mathrm{cm}^{-3}$. Crosses (1) and triangles (2) correspond to the primary and secondary H_{LISM} respectively.

solar wind protons) was developed by Whang et al. [1995]. However, the hydrodynamic system of equations used for the H atoms of the LISM is derived from the Boltzmann equation under the assumption of a Maxwellian distribution function (Whang, [1995]). This last assumption is not correct for neutral atoms because their mean free path is comparable with the characteristic length scale of the problem. Figure 7 demonstrates the last statement. "Effective" temperatures along the Ox and Oz axes are different. This means that the distribution function f_H of H atoms is not Maxwellian. Hydrodynamic equations for H atoms of the LISM were also used by Pauls et al. [1995] to describe the interaction of the solar wind with the two-component LISM gas. In so doing, Pauls et al. [1995] did not take into account the multiple charge exchange, energetic solar wind H atoms (H_{SW}), the gravitational force, and the force of radiation pressure (these combined effects were taken into account in the model by BM93). Finally, the present model is axisymmetric and stationary, although experiments on board Ulysses and other spacecraft have shown that the solar wind velocity and electron number density can be functions of solar latitude and solar activity. Therefore, it is necessary to develop three-dimensional and non-stationary models.

References

Baranov V.B., Space Sci. Rev., **52**, 89 - 120, 1990.

Baranov V.B. and Ruderman M.S., Pis'ma Astron. Zh., **5**, 615 - 619, 1979 (Soviet Astron.Letters).

Baranov V.B. and Malama Yu.G., J. Geophys. Res., **98**, 15,157 - 15,163, 1993.

Baranov V.B. and Malama Yu.G., J. Geophys. Res., **100**, 14,755 - 14,761, 1995.

Baranov V.B. and Zaitsev N.A., Astron. Astrophys., **304**, 631 - 637, 1995.

Baranov V.B., Krasnobaev K.V. and Kulikovsky A.G., Dokl. Akad. Nauk USSR, **194**, 41 - 43, 1970 (Sov. Phys. Dokl., **15**, 791 - 793, 1971, English Translation).

Baranov V.B., Lebedev M.G. and Ruderman M.S., Astrophys. Space Sci., **66**, 441 - 451, 1979.

Baranov V.B., Ermakov M.K. and Lebedev M.G., Sov. Astron. Letters, **7**, 206 - 210, 1981.
Baranov V.B., Lebedev M.G. and Malama Yu.G., Astrophys. J., **375**, 347 - 351, 1991.
Bertaux J.-L. and Blamont J., Astron. Astrophys., **11**, 200 - 217, 1971.
Bertaux J.-L., Lallement R., Kurt V.G. and Mironova E., Astron. Astrophys., **150**, 1 - 20, 1985.
Blum P. and Fahr H., Astron. Astrophys., **4**, 280 - 290, 1970.
Chalov S.V. and Fahr H.-J., Astron. Astrophys., **288**, 973 - 980, 1994.
Chalov S.V. and Fahr H.-J., Planet. Space Sci., **43**, 1035 - 1043, 1995.
Fahr H., Space Sci. Rev., **15**, 483 - 540, 1974.
Fahr H. and Osterbart R., Adv. Space Res., **16**, No. 9 (9)125 - (9),130 1995.
Fahr H. and Ripken H., Astron. Astrophys., **139**, 551 - 554, 1984.
Falle S.A.E.G., MNRAS, **250**, 581 - 596, 1991.
Frisch P., Welty D., York D. and Fowler J., Astrophys. J., **357**, 514 - 523, 1990.
Godunov S.C., Zabrodin A.V., Ivanov M.Ya., Kraiko A.N. and Prokopov G.P., Chislennoe Reshenie Mnogomernych Zadach Gazovoi Dinamiki (in Russian), Nauka, Moscow, 1976.
Gruntman M.A., Sov. Astron. Lett., **8**, 24 - 26, 1982.
Gruntman M.A., Planet. Space Sci., **41** (4), 307 - 319, 1993.
Gruntman M.A. and Morozov V.A., J. Phys. E: Sci. Instrum., **15**, 1356 - 1358, 1982.
Gruntman M.A., Grzedzielski S. and Leonas V.B., in **Physics of the Outer Heliosphere**, edited by S.Grzedzielski and D.E.Page, p.355, Pergamon, New York, 1990.
Gurnett D.A., Kurth W.S., Allendorf S.C. and Poynter R.L.,Science, **262**, 199 - 203, 1993.
Gurnett D.A. and Kurth W.S., Adv. Space Res., **16**(9), 279 - 290, 1995.
Hall D.T., Shemanski D.E., Judge D.L., Gangopadhyay P. and Gruntman M.A., J. Geophys. Res., **98**, 15,185 - 15,192, 1993.
Holzer T., J. Geophys. Res., **77**, 5407 - 5431, 1972.
Hsieh K.C. and Gruntman M.A., Adv. Space Res., **13** (6), 131 - 139, 1993.
Izmodenov V.V., Malama Yu.G. and Lallement R., J. Geophys. Res., 1996 (in press).
Lallement R. and Bertin P., Astron. Astrophys., **266**, 479 - 485, 1992.
Lallement R., Bertaux J.-L. and Clark J.T., Science, **260**, 1095 - 1098, 1992.
Lallement R., Bertin P., Ferlet R., Vidal-Madjar A. and Bertaux J.-L., Astron. Astrophys., **286**, 898 - 965, 1994.
Linsky J.L. and Wood B.E., Astrophys. J., **463**, 254, 1996.
Maher L. and Tinsley B., J. Geophys. Res., **82**, 689 - 695, 1977.
Malama Yu.G., Astrophys. Space Sci., **176**, 21 - 46, 1991.
Parker E., Interplanetary Dynamic Processes, New York, Interscience, 1963.
Pauls H.L., Zank G.P. and Williams L.L., J. Geophys. Res., **100**, **21**, 595 - 604, 1995.
Quemerais E., Lallement R. and Bertaux J.-L., J. Geophys. Res., **98**, 15,199 - 15,210, 1993.
Quemerais E., Malama Yu.G., Sandel B.R., Lallement R., Bertaux J.-L. and Baranov V.B., Astron. Astrophys., 1996 (in press).
Reynolds R.J., in **Physics of the Outer Heliosphere**, edited by S.Grzedzielski and D.E.Page, p.101, Pergamon, New York, 1990.
Richardson J., Belcher J., Lazarus A., Paularena K., Gazis P. and Barnes A., **First ISSI Workshop** "The heliosphere in the local interstellar medium", Bern, Switzerland, November 6 - 10, 1995.
Ripken H. and Fahr H., Astron. Astrophys., **122**, 181 - 192, 1983.
Thomas G. and Krassa R., Astron. Astrophys., **11**, 218 - 233, 1971.
Wallis M., Nature, **254**, 207 - 208, 1975.
Weller C. and Meier R., Astrophys. J., **193**, 471 - 476, 1974.
Whang Y.C., $1-st$ ISSI Workshop "The heliosphere in the local interstellar medium", Bern, Switzerland, 6 - 10 November 1995, Abstracts, p.33.
Whang Y.C., Burlaga L.F. and Ness N.F., J. Geophys. Res., **100**, 17,015 - , 1995.
Witte M., Rosenbauer H., Banaszkiewicz M. and Fahr H.-J., Adv. Space Res., **13** (6), 121 - 130, 1993.

UV STUDIES AND THE SOLAR WIND

J. L. BERTAUX, R. LALLEMENT, E. QUEMERAIS
Service d' Aeronomie du CNRS, Verrieres-le-Buisson, France

Abstract. The solar wind carves a cavity in the flow of interstellar H atoms through the solar system by charge-exchange ionization. The resulting Ly-α sky pattern depends on the latitude distribution of the solar wind flux and velocity. We review how the solar wind characteristics (mass flux latitude distribution) can be retrieved from Ly-α observations, yielding a new remote sensing method of solar wind studies, through UV optical measurements.

1. Introduction

The classical description of an ionization cavity created in the interstellar medium around a star, the Strömgren's sphere, calls for a size r_s (radius of the sphere) which is the distance to the star at which the ionization rate is equal to the recombination rate. However, this description is no longer valid when the relative velocity is significant, and large enough that the time for one H atom to travel over r_s is comparable or smaller than the time of recombination at distance r_s. This fact was recognized quite early (Blum and Fahr, 1970), together with the fact that, for a G type star like our sun, the dominating ionization process would be charge exchange with solar wind protons, rather than EUV photoionization ($\lambda < 912$ Å), which accounts for only 20% of the total ionization. The early modelling of Blum and Fahr (1970) predicted correctly that solar wind plus EUV ionization would create a cavity void of H atoms, elongated in the downstream direction.

All the atoms not yet destroyed by ionization would be illuminated by the sun in the H I resonance line (Lyman α), and would become accessible to observation with space-borne Lα instruments. As a matter of fact, Lα sky maps obtained in 1969 and 1970 with two instruments (one french, one US) on board the OGO-5 US spacecraft proved definitely that the sky Lα emission was generated by interstellar H atoms flowing inside the solar system (Bertaux and Blamont, 1971 ; Thomas and Krassa, 1971). The interstellar wind was thus discovered ; although all the necessary ingredients were available to go one step further, and use the interplanetary Lα radiation as a remote sensing method to study the solar wind, it was not until December 2, 1995, that an instrument dedicated to Lα solar wind remote sensing (SWAN) was launched to space on board SOHO. It should be recognized however that the decision to select the SWAN instrument and add it to the nominal payload of SOHO was taken in late 1987, by a lucid selection committee (whose members are still mostly unknown to us).

Helium atoms can also be ionized by the solar wind. However, the 584 Å resonance emission of He atoms would be sensitive to solar wind very near the sun, where observations are still lacking. Therefore, the present review will concentrate only on the studies of the Lα sky pattern, and what can be derived about the solar wind from these studies. In the next section , the case of an isotropic emission of solar wind will be considered, while the case of an anisotropic solar wind will be discussed in section 3. Solar cycle variation will be adressed in section 4.

Space Science Reviews **78**: 317–328, 1996.

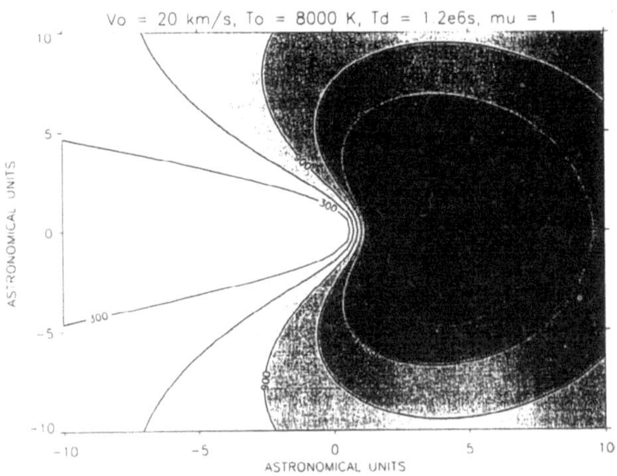

Figure 1. Model distribution of Lα emissivity (in units of number of photons emitted per sec per m3). The values shown correspond to a plane containing the wind axis. The interstellar wind comes from the right side. A Maximum Emissivity Region (MER) lies between 2 and 4 AU, in the upwind direction.

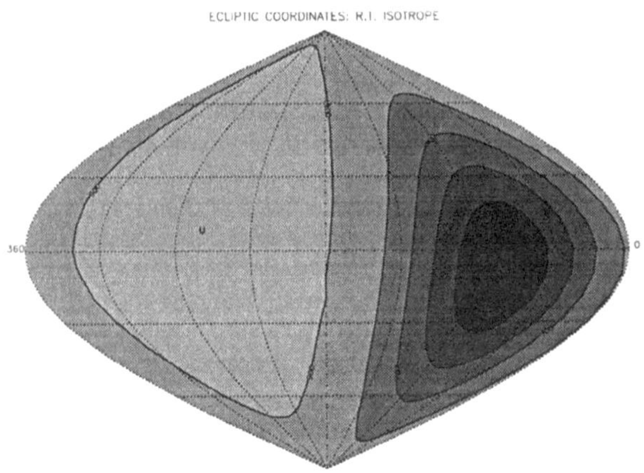

Figure 2. Model distribution of Lα intensity pattern over the sky in ecliptic coordinates. Iso-intensity contours are in Rayleigh; darker on the plot means darker in the sky. The model assumed that the observer was at the position of the Sun, with an isotropic ionization rate. Letters U and D indicate respectively the Upwind (from where the wind blows) and Downwind directions. The pattern is characterized by a dipole-type distribution, with the maximum intensity centered around U. Model parameters are: ratio of Lα radiation pressure to gravitation μ=0.99, ionization lifetime at 1 AU=1.2 10^6 s, interstellar wind velocity +20 km/s and Temperature=8 000 K.

2. The case of spherically symmetric solar wind (isotropic case).

In this section, we consider the case where the ionization rate of one H atom $\beta_t = \beta_{sw} + \beta_{euv}$ at 1 AU is independant of the longitude and heliographic latitude, because both β_{sw} (solar wind charge exchange) and β_{euv} (photoionization rate) are isotropic. Currently admitted values are at 1 AU :

$\beta_{euv} = 6.0\ 10^{-8}\ s^{-1}$

$\beta_{sw} = 6.0\ 10^{-7}\ s^{-1}$ for a proton flux $f_p = 3.\ 10^8\ cm^{-2}\ s^{-1}$ at 1 A.U.

$\beta_t = 6.6\ 10^{-7}\ s^{-1}$

The lifetime $T_D = 1/\beta$ of one H at 1 AU, (1 to 2 x 10^6 seconds) is also frequently used.

To calculate the density distribution of H atoms in the solar system, one must take into account the solar Lα radiation pressure exerting a repulsive force F_s almost equal to the gravitation F_g and the thermal spread of incoming interstellar

H atoms, both factors which were omitted in the early description of Blum and Fahr and introduced later on (Bertaux et al, 1985). The velocity of the wind Vw must also be known. Fortunately, these two parameters T and V_w could be determined by studying the interplanetary Lα line profile carefully, with the technique of a spaceborne Hydrogen absorption cell (Bertaux et al, 1977, 1985). Values of T = 8000K and V_w = 20 km/s were found. As we know now, these parameters are characterizing the interstellar gas at a distance of the sun of 40-60 AU in the upwind hemisphere, well within the heliosphere and may not represent accurately the interstellar gas in the vicinity of the sun before entering the solar system, because of a strong interaction between H neutrals and decelerated interstellar plasma, and, in a smaller extent, decelerated solar wind outside the inner shock (Baranov and Malama, 1993) ; see also Baranov, Zank, Fahr, this book). However, we will ignore these effects in the following for the sake of clarity. Therefore, when parameters T and V_w are used in the following, they refer to parameters of the interstellar flow "at infinity", that is to say far away from the sun, but still within the heliosphere, around 50 AU. Given these interstellar parameters and also the solar parameters μ (Lα radiation pressure) and T_D , the distribution of H density n(r,q) can be calculated in the solar system with the so-called "hot model" (Lallement et al, 1985), which takes into account the thermal spread of H atoms "at infinity". r is the distance to the Sun and q is the angle with the upwind vector. The distribution has a symmetry of revolution about the wind axis. The next modelling step is to calculate the emissivity distribution ϵ(r,q) (figure 1), which is the number of photons emitted per cm^{-3} per sec. In the optically thin approximation, the emissivity is related to the density by :

(eq. 1) $\epsilon(r,q) = g_0\, r^{-2}\, n(r,q)$

with the excitation factor at 1 AU $g_0 = \sigma_l\, F_s = 1.5 \times 10^{-3}\ s^{-1}$ ($\sigma_l = 0.544\ cm^2$ Å , integrated cross section of one H atom at Lα ; F_s is the solar flux at Lα line center).

There is a region where ϵ(r,q) is maximum, and was called for this reason the MER (Maximum Emissivity Region, Lallement et al, 1991). It is located upwind from the Sun for obvious reasons, and relatively nearby to the sun. Its center is at the distance r_{max}, which can be approximated analytically by (Bertaux et al, 1995):

(eq. 2) $r_{max} = 1/2\ \beta_t\ (r_0^2)/V_w$

which is 2.5 AU for $T_D = 1.5 \times 10^6$ sec. This estimate is valid for μ= 1, and therefore depends somewhat on the actual value of the Lα solar flux at line center. In the region of the MER,atoms can approach nearer the sun because they arrive directly from outside, while in the downwind region atoms are much rarefied because they have to travel near the sun before arriving there.

Of course, the only thing which is measured is the interplanetary Lα emission intensity, that is to say the integration along a line of sight of the emissivity across the solar system (in the optically thin approximation). Measurements can be compared to calculated maps of this emission, plotted under the form of iso-intensity contours , as the example shown on figure 2 in Rayleigh unit (1 R= $10^6/4\pi$ photons $cm^{-2}\ s^{-1}\ sr^{-1}$). The main feature of this emission distribution is its dipole character, with a broad maximum around the upwind direction and a minimum around the downwind direction, quite understandable from the distribution of emissivity of fig.1.

Equation (2) shows that the distance rmax to the MER increases with the ionization rate β_t and therefore with solar wind mass flux. As a result, surviving atoms are further out and less illuminated by the sun, and the intensity Iupwind in the direction of the MER is a decreasing function of β_t. Therefore, one first idea to determine β_t is to measure accurately I_{upwind}. Indeed, if the solar flux at line center F_s is known, then μ is fixed, and a modelling exercise shows that (in the optically thin regime, μ, V_w, T known):

(eq. 3) $I_{upwind} \approx n_\infty \; T_D$

Therefore, if n_∞ is known, as well as the absolute calibration of the instrument, one can derive $T_D = (\beta_t^{-1})$ from intensity measurements in the direction of the MER. Unfortunately, there are uncertainties in the knowledge of F_s , n_∞ and usually also in the calibration factor of the instrument, which makes this approach unpractical. Still, time variations of I_{upwind} may be monitored with a photometer, and provided that the variations of F_s are known, one can derive time variations of β_t from equation (2), since n_∞ is constant, though unknown. The timescales of variations of β_t that can be observed can be derived as follows. Obviously the intensity observed when looking at the MER is mainly due to the emissivity at this point, where the lifetime of one H atom is :

(eq. 4) $T_{(rmax)} = T_D \; (r_{max}/r_0)^2 = 1/4 \; T_e^2/T_D$

where we have introduced $T_e = r_0/V_w$, the time needed by one H atom to travel along $r_0 = 1$ AU at the velocity V_w. T_e has a fixed value of 87 days for $V_w = 20$ km/s, while $T_D = 17$ days (for 1.5 x 10^6 s). This formula shows a somewhat paradoxical result : the reaction time of the Lα emission to changes of the ionization rate, as measured by T (rmax), is inversely proportional to the lifetime of one H atom at 1 AU. In other words, fast variations of the solar wind are better reflected on the La emission when the solar wind flux is small, because H atoms are penetrating nearer the sun and react faster.

For values of T_D=1,1.5,2x10^6s, the distances r_{max} are 3.75,2,5 and 1.87 respectively, while the reaction time T (r_{max}) is 160, 107, and 80 days respectively. These crude estimates are made with a series of simplification assumptions but still give a good idea of the time scales involved : relatively short time scale variations of the solar wind ionization rate can be detected by monitoring the upwind emission, because the time variation needs only to be a fraction of the lifetime to be observed. The 27 days variations, if any, could be marginally detected, but more probably would be smoothed out. On the contrary, the 1.3 year period of the solar wind velocity oscillations recently evidenced from IMP-8 and Voyager 2 spacecraft (Richardson et al, 1994) with an amplitude of 100 kms^{-1} might be detectable in the SWAN maps pattern. In particular, a possible latitude variation of this oscillation is an obvious target of interest to the SWAN investigation.

The qualitative discussion above, about time variations, has been recently confirmed by more sophisticated density models computed for a solar cycle variation of μ and T_D (i.e., Kyrölä et al., 1994 ; Bzowski and Rucinski, 1995 ; Summanen, 1996).

In order to derive the value of T_D from measurements, one may get rid of instrument calibration uncertainties and n_∞ imperfect knowledge by considering ratios of intensities in two different directions. It is clear that, the more intense will be the solar wind, the higher will be the ionization rate, the shorter will be T_D, and the longer will be the extension of the cavity. As a result, the La emissivity pattern seen by an observer will change, and in particular the ratio of intensities in two opposite directions $I_{upwind}/I_{downwind}$, or Imax/Imin.

However, the other parameters V_w, T and μ also influence this ratio : curves of modeled intensities along a great circle going through upwind and downwind directions can be of identical shape for various combination of these parameters as was demonstrated by Lallement et al , 1985.

Nevertheless, once V_w and T were determined, as well as μ from H cell measurements on board Prognoz (Bertaux et al, 1985), one could use the observed ratio I_{max}/I_{min} to determine the lifetime T_D. With Prognoz 5 and 6 data, a value of $T_D = 2.5 \times 10^6$s was found , which was somewhat longer than could have been expected from an estimate of the solar wind *in situ* measurements by a number of spacecraft. Ajello et al (1987) found rather an average value of 1.5 x 10^6, based on these direct solar wind measurements. One possibility to explain the discrepancy was to invoke the effect of multiple scattering of Lα photons through the solar system. Indeed, when a full model of radiative transfer was available (Quemerais and Bertaux, 1993), it was found that the upwind intensity Iupwind was slightly increased (by 3% for $n_\infty = 0.1$ cm^{-3}) but that $I_{downwind}$ was more increased (30%). It was confirming earlier results obtained by Keller et al, 1981.

Clearly, radiative transfer could account for a part of the discrepancy, but probably not all of it. In fact, it also depended on the actual value of n_∞ (say, at 50 AU from the Sun), subject to calibration uncertainties as mentionned above.

The parallax method. A third method to derive T_D from Lα intensity observations is to capitalize on the fact that the core of the MER region is relatively nearby from the sun (figure 1). As a result, a spacecraft moving around the sun will see the MER in different directions according to its position in the solar system. As a matter of fact, this parallax effect, observed and measured for the first time from OGO-5 Lα instruments, was the most definitive argument to show that the detected Lα emission was coming from the solar system and was not of galactic origin (Bertaux and Blamont, 1971 ; Thomas and Krassa, 1971) as was thought previously.

From Pioneer Venus Orbiter observations, Ajello et al (1987) used this parallax method to determine the lifetime T_D and arrived at a relatively fair agreement with what can be inferred from direct *in-situ* solar wind mass fluxes and velocities. The orbit of Venus has the advantage of a faster scanning than the Earth's orbit, but the disadvantage of being smaller, therefore reducing the angular parallax effect.

The orbit of Voyager spacecraft was such that, when at 4 AU from the sun in 1978, it was also at right angle from V_w, an ideal position to determine the distance r_{max} by measuring the angular separation between the Sun and the MER with the UVS instrument. However, the sky map which was done at that time did not cover angles to the Sun smaller than about 20 °, preventing to actually see the MER detached from the Sun. Still, a modeling of the MER, compared to the intensity data in the ecliptic plane, could provide an estimate of the lifetime $T_D = 1.0 \times 10^6$ (Lallement et al, 1991), lying in the high side of ionization rates.

The parallax effect is quite obvious when comparing modelled intensity maps. Not only the MER is moving back and forth along the ecliptic plane, but also the area of minimum intensity in the downwind direction is seen to increase and shrink when the Earth is moving from upwind to downwind and back again to upwind position. Even with constant conditions (stationnary H distribution), the Lα maps

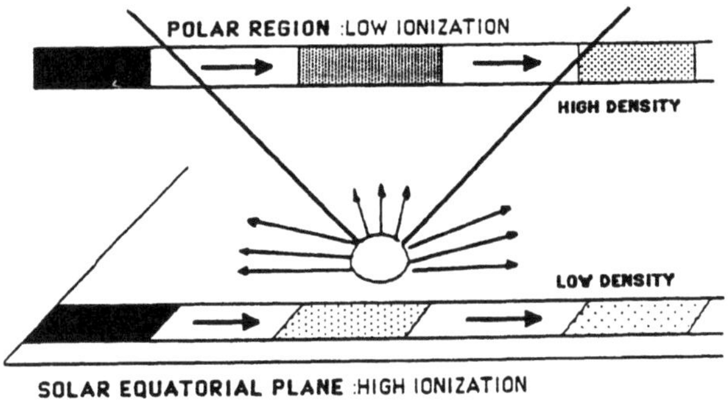

Figure 3. Schematic illustration of the interaction between the hydrogen flow and an anisotropic solar wind, assumed to be smaller at high latitudes. The flow velocity is here assumed to be parallel to the solar equatorial plane for the sake of clarity. Atoms which travel near the solar equatorial plane do suffer higher loss processes by charge exchange with the solar wind than those atoms passing over polar regions (only the North hemisphere is represented here). Though the density difference increases along the flow vector, accurate modelling indicates that the effect of anisotropies are already detectable when looking upwind.

look very dynamic, because the Earth's orbit is in fact immersed into the whole Lα emission region in the solar system.

From the ionization rate to the solar wind mass flux. Once the lifetime T_D and the ionization rate $\beta_t = T_D$-1 has been determined from Lα observations, the solar wind proton flux can be derived by subtracting the photo-ionization rate β_{euv}, and by taking into account the velocity dependence of the charge-exchange reaction rate (eq. 6) σ_{ex}(v)= (7.6 - 2.1 log (0.07217 v))^2 10^{-16} cm^2 (Fite et al., 1962) (v in km s^{-1})

(eq. 7) $\beta_t = \beta_{euv} + f_p$ (1 AU) x σ_{ex}(vsw)

where v can be safely approximated by the solar wind velocity v_{sw}. It requires an independant knowledge of vsw, which is known to vary from 350-400 km/s (slow solar wind) to 500-800 km/s for solar wind originating from coronal holes.

We note here an intrinsic difference between the local photoionization and local charge-exchange rate, relevant to the next section : while the former depends on the whole solar disk (integration for all latitudes), the latter depends only on the solar wind flux at the place considered, and therefore is more prone to latitude variations. In both cases, however, the resulting extinction at a given place must be integrated along the trajectories of H atoms coming to the point considered.

3. The anisotropic case.

The name "anisotropy" refers to variations with heliographic longitudes and latitudes of the characteristics of the solar wind. We restrict immediately to latitude variations, since longitude variations will be smoothed out by the 26 days solar rotation, short with respect to the H lifetime. The idea that solar wind anisotropies would impact on the interplanetary Lα radiation was first published by Joselyn and Holzer (1975). From Figure 3 where the solar ionization is larger at low latitudes it can be understood that, since ionization is cumulated along the trajectories, departures form a spherically symmetric ionization rate (isotropic) will show up more clearly for a spacecraft located downstream from the Earth. Indeed, this was at this privileged position that the effect of solar wind anisotropies were first discovered, by Mariner 10 Ultra-Violet Spectrometer in 1974 (Kumar and Broadfoot, 1978, 1979) and Prognoz Lα photometer in 1977 (Lallement et al, 1985), as shown on figure 4. In order to interpret the data, the latitudinal dependence of the ionization rate β (l,r) was first assumed to follow a harmonic function :

$$\beta(\lambda, r) = \beta_0 \; (1 - A \sin^2\lambda) \; (r_0/r)^2$$

where β_0 is the ionisation rate at 1 AU (Astronomical unit), λ is the solar latitude, A is the anisotropy factor, r_0 is 1 AU and r is the distance to the sun. This harmonic functional dependence has the great advantage that it can be integrated analytically along one H atom trajectory, alleviating the computation time of the forward modelling. On figure 4, the solid curve labeled A = 0.40 is such an harmonic model. The fit to the data is obviously much better than with an isotropic solar wind (A = 0), but it is not perfect.

Some progress was made more recently by the SWAN team in the field of retrieval. Summanen et al (1993) considered a distribution of ionization rate numerically by slices of 10° of latitude (still, symmetric about the solar equator) and re-analyzed the Prognoz data. A better fit was obtained with an ionisation rate decreasing from 0 to +/-30° with large constant plateaus from +/- 30 to 70°. Taking a model proposed by Zhao and Hundhausen (1981) for the latitude dependence of the solar wind velocity V, the solar wind proton flux was found to be as shown in figure 5 for various Prognoz observations. The flux is higher between +/-30° and there are large plateaus from +/- 30 to 70° (Summanen et al ,1993). This overall behavior is strikingly confirmed by Ulysses *in situ* measurements (Smith and Marsden,1995). The proton mass flux measured by the SWOOPS instrument is shown in figure 6 as a function of the heliolatitude λ during the fast latitude scan from South Pole to North (courtesy of Phillips, 1996).

An early presentation of Ulysses SWOOPS results at Solar Wind conference n°8 (Dana Point, June 1996) showed the median value, rather than the average value ; the median is much flatter than the average as pointed out by Bertaux et al (1996a), because of the very large variations encountered near the solar equator. Such a representation made a number of scientists attending SW-8 believe that there were no variation with latitude of the proton mass flux as measured by Ulysses (Marsh 1996, private communication), while the variations are obvious when the average is plotted instead of the median (figure 6).

There is of course a striking similarity between the solar wind retrievals from Prognoz Lα measurements and the in-situ observations of Ulysses, which fully validates the Lα method. It is important to note that in a recently published paper, Scherer and Fahr (1995) have claimed that the departures of La emission measured by Prognoz 5 and 6 from an optically thin modelling can all be explained by considering a more sophisticated modelling of radiation transport of the intensity, and no longer need of solar wind variations with latitude. Bertaux et al. (1996a) have shown that their new model produce completely erroneous results, and therefore their conclusions must be totally discarded. In particular, in the isotropic case, their model produces a minimum of intensity in the upwind direction!

The Lyα groove: Up to now, we have discussed two main features of the Lα pattern : the first feature was the strong dipole asymmetry, with the upwind hemisphere being brigther than the downwind hemisphere. The second feature is the

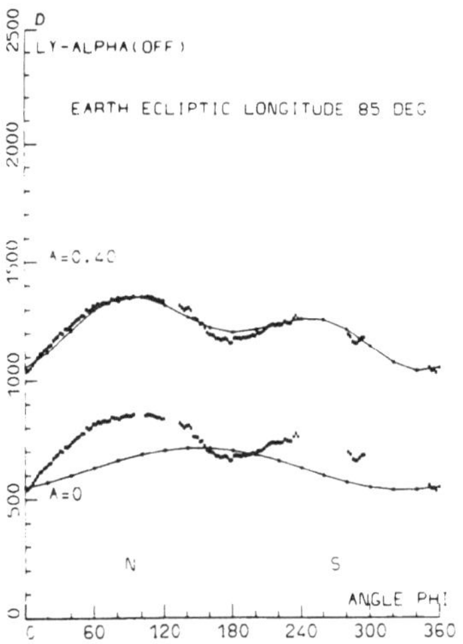

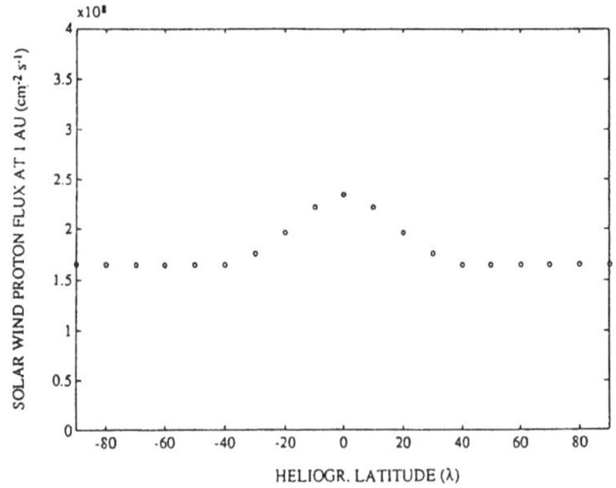

Figure 5. The solar wind flux latitude distribution (assumed symmetric) determined to fit Prognoz data along (after Summanen et al, 1993). It shows an ecliptic enhancement and reduced flux plateaus above 30°.

Figure 4. Prognoz data (points) as a function of spin angle ϕ at 85° earth ecliptic longitude compared to two different models of the solar wind effect. Data points are plotted twice. The ordinate scale in counts per second is valid for the lower set of data points compared to a model assuming an isotropic solar wind ($A = 0$). The same data points are displaced upward by 500 counts/s and are compared to a model including anisotropy of the solar wind characterized by the parameter $A = 0.40$. North and south ecliptic directions are indicated by the letters N and S. In respect to the isotropic model the data points show more Lα emission toward the ecliptic pole, showing that the solar wind was less effective to destroy H atoms at high ecliptic latitudes.

enhancement near the solar poles, when the observing point is downstream from the Sun. A third feature in the pattern of the interplanetary emission has been discovered in the Prognoz data (Bertaux et al, 1996b): a small dip of the emission extending over 30-40 ° across the region of maximum emission, aligned along the ecliptic plane as a groove carved in the emission pattern (Fig 7). This Lα groove is interpreted as the result of enhanced ionization of incoming interstellar H atoms by the solar wind in a thin band of ecliptic latitude, produced by the slow solar wind concentrated in the neutral sheet, flat and thin at the time of the 1976 solar minimum. Such a dip observed by Prognoz was already mentionned in the SWAN proposal to ESA for the SOHO mission (Bertaux et al, 1987, 1989) as a potentially important means of diagnostic for the structure of the solar wind. Indeed, a recent modeling of the Lα sky pattern using the Ulysses solar wind data was able by reproduce the groove (Summanen et al, 1996). The model distribution of intensity for an anisotropic solar wind similar to the wind probed by Ulysses at all latitudes is shown in Fig 8. The large differences from the isotropic Sun case can be seen by comparing with Figure 2. The maximum of intensity is splitted into two maxima which are displaced towards the poles, which produces the groove when the scan planes do contain the poles themselves. This groove is discussed in some details in Bertaux et al (1996b), together with another puzzling observation with Prognoz 6. As a matter of fact we believe that Prognoz-6 data shows the evidence of an assymetry of the ionization rate, with more ionization in the South than in the North, which must have developed in a relatively short time (less than 6 months) between March and September 1977. Here, one caveat is in order. Even if the solar wind were symmetric with respect to the solar equator, the Lα pattern could present a South-North asymmetry, for two reasons : first, the vector V_w is neither in the ecliptic plane, nor in the solar equatorial plane (it deviates from them by 6°). Second, the ecliptic plane deviates by 7.2° from the equatorial plane, and therefore the heliographic latitude of Prognoz was different from 0. It may introduce some geometrical biases that could be eliminated by comparing with anisotropic models containing some North-South asymmetry. The discrete model developped with Summanen et al (1996) is able to handle such a case, but a detailed study remains to be performed. Of course, it is relevant to notice that Ulysses data show more solar wind in the South than in the North : but is this a space or a time variation? Only continuous monitoring of the sky as will be provided by SWAN on SOHO, may bring an unambiguous answer.

4. Variations with the solar cycle

With the clear identification of the groove carved across the MER and supporting Ulysses *in situ* solar wind measurements (both taken near solar minimum) we have developped the following paradigm : Ulysses demonstrates that there is a strong variation of the solar wind mass flux with latitude, the "classical" value (3 x 10^8 protons cm$-2s^{-1}$ found by scores of in-ecliptic instruments flown since several decades) being confined in a narrow band of latitude where the neutral sheet was found, with the rest of space having a smaller flux at 2 x 10^8 protons $cm^{-2}s^{-1}$. Even if the neutral sheet is very narrow, the width of the groove will depend on its particular configuration. In particular, its inclination to the Heliographic equator is very important. It is the so-called tilt angle of the Heliospheric Current Sheet (HCS), identical to the neutral sheet mentionned above, which was calculated from large scale solar magnetic field photospheric measurements by Hoeksema (1991). At the beginning of 1977, it was about 15° only, increasing fast with the development of the solar cycle toward the maximum. Therefore, we may expect a solar cycle variation of the existence of the groove pattern (Fig. 9). Near solar minimum, at the time when the tilt angle is zero, the HCS may be completely flat, and perpendicular to the solar rotation axis,an ideal situation for the creation of the groove inside the MER, because points near the heliographic equator experience permanently a high and slow solar wind mass flux, giving a high ionization rate of incoming interstellar H atoms. When the tilt is increasing, the HCS becomes inclined and warpy, and points in the MER region are experiencing an alternance of slow and fast solar wind, with a resulting smoothed ionization rate. This can possibly explain the difference between the theoretical intensity map of Figure 8 and the data.

The groove may disappear, and when the tilt angle is sufficient, the whole anisotropy Lα pattern might also disappear altogether. Indeed, there are some evidence that anisotropies are less conspicuous near solar maximum.

In addition, from the study of the hydrogen cloud around comet Bennett in 1970, which had an orbit perpendicular to the ecliptic plane, Bertaux et al (1973) derived that the ionization rate was increasing with the heliographic latitude of the comet. We may argue that, if at this time of solar maximum, the HCS tilt is near 90°, a point at the pole will experience continuously a slow and strong solar wind

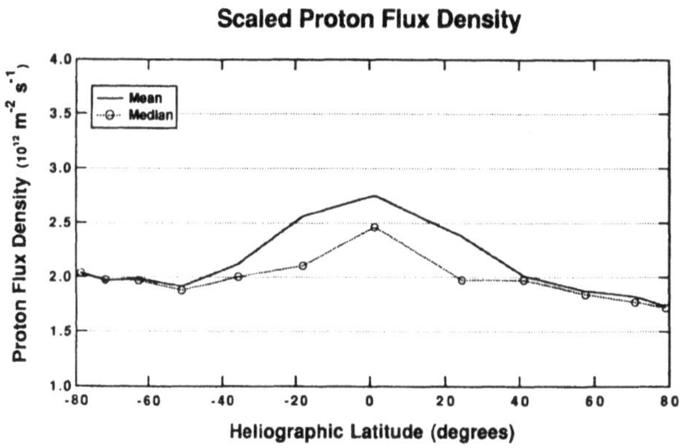

Figure 6. Distribution of the solar wind proton flux as mesured by SWOOPS on board Ulysses during the fast latitude scan of 1994. There is a striking similarity with the profile of figure 5 which was derived from Prognoz data (Courtesy J. Phillips)

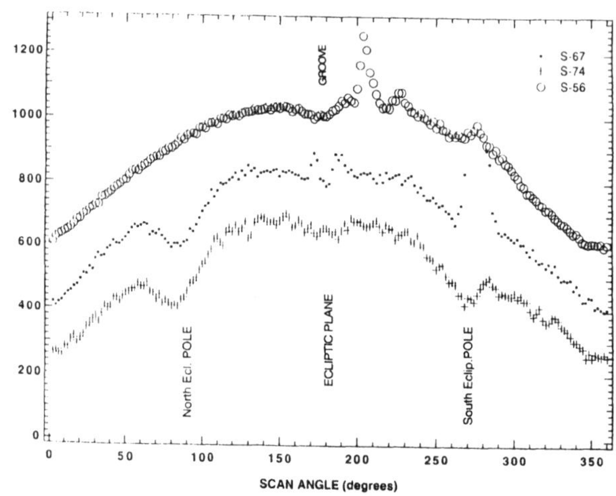

Figure 7. Prognoz-5 intensity at three positions of the earth as a function of scan angle in planes perpendicular to the ecliptic (0 and 180° in ecliptic ; + 90° is North Ecliptic pole, + 270° is South Ecliptic pole). Units are counts/sec.
Curves are displaced by 200 counts/s vertically for clarity. The effect of the H cell absorption on the interplanetary emission is clearly visible for S-67 and S-74 around + 90 and +270. In some cases, there are spikes due to hot stars (205 for S-56, 170-190 for S-67) and also some geocorona around 270°. All curves show a dip of intensity centered on the ecliptic plane, forming a groove in the Lα sky pattern.

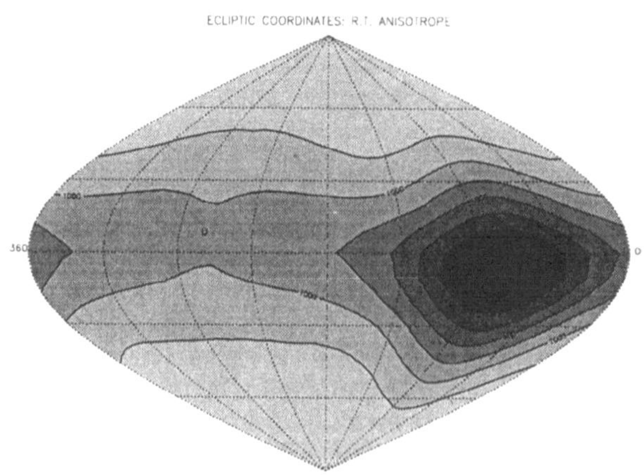

Figure 8. Same as Figure 2, for an anisotropic solar wind similar to the SWOOPS data of Figure 6. The Lα pattern is extremely different from the isotropic Sun map of Figure 2 when observed from this particular location on the upwind side of the earth orbit

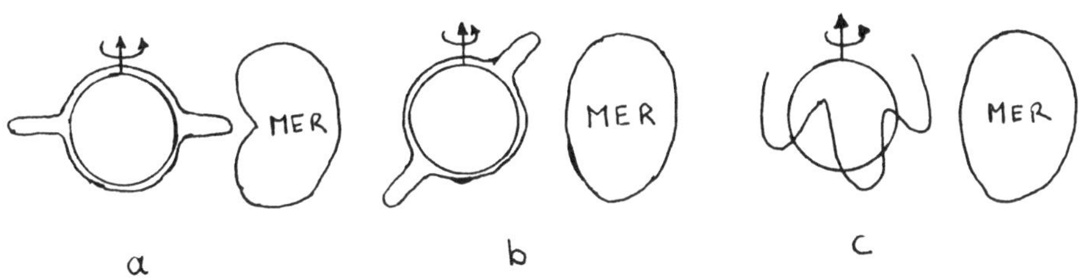

Figure 9. Three possible configurations : a) the neutral sheet is flat, and perpendicular to the sun axis of rotation. This configuration is ideal to produce a pronounced groove in the Lα pattern. b) the neutral sheet is flat, but inclined with respect to the solar axis ; c) the neutral sheet is warpy (exhibiting the "ballerina" type configuration. In both cases b) and c), the groove may be less pronounced, or even disappear.

from the neutral sheet, while in the equator there would be an alternance of fast and slow solar wind, therefore in average a reduced ionization rate.

5. Conclusion

Though almost 20 years old, the Prognoz data have not yet been surpassed in photometric quality, allowing (together with Mariner 10 data) to foster the idea of remote sensing of the solar wind large scale distribution. Ulysses in-situ out-of-ecliptic measurements totally confirmed the picture which was developped from a detailed analysis of Prognoz data and fully validated the Lα method. But one must recognize that both were taken near solar minimum, with the same solar magnetic polarity (two cycles apart). What happens when the solar cycle goes towards its maximum remains to be discovered, thanks to the presence of the SWAN instrument on board the forthcoming SOHO mission, acting as a real "Lyα Solar Wind Mapper". It is perhaps worth to note here three remarks : - if the interstellar wind were blowing, not in the ecliptic plane, but rather from one of the poles, the Lα method would not work. - as shown by Lallement et al (1986), the exact value of the solar wind near the pole is an important boundary condition when one calculates the dynamics of the solar wind expansion in a polar coronal hole. When the solar flux at 1 AU is decreased from 3 x 10^8 to 2 x 10^8 $cm^{-2}s^{-1}$, the maximum temperature required in the corona drops from about 7 x 10^6 K to 3.5 x 10^6 K, relaxing some requirements on the necessary mechanims for coronal heating. - the 3D shape of the heliosphere will also depend on the latitude distribution of the solar wind mass flux. However, it is the momentum flux which counts, rather than the mass flux; and with the much larger velocity (700 km/s) found by Ulysses, outside the HCS, the momentum flux is higher at high latitudes (Smith and Marsden, 1995). The relative increase of the momentum flux from equator to poles can be estimated to be around 40% (Phillips et al, 1995).

References

Ajello J.M., Stewart A.I., Thomas G.E., Graps A., , Astrophys. J. 317, p. 964-986, 1987

Baranov V.B., Malama Yu. G., J. Geophys. Res., 98, A9, 15,157-15,163, 1993

Bertaux J.L., J.E. Blamont, Astron. Astrophys., 11, 200-217 (1971)

Bertaux J.L., J.E. Blamont, M. Festou , Astron. Astrophys., 25, 415-430 (1973)

Bertaux J.L., J.E. Blamont, E.N. Mironova, V.G. Kurt, M.C. Bourgin, Nature, 270, 156-158 (1977)

Bertaux, J.L., Lallement, R., Kurt, V., G., and Mironova, E.N., Astron. Astrophys., 150, pp. 1-20, 1985.

Bertaux J.L., R. Pellinen., E. Chassefiere, E. Dimarellis, F. Goutail, T.E. Holzer, V. Kelha, S. Korpela, E. Kyrola, R. Lallement, K. Leppala, G. Leppelmeier, I. Liede, K. Rautonen and J. Torsti, Original proposal to ESA, 1987 ;

Bertaux et al (same authors) : The SOHO mission ESA SP-1104, 63-68, 1989.

Bertaux J.L., E. Kyrola, E. Quemerais, R. Pellinen, R. Lallement, W. Schmidt, M. Berthe, E. Dimarellis, J.P. Goutail, C. Taulemesse, C. Bernard, G. Leppelmeier, T. Summanen,.., T.E. Holzer, Solar physics, 162, 403 (1995)

Bertaux, J.L., Quemerais E., Lallement, R., 1996a, Sol. Phys., in press

Bertaux, J.L., Quemerais E., Lallement, R., 1996b, G.R.L., in press.

Blum, P. W., and Fahr, H. J., Astron. Astrophys., 4, 280, 1970.

Hoeksema, J.T., Large scale solar and heliospheric magnetic fields, Adv. Space Res., 17, n 1 , pp 15-24, 1991

Joselyn, J.A., and T.E. Holzer, J. Geophys. Res., 80, 903, 1975

Keller H.U., Richter A.K., Thomas G.E., Astron. Astrophys., 102,415 (1981)

Kumar, S., Broadfoot, A.L., Astron. Astrophys., 69, L5, (1978).

Kumar, S., Broadfoot, A.L., Astrophys. J., 228, 302-311, (1979).

Kyrola E.,T. Summanen and P. Raback, Astron. Astrophys., 288, 299-314, 1994.

Lallement R., Bertaux, J.L., F. Dalaudier, Astron. and Astrophys., 150, pp. 21-22 (1985).

Lallement R., Bertaux, J.L., Kurt, V., G., J. Geophys. Res, 90, 1413-1423 (1985).

Lallement R., Holzer T.E., Munro R.H., J. Geophys. Res, 91, 6751- 6759 (1986).

Lallement R. , Bertaux, J.L., , E. Chassefiere, B. Sandel, Astronomy and Astrophysics, 252, 385-401, (1991)

Phillips J.L., S.J. Bame, W.C. Feldman, B.E. Goldstein, J.T. Gosling, CM. Hammond, D.J. McComas, M. Neugebauer, E.E. Scime, S.T. Suess, Science, 268, P.1030-1033,1995

Phillips J.L., Bame S.J., Barnes A., Barraclough B.L., Feldman W.C., Goldstein B.E., Gosling J.T., Hoogeven G.W., McComas D.J., Neugebauer M.,

Suess S.T., Geophys. Res. Lett., 22, 23, 3301 (1995)

Phillips , private communication,1996 Quemerais E. Bertaux J.L., Astron. and Astrophys., 277, 283-301, (1993).

Richardson J.D., K.I. Paularena, J.W. Belcher and A.J. Lazarus, Geophys. Res. Lett. 21, 14, 1559-1560, 1994

Rucinski M. and D. Bzowski, Space Science Rev. 72, 467-470, 1995.

Smith E.J., Marsden R.G., Geophys. Res. Lett., 22, 23, 3297 (1995)

Summanen T., Lallement R., Bertaux, J.L. and Kyrola, E., J. Geophys. Res., 98-A8, pp. 13.215-13.224 (1993)

Summanen T., Astron. Astrophys., in press (1996)

Summanen T., Lallement R., Quemerais E., J. Geophys. Res., in press, 1996

Thomas G.E. and Krassa R.F., Astron. Astrophys., 11, 218, 1971

Zhao X.P., Hundhausen A.J., J. Geophys. Res. 86, 5423, 1981

QUASILINEAR RELAXATION OF PICKUP INTERSTELLAR IONS

A.A. GALEEV and A.M. SADOVSKII
Space Research Institute of Russian Academy of Sciences, 117810 GSP–7 Moscow, Russia

Abstract. The quasilinear relaxation of pickup interstellar helium ions is described in the diffusion shell approximation. It is shown that the Cherenkov damping of Alfvén waves due to their refraction in the nonuniform solar wind could inhibit the complete relaxation of pickup helium ions over the bispherical shell.

1. Introduction

It is well recognized that interstellar pickup ions can excite Alfvén waves (Wu and Davidson, 1972). However until recent *in-situ* measurements of their densities (Geiss, Gloeckler and von Steiger, 1994; Gloeckler *et al.*, 1994; Möbius *et al.*, 1995) it was difficult to assess how the interstellar ion distribution would relax under the action of excited Alfvén waves. The kinetic theory of the relaxation of pickup ion distribution under the action of Alfvén waves differs significantly for cometary (Sagdeev *et al.*, 1986; Galeev *et al.*, 1987) and interstellar (Lee and Ip, 1987) ions. In case of cometary pickup ions the deceleration of solar wind due to the considerable mass loading results in a finite thickness of diffusion shell in velocity space, but the temperature of cometary ions is much lower than that of solar wind protons. For a very low mass density of interstellar pickup ion the thickness of diffusion shell is of the order of their thermal velocity, i. e. negligible in comparison with solar wind velocity, but the thermal spread of hydrodynamic solar wind can not be neglected in comparison with that of interstellar ions. Therefore we account for the interstellar ions thermal velocity to calculate the growth rate of kinetic Alfvén waves (section 2).

The approximation of a thin bispherical distribution (Galeev and Sagdeev, 1988) for the diffusion shell will be used to calculate the reduced growth rate of Alfvén waves and their intensity resulting from the relaxation of pickup ion distribution (section 3). We make these calculations for the parameters of singly charged helium ions that are measured best of all (Gloeckler *et al.*, 1994).

Finally the refraction of Alfvén waves in the nonuniform solar wind results in their Cherenkov damping that could inhibit the complete relaxation of pickup interstellar helium ions in the areas where solar wind gradients are large enough.

2. Alfvén Wave Generation by Pickup He^+ Ions

The dispersion equation for the Alfvén wave generation by interstellar pickup He^+ ions in the solar wind system can be written in the form (Wu and Davidson, 1972):

$$1 - \frac{\omega^2}{k_z^2 v_a^2} + \frac{n_i}{2n_p}\frac{m_p}{m_i}\frac{\omega_{pp}^2}{k_z^2 c^2}\int \frac{2(\omega - k_z u_z - k_z v_z)F_i - k_z u^2 \partial F_i/\partial v_z}{\omega - k_z u_z - k_z v_z \pm \omega_{ci} + i0}\, d^3\mathbf{v} = 0, \quad (1)$$

Space Science Reviews **78**: 329–334, 1996.

where ω and k_z are the frequency and wave vector of Alfvén waves propagating along the magnetic field lines, ω_{pp} the plasma frequency of solar wind protons with the density n_p, ω_{ci} the He^+ ions cyclotron frequency, n_i and F_i the density and velocity distribution function of pickup interstellar helium, v_a the Alfvén wave velocity, u the solar wind velocity, $u_z = -u^2/\sqrt{u^2 + \Omega_\odot^2 r^2}$ the hydrodynamic field aligned velocity of pickup helium ions in the solar wind system, v_z the parallel velocity of pickup helium ions in solar wind system, $\Omega_\odot$ the angular velocity of solar rotation, r the distance from the Ulysses spacecraft to the Sun and c the speed of light.

Using parameters of the solar wind at Ulysses position inside the corotation interaction region (Gloeckler *et al.*, 1994; Möbius *et al.*, 1995; Balogh *et al.*, 1995): $u = 393$ km/s, $n_p = 1.9\ \text{cm}^{-3}$, $T_i = 1.5 \cdot 10^4$ K, $T_p = 4.3 \cdot 10^4$ K, $B = 1.37 \cdot 10^{-5}$ G, $N_i = 6.4 \cdot 10^{-5}\ \text{cm}^{-3}$ (on the bispherical shell only), $r = 4.485$ AU, we calculate the numerical values of parameters: $u_z \approx -80.4\ \text{km/s} \gg v_a = 21.8\ \text{km/s} \sim v_{Tp} = 26.7\ \text{km/s} \gg v_{Ti} = 7.9\ \text{km/s}$, $k_z = 4.2 \cdot 10^{-4}\ \text{km}^{-1}$, $\omega = k_z v_a = 8.9 \cdot 10^{-3}\ \text{s}^{-1} \ll \omega_{ci} = 3.3 \cdot 10^{-2}\ \text{s}^{-1}$. With the help of the above parameters we obtain the Alfvén wave growth rate:

$$\begin{aligned}\gamma_k &= -\frac{\pi^{1/2} n_i}{2n_p}\frac{m_p}{m_i}\frac{\omega_{cp}^2}{\omega}\frac{\omega - k_z u_z \pm \omega_{ci}}{|k_z| v_{Ti}} \exp\left\{-\frac{(\omega - k_z u_z \pm \omega_{ci})^2}{k_z^2 v_{Ti}^2}\right\} \\ &\approx 1.4 \cdot 10^{-2}. \end{aligned} \tag{2}$$

3. Diffusion Shell Approach to the Interstellar Pickup Ion Velocity Distribution and Wave Generation

The interstellar pickup ions initially form a ring perpendicular to the solar wind magnetic field. In the solar wind frame this ring is moving along the magnetic field lines with the velocity $u_z = -u^2/\sqrt{u^2 + \Omega_\odot^2 r^2}$.

As we have shown above such a beam of interstellar pickup ions excites Alfvén waves propagating along the magnetic field lines. The back reaction of Alfvén waves on the pickup ions forces the latter to diffuse in velocity space. In the system of coordinates moving with the Alfvén wave, the wave electric field is zero, and, as a result, the energy of particles interacting with this wave is conserved. Therefore, if the particle in the process of diffusion in velocity space interact only with the waves propagating in the same direction, and therefore having the same phase velocity, their energy is conserved in the system of coordinates moving with the phase velocity of Alfvén waves, i. e., particles diffuse along the spherical shell in velocity space centered on the phase velocity of Alfvén waves.

Taking into account that the pickup ions under the action of Alfvén waves will diffuse from the initial ring in both directions, we expect that pickup ions will

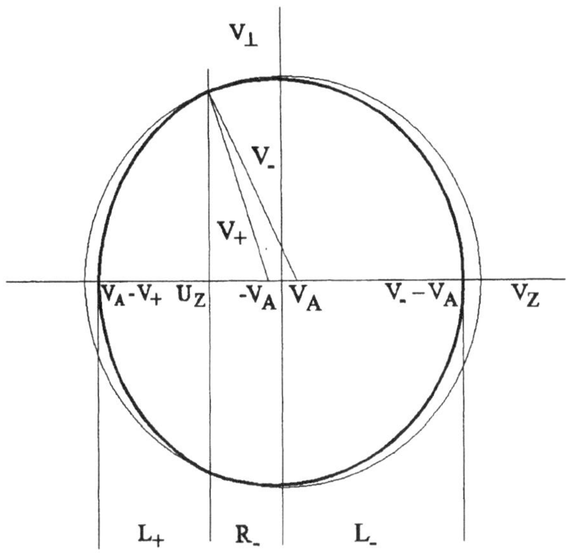

Figure 1. The bispherical distribution of pickup ions

diffuse along the surfaces of two spheres with the centers on the v_z axis at the points $v_z = v_a$ or $v_z = -v_a$. If we draw in Figure 1 the sphere of constant ion energy $v_z^2 + v_\perp^2 = u^2$ through the initial ring of pickup ions, then we see that in order to excite Alfvén waves, the pickup ions have to diffuse towards lower energies, i. e., in the velocity space $v_z < u_z$ and $v_z > u_z$ they diffuse along the spheres centered at $v_z = v_a$ and $v_z = -v_a$, respectively. In other words the initial ring distribution under the action of Alfvén waves transforms into the bispherical distribution shown in Figure 1 (Galeev and Sagdeev, 1988). The value of the wave vector k_z and the sign of polarisation of the waves generated by the pickup ions with the velocity v_z are found from the condition of cyclotron resonance between waves and particles

$$\omega_k - k_z v_z \pm \omega_{ci} = 0, \tag{3}$$

where the signs "+" and "−" refer to the right- and left- hand polarisation respectively (the magnetic field vectors $\mathbf{B}_k^+$ and $\mathbf{B}_k^-$ rotate in the direction of cyclotron motion of electrons and ions, respectively). Having the following ordering of characteristic velocities

$$u \gg |u_z| \gg v_a \gg v_{Ti} \tag{4}$$

in the calculation of expressions for the Alfvén wave growth rate and their intensity we neglect corrections related to small parameters $v_a/|u_z| \ll 1$ and $v_{Ti}/|u_z| \ll 1$. Due to the latter inequality the bispherical shell is assumed infinitely thin. Assuming also that $\omega_k > 0$ we can find from Eq. (3) the sign of k_z and Alfvén wave polarisation

$$k_z = \frac{\pm\omega_{ci}}{v_z - v_{ph}}, \quad k_z = \frac{\omega_k}{v_{ph}} \tag{5}$$

in different regions of the bispherical surface. Here $v_{ph} = \pm v_a$ is the phase velocity of Alfvén waves.

Quasilinear theory of pickup ion relaxation provides the expression for the Alfvén wave growth rate under the assumption of continuous pickup ion production with the rate $\dot{N}_i$ (Galeev *et al.*, 1987):

$$\gamma_k^{\pm} = \frac{2\pi\omega_k\omega_{ci}}{k_z^3}\frac{\dot{N}_i m_i}{|B_k^{\pm}|^2}\left[\left(1 \pm \frac{\omega_{ci}}{k_z u}\right) - 2\eta\left(-\frac{u_z}{u} \pm \frac{\omega_{ci}}{k_z u}\right)\right], \tag{6}$$

where $\eta(x)$ the step function and the Alfvén wave spectral intensity $|B_k^{\pm}|^2/4\pi = I_{R,L}$ can be also found in the diffusion shell approximation (Williams and Zank, 1994) in all areas of velocity space using relations (5) and Figure 1.

1) $-u < v_z < u_z, \quad v_{ph} = v_a, \quad k_z > 0:$

$$\gamma_k^- = \frac{\Gamma_\circ}{I_{L+}}\left(1 - \frac{\omega_{ci}}{k_z u}\right); \quad I_{L+}(k_z) = I_\circ\left(1 - \frac{\omega_{ci}}{k_z u}\right), \tag{7}$$

2) $u_z < v_z < 0, \quad v_{ph} = -v_a, \quad k_z < 0:$

$$\gamma_k^+ = \frac{\Gamma_\circ}{I_{R-}}\left(1 - \frac{\omega_{ci}}{k_z u}\right); \quad I_{R-}(k_z) = I_\circ\left(1 - \frac{\omega_{ci}}{k_z u}\right), \tag{8}$$

3) $0 < v_z < u, \quad v_{ph} = -v_a, \quad k_z < 0:$

$$\gamma_k^- = \frac{\Gamma_\circ}{I_{L-}}\left(1 + \frac{\omega_{ci}}{k_z u}\right); \quad I_{L-}(k_z) = I_\circ\left(1 + \frac{\omega_{ci}}{k_z u}\right), \tag{9}$$

where

$$\Gamma_\circ = \frac{\pi\omega_k\omega_{ci}}{2|k_z|^3}\dot{N}_i m_i, \qquad I_\circ = N_i m_i \frac{v_a\omega_{ci}}{2k_z^2},$$

and the direction of wave propagation is specified by the "+" and "−" sign as subindex of wave spectral intensity.

We see that the spectral intensity of Alfvén waves turns to zero at $\omega_{ci}/k_z \approx \pm u$. However the quasilinear growth rate of the Alfvén waves is constant over the bispherical shell distribution:

$$\gamma_k^{\pm} = \frac{\pi\dot{N}_i}{N_i} = 3.5 \cdot 10^{-5} \quad \mathrm{s}^{-1}, \tag{10}$$

where the pickup He^+ production rate

$$\dot{N}_i = \alpha n_{\mathrm{He}} n_p \sigma_{\mathrm{CE}} u = 7 \cdot 10^{-10} \quad \mathrm{cm}^{-3}\mathrm{s}^{-1}, \tag{11}$$

where $n_{\mathrm{He}} \approx 0.013\ \mathrm{cm}^{-3}$ is the density of interstellar helium (Gloeckler *et al.*, 1994), $\sigma_{\mathrm{CE}} = 2 \cdot 10^{-15}\ \mathrm{cm}^2$ is the charge exchange cross section, and $\alpha = 0.38$ is the fraction of the total production of Helium ions that are injected into the bispherical shell.

4. Damping of Alfvén Waves due to Their Refraction

The damping rate of Alfvén waves was derived by Kotelnicov *et al.* (1991):

$$\gamma_k^D = -\frac{\pi^{1/2}}{4}\frac{k_\perp^2}{|k_z|}v_{Tp}\exp\left\{-\frac{v_a^2}{v_{Tp}^2}\right\}. \qquad (12)$$

To calculate γ_k^D we must evaluate $k_\perp$. Using equation accounting for wave refraction:

$$-\frac{\partial\omega_k^*}{\partial r_\perp}\frac{\partial}{\partial k_\perp}|B_k^\pm|^2 = 2(\gamma_k^D + \gamma_k^\pm)|B_k^\pm|^2, \qquad (13)$$

where $\omega_k^* = \omega_k - k_z u_z$ in the solar wind system, we obtain

$$-\frac{\partial\omega_k^*}{\partial r_\perp} \approx \frac{u_z}{u}k_z\frac{\partial u}{\partial r_\perp} \approx 5.5\cdot 10^{-7}|k_z| \quad \mathrm{s}^{-1}, \qquad (14)$$

where $r_\perp$ is the coordinate perpendicular to the ecliptic plane with the spatial scale of nonuniformity $\sim$ 1 AU. Then we find

$$|B_k^\pm|^2 = \exp\left\{-\frac{Ak_\perp^3}{3}\right\}\left(\frac{I(k_z)}{4\pi} + \phi(k_z)\int_0^{k_\perp}\exp\left\{\frac{Ak_\perp'^3}{3}\right\}dk_\perp'\right), \qquad (15)$$

where

$$A = \frac{2\gamma_k^D}{\partial\omega_k/\partial r_\perp k_\perp^2} > 0 \qquad (16)$$

and

$$-\phi(k_z) = \frac{2\gamma_k^\pm|B_k^\pm|^2}{\partial\omega_k/\partial r_\perp} > 0. \qquad (17)$$

Assuming $I/4\pi \gg \phi(k_z)$ in the region of interest we find from Eq. (20):

$$\frac{k_\perp}{|k_z|} = \left[\frac{6u_z/u\partial u/\partial r_\perp}{\pi^{1/2}v_{Tp}|k_z|}\exp\left(\frac{v_a^2}{v_{Tp}^2}\right)\right]^{1/3} = 5.2\cdot 10^{-3}|k_z|^{-1/3} \quad \mathrm{km}^{-1/3} \qquad (18)$$

and

$$\gamma_k^D = 1.6\cdot 10^{-4}k_z^{1/3} \quad \mathrm{km}^{1/3}\mathrm{s}^{-1} = 1.2\cdot 10^{-5} \quad \mathrm{s}^{-1}. \qquad (19)$$

Here we used typical value of the order of $k_z \sim 4\cdot 10^{-4}$ km^{-1}.

Let us note that in the case of solar wind interaction with comet Halley quasilinear theory predict the formation of plateau at the narrow Cherenkov resonance of magnetosonic and Alfvén waves with cometary ions due to high intensity of waves (Kotelnicov *et al.*, 1991). In our case the width of resonant velocities, derived from the cyclotron resonance condition in Eq. (1) with account for the thermal spread of solar wind velocity is:

$$\Delta v_{zp} = \frac{\Delta k_z}{k_z} v_{Tp} \approx 0.3 v_{Tp}, \tag{20}$$

where $\Delta k_z = k_z v_{Tp}/u_z$. As a result the Alfvén wave intensity is much lower than the energy needed to form a plateau due to very low density of interstellar ions:

$$\frac{I_o k_z}{m_p n_p v_a v_{Tp}} \approx 3 \cdot 10^{-3}. \tag{21}$$

5. Conclusion

Using the measured parameters of the interstellar pickup helium ions we have shown that these ions excite Alfvén waves propagating along the magnetic field lines. Under the action of the excited waves the velocity distribution function of interstellar ions relaxes to the bispherical shell distribution in velocity space with the characteristic time scale of the order 10^4 s. However we also demonstrated that Alfvén wave refraction caused by the latitudinal gradient of the solar wind speed leads to damping of the Alfvén waves with a rate comparable to their growth rate due to production of interstellar helium ions. Therefore it is possible that due to a lager spatial variations of solar wind parameters the damping rate of Alfvén waves can sometime stop the relaxation of pickup ions.

References

Balogh, A., Smith, E. J., Tsurutani, B. T., Southwood, D. J., Forsyth, R. J. and Horbury, T. S.: 1995, *Science* **268**, 1007–1010.

Davidson, R. S.: 1983, in M. N. Rosenblath and R. Z. Sagdeev (Eds.) *Handbook of Plasma Physics, Volume 1: Basic Plasma Physics*, A. A. Galeev and R. N. Sudan (Eds.), North–Holland Publishing Co., Amsterdam – N.-Y. – Oxford, 519–585.

Galeev, A. A., Polyudov, A. N., Sagdeev, R. Z., Szegö, K., Shapiro, V. D. and Shevchenco, V. I.: 1987, *Sov. Phys. JETP* **65**, 1178–1186.

Galeev, A. A. and Sagdeev, R. Z.: 1988, *Astrophys. Space Sci.* **144**, 427–438.

Geiss, J., Gloeckler, G. and von Steiger, R.: 1994, *Phil. Trans. R. Soc. Lond.* **A349**, 213–226.

Gloeckler, G. Geiss, J. and Roelof, E. C., Fisk, L. A., Ipavich, F. M., Ogilvie, K. W., Lanzerotti, L. J., von Steiger, R. and Wilken, B.: 1994, *J. Geophys. Res.* **99**, 17637–17643.

Kotelnicov, A. D., Polyudov, A. N., Malkov, M. A., Sagdeev, R. Z. and Shapiro, V. D.: 1991, *Astron. and Astrophys.* **243**, 546–552.

Lee, M. A. and Ip, W. H.: 1987, *J. Geophys. Res.* **92**, 11041–11052.

Möbius, E., Rucinski, D., Hovestadt, D. and Klecker, B.: 1995, *Astron. and Astrophys.* **304**, 505.

Sagdeev, R. Z., Szegö, K., Shapiro, V. D. and Shevchenco, V. I.: 1986, *Geophys. Res. Lett.* **13**, 85–88.

Williams, L. L. and Zank, G. P.: 1994, *J. Geophys. Res.* **99**, 19229–19244.

Wu, C. S. and Davidson, D. C.: 1972, *J. Geophys. Res.* **77**, 5399–5406.

THE ABUNDANCE OF ATOMIC ^{1}H, ^{4}HE AND ^{3}HE IN THE LOCAL INTERSTELLAR CLOUD FROM PICKUP ION OBSERVATIONS WITH SWICS ON ULYSSES

GEORGE GLOECKLER

Department of Physics and IPST, University of Maryland, College Park, MD 20742, USA,

and

Department of Atmospheric, Oceanic and Space Sciences, University of Michigan, Ann Arbor, MI 48109, USA.

Abstract. Pickup ions measured deep inside the heliosphere open a new way to determine the absolute atomic density of a number of elements and isotopes in the local interstellar cloud (LIC). We derive the atomic abundance of hydrogen and the two isotopes of helium from the velocity and spatial distributions of interstellar pickup protons and ionized helium measured with the Solar Wind Ion Composition Spectrometer (SWICS) on the Ulysses spacecraft between ~2 and ~5 AU. The atomic hydrogen density near the termination shock derived from interstellar pickup ion measurements is 0.115±0.025 cm-3, and the atomic H/He ratio from these observations is found to be 7.7 ± 1.3 in the outer heliosphere. Comparing this value with the standard universal H/He ratio of ~10 we conclude that filtration of hydrogen is small and that the ionization fraction of hydrogen in the LIC is low.

1. Introduction

Pickup ions are created whenever slowly moving neutrals present in the solar system are ionized by charge exchange with the solar wind and solar UV radiation. In the heliosphere the major supply of neutrals is the local interstellar gas which enters the solar system at a relative speed of about 20 to 25 km/s (Lallement *et al.*, 1993; Witte *et al.*, 1993) affected only by the force of gravity and by radiation pressure. These neutrals are ionized at distances sufficiently close to the sun and are embedded (picked up) in the expanding solar wind which sweeps them out of the solar system. Being charged, these newly born, interstellar pickup ions are now subjected not only to a variety of plasma processes that shape their distribution functions, but are also accelerated in the disturbed solar wind typically associated with transient shocks and corotating interaction regions (Gloeckler *et al.*, 1994). At the heliospheric termination shock some fraction of these energetic pickup ions is further accelerated to many tens of MeV to become the so called 'anomalous cosmic rays'. Anomalous cosmic rays were discovered two decades ago (Garcia-Munoz *et al.*, 1973; Hovestadt *et al.*, 1973; McDonald *et al.*, 1974) and recent experimental evidence indicates that they have low charge states (Cummings *et al.*, 1993). Fisk *et al.* (1974) first proposed that anomalous cosmic rays are accelerated interstellar pickup ions. The average radial distance from the sun at which a large fraction of interstellar neutrals has been singly ionized is different for different interstellar atoms, depending on the ionization rates that in turn scale roughly as the first ionization potential of the atom. Helium atoms, whose ionization potential is high, penetrate as far in as ~0.3 AU, whereas hydrogen atoms, and to a lesser degree oxygen, nitrogen and neon atoms are relatively rare at distances closer than several AU from the sun.

The first evidence for the presence of interstellar hydrogen and helium in the heliosphere was provided by measurements of resonantly scattered solar UV in the early

Space Science Reviews **78**: 335–346, 1996.

1970s (Bertaux and Blamont, 1971; Thomas and Krassa, 1971). These and more recent measurements have been used to determine the density, temperature and relative velocity of interstellar hydrogen and helium with respect to the sun (Bertaux *et al.*, 1985; Chassefière *et al.*, 1986; Lallement *et al.*, 1993; Quémerais *et al.*, 1996). Interstellar neutral He has now been detected directly with a novel instrument flown on the Ulysses spacecraft (Witte *et al.*, 1993).

In spite of the dynamical importance of pickup ions, particularly in the outer and distant heliosphere, attempts to detect them directly were unsuccessful until sensitive plasma composition instruments employing time-of-flight technology become available (Gloeckler *et al.*, 1985; Gloeckler, 1990). Using an instrument of this type, interstellar pickup He^+ was detected at 1 AU in 1985 (Möbius *et al.*, 1985). The other pickup ions, namely H^+, $^4He^{++}$, $^3He^+$, N^+, O^+ and Ne^+, were only recently discovered (Gloeckler *et al.*, 1993; Geiss *et al.*, 1994; Gloeckler and Geiss, 1996) using the Solar Wind Ion Composition Spectrometer (Gloeckler *et al.*, 1992) on the Ulysses deep space probe.

Observations of interstellar pickup ions give us a new tool to probe the local interstellar medium, assess conditions in the vast regions beyond the heliospheric termination shock, and study a variety of heliospheric phenomena (see Table 1). While some of this work based on measurements of interstellar pickup ions has already been completed, much more is now in progress or remains to be done in the future. Table 1 summarizes some processes that may be studied and parameters that can be obtained from these new observations.

Table 1. Problems addressed with measurements of interstellar pickup ions

Region	Processes or Parameters
	Relative Velocity and Temperature of He, O
Local	Absolute Neutral Density of H, He
Interstellar	Fractional Ionization of H, He
Cloud	Elemental Composition of Interstellar Gas
Isotopic Ratios of He, Ne	
Heliospheric	Filtration factor for H, O, N
Boundary Layer	Pickup Ion Pressure
	Ionization Rates and Spatial Distribution of Neutral Gas in the Heliosphere
Heliosphere	Heliospheric and Termination Shock Acceleration Processes
	Transport and Plasma Processes

This paper focuses on observations of hydrogen and the two isotopes of helium. Their respective atomic abundance in the local interstellar cloud is derived from detailed observations of the spatial and velocity distributions of pickup protons and ionized helium (both He^+ and He^{++}) at distances between ~2 and ~5 AU with the SWICS instrument on Ulysses.

To interpret observations of interstellar pickup ions (IPIs), it is useful to review briefly some of the basic characteristics of interstellar neutrals and IPIs in the heliosphere obtained from simple models developed in the early 1970s (see, for example, Holzer, 1977, Thomas, 1978, and Vasyliunas and Siscoe, 1976) when the presence of interstellar hydrogen and helium throughout the heliosphere was established from UV observation.

2. Spatial Distribution of Interstellar Neutrals in the Heliosphere

The source material of the IPIs is the local interstellar cloud gas that makes its way deep into the heliosphere. The distribution in space of this gas within the heliosphere is reasonably well understood once the gas passes the heliospheric termination shock. The most appropriate model describing $n_k^*(R,\Theta) = n_k(R,\Theta)/N_k$, the normalized spatial distribution of atoms of type k, in the heliosphere at radial distance R from the sun and angle Θ with respect to the bulk flow direction, is the so-called 'hot model' (e.g. Thomas, 1978). The gas kinetic temperature, the appropriate ionization (loss) rate, $\beta_{k\mathrm{loss}}$, for removing neutrals by ionization, and the ratio of radiation to gravitational forces, μ, are parameters in this model, and N_k is the number density of the gas just inside the termination shock. Of the three generally not well known parameters, uncertainties in the loss rate will have the greatest effect on the spatial distribution of neutrals. In fact, for atoms heavier than hydrogen, radiation pressure is insignificant compared to gravitation ($\mu = 0$) and the gas temperature plays only a minor role.

For hydrogen and some heavier atoms such as oxygen, the density inside the termination shock is believed to be less than the LIC density because of filtration (e.g. Baranov *et al.*, 1993, 1995; Zank and Pauls, 1996) which prevents some of the interstellar gas from crossing the termination shock due to ionization. The filtration factor, $(N_k/N_{k\mathrm{LIC}})$ will depend on the proton density in the LIC and the relevant charge exchange ionization rate. Primarily because of uncertainties in our knowledge of the interstellar proton density, the amount of filtration is not well determined from models.

Filtration is unimportant for helium (Zank, personal communication). Since in addition helium is likely to be unaffected by possible changes in radiation pressure or uncertainties in its temperature, the only relevant model parameter for helium is its loss rate. Fortunately, the He loss rate is small ($< 2 \cdot 10^{-7}\mathrm{s}^{-1}$) implying that at distances of ~5 AU from the sun the helium density will be as high as 90% of its interstellar value. For hydrogen, on the other hand, filtration effects may not be negligible, radiation pressure is comparable to gravitation, and the ratio of thermal to bulk speed is sufficiently large so that the gas temperature becomes also an important parameter. Because the hydrogen loss rate is large, its density is greatly reduced at distances smaller than ~ 3 - 5 AU, and interstellar pickup hydrogen will be virtually undetectable at 1 AU.

3. Model and Measured Velocity Distributions of Pickup Ions

Interstellar pickup ions are created when interstellar atoms are ionized in the heliosphere. The number of IPIs produced is proportional to the ionization (production) rate, $\beta_{k\mathrm{prod}}$, at the time of ionization, and the local density, $n_k(R,\Theta)$, of neutral atoms. While it was often assumed that $\beta_{k\mathrm{prod}} = \beta_{k\mathrm{loss}}$, this is generally not the case as was pointed out by Gloeckler *et al.* (1993). This is because ionization rates vary with time and probably heliolatitude and the instantaneous production rate will not always equal the long-time-averaged loss rate.

The distribution of IPIs in velocity space after pickup is rather complicated and at present not well modeled. However, the basic transformations that take place right after pickup are reasonably well understood. The newly ionized particles will gyrate about the local magnetic field to form a so-called ring distribution in velocity space as they are convected outward with the solar wind at the local solar wind speed, V_{sw}. In the solar wind frame this ring is at $V = |V_{sw}|$ and pitchangle $\cos(\alpha) = \boldsymbol{V}\bullet\boldsymbol{B}/(V|\boldsymbol{B}|)$. After some time τ_r, the ring distribution should relax into a spherical shell distribution (with $V = |V_{sw}|$) due to pitchangle scattering. After some other time τ_s, the shell distributions would broaden due to energy diffusion. By assuming strong pitchangle scattering and little energy diffusion, that is $\tau_r \ll R/V_{sw} \ll \tau_s$ Vasyliunas and Siscoe (1976) found that the distribution function in the solar wind frame is isotropic and can be expressed as

$$f_{vs}(w<1,R,\Theta) = (3/8\pi)V_{sw}^{-4}N_k\beta_{kprod}(R_o^2/R)w^{-3/2}n_k^*(Rw^{3/2},\Theta) \quad w<1 \tag{1}$$

where $w = v_{ion}/V_{sw}$ and $R_o = 1.5\bullet10^{13}$ cm = 1 AU. For $w > 1$ $f(w>1,R,\Theta) = 0$. The strong pitchangle scattering implies adiabatic cooling which is incorporated in eq. (1).

The measured velocity distribution, $F(W)$, of interstellar pickup ions is generally observed in only a limited volume of phase space in the spacecraft frame, and is usually integrated over all directions sampled by the instrument. Furthermore, time averaging over long periods (days to months) is often required to obtain sufficient statistical accuracy. $F(W,R,\Theta)$ is derived from the measured count rate, $r_{m/q}(W,R,\Theta)$, of a given pickup ion (identified by its mass per charge m/q and, in some cases mass) and the measured solar wind speed, with $W = V_{ion}/V_{sw}$, the speed of the ion in the spacecraft frame divided by the solar wind speed,

$$F(W,R,\Theta)_{m/q} = C_{inst}\{r_{m/q}(W,R,\Theta)/\eta_{m/q}(V_{ion})\}\ V_{sw}^{-4}W^{-4}. \tag{2}$$

C_{inst} is the instrument factor (a product of the geometry factor, energy/charge bandwidth and conversion factors) and $\eta_{m/q}(V_{ion})$ is the counting efficiency for that ion at its measured speed V_{ion}. To compare the observed velocity distribution to model predictions (e.g. eq. 1), requires transformation to the spacecraft frame ($\boldsymbol{w} = \boldsymbol{W} - \boldsymbol{V}_{sw}/|V_{sw}|$) and integration of the model distribution function over the instrument view angles, Ω_{inst}

$$F(W, R\,\Theta)_{m/q} = \int_{\Omega_{inst}} d\Omega\, f_{m/q}(w,\theta,\phi,R,\Theta)\,d\theta\, d\phi\,. \tag{3}$$

4. Observations of Pickup Ion Velocity Distributions

Initially pickup ion distributions of He^+ (Möbius *et al.*, 1985) and H^+ as well as He^+ (Gloeckler *et al.*, 1993) were measured in a very limited range of ion speeds corresponding to $W > 1.5$, and appeared to be adequately described by the isotropic velocity distribution function (eq. 1). A sharp drop in phase space density above $W = 2$ was clearly observed. However, when it became possible to measure IPI distribution functions over a broader W range (including speeds well below that of the solar wind, or $W < 1$), distributions that were isotropic in the solar wind frame could no longer describe the measurements (Gloeckler *et al.*, 1995). These new observations forced one to conclude that pitchangle scattering is weak (Gloeckler *et al.*, 1995; Fisk *et al.*, 1996; Möbius, 1996). Since in many instances IPI distributions are likely to be highly anisotropic, pickup ion fluxes, and estimates of interstellar abundances based on observations of distribution functions above $W \sim 1.5$ assuming isotropy, will lead to underestimates of the true atomic densities in the outer heliosphere.

To illustrate these points, we show in Fig. 1 F(W) for pickup hydrogen measured with SWICS on Ulysses (see Gloeckler *et al.* (1992) for the description of the SWICS instrument, and Fig. 3 caption) in the south-polar coronal hole at 2.54 AU. The data were averaged over a 100-day period which was justified because solar wind conditions in these coronal hole regions are extremely steady. The solar wind proton speed was 780 km/s and its density 0.4 cm^{-3}. It is important to note that at these high latitudes the average magnetic field is almost radial and thus nearly aligned with the solar wind flow direction.

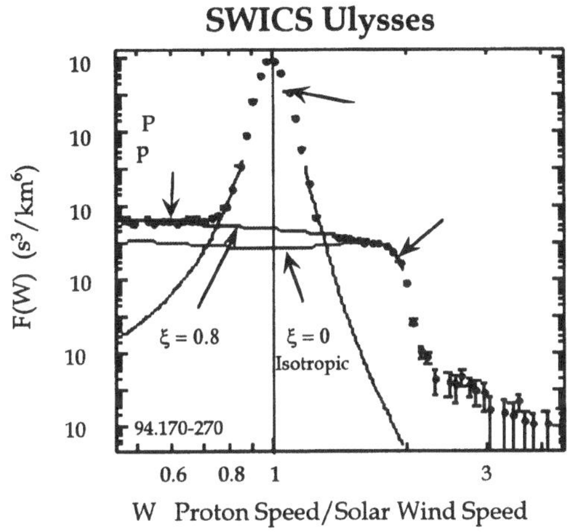

Fig. 1. Phase space density of protons measured with SWICS in the fast solar wind of the south-polar coronal hole. The velocity distribution of the hot pickup protons extends up to the cutoff at twice the solar wind speed and is well separated and distinctly different from the solar wind distribution. Beyond this sharp cutoff the density drops by three orders of magnitude. An isotropic (in the solar wind frame) distribution is ruled out by the observations. Counts accumulated below $W = 1$ in the sun sector have not been included because of possible instrument background due to scattered solar wind protons.

Several characteristics of pickup ions can be established from this proton spectrum measured above the solar wind peak. First, the pickup proton spectrum extends all the way to the predicted cutoff at $W = 2$. This implies that there is little if any thermalization of pickup ions, that is, the coupling between the solar wind plasma and pickup ions is so weak that little energy is exchanged as was first discussed by Feldman *et al.* (1974). Second, the drop-off of phase space density beyond the cutoff at $W = 2$ (corresponding to $w = 1$ in the solar wind frame) is very sharp, implying little if any energy diffusion. Thus, one can conclude that in the unperturbed solar wind, energy diffusion is negligible and $\tau_s \gg R/V_{sw} \approx 10^6$ sec. The more than three orders of magnitude smaller phase space density above $W = 2$ is probably a combination of a residual suprathermal tail and accidental low-level instrument background.

The anisotropy in the solar wind frame velocity distribution of the hydrogen IPI was revealed when its F(W) could be measured at speeds well below V_{sw} ($W < 1$), that is for protons in the solar wind frame moving in the direction towards the sun. The curve labeled $\xi = 0$ was obtained from eq. (3) using eq. (1) for f(w). Clearly, this curve falls a factor of three to four below the observations and is thus inconsistent with the measurements. To account for the observed excess of pickup protons below $W = 1$ it is necessary to use anisotropic solar wind frame distributions. The curve labeled $\xi = 0.8$ shows an example of an anisotropic distribution of the form

$$f(w<1,R,\Theta,\theta) = f_{vs}(w<1,R,\Theta)\{1+\xi\cos(\theta)\} \tag{4}$$

where θ is the angle between the velocity, $\boldsymbol{w}$, of the pickup ions in the solar wind frame and the inward radial direction. The anisotropy is the result of weak pitchangle scattering (Gloeckler *et al.*, 1995; Möbius, 1996) and its magnitude, ξ, depends not only on the pitchangle scattering mean free path, λ, but also on the average time between the creation of the pickup ion and its detection. Thus, ξ is likely to be larger for pickup H^+ than for He^+ because most protons observed at several AU have been produced relatively recently and thus had little time to isotropize, while most pickup He was ionized closer to the sun and hence had additional time to become more isotropic. Weak pitchangle scattering ($\lambda \approx 1$ AU) implies that the coupling between pickup ions and the ambient solar wind is weak, and that the assumption $\tau_r \ll R/V_{sw} \approx 10^6$ sec no longer holds.

With τ_r comparable to the solar wind convection time of $\sim 10^6$ sec, the exact solution for $f(\boldsymbol{w})$ becomes difficult and is currently not available (Fisk *et al.*, 1996; Möbius, 1996). However, we now believed that in the case when the magnetic field is nearly radial, the ring distribution of newly created pickup ions in the inward ($\boldsymbol{w}\bullet \boldsymbol{V}_{sw} < 1$) hemisphere of phase space initially decays with some time constant τ_1. Then, with scattering suppressed near 90° pitchangles, as seems to be indicated by pickup proton observations (Fisk *et al.*, 1996), the phase space density in the outward ($\boldsymbol{w}\bullet \boldsymbol{V}_{sw} > 1$) hemisphere becomes nearly isotropic but is reduced compared to that in the inward hemisphere by an amount that increases with λ. Adiabatic cooling still persists (Fisk *et al.*, 1996). Lacking an exact solution we use eq. (4) which appears to be an adequate approximation to the true anisotropic solar wind frame distribution function, $f(\boldsymbol{w})$ (see Fig. 1).

In the special case when the magnetic field is perpendicular to the solar wind flow direction (i.e. $\cos(\alpha) \approx 0$), as is often the case at radial distances beyond ~ 5 AU in the ecliptic plane, the phase space density in the two hemispheres should be about equal and no pronounced inward/outward anisotropy should be present.

5. The Hydrogen to Helium Ratio Near the Termination Shock

The spatial distribution of pickup ion fluxes, J_k (or densities) is approximately determined by the spatial distribution of interstellar neutrals $N_k n_k^*(\rho,\Theta)$ and the relevant production rates, $\beta_{k\mathrm{prod}}$,

$$J_k(R,\Theta) = N_k \beta_{k\mathrm{prod}} (R_o/R)^2 \int_0^R n_k^*(\rho,\Theta) d\rho \,. \qquad (5)$$

While $n_k^*(\rho,\Theta)$ is not directly measured as a function heliocentric radius R it may be estimated from model predictions.

In Fig. 2 is shown the spatial gradient of pickup hydrogen measure with SWICS on Ulysses during the high-latitude pass over the south-polar regions of the sun. In order to minimize effects due to the large anisotropies in the distribution function, the H^+/He^+ flux ratio is plotted versus heliocentric distance. The dependence of this ratio on the ionization rates is shown in eqs. (6a) and (6b) below.

$$J_H(R)/J_{He}(R) = \{(N_H \beta_{H^+\mathrm{prod}})/(N_{He}\beta_{He^+\mathrm{prod}})\} \bullet \mathbf{G}(R,\beta_{H\mathrm{loss}},\beta_{He\mathrm{loss}}), \qquad (6a)$$

$$\mathbf{G}(R,\beta_{H\mathrm{loss}},\beta_{He\mathrm{loss}}) = \int_0^R n_k^*(\rho,\beta_{H\mathrm{loss}}) d\rho \Big/ \int_0^R n_k^*(\rho,\beta_{He\mathrm{loss}}) d\rho \,. \qquad (6b)$$

During solar minimum conditions when these measurements were made, ionization loss rates were low (Rucinski *et al.*, 1996) and thus a larger fraction of neutrals, especially helium, made their way to heliocentric distances of less than 5 AU. For helium, the fraction was over 0.9 and therefore even large uncertainty in its loss rate would result in small error in n_k*. In computing **G**(R) we used the 'hot model' for protons with $\mu = 1$, gas temperature of 9000 K and an average total production rate of $3.3 \cdot 10^{-7}$ s^{-1}, the largest part of which was derived from the simultaneously measured solar wind proton flux. For helium we neglected radiation pressure ($\mu = 0$) and assumed $\beta_{He^+prod} = 0.7 \cdot 10^{-7}$ s^{-1}. The other parameters are given in the caption of Fig. 2.

Changing the loss rate for helium by ±30% had little effect on the shape of **G**(R) which is determined primarily by the hydrogen loss rate, β_{Hloss}. The height of the gradient curve in Fig 2. is proportional to (a) the product of the N_H/N_{He} density ratio at the termination shock, (b) the ratio of the respective production rates, and (c) the ratio of the hydrogen-helium anisotropy amplitudes. The best fit to the gradient measurements is for $\beta_{Hloss} = (5.5 \pm 0.7) \cdot 10^{-7}$ s^{-1} and $N_H/N_{He} = 7.3^{+2.2}_{-1.7}$. A more accurate determination of the N_H/N_{He} ratio comes from comparison of the observed velocity distributions of H^+ and He^{++} (rather than He^+) above W ~1.4. Using this method Gloeckler *et al.* (1996) obtained $N_H/N_{He} = 7.7 \pm 1.3$ from SWICS observations in the high-latitude coronal hole with $\beta_{Hloss} = 5.5 \cdot 10^{-7}$ s^{-1}. It is important to note that in their study the production rates of both H^+ and He^{++} were determined and thus, the production rate of He^+ was also obtained. The values of the production rates of H^+, He^+ and He^{++} were found to be $(3.3 \pm 0.3) \cdot 10^{-7}$ s^{-1}, $(0.65 \pm 0.07) \cdot 10^{-7}$ s^{-1}.and $(0.02 \pm 0.003) \cdot 10^{-7}$ s^{-1} respectively.

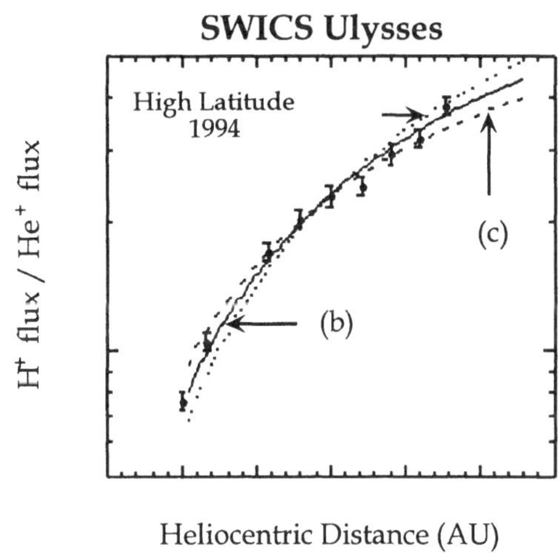

Fig. 2. Ratio of H^+ to He^+ pickup ion fluxes versus heliocentric distance. The flux ratio is derived from the observed phase space densities of H and He for W between 1.75 and 2.0 and an assumed anisotropy ratio for H^+/He^+ of ~1.2. Curves (a), (b) and (c) are model predictions of the flux ratio with β_{Hloss} = $6.5 \cdot 10^{-7}$ s^{-1}, $5.5 \cdot 10^{-7}$ s^{-1}, and $4.5 \cdot 10^{-7}$ s^{-1}, and N_H/N_{He} density ratios (near the termination shock) of 10.5, 7.3 and 5.0 respectively. Each data point represents a one-month average.

6. The $^3He/^4He$ Abundance Ratio in the Local Interstellar Cloud

The primordial abundances of the light elements and their isotopes provide essential information on processes of nucleosynthesis that occurred in the Big Bang (e.g. Schramm, 1993; Geiss, 1993). At present, the best estimates of the universal baryon/photon ratio, a fundamental constant of cosmology and the theory of matter, are extrapolations to primordial times of light-element abundances measured in the protosolar cloud and in the

Milky Way. The first direct measurement of the $^3He/^4He$ abundance ratio in the LIC based on the isotopic analysis of interstellar pickup helium (Gloeckler and Geiss, 1996) gives a value of $(2.25 \pm 0.65) \cdot 10^{-4}$, and indicates little change in this ratio during the past $4.5 \cdot 10^9$ years. The observations are summarized in Fig. 3. Because all known physical processes that transform interstellar atoms into pickup ions found well inside the heliosphere are very likely to affect isotopes of the same element in similar fashion, the isotopic ratio of pickup ions measured close to the sun is the local interstellar cloud isotopic abundance.

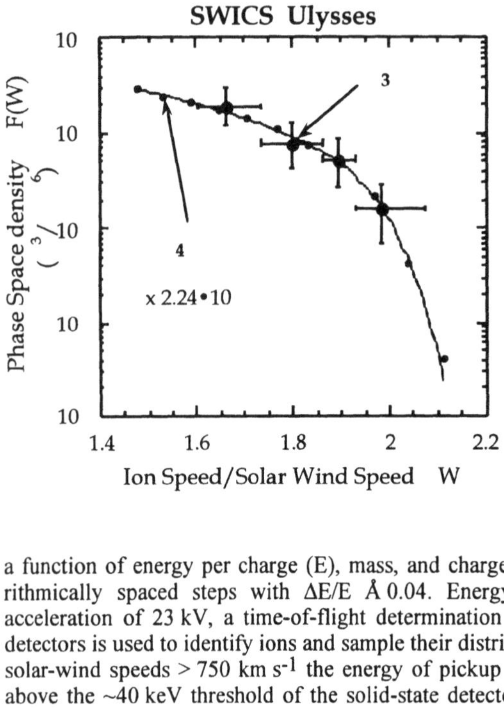

Fig. 3. (from Gloeckler and Geiss, (1996)). The measured distribution functions, $F_m(W)$, of interstellar pickup $^3He^+$ and $^4He^+$ (times $2.24 \cdot 10^{-4}$) versus the normalized ion speed, W. The phase space densities of $^3He^+$ (m=3) and $^4He^+$ (m=4) were computed from their respective efficiencies $\eta_m(W)$, and count rates $r_m(W)$, (with $^3He^+$ corrected for a 10% background). $F_m(W) = 4541(r_m/\eta_m)(438/V_{sw})^4 W^{-4}$. Despite relatively large statistical uncertainties in the four values of the $^3He^+$ phase space densities, the similarity of the two spectral shapes is striking. No $^3He^+$ was observed above W Å 2, consistent with the expected cutoff seen in the distribution functions of pickup ions. Below W Å 1.5 the probability for triple coincidence becomes extremely small. SWICS measures the intensity of ions as a function of energy per charge (E), mass, and charge state, from 0.6 to 60 keV e^{-1} in 64 logarithmically spaced steps with $\Delta E/E$ Å 0.04. Energy-per-charge analysis, followed by post-acceleration of 23 kV, a time-of-flight determination and energy measurement with solid-state detectors is used to identify ions and sample their distribution functions once every 13 minutes. At solar-wind speeds > 750 km s^{-1} the energy of pickup $^3He^+$ with speeds $>1.6 \cdot V_{sw}$ was typically above the ~40 keV threshold of the solid-state detectors, and the probability of obtaining triple coincidence analysis was thus high. Only triple coincidence data were used to obtain the velocity distribution functions of interstellar $^3He^+$ and $^4He^+$ shown here.

7. Summary and Conclusions

We have shown how measurements of interstellar pickup ions give new information on atomic abundances in the local interstellar cloud, as well as important insight on the interaction of low energy particles with the ambient solar wind. Far from the sun, pickup ions become dynamically important (Whang and Burlaga, 1996; Ness *et al.*, 1996) and are likely to affect the structure of the outer heliosphere and the termination shock. There is no doubt that pickup ions are accelerated (Gloeckler *et al.*, 1994) in the turbulent solar wind that often prevails in the near ecliptic region of the heliosphere as shown in Fig. 4. These pre-accelerated pickup ions are most likely the seed population of the anomalous cosmic rays, and detailed information on the velocity distribution of these pickup ions is essential for determining termination shock acceleration parameters, such as ion injection efficiencies.

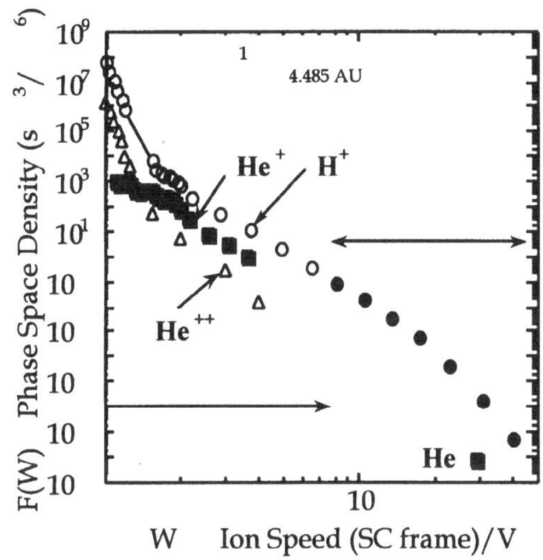

Fig. 4. (from Gloeckler *et al.*, (1994)). Velocity distribution functions of H^+(circles), He^+ (squares) and He^{++} (triangles) in the spacecraft frame. Data below $W \sim 8$ (0.6 to 60 keV/e) are obtained with SWICS, while those above $W \sim 8$ are measured with the HI-SCALE instrument (Lanzerotti *et al.*, 1992). Interstellar pickup H^+ appears as a bump on the spectrum below $W = 2$. The accelerated solar wind H^+ and He^{++} ($1.1 < W < 1.5$) have steep power law spectra with respective indices of $\sim$-22 and $\sim$ -30. Accelerated pickup H^+ and He^+ ($W > 2$) have identical spectral shapes that are less steep than that of accelerated solar wind He^{++} above $W \sim 1.5$ which is a power law of index $\sim$ -8.

The light-element and isotope abundances in the LIC address fundamental questions of Big Bang nucleosynthesis and galactic chemical evolution. For example, the first measurements of the LIC $^3He/^4He$ abundance ratio are already placing constraints on how much ^3He is produced by low mass stars. The accuracy to which this ratio is determined is currently limited by low statistics. This will improve as additional data are acquired by a number instruments that can make these measurements. It should be possible to measure the $^{22}Ne/^{20}Ne$ abundance ratio from SWICS observations in the near future.

Our ability to determine accurate atomic abundances in the LIC is currently limited by systematic uncertainties due to poor knowledge of ionization rates, in particular the loss rates that govern the amount of depletion of the interstellar gas in the inner heliosphere where pickup ion measurements are made. These loss rates are time dependent, vary with solar activity (Rucinski *et al.*, 1996) and may well be different in the high-latitude regions of the heliosphere, thus producing additional asymmetries in the spatial distribution of interstellar neutrals. Present uncertainties could be reduced with direct UV measurements (now done on SOHO with the Celias EUV He line monitor), or constrained by detailed observations of the spectral shapes and gradients of pickup ions, especially H^+, as discussed in this paper. The last method is judged to be the more reliable for obtaining these unknown parameters, especially for hydrogen, because it is more direct and does not require a-priori knowledge of how to integrate and over what time intervals to average solar wind or UV fluxes generally measured at 1 AU near the ecliptic plane.

We are in better shape when it comes to determining production rates. For H^+ the dominant part of the ionization is due to charge exchange with solar wind protons and this part is continuously measured. For He^+ the dominant part is due to photoionization which was not measured routinely until the launch of SOHO in late 1995. However, in their determination of the H/He ratio, Gloeckler *et al.* (1996) were able to use SWICS observations of pickup He^{++} produced almost entirely by double charge exchange with solar wind alpha particles. The solar wind alpha particle flux, just as the solar wind proton flux, was also measured simultaneously with the IPIs and thus the production rate of pickup He^{++} was known. This removed the large uncertainty introduced when determining the atomic abundance of He from measurements of pickup He^{+} (Möbius, 1996) whose production rate was not directly measured.

In Fig. 5 we present atomic abundances of hydrogen in the outer (> ~50 AU) heliosphere and helium in the local interstellar cloud. The helium density, labeled 'Witte' is based on direct observations of atomic helium with the GAS instrument on Ulysses (Witte *et al.*, 1993; Geiss and Witte, 1996) and falls in the middle of the SWICS range. The helium density labeled 'Möbius', is derived from 1 AU observations of pickup He^{+} (Möbius *et al.*, 1995) corrected for distribution function anisotropy effects (Möbius, 1996). The Möbius results are marginally consistent with the SWICS determination, but fall outside the limits of the Witte *et al.* (1993) range of values, possibly because of systematic uncertainties in the He^{+} production rate. The range of hydrogen densities (0.135 ± 0.025 cm^{-3}) determined from Lyman α emission data (see Quémerais *et al.* (1996) for an excellent review of these observations) overlaps nicely with the SWICS hydrogen abundance. The interval of consistency of the three independent methods -- pickup ions, direct measurements of neutral helium and UV emission -- is the small dark-shaded area.

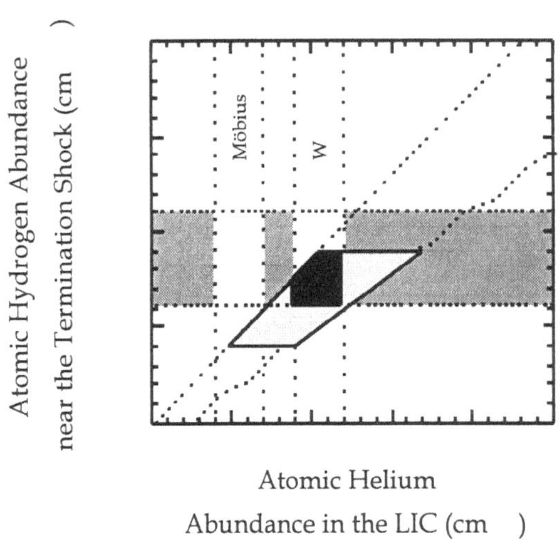

Fig. 5. Hydrogen versus helium density values derived from interstellar pickup ions, direct neutral helium measurements and UV observations. Ranges of four sets of observations are indicated by the various shaded regions. The small dark-shaded area is the region of consistency between abundances derived from pickup ion observations with SWICS (Gloeckler *et al.*, 1996), UV observations (Quémerais *et al.*, 1996) and atomic helium measurements (Witte *et al.*, 1993).

In Table 2 we list current best values for the abundance of the light elements and their isotopes (from Fig. 5) and the ionization rates for hydrogen and helium that were obtained from measurements of pickup ions with SWICS. The loss rates for hydrogen are determined to be different during the two phases of the solar cycle. However, in each case they fall well within the range given by Rucinski *et al.* (1996).

The hydrogen density near the termination shock (N_H) obtained from a combination of pickup hydrogen and UV observations, is 0.125 ± 0.015 cm^{-3}, and the neutral helium density ($N_{He}=He_I$) in the LIC, from direct detection of neutral helium, is 0.0155 ± 0.0015 cm^{-3}, giving a N_H/He_I ratio of 8.1 ± 1.3. Comparing this with the standard universal H/He abundance ratio of ~10 (Anders and Grevesse, 1988) implies little filtration of hydrogen and relatively low proton densities in the local interstellar cloud. A more precise determination of these parameters was made by Gloeckler *et al.* (1996). From their analysis of pickup ion measurements in combination with other observation and model calculations they find that the total hydrogen density in the LIC is ~0.26 cm^{-3}, the filtration factor for hydrogen is ~0.6, and the fractional ionization of H and He is ~0.15 and ~0.4 respectively.

Table 2. Hydrogen and helium densities, abundance ratios and ionization rates in the outer heliosphere and the local interstellar cloud

Quantity	Density (cm^{-3}), or *density ratio*	Production Rate $\cdot(10^{-7}\ s^{-1})^*$ 1991	Production Rate $\cdot(10^{-7}\ s^{-1})^*$ 1994	Loss Rate $\cdot(10^{-7}\ s^{-1})^*$ 1991	Loss Rate $\cdot(10^{-7}\ s^{-1})^*$ 1994
N_H	$0.125 \pm 0.015^{\S}$	$3.1\pm0.3^{\#}$	3.3 ± 0.3	$8.5\pm1.0^{\#}$	5.5 ± 0.5
He_I	$0.0155 \pm 0.0015^{\dagger}$	---	$0.65\pm0.10^{\#}$	---	0.6 ± 0.1
N_H/He_I	*$8.1 \pm 1.3^{\S}$*				
$(^3He/^4He)_{LIC}^{\ddagger}$	*$(2.25 \pm 0.65)\cdot10^{-4}$*				

§Combination of UV (Quémerais *et al.*, 1996) and pickup ion (Gloeckler *et al.*, 1996) results.
†From Witte *et al.* (1993) and Geiss and Witte (1996). ‡From Gloeckler and Geiss (1996).
*μ=1 was used for hydrogen in 1991 and 1994. #From Gloeckler *et al.* (1996).

Acknowledgments

I am very grateful to the many individuals at the University of Maryland, the University of Bern, the Technical University of Braunschweig, the Max-Planck-Institut für Aeronomie, and the Goddard Space Flight Center who have contributed to the success of the SWICS experiment during the many years of its development. Of particular benefit to me have been the many insightful discussions with Johannes Geiss and Len Fisk. I thank Christine Gloeckler for her help with data reduction and preparation of the manuscript, and Daniel Rucinski for making his hot model subroutine available to me. This work was supported by NASA/JPL contract 955460.

References

Anders, E. and Grevesse, N.: 1988, *Geochim. Cosmochim. Acta* **53**, 1197.
Baranov, V. B. and Malama, Yu. G.: 1993, *J. Geophys. Res.* **98**, 15157-15163.
Baranov, V. B. and Malama, Y. G.: 1995, *J. Geophys. Res.* **100**, 755-761.
Bertaux, J.-L. and Blamont, J. E.: 1971, *Astron. Astrophys.* **11**, 200.
Bertaux, J.-L., Lallement, R., Kurt, V. G., and Mironova, E. N.: 1985, *Astron. Astrophys.* **150**, 1.
Chassefière, E., Bertaux, J.-L., Lallement, R., and Kurt, V. G.: 1986, *Astron. Astrophys.* **160**, 299.
Cummings, J. R., Cummings, A. C., Mewaldt, R. A., Selesnick, R. S., Stone, E. C., and von Rosenvinge, T. T.: 1993, *Geophys. Res, Lett.* **20**, 2003.
Feldman, W. C., Asbridge, J. R., Bame, S. J., and Kearney, P. D.: 1974, *J. Geophys. Res.* **79**, 1808.
Fisk, L. A., Koslovsky, B., and Ramaty, R.: 1974, *Astrophys, J. Lett.* **190**, L35.
Fisk, L. A., Schwadron, N. A., and Gloeckler, G.: 1996, *Geophys. Res. Lett.*, (in press).

Garcia-Munoz, M., Mason, G. M., and Simpson, J. A.: 1973, *Astrophys. J. Lett.* **182**, L81.
Geiss, J.: 1993, in *Origin and Evolution of the Elements* (eds. Prantzos, N, Vangioni-Flam, E. & Cassé, M.) 89-106 (Cambridge University Press).
Geiss, J., Gloeckler, G., Mall, U., von Steiger, R., Galvin, A. B., and Ogilvie, K. W.: 1994, *Astr. Astrophys.* **282**, 924-933.
Geiss, J. and Witte, M.: 1996, *Space Sci. Rev.*, this issue.
Gloeckler, G., *et al.*: 1985, *IEEE Trans. on Geosci. and Remote Sensing, GE-23*, **3**, 234-240.
Gloeckler, G.: 1990, *Rev. Sci. Instrum.* **61**, 23613-23620
Gloeckler, G., *et al.*: 1992, *Astr. Astrophys. Suppl. Ser.* **92,** 267-289.
Gloeckler, G., *et al.*: 1993, *Science* **261**, 70.
Gloeckler, G., *et al.*: 1994, *J. Geophys. Res.* **99**, 17,637-17,643.
Gloeckler, G., Schwadron, N. A., Fisk, L. A., and Geiss, J.: 1995, *Geophys. Res. Lett.* **22**, 2665-2668.
Gloeckler, G. and Geiss, J.: 1996, *Nature* **381**, 210.
Gloeckler, G., Geiss, J, and Fisk, L. A.: 1996, submitted to *Nature*.
Holzer, T. E.: 1977, *Rev. of Geophys. and Space Phys.* **15**, 467.
Hovestadt, D., Vollmer, O., Gloeckler, G., and Fan, C. Y.: 1973, *Phys. Rev. Lett.. J.* **31**, 650.
Lallement, R., Bertaux, J.-L., and Clarke, J. T.: 1993, *Science* **260**, 1095.
Lanzerotti. L. J., *et al.*: 1992, *Astron. Astrophys. Suppl. Ser.* **92**, 349-365.
McDonald, F. B., *et al.*: 1974, *Astrophys. J. Lett.* **185**, L105.
Möbius, E., Rucinski, D., Hovestadt, D., and Klecker, B.: 1995, *Astron. Astrophys.* **304**, 505.
Möbius, E. *et al.*: 1985, *Nature* **318**, 426.
Möbius, E.: 1996, *Space Sci. Rev.*, this issue.
Ness, N. F., *et al.*: 1996, *Space Sci. Rev.*, this issue.
Quémerais, E., Bertaux, J._L., Sandel, B. R., and Lallement, R.: 1996, *Astron. Astrophys.* **290**, 961.
Rucinski, D., Cummings, A. C., Gloeckler, G., Lazarus, A. J., Möbius, E. and Witte, M.: 1996, *Space Sci. Rev.*, this issue.
Schramm, D. N.: 1993 in *Origin and Evolution of the Elements* (eds. Prantzos, N., Vangioni-Flam, E. & Cassé, M.) 112-131 (Cambridge University Press).
Thomas, G. E. and Krassa, R. F.: 1971, *Astron. Astrophys.* **11**, 218.
Thomas, G. E.: 1978, *Ann. Rev. Earth Planet. Sci.* **6**, 173.
Vasyliunas, V. M. and. Siscoe, G. L.: 1976, *J. Geophys. Res.* **81**, 1247.
Whang, Y. C. and Burlaga, L. F.: 1996, *Space Sci. Rev.*, this issue.
Witte, M., Rosenbauer, H., Banaszkiewicz, M. and Fahr, H. J.: 1993, *Adv. Space Res.* **13**, (6)121.
Zank, G. P. and Pauls, H. L: 1996, *Space Sci. Rev.*, this issue.

PHYSICS OF INTERPLANETARY AND INTERSTELLAR DUST

EBERHARD GRÜN
Max-Planck-Institut für Kernphysik, Heidelberg, Germany

JIRI SVESTKA
Prague Observatory, Prague, Czech Republic

Abstract. Observations of dust in the solar system and in the diffuse interstellar medium are summarized. New measurements of interstellar dust in the heliosphere extend our knowledge about micron-sized and bigger particles in the local interstellar medium. Interplanetary grains extend from submicron- to meter-sized meteoroids. The main destructive effect in the solar system are mutual collisions which provide an effective source for smaller particles. In the diffuse interstellar medium sputtering is believed to be the dominant destructive effect on submicron-sized grains. However, an effective supply mechanism for these grains is presently unknown. The dominant transport mechanisms in the solar system is the Poynting-Robertson effect which sweeps meteoroids bigger than about one micron in size towards the sun. Smaller particles are driven out of the solar system by radiation pressure and electromagnetic interaction with the interplanetary magnetic field. In the diffuse interstellar medium coupling of charged interstellar grains to large-scale magnetic fields seem to dominate frictional coupling of dust to the interstellar gas.

1. Introduction

Dust in interplanetary space has various appearances and can be detected and analyzed by a number of techniques. Comprehensive surveys of radio meteor orbits observed by the Harvard-Smithsonian radar have been published by Southworth and Sekanina (1973). Interplanetary dust particles collected in the atmosphere give cosmochemical information (composition and structure, Brownlee, 1985 and Bradley, 1988). The size distribution of dust at $1AU$ distance from the sun is documented in lunar microcrater records (Grün et al., 1985a). Zodiacal light observations form the Earth and from spacecraft (Leinert and Grün, 1990, Levasseur-Regourd, 1980) demonstrate the spatial distribution of dust in the inner solar system. Dust in the asteroid belt has been identified by its thermal emission (Hauser, 1984). Measurements of dust particles by in situ detectors on interplanetary spaceprobes give evidence of new populations of dust (e.g. interstellar dust and dust emitted from the Jovian system, Grün et al. 1993, Baguhl et al., 1995) which are not easily observed at the Earth's distance.

Gravity of the sun dominates all forces on bigger than micron-sized interplanetary meteoroids. Smaller dust particles are increasingly more affected by solar radiation pressure which can exceed gravity for dust particles below $1\mu m$ in size. For tenth micron-sized dust grains electromagnetic interaction

Space Science Reviews **78**: 347–360, 1996.

with the interplanetary magnetic field exceeds gravity by more than a factor 10. Mutual collisions among meteoroids determine the life-time of bigger than tenth millimeter-sized grains and generate smaller particles by fragmentation. Smaller grains are transported towards the sun by the Poynting-Robertson effect where they evaporate.

Little is known about dust in the diffuse interstellar medium which surrounds the solar system. Dust is observed in the diffuse interstellar medium by extinction of star light (Mathis, 1993) and by absorption features in stellar spectra (Jenniskens and Desert, 1993). Scattering of light by dust in the vicinity of near-by stars and thermal emission from interstellar cirrus (Low et al., 1984) give direct evidence of dust in the diffuse interstellar medium. Polarization measurements of interstellar extinction indicates the existence of elongated dust grains which are aligned by large-scale interstellar magnetic fields (Whittet, 1992).

In interstellar space the effects which need to be understood are interactions with: the radiation field, the ambient plasma, energetic particles, the magnetic field, the interstellar gas and grains. Sputtering by energetic particles in a high temperature plasma and mutual collisions are the main loss processes of interstellar grains. Gas drag and magnetic drag of charged dust particles are the dominant dynamic effects in the diffuse interstellar medium.

In-situ dust measurements in the heliosphere open a new window of information about the local interstellar medium. Micron-sized interstellar dust particles have been identified by the Ulysses and Galileo dust instruments inside the orbit of Jupiter (Grün et al., 1994, Baguhl et al., 1995). There are indications that even bigger interstellar grains pass through the solar system. Pioneer 10/11 detected a flux of particles approximately a few microns in size (Humes, 1980) from the opposite hemisphere than the local interstellar gas flux. Recent radio meteor observations indicate a population of extremely fast ($> 100 km/s$) meteor particles of obviously interstellar origin (Taylor et al., 1996). All these new observations should eventually lead to the determination of the physical and chemical properties and the size distribution of local interstellar dust.

In the following chapters we discuss different processes which affect interplanetary and interstellar dust. We start with a brief description of the sources of dust in interstellar and interplanetary space and continue to discuss life-time and dynamic effects. We describe the basic effects and show their relevance in both scenarios. This review is not considered to be complete in the sense that it describes all processes which affect interplanetary and interstellar dust but it concentrates on the dominant effects which are manifested by observational evidences. With this paper a comparison is attempted between the interplanetary and interstellar environment for the mutual benefit of both fields of study.

2. Sources of dust in interplanetary and interstellar space

Dust in the solar system is characterized by short life times, ($< 10^6$ years). Several effects destroy dust grains and disperse the material in space. Therefore, interplanetary dust must have contemporary sources: bigger objects like meteoroids, comets and asteroids.

Obvious sources of interplanetary dust are comets and asteroids. In the inner solar system comets shed large amounts of dust into interplanetary space. However, it can easily be shown that most of this dust is rapidly lost to interstellar space due to the action of radiation pressure. Bands of increased thermal radiation detected by the IRAS satellite indicate the presence of mm-sized particulates in the asteroid belt. However, in-situ detectors on board of the Pioneer 10 and 11 as well as the Galileo and Ulysses spacecraft did not find significant enhancements of small micron-sized dust in the asteroid belt. Where does interplanetary dust come from? The clue was given by Whipple 1967 and Dohnanyi 1969 who pointed out that cometary and asteroidal matter is efficiently injected into the solar system as bigger (millimeter- to meter-sized) meteoroids which are subsequently ground-up by mutual collisions. Grün et al. (1985a, b) demonstrated how the observed size distribution of meteoroids can be quantitatively explained by this effect.

There is a continuous stochastic stirring of asteroid orbits by the gravitational pull of the planets, especially of Jupiter. Dust in the asteroid belt originates from mutual collisions among asteroids. Small, micron-sized meteoroids produced in such a collision will be immediately removed by solar radiation pressure from their place of origin and will be swept by Poynting-Robertson effect towards the sun.

Comets produce dust and bigger meteoroids by a quite different process: Sublimation of the cometary ices drag particulates embedded within the cometary nucleus into interplanetary space. Most dust visible in the coma and in the tail consists of micron-sized dust particles which leave the solar system on hyperbolic orbits due to the action of radiation pressure. However, we know that also much bigger objects are released from comets. Meter-sized fireballs which hit the Earth's atmosphere demonstrate by their trajectories clearly their cometary origin. Splitting of comets give testimony that substantial fractions of kilometer-sized nuclei are left behind in interplanetary space.

Besides the ejection of dust into interstellar space from our own planetary system there are various other ways of dust formation and injection into the interstellar medium. Most visible stardust is born in stellar outflows especially from cool high luminosity stars (M- and C- giants, and supergiants, Jones and Tielens, 1994). Planetary nebulae, novae and supernovae provide the rest. Carbon and silicate grains and metal oxides like Al_2O_3 have been identified in the outflowing material (see e.g. the review by Dorschner and

Table I
Phases of the diffuse interstellar medium. HI is neutral and HII is ionized hydrogen.

phase	temperature (K)	density ($n(H)/cm^3$)
cold neutral matter	30 – 80	20 – 50 (HI)
warm neutral matter	5000 – 8000	0.1 – 1.0 (HI,HII)
thin hot ionized matter	$10^5 - 10^6$	$10^{-3} - 10^{-2}$ (HII)

Henning, 1995). However, not all of the heavy elements is condensed in the dust: e.g. much of the carbon is contained in gaseous CO. Refractory carbon phases are amorphous and hydrogenated amorphous graphite and polycyclic aromatic hydrocarbon (PAH, Allamandola et al., 1987). PAHs range from macro- molecules to nanometer-sized particulates.

The diffuse interstellar medium appears in three distinct phases (McKee and Ostriker, 1977): dusty gas clouds consisting out of a cold neutral core and warm neutral envelopes are embedded in thin hot ionized matter. Typical parameters of the diffuse interstellar medium are summarized in Table I.

The sizes of interstellar grains which are observable by astronomical means range from nanometer-sized grains over classic grains ($0.1\mu m$) to very big grains ($10\mu m$) in some interstellar environments. For classic grains a power law $n(s) \propto s^{-3.5}$ is postulated with a sharp cut-off at $s = 0.2\mu m$ (Mathis et al., 1977).

There is indirect evidence that grain growth partially balances grain destruction in the diffuse interstellar medium (Draine, 1990). Observations of interstellar absorption lines (Whittet, 1992, Sofia et al., 1994) show that refractory elements (like e.g. Al, Ca, Fe, Ni, Si, Mg) are underabundant with respect to cosmic abundance (Anders and Grevesse, 1989). Draine (1990) suspects that much of the dust observed in the diffuse interstellar medium was condensed in molecular clouds and not in stellar outflows. Rapid exchange with the diffuse interstellar medium supplied enough dust to balance the losses expected to occur.

The local interstellar medium in the solar neighborhood consists of a local bubble of hot ionized matter with about $50pc$ radius. Embedded in it are clouds of local fluff of warm neutral matter of a few pc dimension. From stellar absorption measurements (Bertin et al., 1993) and in-situ measurements by the Ulysses spacecraft (Witte et al., 1993) we know that the solar system is moving at $26km/s$ through interstellar medium of $7000K$ and density $n(H) = 0.1cm^{-3}$.

3. Grain destruction

MUTUAL COLLISIONS

In interstellar space mutual collisions are not believed to play a major role (Tielens et al., 1994), although fragmentation of bigger interstellar grains could provide an efficient source for the much more abundant smaller dust particles. On the contrary, collisions determine the lifetime of interplanetary meteoroids and they are the main process by which micrometeoroids are generated. It has been recognized since long that meteoroids have a continuous distribution of sizes (and masses) ranging from km-sized asteroids to submicron-sized dust particles. The flux of interplanetary meteoroids is generally given in the form of the cumulative meteoroid flux $F(m)$, which is the number of meteoroids with masses bigger than or equal to mass m which impact $1m^2$ each second. It is related to the differential flux $f(m)$ of particles in the mass range m to $m + dm$

$$F(m) = \int_m^\infty f(\tilde{m})d\tilde{m} \tag{1}$$

In collisions between interplanetary meteoroids the impact speed $v(r)$ is usually high enough to result in fragmentation of one or both particles. We approximate the average impact speed $\langle v(r)\rangle \propto r^{-1/2}$ as given by Kepler's law. In such a collision the smaller projectile always gets destroyed. A catastrophic collision, i.e. fragmentation of both particles, only occurs, if the mass ratio of target and projectile is not too large, i.e. if

$$m_{projectile} \geq \frac{1}{\Gamma(v)} M_{target} \tag{2}$$

where the maximum allowable mass ratio depends on velocity and was found experimentally for basalt to bc $G(v) = 500v^2$ (km/s, Fujiwara et al. 1977). This may or may not also be typical for interplanetary particles. If the projectile mass is smaller, the larger particle is eroded by the impact cratering process, a kind of sputtering on a macroscopic scale. Dohnanyi (1970) showed that for the meteoroids of concern to us ($m < 10^{-3} kg$) erosive collisions are much less important than catastrophic collisions, so that we may limit our discussions to the latter case.

The rate of catastrophic collisions $C(m, r)$ of a meteoroid of mass m at heliocentric distance r can now be calculated by adding the probabilities that the meteoroid will encounter during the following second a projectile particle large enough to disrupt the meteoroid. This integral has to be taken over all projectile masses m_p larger than the minimum mass given in eqn. (2):

$$C(m, r) = \int_{m/\Gamma(v(r))}^\infty \sigma \cdot f(m_p, r)dm_p \tag{3}$$

Here $f(m_p, r) = n(m_p, r)\langle v(r)\rangle$ is the flux of particles of mass m at heliocentric distance r. The cross section is taken as the total area inside the circle where the particles touch, $\sigma = \pi(s + s_p)^2$, where s and s_p are the radii of the meteoroid and projectile particles, respectively.

The collisional lifetime for this particle then is defined as the reciprocal of the average collision rate,

$$\tau_C = \frac{1}{C(m, r)} \tag{4}$$

In interplanetary space the lifetime of particles with masses $> 10^{-5}g$ is dominated by collisions. The mass of particles destroyed by collisions is not really lost but reappears in the form of small fragments, constituting a gain of particle in those size ranges. Comparing gains and losses, the net effect of collisions is to produce dust particles ($m < 10^{-5}g$, radii $< 100\mu m$) at the expense of the larger meteoroids (Grün et al., 1985a). This means that at $1AU$ the spatial density of interplanetary dust increases by 10% in 3000 years, at $0.1AU$ even in only 30 years. Most of the interplanetary dust is produced by collisions of larger meteoroids at a rate of about $10t/s$ inside $1AU$, which present a reservoir continually being replenished by disintegration of comets or asteroids. Part of it ($\approx 1t/s$) is lost by evaporation after being driven close to the sun by the Poynting-Robertson effect. The remainder is transformed by collisions to beta-meteoroids ($\approx 9t/s$) which is blown out of the solar system (Grün et al., 1985b).

SPUTTERING

In the diffuse interstellar medium sputtering is considered to be the main loss process for interstellar dust, whereas, in interplanetary space sputtering plays only a minor role. Sputtering is the process when an individual atom or ion collides with a solid body and removes from it one or more ions or atoms. The sputtering yield $Y(E)$ is the average number of atoms or ions released by impact of a single atom or ion with an energy E. Sputtering yields, especially for lower energy atoms or ions, are rather uncertain, and new laboratory experiments with appropriate targets and projectiles are desirable. Sputtering of dust particles by ions will be investigated in a near future by an experimental set-up for studies of dust electric charging by electrons and ions described in Svestka et al. (1993) and Cermak et al. (1995).

The number of sputtered ions is generally much smaller than the number of atoms. The threshold energy, i.e., the minimum energy of incoming particle required for sputtering, is approximately four times higher than the sublimation energy, e.g., for metals equal 5 to $40eV$ (Barlow, 1978). Most important for sputtering are atoms (or ions) of hydrogen or helium. The abundance of helium is about an order of magnitude smaller than that of hydrogen, but the sputtering yield of helium atoms is about an order of magnitude higher compared to hydrogen. Therefore, relative importance of sputtering by hydrogen and helium atoms is about the same. Sputtering yields of atoms of the CNO group are about an order of magnitude higher compared to helium atoms, but their relative abundance compared to hydrogen is only about 10^{-3} and so they contribute to the total sputtering rate only by about 2.5%.

In a hot gas where plasma particles are moving with thermal velocities we speak about "thermal sputtering". This process may be important for the destruction of particles in a hot gas. In the case of the local interstellar medium, where we find temperatures of $7000K$ to 10^6K and densities of hydrogen atoms 0.1 to $1cm^{-3}$, we can use sputtering rates estimated by Draine and Salpeter (1979). In the local cloud with a temperature not higher than 10^4K, thermal sputtering rates are negligible. Sputtering yields of hydrogen and helium atoms of energies $100eV$ incident on graphite are equal to about 0.01, the yield of oxygen atoms is 0.1. In the case of silicates the yield for hydrogen is equal to 0.01, for helium 0.1, and for oxygen 0.3. In the local bubble with temperature 10^6K and hydrogen number density 0.1 the erosion rates of silicate, graphite and iron grains due to the thermal sputtering are equal to approximately $3 \cdot 10^{-12} cm\ yr^{-1}$. It means that the lifetime of $0.01\mu m$ grains against the thermal sputtering would be about $3 \cdot 10^5$ years, the lifetime of $0.1\mu m$ grains $3 \cdot 10^6$ years. Erosion rates due to thermal sputtering are proportional to the hydrogen density and, therefore, the lifetimes against the sputtering are inversely proportional to the hydrogen density.

For estimates of the hydrogen and helium sputtering yields of candidate grain materials see also Barlow (1978) and Tielens et (1994), for reviews of experimental data on sputtering see Andersen and Bay (1981) and Betz and Wehner (1983), and for a theory of the sputtering process see Sigmund (1969; 1981).

4. Grain Transport

Interplanetary dust dynamics

In this paragraph we review the basic dynamic effects which act on meteoroids in interplanetary space. All dust particles in space feel the gravitational pull of the sun. Table II gives values for the solar gravitational force on particles of different masses and compares them with other forces acting on particles in interplanetary space. For particles with masses $> 10^{-8}g$ this is by far the dominating force. As a consequence meteoroids move on Keplerian orbits which are conic sections with the sun in the focus – other forces

Table II

Main forces on dust in interplanetary space at $5AU$, radius for spherical particles, absorbing particle, density $1000 kg/m^3$, $5V$, $v_{rel} \approx 400 km/s$, $B \approx 1nT$, $\alpha \approx 80°$.

mass (kg)	10^{-20}	10^{-17}	10^{-14}	10^{-11}	10^{-8}
radius (μm)	0.01	0.1	1	10	100
F_{grav} (N)	10^{-24}	10^{-21}	10^{-18}	10^{-15}	10^{-12}
F_{rad}/F_{grav}	0.5	2	0.5	0.05	$5 \cdot 10^{-3}$
charge-to-mass ratio (C/kg)	600	6	0.06	$6 \cdot 10^{-4}$	$6 \cdot 10^{-6}$
F_L/F_{grav}	2000	20	0.2	$2 \cdot 10^{-3}$	$2 \cdot 10^{-5}$

are only small disturbances. Certainly, all observations of big particles are compatible with such orbits.

The force F_{rad} exerted by the solar radiation on dust in interplanetary space decreases with the inverse square of the distance to the sun, i.e. this is the same dependence as the gravitational pull F_{grav}. Therefore, the ratio of both forces F_{rad}/F_{grav} is constant everywhere in interplanetary space and it is only dependent on material properties (Burns et al., 1979). This ratio is generally termed β:

$$\frac{F_{rad}}{F_{grav}} = \beta = 5.7 \cdot 10^4 \frac{\langle Q_{pr} \rangle}{\rho s}, \tag{5}$$

where $\langle Q_{pr} \rangle$ is the efficiency factor for radiation pressure on the meteoroid averaged over the solar spectrum, s is the radius of a spherical particle, and ρ is its density. E.g. for $s = 10^{-6} m$, and $\rho = 100 kg/m^3$ follows $\beta = 0.57$. For big particles $\langle Q_{pr} \rangle$ is of the order of 1, depending somewhat on the material properties, but it decreases for particles smaller than the effective wavelength of sun light. As a consequence β increases for smaller s values and reaches its maximum value between 0.1 and 1 microns. The maximum value is about 0.5 for non-absorbing dielectric materials and increases with increased absorptivity; it reaches values of 3 to 10 for metallic particles.

There are important consequences for the dynamics of small particles because of the radiation pressure. Small particles which are generated from big particles (e.g. ejection from comets or impact ejecta from meteoroids or asteroids) carry specific kinetic energy of their parents but find themselves in a reduced potential field of the sun. As a consequence they move on different orbits than their parents. E.g. a dust particle with radiation pressure constant b which is released from a big parent object on an eccentric orbit (eccentricity e_p) at perihelion will leave the solar system on a hyperbolic orbit if

$$\beta > \frac{1}{2}(1 - e_p) \tag{6}$$

It can be seen that even for a parent object on a circular orbit the ejected dust grain will move on an unbound hyperbolic orbit if its β value is only 0.5. In the data from Pioneers 8 and 9 spaceprobes (Berg and Grün, 1973) a flow of beta meteoroids leaving the solar system on hyperbolic orbits was first detected.

Besides the direct effect of radiation pressure on the trajectories of small dust grains there is also the more subtle Poynting-Robertson effect. It is caused by radiation pressure which is not perfectly radial on a moving dust particle but has a small component acting opposite to the particle motion. This drag force leads to a loss of angular momentum and orbital energy of the orbiting particle. The time, τ_{PR}, for a particle on a circular orbit to spiral to the sun is

$$\tau_{PR} = 2.2 \cdot 10^{13} \frac{s\rho}{Q_{pr}} \left(\frac{r}{r_0}\right)^2, \tag{7}$$

with $r_0 = 1AU$. E.g for $s = 10^{-2}m$, $\rho = 1000kg/m^3$, $\tau_{PR} = 2.2 \cdot 10^{14}s$ or $7 \cdot 10^6$ years.

All meteoroids in interplanetary space are electrically charged. Several competing charging processes determine the actual charge of a meteoroid: collection of plasma ions and electrons and emission of secondary and photoelectrons. Because of the predominance of the photoelectric effect in interplanetary space meteoroids are mostly charged positive at a potential of a few Volts. The outward (away from the sun) streaming solar wind carries a magnetic field from the sun. The polarity of the magnetic field can be positive or negative depending on the polarity at the base of the field line in the solar corona which varies spatially and temporary. The Lorentz force on a charged dust particle near the ecliptic plane is mostly either up- or downward depending on the polarity of the magnetic field. Submicron-sized grains are carried by the magnetic field out of the solar system. Since the polarity changes due to the sector structure near the ecliptic at a frequency much faster than the orbital period of bigger interplanetary dust particles the net effect of the Lorentz force is small. Only secular effects on the bigger zodiacal particles are expected to occur (Morfill et al., 1979, 1986) which could have an effect on the symmetry plane of the zodiacal cloud close to the sun. Zodiacal light observations (Leinert et al., 1980) show such an effect on the symmetry plane but there are other explanations as well.

The overall polarity of the solar magnetic field changes with the solar cycle of 11 years. For one solar cycle positive magnetic polarity prevails in the northern and negative polarity in the southern solar hemisphere. Submicron-sized interstellar particles which enter the solar system are either deflected towards the ecliptic plane or away from it depending on the overall polarity of the magnetic field. Therefore, interstellar particles are either prevented (during one solar cycle) from reaching the inner solar system or are concentrated (in the other solar cycle) in the ecliptic plane.

DUST DYNAMICS IN THE DIFFUSE INTERSTELLAR MEDIUM

In the local interstellar medium we find temperatures from $7000K$ to 10^6K and gas number densities 0.1 to 1 (Holzer et al., 1989; Lallement et al., 1994). Threshold energies for secondary electron emission under ion impacts are $\approx 100eV$, so that we can neglect this process. Potentials will be basically determined only by interactions with electrons and UV radiation. We will confine further our discussion to grains of radii $\geq 0.01\mu m$. In the case of the mentioned temperatures and grain radii, the ranges of electrons and ions are smaller than the grain dimensions, so that it is not necessary to consider a possibility of penetration of electrons or ions through grains with subsequent electron emission from the exit side (Svestka and Grün, 1991; Chow et al., 1994).

At a given potential non-spherical grains can carry more electric charges than spherical grains of the same mass; they can have higher charge-to- mass ratio compared to spherical grains. For example, if we keep in the case of conducting prolate and oblate spheroids the potential and mass constant, the total charge increases with the increasing ratio of the major to the minor axis. Calculations of charges on non-spherical grains of more complicated shapes have just started (Svestka et al., 1996).

In a Maxwellian plasma with electron and ion energy distributions at the same temperature, the velocity and, therefore, the charging current of electrons is higher - grains are charged negatively. As a result, ions are attracted and electrons repelled until both charging currents become equal in the magnitude and equilibrium negative charge is reached.

From the conservation of energy and momentum of electrons and ions, and under the assumption that the radius of a grain is much smaller than the Debye length in the surrounding plasma, the collisional cross- section σ can be calculated for an electron or ion with charge q and kinetic energy E (in infinity) to collide with a grain of radius, s, and charge, Q, $\sigma = \pi s^2(1 - qU/E)$, where $U = Q/(4\pi\epsilon_0 s)$ is the surface electrostatic potential of a spherical grain and permittivity $\epsilon_0 = 8.859 \cdot 10^{-12} C/Vm$. If we integrate this cross-section over a Maxwellian velocity distribution of electrons or ions with temperatures T, masses m, and number densities n, we obtain for a charging currents $J = J_0 \exp(-qU/kT)$, for $qU > 0$, and $J = J_0(1 - qU/kT)$, for $qU < 0$, where J_0 is the current on a grain with zero charge (k is Boltzmann constant). In a fully ionized hydrogen plasma with Maxwellian distributions of electrons and ions at equal temperatures T, the equilibrium potential U is independent of the radius of a grain and it is given by $U = -2.51kT$ (Spitzer, 1941).

Interactions of electrons of sufficiently high energies with a grain leads to secondary electron emission. The shape of the secondary electron emission yield $\delta(E)$ (number of electrons released by impact of a single electron of energy E) normalized to the maximum yield, δ_{max}, at the corresponding energy E_{max} is approximated by

$$\delta(E) = 7.4\ \delta_{max} \frac{E}{E_{max}} \exp(-2 \left(\frac{E}{E_{max}}\right)^{\frac{1}{2}})$$

(Sternglass, 1954). δ_{max} is of the order of one for metals and semiconductors, for insulators it is higher (2 to 30, Bruining, 1954). According to Suszcynsky et al. (1992), δ_{max} of various ices equals 2 to 7. E_{max} is equal to 300–2000eV (higher δ_{max} implies higher E_{max}). Threshold energy of primary electrons to produce secondary ones is about 4 to 7eV. It is expected that the energy distribution of secondary electrons is Maxwellian with mean energy of 1 to 5eV. The yield, $\delta(E)$, increases with decreasing grain size (Chow et al., 1993). At energies above a few hundred eV, the yield $\delta(E)$ becomes > 1 and, therefore, the process of secondary electron emission is very important for charging of grains in a plasma of temperature $T \geq 10^6 K$.

UV photons, which are relatively abundant in interstellar space, release photoelectrons from a grain: Photon absorption leads to excitation of an electron to an energy level with enough energy to escape from the grain. Photoemission is very important charging mechanism for grains surrounded by a gas of temperature $T < 3 \cdot 10^5 K$. The photoelectric yield $Y(h\nu)$ is defined as the number of photoelectrons released by absorption of a single UV photon with energy $h\nu$. In the case of very small grains the yield can be strongly enhanced (Schleicher et al., 1994). From several theoretical estimates and observations, Drain (1978) fitted the "average" galactic UV radiation for photons of energies $h\nu = 5 - 13.6 eV$ by $F(h\nu) = (1.658 \cdot 10^6 (h\nu/eV) - 2.152 \cdot 10^5 (h\nu/eV)^2 + 6.919 \cdot 10^3 (h\nu/eV)^3)$ photons $cm^{-2} s^{-1} sr^{-1} eV^{-1}$. As a result, he obtained for a flux of photoelectrons $J_{phe} = 2.4 \cdot 10^6 Q_{abs} cm^{-2} s^{-1}$, for $U < 0$, $J_{phe} = 2.4 \cdot 10^6 Q_{abs} (1 - U/5.6V)^3 cm^{-2} s^{-1}$, for $0 < U < 5.6V$, and $J_{phe} = 0$, for $U > 5.6V$, where the average UV absorption efficiency Qabs is an approximate fit for silicates and graphite.

In the local cloud we assume an ionization degree of 0.1 and average ultraviolet background. From calculations of Draine (1978) follows the equilibrium surface electrostatic potentials of graphite or silicate dust grains of radii 0.01 to 0.1μm will be very low and positive, equal to +0.5 to +1V. In the local bubble with a temperature of about $10^6 K$ and a hydrogen number density of about $0.1 cm^{-3}$ the equilibrium surface electrostatic potentials of dust grains of radii 0.01 to 0.1μm are again positive, higher compared to the local cloud. They depend on the material of a grain. The potentials of graphite grains are equal to about +5V, potentials of silicate grains +12 to +13V, potentials of iron grains about +6V.

Table III

Gyroradius r_{gy} of charged particles in interplanetary and interstellar space. Surface potentials of 5 and $0.5V$, speeds of 400 and $5km/s$, and magnetic fields of 1 and $0.5nT$ have been assumed for interplanetary and interstellar conditions, respectively. $1AU = 1.5 \cdot 10^{11} m$, $1pc = 2 \cdot 10^5 AU$.

Size (μm)	mass (kg)	charge (C)	r_{gy} (AU)
interplanetary condition			
0.1	10^{-17}	$6 \cdot 10^{-17}$	500
1	10^{-14}	$6 \cdot 10^{-16}$	$5 \cdot 10^4$
10	10^{-11}	$6 \cdot 10^{-15}$	$5 \cdot 10^6$
interstellar condition			
0.1	10^{-17}	$6 \cdot 10^{-18}$	120
1	10^{-14}	$6 \cdot 10^{-17}$	$1.2 \cdot 10^4$
10	10^{-11}	$6 \cdot 10^{-16}$	$1.2 \cdot 10^6$

Charged dust couples to the ambient magnetic field. We will compare this effect in two different environments, in the local interstellar medium and in the heliosphere. In interplanetary space dust is charged to approximately $+5V$, and the typical magnetic field at $5AU$ distance is primarily azimuthal with a strength of about $1nT$. In interstellar space we assume a potential of $+0.5V$. The interstellar magnetic field is largely unknown but has been estimated to be $0.5 \pm 0.3nT$ (Holzer, 1989). The speed of the dust particle relative to the magnetic filed is assumed to be about $5km/s$. The gyroradius r_{gy} in a magnetic field B of a particle with charge Q traveling with speed v is $r_{gy} = mv/QB$.

In Table III we have compared the gyroradii for dust particles in interplanetary and interstellar space, respectively. It can be seen that in interplanetary space the resulting gyroradii are huge compared to the extend of the heliosphere ($\approx 100AU$), even for $0.1\mu m$-sized particles. To the contrary, in interstellar space smaller than micron-sized dust particles couple to the magnetic field on a distance scale which is short compared to the extend of the local cloud (a few pc).

Besides coupling to the magnetic field interstellar grains couple to the ambient gas via collisions. The dust–gas–coupling length can be estimated from the distance l a dust particle travels through the gas until it has traversed a column density of gas with a similar mass as that of the particle $l \approx 4\pi\rho_d/3m_g n_g$ (Egger et al., 1995). Using $\rho_d = 1000kg/m^3$ a frictional coupling length of $l \approx 300 \cdot s_\mu pc$ is obtained, where s_μ is the dust size in microns. This length scale is larger than the extend of the local interstellar cloud even for $0.1\mu m$-sized particles. This implies that the gas and dust are not in frictional equilibrium. We conclude that in the diffuse interstellar medium electromagnetic coupling is much more important than frictional coupling. Interstellar dust is rather coupled to the large-scale galactic magnetic field than to the gas clouds in the diffuse interstellar medium. However, the ionized component of the interstellar medium provides the medium for coupling the neutral component to the galactic fields as well. Only in the cores of molecular clouds the gas density is high enough in order to couple directly interstellar dust to the gas.

New observations and updated theoretical considerations will eventually allow us to address important questions concerning the local interstellar medium:

1. Interstellar dust grains in the heliosphere are probes of the local interstellar medium: e.g. their speed dispersion is a function of the coupling to interstellar gas and interstellar magnetic field. Determinations of the speed dispersion may allow us to determine this coupling strength.
2. What is the size distribution of interstellar dust, which particles carry the most mass and what is the role of mutual collisions for the generation of submicron-sized grains?
3. Interstellar dust is considered to contain a major fraction of the condensable interstellar elementar inventory. It is generally believed that the large scale elementar average composition has to have cosmic abundance. How large can be local deviations from the average value?

References

Allamandola L.J., Tielens A.G.G.M., Barker J.R.: 1987, in "Interstellar Processes", ed. D.J. Hollenbach, H.A. Thronson, (D. Reidel Co., Dordrecht), p. 471.

Anders, E. and Grevesse, E.: 1989, *Geochim. Cosmochim. Acta* **53**, 197–214.

Andersen, H.H. and Bay, H.L.: 1981, in "Sputtering by Particle Bombardment I", ed. R. Behrisch (Springer-Verlag, New York), p. 145.

Baguhl, M., Grün, E., Hamilton, D.P., Linkert, G., Riemann, R., and Staubach P.: 1995, *Space Sci. Rev.* **72**, 471–476.

Barlow, M.J.: 1978, *Monthly Not. Roy. Astr. Soc.* **183**, 367.

Berg, O.E. and Grün, E.: 1973, Space Research XIII, (eds. M. J. Rycroft and S. K. Runcorn), Akademie-Verlag, Berlin, pp. 1047–1055.

Bertin, P., Lallement, R., Ferlet, R., and Vidal-Madjar, A.: 1993, *A&A* **278**, 549

Betz, G. and Wehner, G.K.: 1983, in "Sputtering by Particle Bombardment II", ed. R. Behrisch (Springer-Verlag, New York), p. 11.

Bradley, J.P.: 1988, *Geochim. Cosmochim. Acta* **52**, 889–900.

Brownlee, D. E.: 1985, *Ann. Rev. Earth Planet. Sci.* **13**, 147–173.

Bruining, H.: 1954, in "Physics and Applications of Secondary Electron Emission" (Pergamon Press, London).

Burns, J.A., Lamy, P.L., and Soter, S.: 1979, *Icarus* **40**, 1–48.

Cermak, I., Grün, E., and Svestka, J.: 1995, *Adv. Space Res.* **15**, (10)59–(10)64.

Chow, V.W., Mendis, D.A., and Rosenberg, M.: 1993, *J. Geophys. Res.* **98** (A11), 19065.

Chow, V.W., Mendis, D.A., and Rosenberg, M.: 1994, *IEEE Trans. Plasma Sci.* **22**, 179.

Dohnanyi, J.S.: 1969, *J. Geophys. Res.* **74**, 2531–2554.

Dohnanyi, J.S.: 1970, *J. Geophys. Res.* **75**, 3468–3493.

Dorschner, J. and Henning, T.: 1995, *Astron. Astrophys. Rev.* **6**, 271–333.

Draine, B.T.: 1990, in "The Evolution of the Interstellar Medium", ed. L. Blitz, ASP Conf. Ser. 12, p. 193.
Draine, B.T.: 1978, *Astrophys. J. Suppl.* **36**, 595.
Draine, B.T. and Salpeter, E.E.: 1979, *Astrophys. J.* **231**, 77.
Egger, R., Freyberg, M.J., and Morfill, G.E.: 1995, *Space Sci. Rev.*
Fujiwara, A., Kamimoto, G., and Tsukamoto, A.: 1977, *Icarus* **31**, 277–288.
Grün, E., Zook, H. A., Fechtig, H., and Giese, R. H.: 1985a, *Icarus* **62**, 244–272.
Grün, E., Zook, H. A., Fechtig, H., and Giese, R. H.: 1985b, in "Mass Input into and Output from the Meteoritic Complex in Properties and Interactions of Interplanetary Dust", eds. R.H. Giese and P. Lamy, Reidel, Dordrecht, pp. 411–415.
Grün, E., Zook, H. A., Baguhl, M., *et al.*: 1993, *Nature* **362**, 428–430.
Grün, E., Gustafson, B.Å.S., Mann, I., Baguhl, M., Morfill, G. E., Staubach, P., Taylor, A., and Zook, H. A.: 1994, *Astron. Astrophys.* **286**, 915–924.
Hauser, M.G., Gillett, F.C., Low, F.J., *et al.*: 1984, *Astrophys. J.* **278**, L15–L18.
Holzer, T.E.: 1989, *Ann. Rev. Astron. Astrophys.* **27**, 199.
Humes, D.H.: 1980, *J. Geophys. Res.* **85**, 5841.
Jenniskens, P. and Desert, F.X.: 1993, *Astron. Astrophys. Suppl.* **106**, 39.
Jones, A.P. and Tielens, A.G.G.M.: 1994, in "The Cold Universe", eds. T. Montmerle, C.J. Lada, I.F. Mirabel, J. Tran Than Van (Edition Frontiers, Gif-sur-Yvette) p. 35.
Leinert, Ch., Hanner, M., Richter, I., and Pitz, E.: 1980, *Astron. Astrophys.* **82**, 328–336.
Leinert, Ch. and Grün, E.: 1990, in "Physics of the Inner Heliosphere I", eds. R. Schwenn, and E. Marsch, Springer-Verlag, Berlin, pp. 207–275.
Levasseur-Regourd, A.-Ch. and Dumont, R.: 1980, *Astron. Astrophys.* **84**, 277–279.
Low, F. J., Beintema, D. A., Gautier, T. N., *et al.*: 1984, *Astrophys. J.* **278**, L19–L22.
Lallement, R., Bertin, P., Ferlet, R., Vidal-Madjar, A., and Bertaux, J.L.: 1994, *Astron. Astrophys.* **286**, 898.
Mathis, J.S.: 1993, *Rep. Prog. Phys.* **56**, 605.
Mathis, J.S., Rumpel, W., and Nordsiek, K. H.: 1977, *Astrophys. J.* **217**, 425.
McKee, C. F. and Ostriker, J. P.: 1977, *Astrophys. J.* **218**, 148.
Morfill, G.E. and Grün, E.: 1979, *Planet. Space Sci.* **27**, 1283–1292.
Morfill, G. E., Grün, E., and Leinert, Ch.: 1986, in "The Sun and the Heliosphere in Three Dimensions", ed. R. G. Marsden, Reidel, Dordrecht, pp. 455–474.
Schleicher, B., Burtscher, H., and Siegmann, H.C.: 1994, *Applied Phys. Lett.* **63** (9), 1191.
Sigmund, P.: 1969, *Phys. Rev.* **184**, 383.
Sigmund, P.: 1981, in "Sputtering by Particle Bombardment I", ed. R. Behrisch (Springer-Verlag, New York), p. 9.
Spitzer, L., Jr.: 1941, *Astrophys. J.* **93**, 369.
Sternglass, E. J.: 1954, Sci. Paper 1772, Westinghouse Res. Lab., Pittsburgh.
Suszcynsky, D.M. and Borovsky, J.E.: 1992, *J. Geophys. Res.* **97** (E2), 2611.
Svestka, J. and Grün, E.: 1991, in "Origin and Evolution of Interplanetary Dust", eds. A.C. Levasseur-Regourd and H. Hasegawa (Kluwer Acad. Publ., Dordrecht), p. 367.
Svestka, J., Cermak, I., and Grün, E.: 1993, *Adv. Space Res.* **13**(10), 199.
Svestka, J., Auer, J., Grün, E., and Baguhl, M.: 1996, Proc. IAU 150, in press.
Sofia, U.J., Cardelli, J.A., and Savage, B.D.: 1994, *Astrophys. J.* **430**, 650.
Southworth, R.B. and Sekanina, Z.: 1973, in "Physical and Dynamical Studies of Meteors", NASA CR-2316.
Taylor, A.D., Baggaley, W.J., and Steel, D.I.: 1996, *Nature* **380**, 323–325.
Tielens, A.G.G.M., McKee, C.F., Seab, C.G., and Hollenbach, D.J.: 1994, *Astrophys. J.* **431**, 321.
Whipple, F. L.: 1967, in "The Zodiacal Light and the Interplanetary Medium" (J. L. Weinberg, ed.), NASA SP-150, U. S. Govt. Prt. Off., Washington, D. C., pp. 409–426.
Whittet, D.C.B.: 1992, in "Dust in the Galactic Environment" (Inst. of Physics Publ., Bristol) p. 193.
Witte, M., Rosenbauer, H., Banaskiewicz, M., and Fahr, H.: 1993, *Adv. Space Res.* **13**, (6)121–(6)130.

RELATIONS BETWEEN ISM INSIDE AND OUTSIDE THE HELIOSPHERE

R. LALLEMENT
Service d' Aéronomie du CNRS, Verrières le Buisson, France

Abstract. Thanks to remarkable new tools, such as the Goddard High Resolution Spectrograph (GHRS) on board the HST and the EUVE spectrometer on the interstellar side, and Ulysses particle detectors on the heliospheric side, it is possible now to begin to compare abundances and physical properties of the interstellar matter outside the heliosphere (from absorption features in the stellar spectra), and inside the heliosphere (from "in situ" or remote detection of the interstellar neutrals or their derivatives, the pick-up ions or the Anomalous Cosmic Rays detected by the two Voyager spacecraft).

Ground-based and UV spectra of nearby stars show that the Sun is located between two volumes of gas of different heliocentric velocities V and temperatures T (see also Linsky et al, this issue). One of these clouds has the same velocity (V= 25.6 km s^{-1} from λ= 255°and β=8°) and temperature (6700 K) as the heliospheric helium of interstellar origin probed by Ulysses, and is certainly surrounding our star (and then the Local Interstellar Cloud or LIC). This identification allows comparisons between interstellar constituents on both sides of the heliospheric interface.

Ly-alpha background data (absorption cell and recent HST-GHRS spectra) suggest that the heliospheric neutral H velocity is smaller by 5-6 km s^{-1} than the local cloud velocity, and therefore that H is decelerated at its entrance into the heliosphere, in agreement with interaction models between the heliosphere and the ISM which include the coupling with the plasma. This is in favor of a non negligible electron density (at least 0.05 cm^3). There are other indications of a rather large ionization of the ambient ISM, such as the ionization equilibrium of interstellar magnesium and of sodium. However the resulting range for the plasma density is still broad.

The heliospheric neutral hydrogen number density (0.08 - 0.16 cm^{-3}) is now less precisely determined than the helium density (0.013-0.017 cm^{-3}, see Gloeckler, Witte et al, Mobius, this issue). The comparison between the neutral hydrogen to neutral helium ratios in the ISM (recent EUVE findings) and in the heliosphere, suggests that 15 to 70% of H does not enter the heliosphere. The comparison between the interstellar oxygen relative abundance (with respect to H and He) in the ISM and the heliospheric abundance deduced from pick-up ions is also in favor of some filtration, and thus of a non-negligible ionization.

For a significant ISM plasma density, one expects a "Hydrogen wall" to be present as an "intermediate" state of the interstellar H around the interface between "inside" and "outside". Since 1993, the two UVS instruments on board Voyager 1 and 2 indeed reveal clearly the existence of an additional Ly-alpha emission, probably due to a combination of light from the compressed H wall, and from a galactic source. On the other hand, the decelerated and heated neutral hydrogen of this "H wall" has recently been detected in absorption in the spectra of nearby stars (see Linsky, this issue).

1. Velocity and temperature inside and ouside the heliosphere

1.1. Velocity and temperature of the Local Cloud

The study of the local interstellar cloudlets by means of stellar spectroscopy requires high resolution and high sensitivity. This was up to the Hubble Space Telescope

Space Science Reviews **78**: 361–374, 1996.

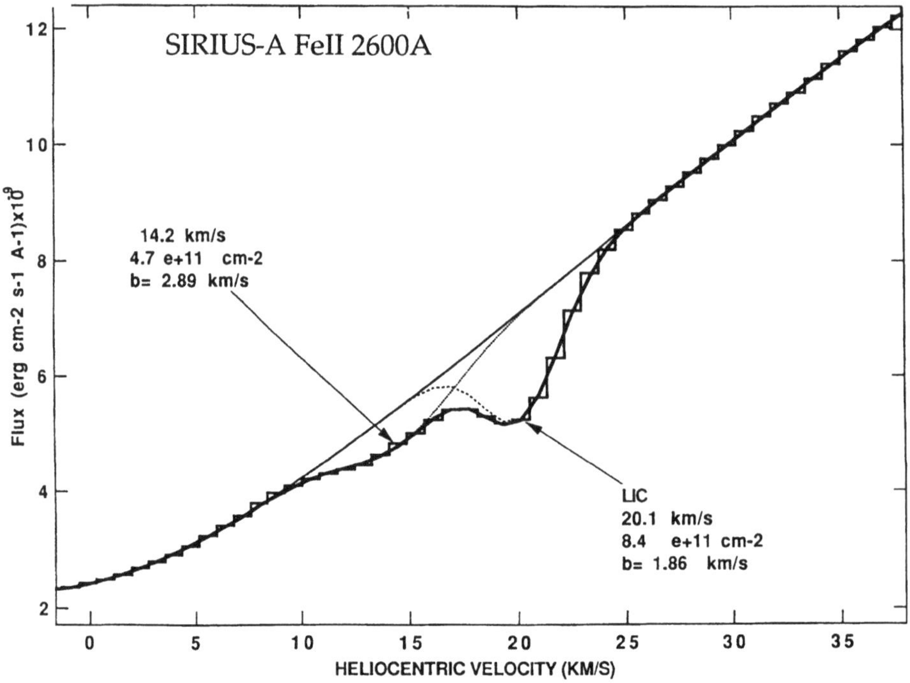

Figure 1. GHRS Ech-A spectrum of Sirius around the FeII 2600 Å resonance line. Two clouds are detected along the 2.7 pc L-O-S, the LIC and a companion cloud. b is the sum of thermal and turbulent widths. b_{LIC} =1.86 km/s (apparent temperature 11,000K) corresponds to T=7000K plus 1.5 km/s turbulence

(HST) launch achievable only from ground. Unfortunately, interstellar lines in the visible are weak. Nonetheless, several masses of gas have been detected very close to the Sun, and the Doppler Triangulation Method (DTM) was successfully applied to two nearby parcels of gas. The DTM is used to derive three dimensions motions and is based on a very simple principle: assuming a cloudlet is moving like a solid body, the Doppler shifts of its absorption lines in several target stars spectra, are simply the projections of the velocity vector onto the star directions, and then they obey simple relationships.

First, ground based spectra obtained at the Observatoire de Haute-Provence in France with the Aurelie spectrograph (same observatory as for the recent first detection of an extra-solar planet by Mayor and Quéloz (1995) from Observatoire de Geneve, with a spectrograph built by the same french team...), have allowed the derivation of the 3D motion of a very close mass of gas, called initially the AG (for anti-galactic) cloud detected in the second and third galactic quadrants (Lallement and Bertin, 1992). This cloudlet flows with respect to the Sun at 25.7 km s^{-1} towards the galactic longitudes l=186 °, b= 16 ° (λ= 75 °, β= -8°). Independently and earlier, ESO (European Southern Observatory) La Silla and CFHT (Canada-France-Hawaii Telescope) observations had provided a velocity vector for another very close mass of gas (called the G cloud) in the first and fourth quadrant (Lallement & al, 1990). This velocity vector is only slightly different from the previous vector (29.4 km s^{-1} heliocentric velocity modulus and less than 3 degrees of difference in direction).

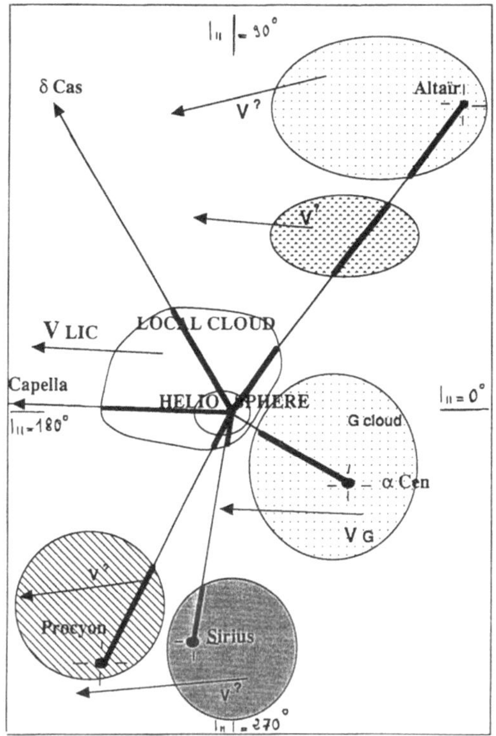

Figure 2. Schematic view of the cloudlets in the Sun vicinity. It is assumed that the Sun is embedded in the 26 km/s cloudlet (AG cloud) in agreement with Ulysses neutral helium measurements. The heliosphere is represented dilated about two hundred times. The Sun is close to the Local Cloud boundary in the direction of α Cen. This star shows only one main absorption at the G cloud velocity. The 3D velocity vectors are known for the LIC and G only.

During the last four years, the GHRS on board the HST has been the best tool for local matter studies, because it opens the access to the strong UV lines. The GHRS spectra, of unprecedented quality, reveal absorption lines in the lines-of-sight to all the closest stars. An exemple is given in Fig. 1, showing the detection of two lines towards Sirius.

The DTM, when applied to the whole set of visible and UV spectra, shows a perfect agreeement in all directions for the aforementioned velocity vector AG (Lallement et al, 1995). As one would expect for the cloud surrounding the Sun, for each star one of the observed lines has a Doppler shift at the exact projection of this velocity vector (e.g., in Fig. 1, the redshifted absorption is due to the AG cloud, and the other one to a second cloud closer to Sirius). There is, however, one exception to this excellent agreement. Surprisingly this exception is for α Cen, the closest target star. Towards this star, the projected velocity of the apparently unique cloud detected in the MgII and FeII lines is slightly larger than predicted, by about 2 km s^{-1}. This Doppler shift is in fact in perfect agreement with the second close cloud, the "G" cloud. At least a part of the G cloud is thus within

1.3 pc. Now, is the Sun inside the AG or the G cloud? If the Sun is embedded in the AG cloud detected in all directions (except α Cen), then there is necessarily gas at the AG velocity along the Sun- α Cen line-of-sight (L-O-S). Because such a gas is undetected, it means that the Sun is extremely close to the boundary of the cloud in the direction of α Cen, and the corresponding column-density is correspondingly small. A small quantity of gas at the AG velocity can indeed be hidden in the main faster component G. Estimates give a very low upper limit of about 2. 10^{16} cm^{-2} as the maximum hidden column of H at the AG velocity (Lallement & al, 1995). This implies that the AG boundary is distant from the Sun by a fraction of pc only. If the Sun were in the faster G cloud, then we would have to face the same problem, i.e. why is the fastest gas not detected in any of the other targets? A tempting way to solve the problem is to consider that the differences in the velocities of the LIC and the G cloud are so small (3-4 km s^{-1} difference between the two speeds) that we probably see a unique mass of gas with an internal velocity gradient. However, first, the Doppler shifts suggest a clear separation into two entities, and, second, these two masses of gas have different temperatures and abundances. The most conclusive data on the discrimination between the two LIC and G clouds are indeed the recent observations of α Cen of Linsky & Wood, 1996, which have demonstrated that the temperature of the faster G cloud (5400K) is definitely lower than the LIC temperature of 6700K they have mesured for the LIC towards Procyon and Capella (Linsky & al, 1993,1995). Moreover, it appears that relative abundances also vary between the two media.

A schematic view of the local "galactography" is drawn in Fig 2 (galactic plane section). Here the number of clouds and their relative positions are realistic, but one does not know whether or not these cloudlets are really separated by empty space (in fact very tenuous hot gas) or if they belong to a unique extended volume, with velocity and temperature jumps at discontinuity surfaces responsible for the apparent existence of different cloudlets. Although there is no pressure equilibrium between the hot and the warm gas, the former description (separate cloudlets) is generally prevailing in the interstellar community.

1.2. VELOCITY AND TEMPERATURE IN THE HELIOSPHERE

1.2.1 Helium velocity and temperature

The neutral helium is negligibly coupled to the low density plasma of the interstellar medium and of the heliospheric boundary at spatial scales of the order of the size of the heliosphere. This means that the interstellar helium flow keeps its initial properties when entering the heliospheric interface, and is perturbed only by the supersonic solar wind and the solar UV light when approaching within a few A.U. from the Sun. The GAS experiment on board Ulysses has measured in situ the interstellar neutral helium. The initial (i.e. before perturbations by the Sun) velocity and temperature of the flow have been determined by Witte et al (1993) and are in excellent agreement with the velocity vector and the temperature of the AG cloud. The G cloud velocity and temperature are outside the possible ranges for these parameters quoted by Witte. This implies almost certainly that the Sun is located in the AG cloud, and in what follows the AG cloud will be identified with the LIC (with the Sun being located very close to the boundary of the LIC in the direction of α Cen, see 1.1). The Ulysses results on helium do not agree very well with previous determinations of V and T from helium resonance glow (Dalaudier et al, 1985; Chassefiere et al, 1988). The reasons for the discrepancies are possibly linked to the solar line (HeI 584 A) which illuminates He atoms by resonance scattering, because the helium emission is very sensitive to the width, the shape and the shift of the solar line, or to the lack of a precise enough representation of the electron impact ionization. Further work is needed to solve these problems, and the SOHO/SUMER measurements of the solar helium line profile should help significantly.

1.2.2 Hydrogen velocity and temperature

The interplanetary H Lyα emission produced by resonance scattering of the solar line, when observed from the earth's orbit, is almost entirely coming from within 50 A.U. of the Sun, essentially due to the solar flux decrease with heliocentric distance. The Doppler shifts and the widths do reflect directly the bulk motion and the velocity distribution. Ignoring the perturbations at large distances,the H flow and its emission pattern can be represented at first order by the classical models which assume a homogeneous flow "at infinity", characterized by a velocity V and a temperature T . For these models, infinity means at about 50 A.U., i.e. outside the ionization cavity carved around the Sun, i.e. before any significant perturbation due to the Sun and the supersonic solar wind. In the framework of such a "classical" hot model, hydrogen cell data at the Lyα resonance line, recorded on board the Prognoz 5 and 6 satellites, were best fitted with a velocity of about 20 km s^{-1} and a transverse (i.e. perpendicular to the flow direction) temperature of about 8000K, as also did an HST-GHRS spectrum recorded during an observation of planet Mars (Lallement et al, 1993).

In light of the helium data, this suggests a modification of the velocity distribution of the H flow with respect to interstellar conditions. Neutral hydrogen is strongly coupled to the plasma via charge-exchange interactions, and one could expect such modifications if the LIC gas is partially ionized.

Recently, spectra of the Ly-alpha emission recorded by the GHRS (P.I. J. Clarke) have provided the best quality spectra of the H glow ever obtained, after Copernicus and IUE (Adams and Frisch 1977, and Clarke et al, 1984, 1995). The observations combine high sensitivity, high resolution, and a good separation of the earth and interplanetary emissions. Spectra were recorded for the upwind and crosswind (at right angle with the flow) directions, and preliminary comparisons with classical models have been done. For those two directions, the emitting gas is rather close to the Sun and corresponds to atoms which have traveled close to the stagnation line. For the crosswind spectrum, it has been possible to find a reasonable fit to the data with a classical model of an initial homogeneous flow (i.e. without any coupling with the plasma) at a temperature of about 8000K. This is in agreement with an absence of velocity dispersion in this direction and with Prognoz 5 and 6 H cell measurements . For the upwind spectrum (Fig. 3), it appears impossible to fit the data with an homogeneous flow at an initial velocity of 26 km/s (identical to the LIC and helium velocity), unless the deceleration due to the radiative pressure is very large, which is not in agreement with the crosswind data. The data imply a lower velocity "at infinity" of about 19-20 km/s (Lallement et al, 1996). This is in agreement with Fig. 4, which represents Baranov's two-shock model results (Baranov and Malama, 1993) . This figure displays the radial bulk velocity of the H flow along the Sun-upwind axis for an interstellar plasma density of 0.1 cm^{-3}. This bulk velocity represents the mean flow of both the primary (initial unperturbed

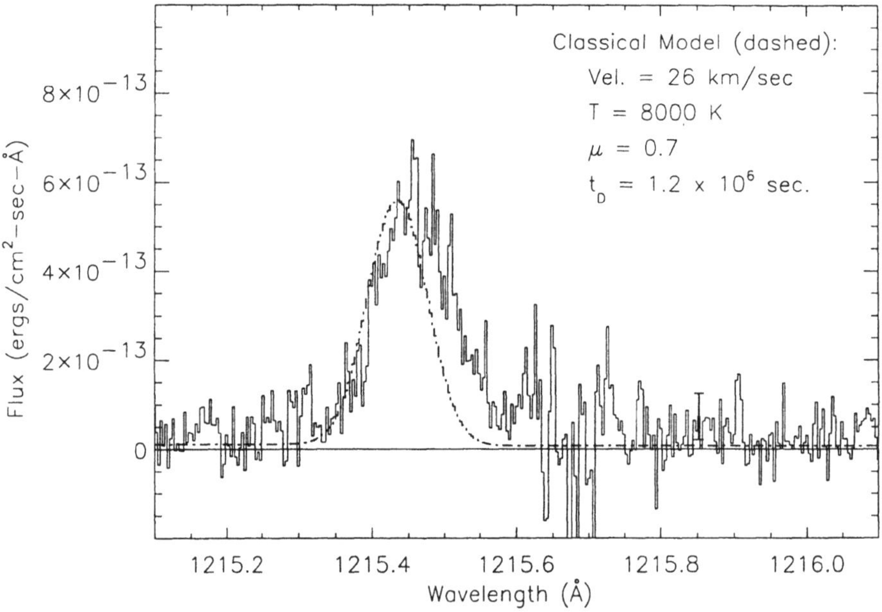

Figure 3. HST/GHRS spectrum of the interplanetary-interstellar Lyα emission. A classical model for an initial velocity of 26 km/s has been superimposed. The observed spectrum is shifted towards lower heliocentric velocities. The adjustment requires an initial bulk velocity of 19-20 km s^{-1}, or an improbable very large value of the radiation pressure. The observed line width is also apparently broader than the model predicts. (Courtesy J.T. Clarke and collaborators)

atoms) and secondary atoms (resulting from charge-exchange between the neutrals and the decelerated protons). In the region between 50 and 100 A.U. from the Sun, before the flow begins to be perturbed by the Sun, the bulk velocity in the model is smaller than the initial velocity of 26 km/s by about 5-6 km/s. The model also predicts a significant velocity dispersion in this direction, due to the differentiation into two flows, primary and secondary atoms , while perpendicular to the flow, the velocity dispersion is negligible. The higher the interstellar plasma density, the larger the proportions of secondary atoms and the bulk velocity decrease. At the same time, for an observer located close to the Sun, the apparent density "at infinity" (at, say, 50-70 AU), is smaller than the true interstellar density, due to the exclusion from the heliosphere of a fraction of the H atoms (see 2.1.1).

2. Densities

Interstellar number densities can be obtained by assuming the measured column-density is homogeneously distributed along the line-of-sight. This can be a very crude approximation, because one does not know where the clouds boundaries are, especially when there are two or more components. Relative abundances, however, deduced from column-density ratios, are excellent starting points for comparisons.

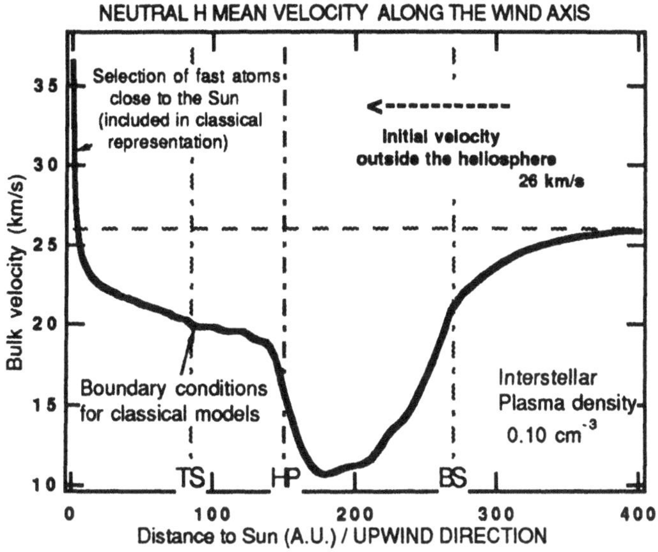

Figure 4. 2-shock model results of Baranov and Malama (1993): combined primary and secondary H atoms radial bulk velocity along the Sun-Wind axis on the upwind side.

2.1. Neutral species

2.1.1 Helium and Hydrogen

The neutral helium number density in the heliosphere has been recently determined to be very likely around 0.013-0.017 cm^{-3} (see Gloeckler, Witte et al, Mobius in this issue). Because helium is theoretically unchanged when flowing through the interface, this is also the interstellar number density. On the other hand, recently Dupuis et al (1995) have derived from EUVE observations of nearby white dwarfs, below the H ionization threshold at 91.2 nm, that the neutral H to neutral He ratio is 12-18.5 in the nearby medium, with the important result that helium is always equivalently or more ionized than H. This and the aforementioned helium measurements strongly suggest that in the Local Cloud just outside the heliosphere the neutral H density is in the range between 0.18 and 0.31 cm^{-3}.

Neutral H density measurements inside the heliosphere have been derived up to now from Lyα glow analysis, but, even when calibration-independent estimates are possible, they unfortunately depend on the choices of solar parameters. In particular, when modelling the Lyα emission, for the same number of detected Lyα photons, the lower the actual ionization, the lower the derived density. The latitudinal dependences of the solar wind charge-exchange rate and the solar Lyα flux are not known with good precision. In addition, temporal variations are difficult to take into account (Rucinski & Bzowski, 1995). The most probable density is in the interval (0.08-0.16 cm^{-3}), with the high values having been recently derived from radiative transfer calculations close to the Sun (0.16 cm^{-3}, Quemerais et al, 1994) and at large distance from the Sun (0.15 cm^{-3}, Quemerais & al,1995,1996), and the low value from recent Ulysses measurements of pick-up protons (Gloecker

& al, 1993). A compilation of the density estimates has been given in Quemerais & al, 1994. As it can be seen, unlike helium, there is apparently no convergence yet for the hydrogen density. However, it must to be noted that the high value of Quemerais & al, 1994 is probably overestimated, due to an overestimate of the solar wind ionization rate. As a matter of fact, Ulysses measurements have revealed that the proton fluxes from above 25 °solar latitude are rather small (Phillips & al, 1995), and Summanen et al (1996) have shown that the subsequent average ionization of H is significantly lower than previously predicted. As a consequence the derived density could be lower by 20% (n_H= 0.13 cm^{-3}) or more. It must also be noted that during the last years the density has been revised upward, to correct for an earlier use of a very low ionization rate, incompatible with solar wind data (Quemerais & al, 1994, Summanen & al, 1996). Now, a fraction of this correction appears to be unnecessary, and we go back to smaller values of the density. The Voyager value derived at large distance, which is also in the upper part of the interval, is almost independent of the solar parameters, but it may be somewhat larger than the inner heliosphere value due to the "hydrogen wall" density increase (see 3.), and can be considered as an upper limit. It is a pity that after so many measurements such an important parameter is so uncertain. Hopefully extensive pick-up hydrogen measurements will help to reduce the uncertainty.

The comparison between the inside density (lower than 0.15 cm^{-3}) and the interstellar density (larger than 0.18 cm^{-3}), suggests a density decrease along the heliospheric interface crossing, ranging from a small (15 %) to a significant (70 % if using the extreme values) effect. According to interface models, either subsonic or supersonic, one expects such a filtration at the heliospheric interface (1.2.2).

2.1.2 Oxygen vs Hydrogen, Oxygen vs Helium, Carbon vs Oxygen

A species of particular interest for its coupling with the plasma is atomic oxygen. The penetration of oxygen has been calculated recently by Fahr & al, (1995) and Izmodenov & al (1996). These authors have shown that taking into account the reverse charge-exchange between O ions and neutral H implies a filtration of oxygen much weaker than previously thought (Fahr, 1991). In Fig. 5 are displayed the neutral oxygen densities for three directions (upwind, crossind, downwind) resulting from a Monte-Carlo simulation of the O and O ions flows through a Baranov two-shock self-consistent interface (Izmodenov & al, 1996). It can be seen that for an interstellar electron density of 0.1 cm^{-3}, about 70% of the initial neutral oxygen penetrates close to the Sun. This number however may be slightly decreased if electron impact is taken into account (Rucinski, 1996). The recent pick-up ion measurements of Geiss & al, 1995 with the SWICS instrument on board Ulysses can be compared with the (also recent) interstellar oxygen abundances measured with the GHRS (Linsky & al, 1993) in the LIC towards Capella. More precisely, hydrogen is taken here as a reference for both inside and outside the heliosphere. O and H atom LIC column-densities have been derived by Linsky towards Capella, providing the (neutral O/neutral H) ratio, which at equilibrium is nearly equal to the (O ion/H ion) ratio. Assuming the (low) value of 0.14 cm^{-3} for H in the LIC (see discussion above), and using the Linsky & al interstellar relative abundances, the Izmodenov & al model predicts an O ion flux at about 5 A.U. higher by 30% than the measurements (Gloeckler & al, 1993), for a plasma density of 0.1 cm^{-3}. This suggests that the filtration of O is probably higher than in the chosen case of a plasma density of 0.1 cm^{-3}. If the neutral H density in the LIC is larger,

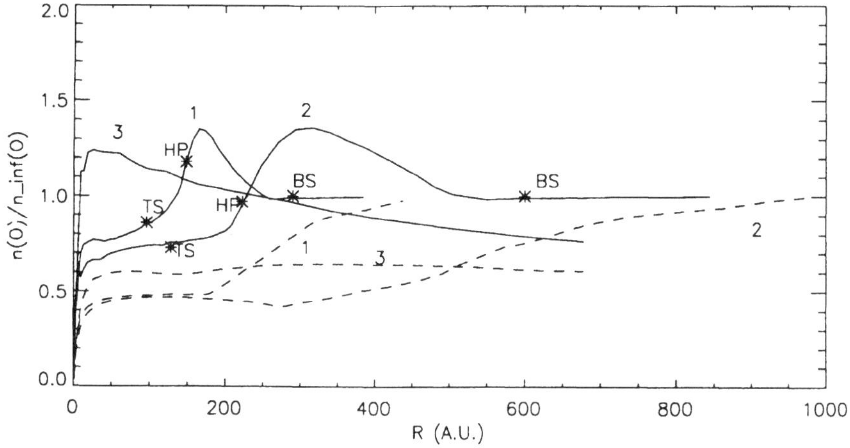

Figure 5. Neutral oxygen densities from a Monte-Carlo simulation of the O and O^+ ions flow through a Baranov-type interface (Izmodenov & al, 1996). Dashed lines correspond to the absence of reverse charge-exchange and are shown for comparison. Labels 1, 2,3 correspnd to upwind,crosswind and downwind directions respectively.

as is likely, and the oxygen density varies proportionally, then the Ulysses data imply an even larger filtering. This is in agreement with the oxygen/helium ratio measured by Geiss et al, 1994. For reasonable values of the ionization rates, the oxygen/helium ratio in the heliosphere (Geiss & al, 1994) is found to be much smaller than the solar system O/He ratio (Fahr & al,1995). Now, comparing with the more appropriate interstellar O/He ratio (derived from both Linsky (H/O) and EUVE (H/He) measurements), instead of the solar system ratio, Izmodenov & al (1996) suggest that there is a significant filtration of O, by a factor of 2 or more. This is more than the model prediction for n_e=0.1 cm^{-3}. As a conclusion, the heliospheric oxygen abundance does not preclude interstellar electron densities larger than (if electron impact ionization is neglected) or of the order of (when including electron impact) 0.1 cm^{-3}. Finally, the relative abundances of heliospheric carbon and oxygen deduced from the ACR measurements by Voyager (e.g. Cummings & Stone, 1990) have been analysed by Frisch (1994,1995) as evidence for a significant electron density in the ISM, using ionization equilibria of these species in the ISM. This study makes the assumption that O and C are not substantially filtered, or that they are equally filtered. The Fahr & al, (1995) and Izmodenov & al (1996) calculations show that the former assumption is valid for oxygen. However, there is still a lack of carbon flow simulations, to refine these carbon-oxygen comparisons and the deduced electron density.

2.2. Electron density in the LIC

The electron number density in the LIC can be derived from abundances ratios of ions of different ionization states. One of the commonly used couples is neutral magnesium (MgI) and singly ionized magnesium (MgII). As a matter of fact, neutral magnesium is formed through recombination of Mg^+, and destroyed mainly by photoionization. If equilibrium conditions prevail, the electron density is derived as a function of the photoionization rate (rather well measured for Mg^+) and the gas temperature (also now rather well constrained) (Frisch, 1995). However, the MgI

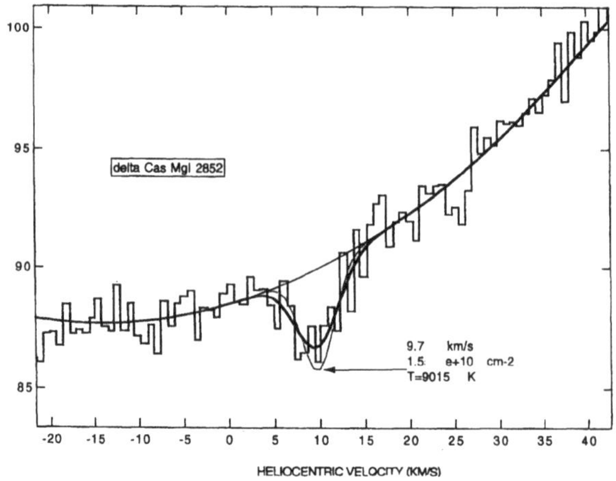

Figure 6. The very small magnesium absorption line detected toward δ Cas. The comparison of the neutral magnesium column-density with the (independently measured) singly ionized magnesium column-density allows to infer the average electron density in the Local Cloud.

line is extremely small for such small diffuse clouds, and very difficult to detect. The first target which was used for such a purpose is Sirius. A very small neutral magnesium line has been detected at the LIC Doppler shift, allowing a comparison between neutral and singly ionized magnesium abundances in the LIC. Contrary to previous estimates, the derived electron density was found to be surprisingly large, of the order of 0.2 to 0.6 cm^{-3} for temperatures of the order of 7000-10000K (Lallement & al, 1994). Electron densities higher than about 0.3 cm^3 are precluded, because if the ambient density were so huge, the heliosphere would be small enough for the Voyager to have already crossed the termination shock. Such a high ionization is also difficult to explain if one considers the known sources of ionization (Cox and Reynolds, 1987, Cheng and Bruhweiler, 1990, Jelinsky et al, 1994, Vallerga and Welsh, 1995). But note that the helium ionization is also difficult to explain.

A small neutral magnesium line has also been detected in the spectrum of δ Cas as shown in Fig 6. The derived Mg/Mg^+ ratio of about 400 is a factor of two larger than the Sirius ratio, implying an electron density smaller by a factor of two, and suggesting an ionization gradient inside the LIC. Such a gradient is certainly plausible, according to the models, but still the inferred ionization degree is larger than expected, because the derived range for the electron density, $n_e = 0.11 - 0.50cm^{-3}$, is very high..

An independent way to estimate the electron density is the use of the sodium ionization equilibrium. It can be obtained by comparing neutral sodium (deduced from ground-based spectra) and singly ionized sodium (deduced from neutral H data, because the sodium depletion is rather well constrained and almost all the sodium is singly ionized). Such a study yields an electron density of 0.04-0.25 cm^{-3} towards δ Cas, i.e. significantly smaller than the density derived from magnesium (Lallement and Ferlet, 1996). These different attempts to measure the electron density have a common result, which is the non-negligible ionization of the LIC. However, the error on the local plasma density is still rather high, even if one

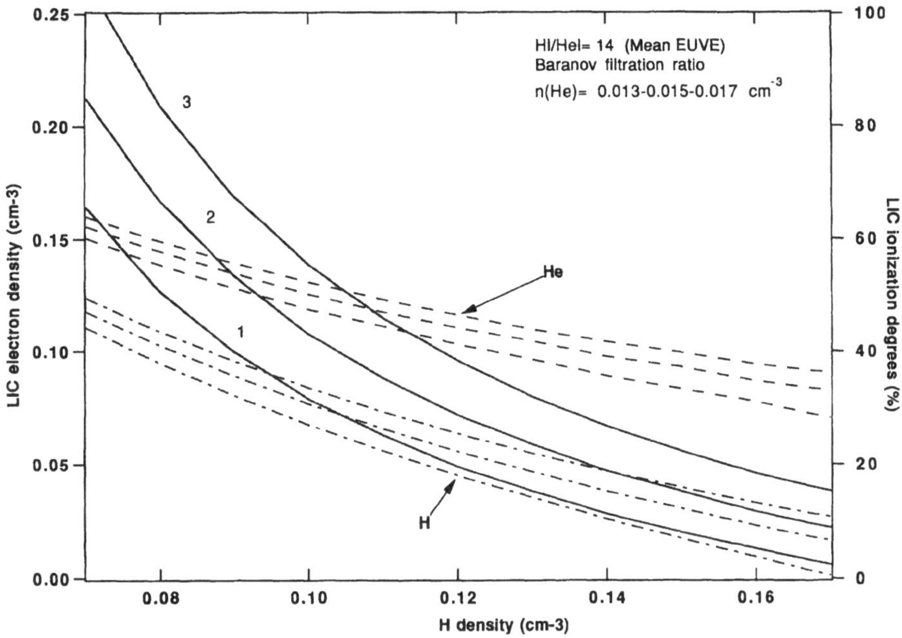

Figure 7. The electron density (solid curves) and ionization degrees of H (dot-dashed curves) and He (dashed curves) in the LIC assuming the mean EUVE neutral H/neutral He ratio of 14 and the Baranov filtration factor dependence for neutral H, as a function of the heliospheric neutral H density. Three values of the neutral helium heliospheric densities are considered: from upper to lower curves, n(He)= 0.013, 0.015, 0.017 cm^{-3}.

considers optimistically that neutral H deceleration and LIC measurements favour a common value around 0.1 cm^{-3}.

Playing with the ranges for the hydrogen and helium number densities derived from neutral detection, pick-up ions and the Ly-alpha glow, the EUVE HI-HeI ratio, and the Baranov filtration factor (interpolated from models for 0.05, 0.1 and 0.21 e_{-} cm^{-3}), one can connect the plasma density and the H and He ionization degrees in the ISM to the heliospheric densities. Fig. 7 shows such estimates. It can be seen that the precise knowledge of the H density in the inner heliosphere (in abscissa) is crucially missing. It would allow us to distinguish between the medium ionization (0.03 e^{-} cm^{-3}) and the high ionization (0.22 e_{-} cm^{-3}) cases. On the other hand, assuming $n^{e} = 0.1$ cm^{-3} as suggested above, one infers minimum ionization degrees of 33 and 44% respectively for H and He, which is quite large compared to the values measured by Dupuis et al (1995) towards nearby white dwarfs with the EUVE.

3. Between inside and outside: the hydrogen "wall"

Since 1993, a series of Voyager UVS observations have been scheduled to search for the signature of the layer of compressed neutral hydrogen at the periphery of the heliosphere predicted by theoretical calculations taking into account neutral-plasma coupling in this area. There is definitely an excess of Lyα intensity from

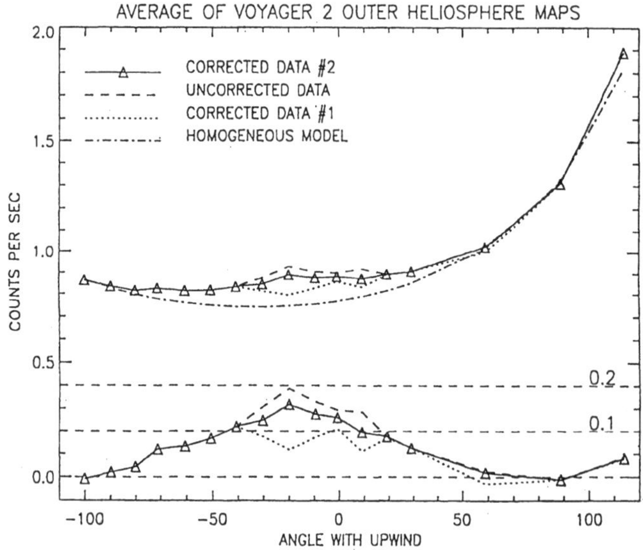

Figure 8. Ly-alpha intensities recorded in 1993 and 1994 by the Voyager 2 UVS instrument as a function of the angle between the line-of-sight and the upwind direction, compared with theoretical intensities for a density distribution corresponding to the "classical" model without interface. There is an excess of emission centered on the upwind region, which by chance is within 18 deg from the galactic center direction. The bottom curve is the difference between the data and the model.

the upwind side (the "nose" of the heliosphere), as shown in Fig 8. However, it is at present clear that a fraction of the detected excess at Lyα has a galactic origin (Quemerais & al, 1995, 1996). As a matter of fact, by chance, the galactic center is within a few degrees from the direction of the heliocentric motion of the interstellar gas. More specifically, radiative transfer models do predict a part of the excess, but not the observed pattern with a marked "bump" in the direction of the galactic center. Multiple scattering would smear out such a "bump". Incidentally, the galactic longitude range of the "bump" does correspond to the Sagittarius spiral arm, but whether or not this is the source of the emission remains to be demonstrated. Fortunately, the two sources should be disentangled in the future. The solar flux is expected to decrease at Ly-alpha by almost a factor of two during the next four years. The H wall Lyα excess will vary in proportion, because it is illuminated by the Sun, while the galactic emission will remain constant. Hopefully future scans will allow us to measure the respective contributions of the two sources. If so, the "height" of the "wall" should be a valuable diagnostic of the location of the interface and the plasma density.

The H "wall" has indeed been detected, not in emission but in absorption, in the spectrum of αCen by Linsky and Wood (1996). This new findings is detailed in this volume by Linsky.

4. Conclusions

Our Sun is located between two interstellar "cloudlets" of slightly (but definitely) different velocities (25.7 and 29.4 km s^{-1} heliocentric velocities), temperatures (6700K and 5400 K) and abundances. According to the Ulysses interstellar neutral helium direct measurements, the Sun is located on the warmer/slower side, i.e. the LIC is the first "cloudlet". If we transform the neutral H (or other elements) columns along the lines-of-sight to the nearby stars, into a distance to the surface of the cloud by using the H number density, then the distance already covered by the Sun in the LIC is found to be about 2 to 3 pc, for a travel time of 75,000 to 110,000 yrs. The maximum remaining time in the LIC before entering the colder cloud , or maybe hot gas separating the two cloudlets, can be estimated from the maximum column of LIC gas possibly hidden in the G cloud line in the spectrum of α Cen. It seems very small, between 1100 and 1600 yrs assuming LIC Mg and Fe abundance, (a distance of around 1/30 pc), but this is an upper limit. It may be interesting to estimate what kind of consequences leaving the LIC would have for our heliosphere and the cosmic ray fluxes reaching our planet in the two different situations (colder and faster cloud, or tenuous hot gas).

According to the ionization balance of magnesium on the one hand (HST), and calcium-sodium on the other (ground telescopes and HST), the local clouds seem to be significantly ionized (interstellar electron density of the order of 0.1 cm^{-3} or more), in agreement with recent EUVE findings of an non-negligible helium ionization. An interesting finding of the EUVE is that helium is more highly ionized than hydrogen, with an HI/HeI ratio of 12-18. Making use of this observed range, the neutral hydrogen number density in the LIC should be within the interval 0.18-0.31 cm^{-3}, i.e. 12 to 18 times the neutral helium density of 0.013-0.017 cm^{-3} (derived from neutral and pick-up helium measurements by AMPTE and Ulysses). Because the neutral H density in the inner heliosphere seems to be lower (0.08 -0.15 cm^{-3}, see text), there is evidence for a filtering of the neutral H across the interface.

The GHRS Ly-alpha glow data suggest a deceleration and a dispersion of the interstellar neutral H flow in the inner heliosphere, as expected theoretically from coupling with the plasma at the heliospheric interface, compatible with an ambient interstellar proton density of the order of 0.1 cm^{-3} in the framework of the Baranov model. Hopefully the SOHO/SWAN experiment will bring new constraints. There is still an ambiguity in the relative proportions of the heliospheric (hydrogen wall) and galactic intensities as the sources of the emission detected by the Voyager UVS since 1993 on the upwind side of the heliosphere. Hopefully observations during the solar cycle ascending phase will help to disentangle the two emissions. The resolution of the two sources would result in strong constraints on the LIC plasma density.

A very good illustration of what can be done now in terms of establishing conditions inside and outside the heliosphere is the case of interstellar oxygen. Comparing the interstellar neutral oxygen abundance (relative to hydrogen and helium) measured in the LIC with the HST, with the pick-up oxygen ion data of Ulysses, and using a simulation of the neutral oxygen and oxygen ion flow through the heliospheric interface, strongly suggests that oxygen is non negligibly filtered by the interface. Because absolutely all the elements for this comparison , i.e. all sets of data (HST, EUVE, Ulysses...) and the simulations themselves are very recent (1995,1996), it gives a perfect illustration of the activity and the potential advances in the field to which this meeting is devoted.

Acknowledgements: Many thanks to John Clarke and collaborators for allowing inclusion of HST spectrum of interplanetary Ly-α not yet published and to Vladimir Baranov and Youri Malama for the results of their model calculations.

References

Adams T.F., Frisch P.C. 1977, ApJ, 212, 3000
Baranov V.B, Malama Y. 1993, JGR, 98, 15157
Bertin P., Lallement R., Ferlet R., Vidal-Madjar A., Bertaux J.L., 1993, J. Geophys. Res., 98, A9, 15193
Cheng K.P., Bruhweiler F.C, 1990, ApJ 364, 673
Clarke J.T., Bowyer S., Fahr H., Lay G. 1984, Astronomy and Astrophysics, 139, 389
Clarke J.T., Lallement R., Bertaux J.L., Quemerais E. 1995, ApJ, 448, 893
Cox D.P., Reynolds R.J., 1987, ARA&A 25, 303
Cummings A. C., Stone E.C., 1990, 21 st Int. Cosmic Ray Conference 6, 202
Dupuis J., Vennes S., Bowyer S., ApJ,455, 574
Fahr, H.J.,1991, A&A 241, 251
Fahr, H.J., Osterbart, O. and Rucinski, D. 1995, A & A, 294, N 2, 584
Ferlet R., Lallement R., Vidal-Madjar A., 1986, Astronomy and Astrophysics, 163, 204
Frisch P.C., 1994, Science, 265, 1443
Frisch P.C., 1995, Space Science Reviews, 72, 499
Geiss & al, 1994, Astron. Astrophys. ,282, 924
Gloeckler & al, 1993, Science, 261, 70
Gloeckler G, 1996, this issue
Jelinsky, P., Vallerga, J. and Edelstein, J., 1994, ApJ, 442, 653
Lallement R., Ferlet R., Vidal-Madjar A., Gry C., 1990, in "Physics of the Outer Heliosphere", (Varsaw, Sept 89), Grzedzielski and Page, Edts, Pergamon.
Lallement R., Bertin P., 1992, Astronomy and Astrophysics, 266, 479
Lallement R., Bertaux J.L., Clarke J.T., 1993, Science, 260, 1095
Lallement R. Ferlet R., Lagrange A.M., Lemoine M., Vidal-Madjar A., 1995, Astronomy and Astrophysics, 304 (2), 461
Lallement R., Bertin P., Ferlet R., Vidal-Madjar A., Bertaux J.L., 1994, A&A, 286, 898
Lallement R., Clarke J.T., Malama Y., Quémerais E., Baranov V.B., Bertaux J.L., Fahr H.J., Paresce F., in "Science with the Space Telescope", in press.
Lallement R. Ferlet R., 1996, Astronomy and Astrophysics, in press
Linsky J.L., Brown A., Gayley K. et al 1993, ApJ,402, 694
Linsky J.L., et al 1995, ApJ, 451, 335
Linsky J.L., Wood B.E., 1996, ApJ, vol 462
Mayor M., Queloz D., 1995, Nature, Vol 378, 355
Mobius E., 1996, this issue
Phillips J.L., Bame S.J., Feldman W.C.,&al, 1995, Geophys. Res. Let. 22, 23, 3301
Quémerais E., Bertaux J.L., Sandel B.R., Lallement R., 1994, 290, 941
Quémerais E., Sandel B.R., Lallement R., Bertaux J.L.,1995, A&A, 299,249
Quémerais E., Malama Y., Sandel B.R., Lallement R., Bertaux J.L., Baranov V.B., 1996, A&A, 308, 279
Rucinski D., and Bzowski M., 1995, Space Sci. Rev. ,72, 467
Rucinski D., 1996 (private communication)
Summanen T., Lallement R., Quémerais E., 1996, JGR, in press
Vallerga, J. and Welsh, B. ,1995, ApJ, 444, 202
Witte M. et al, 1993, Adv. Space Res., 6, (13), 121
Witte M. Banaskiewicz M., Rosenbauer H., 1996, this issue

THE LOCAL INTERSTELLAR MEDIUM VIEWED THROUGH PICKUP IONS, RECENT RESULTS AND FUTURE PERSPECTIVES

E. MÖBIUS
Dept. of Physics and Inst. for the Study of Earth, Oceans and Space
University of New Hampshire, Durham, NH 03824, U.S.A.

Abstract. Over the last 10 years the experimental basis for the study of the very local interstellar medium (VLISM) has been substantially broadened by the direct detection of pickup ions and of neutral helium. The strength of these methods lies in the local measurement of the particles. By scanning the gravitational focusing cone of the interstellar wind, a consistent set of interstellar helium parameters, neutral density, temperature and relative velocity, has been derived. However, the accuracy of these parameters is still hampered by uncertainties in some of the crucial ionization rates and in the pickup ion transport. Recent observations have shown that the scattering mean free path of pickup ions is comparable with the large scale variation of the interplanetary magnetic field (IMF) in the inner heliosphere. This requires a substantial modification in the modeling of the ion distribution and more detailed measurements, tasks that can be addressed in the near future.

1. Introduction

Due to the relative motion of the Sun with respect to the VLISM an interstellar wind blows through the solar system, and thus neutral interstellar material becomes accessible to local detection techniques. This is of very general interest in astrophysics, because of the direct link between in-situ measurements of the interstellar gas, that can be more powerful in detail, with remote sensing techniques, that reach farther out. For the first time a set of benchmark parameters can be derived for the VLISM that can be put into perspective with the "local bubble" and its interaction with nearby star systems. Still the interstellar parameters outside the solar system have to be inferred from data taken in the inner heliosphere (typically at distances of 1 - 5 AU), and thus a detailed modeling of the transport of the interstellar gas into the heliosphere including other related processes is necessary for the interpretation. The incoming interstellar neutrals are ionized by solar EUV radiation, by charge exchange with solar wind ions, and by electron collisions, whereby a cavity is created in the interstellar gas cloud. Except in the case of hydrogen where radiation pressure is important the Sun acts as a huge gravitational lens creating a focusing cone with a substantial density increase of the interstellar gas on the downwind side. The density increase and the width of the cone depend sensitively on the temperature and relative velocity of the interstellar gas cloud as well as on the mass of the species. It is this structure which provides the main experimental handle on the interstellar gas temperature and the relative velocity. For overviews on this topic the reader is referred to the reviews of, e.g., Axford (1972) and Holzer (1989).

Since the late 1960's back-scattering of solar UV radiation has been employed to map the interstellar gas distribution in the inner heliosphere (e.g. reviews by Chassefière *et al.*, 1986; Lallement 1990). This technique is limited to the two most abundant species of the interstellar medium, hydrogen and helium. The interpretation of sky maps of backscattered Lyman-α and He I 58.4 nm photons, based on simulated distributions of the interstellar gas, has provided us with VLISM parameters with relatively wide error bars (compilation by Möbius, 1993). They also seemed to suggest that the temperatures and relative velocities for H and He differ significantly (e.g. Dalaudier *et al.*, 1984), a puzzle, which has not been completely resolved to date. In addition, there is an ongoing

Space Science Reviews **78**: 375–386, 1996.

debate about the effects of a heliospheric boundary layer, created by the deflection of interstellar plasma around the heliosphere, which can drag a substantial fraction of the neutral gas along due to charge exchange (e.g. Fahr and Ripken, 1984; Baranov *et al.*, 1991). This may explain an apparent depletion of hydrogen over helium in the interstellar gas as compared with the solar system abundances. Because of the different cross sections this drag effect will be significant for H and O, while unimportant for He (Bleszynski, 1987), and thus would lead to a depletion of H and O with respect to the other species inside the heliosphere. However, this effect does not seem to be significant in the composition of the anomalous component of cosmic rays (Cummings and Stone, 1987), which consists of interstellar material ionized within the heliosphere and then accelerated at the solar wind termination shock.

Over the last decade, the capabilities for studying the VLISM have been vastly improved by the advent of new in situ techniques. Möbius *et al.* (1985a) reported the first detection of interstellar He^+ pickup ions in the solar wind. More recently, the direct capture of neutral interstellar helium was accomplished (Witte *et al.*, 1993). Furthermore, the detection of interstellar H^+ (Gloeckler *et al.*, 1993), N^+, O^+, and Ne^+ pickup ions has been reported (Geiss *et al.*, 1994), as well as that of C^+ which stems mostly from local sources (Geiss *et al.*, 1996). Using the spatial distribution of pickup ions a comprehensive determination of the interstellar He parameters has been carried out by Möbius *et al.* (1995a, 1996). They pointed out that production and transport of pickup ions need to be understood in more detail before the remaining questions on the interstellar parameters and the heliospheric boundary can be settled. Since any uncertainty in the ionization rates of the interstellar gas translate into inaccuracies in the determination of the interstellar gas parameters, a direct simultaneous measurement of these rates is necessary. Transport effects can modify substantially the pickup ion over the neutral gas distribution, which then would skew the determination of the temperature and flow velocity. In addition, the density determination can be heavily biased by anisotropic pickup ion distributions.

We will start with a summary of the method to determine the interstellar parameters from pickup ion distributions and a brief comparison with the other methods. Then we will discuss in some detail the uncertainties in the ionization rates and the pickup ion transport followed by an empirical correction for transport effects. The paper will be concluded with potential improvements through new data from upcoming spacecraft missions in conjunction with improved modeling of the pickup ion distributions.

2. Interstellar Gas Parameters from Pickup Ion Distributions

As mentioned in the introduction, a careful modeling of the influx of neutral interstellar atoms into the heliosphere is needed in order to extrapolate the parameters in the undisturbed interstellar medium from local observations in the inner heliosphere. The reason is two-fold: 1) a reduced density of the neutral gas is observed in the inner heliosphere due to ionization losses during the approach of the Sun and 2) the spatial and temporal variations of the neutral gas density over the observation region contain valuable information that can be utilized to infer the parameters. The modeling of the observations ultimately consists of two parts, a simulation of the neutral gas distribution, which is common for all observation methods, and a specific simulation of the observables. These are the intensity distribution of backscattered UV light across the sky, the velocity distribution of pickup ions at the observer's location or the neutral atom fluxes as a function of looking direction and relative velocity between spacecraft and gas.

A schematic representation of the neutral helium distribution in the inner heliosphere is shown in Fig. 1 with sample trajectories of inflowing gas under the influence of the Sun's gravitation. A suitable simulation is based on a so-called hot model of the interstellar gas in which the incoming neutrals move on Keplerian orbits through the solar system. It starts with a shifted Maxwellian distribution at infinity that is characterized by the gas temperature and its velocity relative to the Sun (e.g. Fahr, 1971; Wu and Judge, 1979). Using Liouville's Theorem the resulting velocity distributions for any location inside the heliosphere can be calculated neglecting ionization loss. The latter effect is taken into account separately as an accumulative loss along the trajectories. The resulting neutral gas distribution constitutes the basis for all in-situ observations.

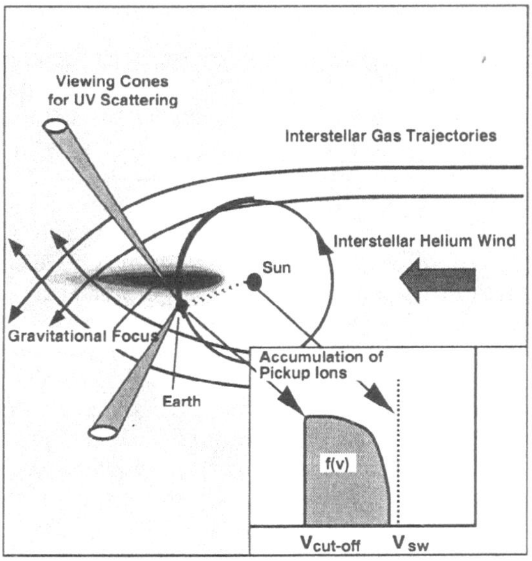

Fig. 1: Comparison of the diagnostics of the interstellar gas in the solar system by different methods. The use of backscattered solar UV light requires a deconvolution of the line-of-sight integral measurement into the spatial neutral gas distribution (shaded cones extending from the observation point). Energy spectra of pickup ions contain a differential information on the neutral gas density along the observer-sun line (inset with a typical pickup ion spectrum).

UV backscattering measurements provide an integral information along the line-of-sight (indicated by viewing cones in Fig. 1) that contains the neutral gas distribution and the spatial and temporal intensity distribution of the solar flux in the resonance line chosen for the observation. By scanning the sky a roughly spherical volume with the extent of a few AU can be covered by the observations. As an additional complication for helium the absorption line of the interstellar gas is slightly detuned with respect to the solar emission line due the relative velocity of the gas. Because of the narrow line width a variable Doppler dimming of the backscattered light has to be taken into account depending on the velocity of the gas in the field-of-view. The sensitivity of the backscatter method to this effect can be seen from the fact that a potential shift of the original solar line position and potential peculiarities of the profile of the HeI 58.4 nm line have been invoked to explain the apparent differences of the reported He temperature and velocity over that of H (Chassefière *et al.*, 1988).

Although pickup ions are accumulated along the spacecraft-sun line a more or less differential measurement can be achieved with this method, if rapid pitch-angle scattering of the ions to isotropy in the solar frame occurs after the initial pickup of the newborn ions by the interplanetary magnetic and convection electric fields (Vasyliunas and Siscoe, 1976). In this case, which is equivalent to a scattering mean free path length small compared with the typical scale length of accumulation, the emerging spherical shells in velocity space shrink due to adiabatic cooling while convected with the radially expanding solar wind. Therefore, a radial cut through the pickup ion velocity distribution centered on the solar wind reflects the production rate $S(\boldsymbol{R},t) = N(\boldsymbol{R},t)$ $\beta^{+}(\boldsymbol{R},t)$, where $N(\boldsymbol{R},t)$ and $\beta^{+}(\boldsymbol{R},t)$ represent the neutral gas density and the ionization rate as summed over all relevant ion production processes. A radial cut of the velocity distribution can be realized most easily in the ion energy spectrum as observed in the solar wind direction (shown as insert in Fig. 1). When taken over a substantial fraction of a spacecraft mission, the combined pickup ion energy spectra map out a 2-dimensional cut through the interstellar gas distribution in the plane of the spacecraft trajectory. The primary result is a profile of the ion production rate which requires an accurate knowledge of the ionization rate β^{+} in order to derive the neutral gas density distribution $N(\boldsymbol{R},t)$ from these measurements.

The most direct method to measure the neutral gas density, of course, is the detection of neutral atoms themselves (Witte *et al.*, 1993). It is based on the fewest assumptions and has the potential to become a very precise tool for the determination of velocity, temperature and density of the interstellar gas. However, at this time the method is limited to situations where the spacecraft and the gas velocity add up favorably and can only be employed successfully to helium, both because of the detection threshold of the sensor. Major instrumental developments are needed, before it can be expanded to the full spectrum of interstellar species. Yet for helium this leads to the very fortunate situation that three independent methods can be applied and compared. An instant cross-calibration is possible and therefore should lead to a benchmark parameter set for interstellar helium within the near future. Because helium is affected only very little by potential filtering at the heliospheric boundary, this will provide a solid baseline for further investigations of the VLISM and its interaction with the heliosphere. At this time the temperature, velocity and density for He have been determined, and first density values for H, O, N, and Ne are available. With the existing and upcoming spacecraft missions it will be possible to derive all three parameters for all key species of the interstellar medium and thus provide the first comprehensive in-situ values for interstellar material that can be compared with results from remote sensing of the solar neighborhood. To prepare for this task a critical review of the current results, their error sources and potential improvements is a timely task.

3. Significance of the Ionization Rates for the VLISM Parameters

As has been pointed out above the knowledge of the ionization rate is of critical importance for the accuracy with which the interstellar parameters can be derived from pickup ion observations. In this context, it is important to distinguish between the "production" rate β^{+} of pickup ions and the "destruction" rate β^{-} of neutrals. The production rate represents the instantaneous ionization rate at the time of the creation of pickup ions and is sensitive to short term variations on a scale of days or less. Errors in its knowledge have to be directly factored into the accuracy of the resulting neutral gas density. The destruction rate is incorporated into the calculation of the local neutral gas density at the time of observation $N(\boldsymbol{R},t)$ and represents the "average" rate of the long-term, continuous destruction of the inflowing atoms accumulated over periods of several months to a few years into the past, depending on the species. A more detailed discussion of the dual role of the ionization rate was given in Möbius *et al.* (1995a). They concluded that a significant source of error (of at least 25%) in the interpretation of the pickup ion data was the lack of continuous measurement of the ionization rates, which necessitated the use of indirect proxies for the solar EUV flux, such as the 10.7

cm radio flux. In addition, the lack of a reliable long term record of the solar EUV flux impacts the extrapolation of the undisturbed VLISM conditions from local gas density determinations. This becomes particularly important during times of increased solar activity during the months preceding the observation. Variations in the temporal profile of the ionization rates can modify substantially the spatial distribution of interstellar neutral gas in the inner heliosphere (Rucinski and Bzowski, 1995a, b). Therefore, the past history of solar activity with the related EUV emission and of the solar wind flux become important data sets for the analysis of the interstellar gas.

Whereas the simultaneous measurement of the solar wind flux has already been employed by Gloeckler (1996) to provide the necessary information on production rates for interstellar pickup ions which are preferentially created by charge exchange, such as H^+ and He^{2+}, the direct monitoring of solar EUV, which dominates the production of, e.g. He^+ and Ne^+, has yet to be performed. In addition, there has never been any direct long term monitoring of the solar EUV flux, which is crucial for obtaining an accurate destruction rate. The SOHO mission, for the first time will provide simultaneous measurements of pickup ion fluxes and the corresponding photoionization rates of neutrals. The instrumentation for the interplanetary plasma observations is augmented with an EUV monitor. While providing an absolute cross-calibration for the Sun-observing UV instruments on SOHO, it will also provide direct data on the photoionization rate. The combination of a photodiode with an Al filter defines the spectral response such that the wave length range relevant for the ionization of He and Ne is covered by the EUV monitor (Hovestadt *et al.* 1995). An interference filter allows the separate monitoring of the He II 30.4 nm line, which contributes up to ≈ 50% to the ionization rate and has a stronger variation with solar activity than the EUV continuum. Combined with the measurement of the total solar wind flux by the proton monitor in the same instrument package this will allow a precise monitoring of the ionization rates for all key interstellar species and will substantially narrow the uncertainties in the interpretation of the pickup ion data. A detailed discussion of the relevant ionization rates and their determination may be found in the paper by Rucinski *et al.* (1996).

4. Significance of Ion Transport for Pickup Ion Distributions

The current interpretation of pickup ion distributions and its translation into interstellar gas parameters contains several assumptions that have simplified the analysis. They seemed valid for the early pickup ion results. However, according to more recent observations and analysis modifications are required. The simplifying assumptions can be maintained as long as the mean free path length for pitch-angle scattering is short compared with any other relevant scale length in the evolution of the observed pickup ion distributions. Firstly, a short mean free path, or the fact that pitch-angle scattering is much faster than all other processes, allows that pickup ion distributions can be taken as nearly isotropic in the solar wind frame (e.g. Vasyliunas and Siscoe, 1976). Therefore, an accurate determination of the total pickup ion flux and thus the local neutral gas density is possible, even from a fraction of the distribution. Any deviation from this assumption may require a substantial correction of the resulting neutral gas density. Secondly, because pitch-angle scattering modifies the velocity component of the pickup ions along the magnetic field line, it represents also the major spatial diffusion process. The assumption of very efficient pitch-angle scattering therefore implies that spatial diffusion takes place with a short mean free path ($\lambda \approx 0.1$ AU). In this case the interstellar neutral gas distribution is mapped by pickup ions, with only minor modifications, along the spacecraft-sun line through the convection with the solar wind velocity. For a larger mean free path spatial diffusion will wash out structures. For example, the gravitational focusing cone will appear wider and flatter when observed in the "light" of pickup ions. In essence a slower bulk flow velocity and a higher temperature of the interstellar medium are implied by this effect than are realistic.

Recently, there have been indications that the mean free path for diffusion of pickup ions can be much larger, on the order of 1 AU, which is comparable with the size of the inner heliosphere. Such conclusions were drawn, for example, from studies of anisotropies in the velocity distribution of pickup H^+ in unperturbed high-speed solar wind (Gloeckler *et al.*, 1995). The significant reduction of the pickup ion flux during radial IMF conditions in comparison with pickup during perpendicular IMF (Möbius *et al.*, 1995b) can also be explained most naturally in terms of a larger mean free path. This reduction is more pronounced closer to the cut-off energy, i.e. where freshly created ions need the longest time to be transported by pitch-angle scattering.

A larger mean free path has profound implications on the shape of the gravitational focusing cone as seen with pickup ions. Depending on the actual IMF direction and its variation with time, pickup ions are transported across the cone structure which leads to a wider cone with reduced density enhancement. Such cross-cone transport would explain the higher He temperature and lower relative velocity reported by Möbius et al. (1995a) from the pickup ion distribution in comparison with the results from neutral gas measurements (Witte *et al.*, 1993) and from UV absorption lines in the VLISM (Bertin *et al.*, 1993). In a heuristic approach let us now assume a spread of the focusing cone due to spatial diffusion. Modifying the ideal model cone of the neutral gas by a convolution with the diffusive width leads to a reasonable agreement of the pickup ion results with the other two methods for both temperature (Fig. 2 a) and relative velocity (Fig. 2 b) when a mean free path of $\lambda \approx 0.6 - 0.9$ AU is assumed. The reasonable result of this empirical correction to the interstellar gas parameters as derived from He pickup ion measurements lends additional support for a large mean free path.

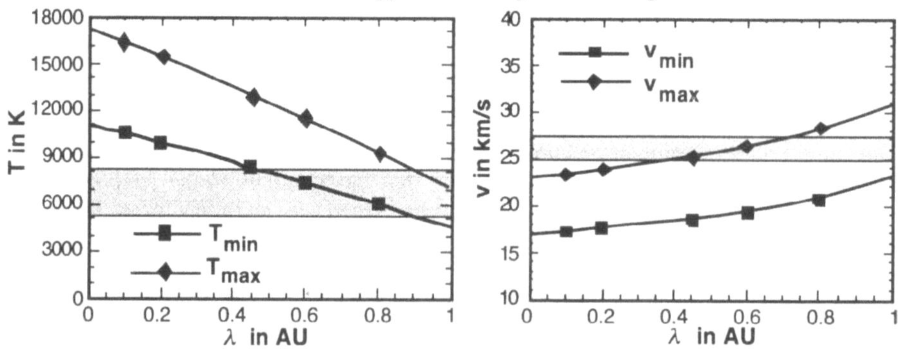

Fig. 2: (Left) Temperatures of interstellar He as derived from the pickup ion distribution across the focusing cone, assuming spatial diffusion of the ions with λ varying from 0 - 1 AU (Möbius et al., 1996). The shaded bar indicates the values derived from neutral gas detection (Witte et al., 1993) and from UV absorption lines (Bertin et al., 1993). (Right) The same representation as (Left) for the bulk velocity of interstellar He.

In addition, the crucial assumption of a nearly isotropic distribution in the solar wind frame no longer holds. Because of instrumental constraints the pickup ion flux and thus the source neutral density has been traditionally derived from the anti-sunward portion of the distribution. Except in cases when the IMF is oriented almost perfectly perpendicular to the solar wind direction the revised view of the pickup ions predicts a substantial reduction of the flux in the anti-sunward hemisphere of the distribution. In the solar wind frame all pickup ions are injected with a velocity $-v_{sw}$. For radial magnetic field conditions all ions will have to be transported to the anti-sunward hemisphere of the velocity distribution solely by pitch-angle scattering. This forms a natural explanation for the much lower pickup ion fluxes that have been observed during mostly radial IMF conditions (Möbius et al, 1995a). As a consequence, this also calls for a revision of the neutral gas density as derived from the pickup ion measurements. Because of the observed flux variation with the magnetic field direction, only time periods with field orientations of $\Theta_{IMF} = 45° - 90°$ with respect to the solar wind were selected for the study of the interstellar gas parameters by Möbius *et al.*. (1995a). However, even this conservative restriction of the data is not sufficient to derive an accurate neutral gas density from the anti-sunward hemisphere of the pickup distribution. A correction of the result seems necessary. Let us use a comparison between periods with more or less radial magnetic field direction to estimate the correction factor heuristically. In fact the radial cases, for which a typical reduction of 50% over the average flux during oblique magnetic field is observed, encompass a range of magnetic field directions $\Theta_{IMF} = 0° - 45°$. Assuming a perfectly isotropic distribution for $\Theta_{IMF} = 90°$ and a linear variation of the flux with angle, the correction factor α between the average neutral density $n_{average}$ as derived earlier and the corrected density n_o ($n_o = \alpha \cdot n_{average}$) can be easily estimated from Fig. 3. If we assume that the measurements represent the centers of the two angular ranges, this leads to $\alpha = 1.25$. Because an IMF orientation along the Parker spiral ($\Theta_{IMF} = 45°$) is more likely than the two extremes, our measurements maybe biased even further, which would lead to a steeper variation with the angle.

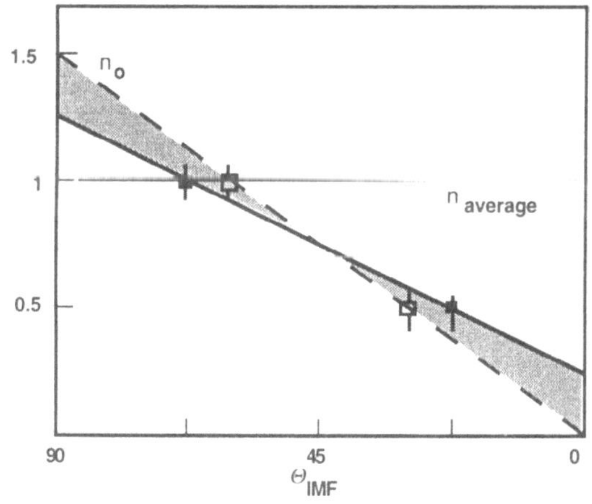

Fig. 3: Estimate of the variat-ion of the average neutral gas density $n_{average}$ with magnetic field angle as derived from the pickup ion flux in the anti-sunward hemi-sphere of the distribution. The symbols indicate the observations for the two angular ranges (45° - 90° and 0° - 45°; filled, if the average direction is centered; open, if the average is closer to 45° in each case.

On the other hand we must assume that even for a perfectly radial magnetic field some pickup ions are scattered into the anti-sunward hemisphere. Therefore, α has to be < 1.5. Taking the middle ground we adopt a correction factor $\alpha = 1.3 - 1.4$ in the absence of further detailed information. Applying this correction to the results by Möbius *et al.* (1995a) leads to a neutral helium density of $n_{He} = 0.9 - 1.2 \cdot 10^{-2}$ cm^{-3}. The modified set of parameters of the VLISM after all empirical corrections for effects of a large mean free path are compiled in Table 1 together with the pickup ion result reported by Gloeckler (1996) and the parameters derived from UV scattering and absorption as well as from direct neutral gas measurements. The resulting temperature and velocity are in relatively good agreement with the other results. The density still seems somewhat low compared with the pickup ion result by Gloeckler (1996) and the neutral gas result by Witte (1996), but it should be pointed out that an apparent agreement within a bandwidth of 30% around the average between the three independent measurement points is a great achievement, given the remaining uncertainties as pointed out in this paper.

TABLE 1

Interstellar He parameters from pickup ion measurements and other methods

Method	n_{He} [cm^{-3}]	T_{He} [K]	v_{rel} [km/s]	Reference
pick-up ions	0.009 - 0.012	4800 - 7200	23 - 30.5	*Möbius et al. 1995a, 1995b*
($\lambda \approx 1$ AU)	0.015	-	-	*Gloeckler et al. 1993, 1994*
UV	0.015 - 0.02	16000	24 - 30	*Dalaudier et al. 1984*
scattering	0.005 - 0.015	8000	19 - 24	*Chassefiere et al. 1988*
neutrals	≈ 0.014	5200 - 8200	25 - 27	*Witte et al. 1993*
absorption	-	7000 ±200	25.7 ±1	*Bertin et al. 1993*

5. Improvements in the Modeling and Observations of Pickup Ions

With several consistent indications that the mean free path for pitch-angle scattering of pickup ions may be of the order of 1 AU a much more rigorous approach has to be taken for the pickup ion distribution and related transport effects, starting with the transport equation for ions in interplanetary space (Roelof, 1969):

$$\frac{df}{dt} = \underbrace{\frac{V}{\lambda}\frac{\partial f}{\partial \mu}\left[\left\{1 - \mu^2\right\}\frac{\partial f}{\partial \mu}\right]}_{(1)} + \underbrace{\frac{1}{B}\frac{\partial B}{\partial s}\left\{1 - \mu^2\right\}\frac{\partial f}{\partial \mu}}_{(2)} + \underbrace{\frac{1}{3}\nabla V p\frac{\partial f}{\partial p}}_{(3)}$$

For observations at 1 AU λ (important in term (1)) is of the same order as

- the length scale for adiabatic focusing (term (2))
- the length scale for adiabatic deceleration of the ions (term (3)) and
- the typical scale for the accumulation of the ions from the Sun to the spacecraft

TABLE 2

Pickup ion transport effects

Effect	Scale	Old Model	New Model
pitch-angle scattering	λ	0.1 AU	1 AU
adiabatic deceleration	≈ 1 AU	isotropic cooling (time ordering)	deceleration $\perp B$ (same time scale)
adiabatic focusing	≈ 1 AU	small correction	strong anisotropic filling of angle space
modification of pitch-angle in *B* curvature	a few AU	unimportant	trajectory along curved *B* field necessary

These processes and their relative importance are listed in Table 2, comparing situations when λ is small compared with the typical scale length of gradients in the given geometry and when both scales are of the same order of magnitude.

In the first case pitch-angle scattering can be treated independently of the remaining effects and only adiabatic cooling is the first order effect that survives. In the second case all effects have to be weighted equally, and the large scale curvature of the magnetic field also has be taken into account. Therefore, depending on the IMF direction the pickup distribution will evolve initially in the following way: During radial field conditions a substantial pile-up of newly ionized particles, which are initially at rest, will accumulate in the sunward hemisphere of the velocity distribution. Under these circumstances the ions are not picked up by the magnetic field, and it takes a typical time $\tau = \lambda / v_{sw}$ until a substantial scattering towards 90° pitch-angle has occurred. Therefore, the highest differential pickup ion density is found close to $v = 0$ in the rest frame. Because of inefficient scattering adiabatic deceleration will act mainly on the velocity component perpendicular to the magnetic field, which leads to a cigar-shaped distribution (Fig. 4a). Because the magnetic mirror force acts on the same scale as pitch-angle scattering adiabatic focusing may now play a major role in bringing the ions from the sunward into the anti-sunward hemisphere of the distribution. In addition, the pitch-angle of a particle will change on its way out due to the gradual turn of the field according to its garden hose angle

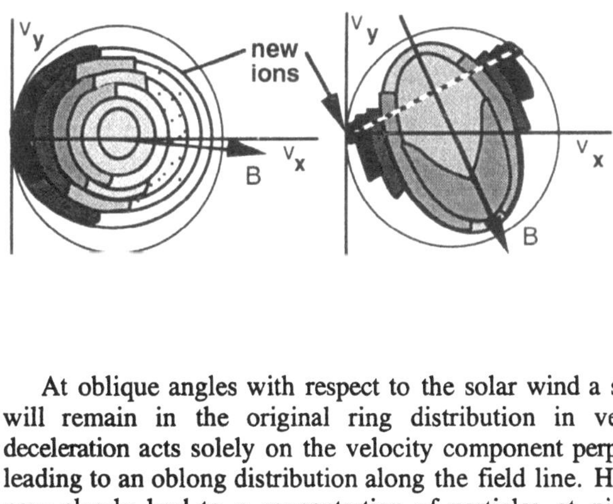

Fig. 4: (Left) 2D cut of a pick-up ion distribution in radial IMF for large λ. The shape of the rings with different shading indicates qualitatively the density distribution as expected from the effects of adiabatic deceleration perpendicular to B and adiabatic focusing. (Right) Similar representation for oblique magnetic field.

At oblique angles with respect to the solar wind a substantial fraction of the ions will remain in the original ring distribution in velocity space. Again adiabatic deceleration acts solely on the velocity component perpendicular to the magnetic field leading to an oblong distribution along the field line. However, adiabatic focusing may now clearly lead to a concentration of particles at pitch-angles in the anti-sunward direction along the field line (Fig. 4b). As can be seen from the schematics a substantially more structured velocity distribution evolves from pickup with a large scattering mean free path than considered before. However, with the phase space resolution and coverage of the earlier instruments these effects were invisible. For example, AMPTE SULEICA (Möbius *et al.*, 1985b) with its eight 45° sectors in azimuthal angle and only a 40° fan in polar angle averaged over the features which would distinguish between the different effects and cut off the complete sunward hemisphere. The respective binning and coverage of SULEICA is indicated in Fig. 5a.

In order to study these transport effects and to accurately determine the mean free path a complete mapping of the full distribution with much improved phase space resolution including the important pile-up region at low velocities is necessary. The CODIF instrument on Cluster (Rème *et al.*, 1993) will provide angular resolution in polar angle of 22.5° (8 pixels over 180°) and 22.5° or 11.25° (16 or 32 sectors) in azimuth over one spacecraft spin (depending on the instrument mode). The instrument also covers the full energy range from spacecraft potential to 40 keV/charge so that the low velocity part of the distribution function is included. The binning and coverage of CODIF is shown in Fig. 5b. It is worth emphasizing that the count rate of an electrostatic analyzer scales as E^2 for a given phase space density, which makes the coverage of the sunward hemisphere of the pickup distribution a challenge. However, the geometric factor of CODIF is approximately 4 times larger than that of SULEICA for the same viewing angle. For the large scale structures of the interstellar gas CODIF can take advantage of another benefit of the Cluster program: the data from all 4 satellites can be accumulated, thus increasing the counting statistics by another factor of 4. In addition, there is no loss of efficiency at low energies in the time-of-flight section of the instrument because of the post-acceleration of the ions. With SULEICA typically a 10 minute integration was sufficient for a spectrum between the solar wind energy and the cut-off. CODIF will accumulate a distribution function for $E \geq 0.1\ E_{sw}$ within one hour with the same counting statistics at the low energy limit as with SULEICA in 10 minutes at the solar wind energy of He^+. This will leave a relatively small portion of velocity space uncovered. The SWICS instrument on Ulysses can cover the distribution only for $E > 0.6$ keV/Q, and it requires a long integration time (of the order of 30 days) for the accumulation of a H^+ pickup ion distribution in its sunward hemisphere (Gloeckler *et al.* (1995). This is problem is even more pronounced for a similar study of the He^+ distribution.

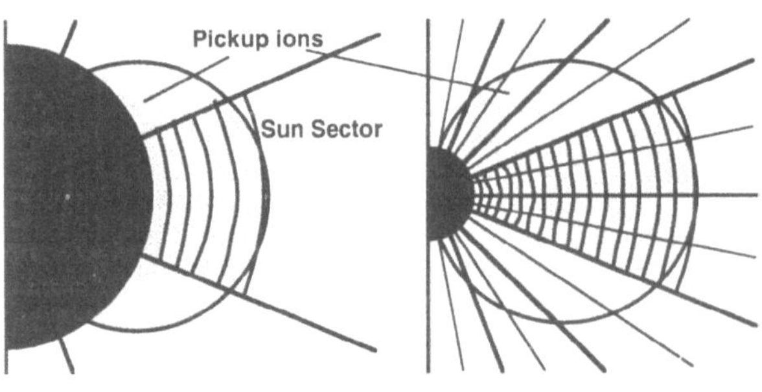

Fig. 5: (Left) Sectors and energy steps of AMPTE SULEICA. The low-energy cut-off of the instrument is marked by dark shading and the phase space occupied by pickup ions by light shading. (Right) Similar representation for Cluster CODIF (sectors in elevation: heavy lines; additional sectors in azimuth: thin lines).

The lack of efficient pitch-angle scattering also has profound effects on the transport of pickup ions in the vicinity of the gravitational focusing cone. A schematic representation of the expected differences in the transport of different parts of the distribution function close to the cone is given in Fig. 6. Cuts through the velocity distribution are shown with expected maxima in the phase space distribution in relation to the location of the focusing cone. The bulk of the ions that remains close to the original pickup ring perpendicular to the field will, in the case of a Parker spiral configuration, move eastward with respect to the source. Therefore, we can expect to observe a maximum in this portion of the distribution, after Earth has passed the focusing cone in December. Adiabatic focusing tends to fill the anti-sunward portion of the distribution along the field line. Ions in this part of the distribution move mainly along the field lines and therefore are transported westward with respect to the cone. As a consequence we may expect a maximum along the field lines during the time period before Earth crosses the cone in late November. Convection with the solar wind velocity occurs for ions at 90° pitch-angle, where pickup ion distributions near the nominal center of the cone may have a maximum. Because Cluster will initially be launched into the midnight direction, the current launch schedule with a summer launch would bring the satellites into the solar wind when Earth crosses the focusing cone. This scenario would then provide the best coverage of the region around the cone with the ideal instrumentation.

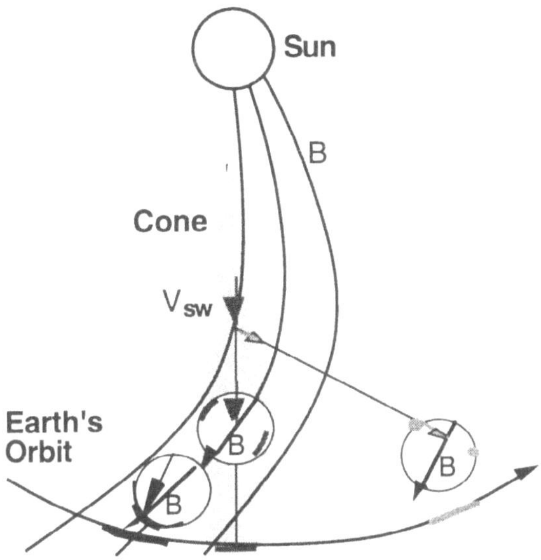

Fig. 6: Schematic representat-ion of transport effects near the focusing cone for pickup ions with different pitch-angles. The expected maxima in the velocity distributions and the locations of their arrival at Earth's orbit are indicated by different shading. The main direction for their transport out of the cone is indicated by a thick arrow with the respective shading in each distribution.

6. Summary and Outlook

The results compiled in this paper show that a consistent set of interstellar gas parameters for helium can now be derived from pickup ion measurements. After correcting for expected ion transport effects in interplanetary space the densities for the current in-situ measurements of the interstellar gas density are compatible within 30%, and the temperature and velocity agrees with neutral gas and UV absorption results. The remaining error bars for results derived from He^+ are related to uncertainties in the ionization rates and only a cursory knowledge of the transport effects. With the launch of SOHO in December 1995 and the upcoming launch of Cluster in summer 1996 a complement of instruments will become available to address both remaining shortcomings. We can expect to bring the results from He^+ pickup ions to an accuracy level that will be comparable with that of a direct neutral gas measurement. With several different complementary methods available it is the time to set a benchmark value for interstellar helium in the solar neighborhood. This will also set a cornerstone for using pickup ions as a general technique for all species of the interstellar gas that penetrate into the solar system. With a well established modeling of the parameters from pickup ion measurements a similar accuracy can be expected for the other elements and a multi-parameter comparison with remote sensing of the neighborhood can be initiated.

Acknowledgements

The author is grateful for many helpful and stimulating discussions with G. Gloeckler, P. A. Isenberg, M. A. Lee, and D. Rucinski. Part of the work towards this paper originates in a very fruitful visit of the author at the Space Research Centre in Warsaw. The work was supported by NSF Grant INT-9116367, NASA Grant NAGW-2579, and Grant 2 P 304 018 04 from the Committee for Scientific Research, Poland.

References

Axford, W.I.: 1972, in C.P. Sonnett, P.J. Coleman, Jr., and J.M. Wilcox (eds.), *The Solar Wind*, NASA SP, **308**, 609.
Baranov, V.B., Lebedev, M.G., and Malama, Yu. G.: 1991, *Astrophys. J.*, **375**, 347.
Bertin, P., Lallement, R., Ferlet, R., and Vidal-Madjar, A.: 1993, *J. Geophys. Res.*, **98**, 15193.
Bleszynski, S.: 1987, *Astron. Astrophys.*, **180**, 201.
Chassefière, E., Bertaux, J.L., Lallement, R., and Kurt, V.G.: 1986, *Astron. Astrophys.*, **160**, 229.
Chassefière, E., Dalaudier, F., and Bertaux, J.L.: 1988, *Astron. Astrophys.*, **201**, 113.
Cummings, A., and Stone, E.C.: 1987, *Proc. 20th Intl. Cosmic Ray Conf.*, **SH3**, 413.
Dalaudier, F., Bertaux, J.L., Kurt, V.G., and Mironova, E.N.: 1984, *Astron. Astrophys.*, **134**, 171.
Fahr, H.J.: 1971, *Astron. Astrophys.*, **14**, 263.
Fahr, H.J., and Ripken, H.W.: 1984, *Astron. Astrophys.*, **139**, 551.
Geiss, J., et al.: 1994, *Astron. Astrophys.*, **282**, 924.
Geiss, J., Gloeckler, G., and von Steiger, R.: 1996, *Space Sci. Rev.*, this issue.
Gloeckler, G., et al.: 1993, *Science*, **261**, 70.
Gloeckler, G., Fisk, L.A., and Schwadron, N.: 1995, *Geophys. Res. Lett.*, **22**, 2665.
Gloeckler, G.: 1996, *Space Sci. Rev.*, this issue.
Holzer, T.E.: 1989, *Ann. Rev. Astron. Astrophys.*, **27**, 129.
Hovestadt, D., et al.: 1995, *Solar Physics*, **162**, 441.
Lallement, R.: 1990, in S. Grzedzielski, D.E. Page (eds.), *Physics of the outer heliosphere*, COSPAR Coll. Series, 1, 49.
Möbius, E., et al.: 1985a, *Nature*, **318**, 426.
Möbius, E., et al.: 1985b, *IEEE Trans. on Geosci. and Remote Sens.*, **GE-23**, 274.
Möbius, E., Klecker, B., Hovestadt, D., and Scholer, M.: 1988, *Astrophys. Space Sci.*, **144**, 487.
Möbius, E.: 1993, in Landolt-Börnstein, *New Series VI.3a*, 184.
Möbius, E., Rucinski, D., Hovestadt, D., and Klecker, B.: 1995a, *Astron. Astrophys.*, **304**, 505.
Möbius, E., Lee, M.A., Isenberg, P.A., Rucinski, D., and Klecker, B.: 1995b, *EOS Trans.*, **76**(17), S229.
Möbius, E., Rucinski, D., Isenberg, P.A., and Lee, M.A.: 1996, *Ann. Geophys.*, **14**, 492.
Rème, H., et al.: 1993, *ESA-SP*, **1159**, 133.
Roelof, E.C.: 1969, in H. Ögelman and J.R. Wayland (eds.), *Lectures in High Energy Astrophysics*, NASA-SP, **199**, 111.
Rucinski, D., and Bzowski, M.: 1995a, *Astron. Astrophys.*, **296**, 248.
Rucinski, D., and Bzowski, M.: 1995b, *Adv. Space Res.*, **16**(9), 121.
Rucinski, D., et al.: 1996, *Space Sci. Rev.*, this issue.
Vasyliunas, V., and Siscoe, G.L.: 1976, *J. Geophys. Res.*, **81**, 1247.
Witte, M., Rosenbauer, H., Banaszkiewicz, M., and Fahr, H.: 1993, *Adv. Space Res.*, **13**(6), 121.
Witte, M., Banaszkiewicz, M., and Rosenbauer, H.: 1996, *Space Sci. Rev.*, this issue.
Wu, F.M., and Judge, D.L.: 1979, *Astrophys. J.*, **231**, 594.

MOMENT EQUATION DESCRIPTION OF INTERSTELLAR HYDROGEN

Y. C. WHANG
Catholic University of America, Washington, D. C.

Abstract. The flow of interstellar hydrogen in the heliosphere can be studied using the moment equation approach. The Boltzmann equation is integrated over the velocity space to obtain the moment equations, the moment equations are then solved directly for the flow conditions. We present a closed system of moment equations. This approach can include anisotropic pressure when the distribution function is distorted into skewed ellipsoid.

1. Introduction

Interstellar hydrogen can enter the heliosphere, and may become ionized by photoionization and by charge exchange with solar wind protons to produce pickup protons. The charge exchange between solar wind protons and interstellar hydrogen atoms produces hydrogen atoms of solar origin, these hydrogen atoms of solar origin make a negligible contribution to the population of neutral hydrogen inside the heliosphere. The bulk speed of hydrogen atoms of solar origin is of the order of the solar wind speed that is ~20 times the bulk speed of interstellar hydrogen. Because of the distinct difference in bulk speed, we treat hydrogen atoms of interstellar origin and hydrogen atoms of solar origin as two kinds of neutral hydrogen.

It is desirable to understand the variation of the number density N_H, the bulk velocity $\mathbf{V}_H$, and the temperature T_H of interstellar hydrogen in the heliosphere. These macroscopic properties can be calculated from the zeroth, first, and second moments of the distribution function of interstellar hydrogen f_H. Two approaches can be employed in studying the variation of N_H, $\mathbf{V}_H$, and T_H: the microscopic kinetic theory approach and the macroscopic moment equation approach.

The kinetic theory approach is to find the solution of the distribution function f_H from the Boltzmann equation in the first step. The typical method is to evaluate the solutions for enormous amount of detailed information about the distribution function f_H over the three-dimensional velocity space at each point of interest in the physical space, then calculate the volume integrals in the velocity space in the second step for the moments of f_H to obtain N_H, $\mathbf{V}_H$, and T_H at that point. Numerical solutions are obtained in both steps. Using the kinetic theory approach, the problem has been well studied by many authors [Fahr 1979, 1986; Fahr and Ripken 1984; Lallement et al. 1985; Osterbart and Fahr 1992; Ripken and Fahr 1983; Thomas 1978; Vasyliunas and Siscoe 1976; Wu and Judge 1979]. We propose to study the flow of interstellar hydrogen in the heliosphere using moment equations. The alternative moment equation approach simply reverses the order of two mathematical operations. The first step is to analytically integrate the Boltzmann equation over the velocity space to obtain the moment equations, then solve of the moment equations directly for the flow conditions.

Space Science Reviews **78**: 387–392, 1996.

Analytical integration results are obtained in its first step and numerical solutions are obtained in its second step.

2. Moment equations of the Boltzmann equation

The Boltzmann equation describes the time rate of change of f_H following the trajectory of a hydrogen atom in a six-dimensional phase space as due to the rate at which f_H is altered by the ionization process:

$$\frac{\partial f_H}{\partial t} + \mathbf{v}\cdot\nabla f_H + (1-\mu)\nabla\left(\frac{GM}{R}\right)\cdot\nabla_v f_H = -\beta f_H \tag{1}$$

Here G is the gravitational constant, M the solar mass, R the heliocentric distance, $\mathbf{v}$ the velocity of hydrogen atom, and μ the ratio of the repulsion force by solar radiation to the gravitational attraction force on a hydrogen atom; the total ionization rate per hydrogen atom β is

$$\beta = v_0\left(\frac{R_0^2}{R^2}\right) + \sigma N_S V \tag{2}$$

where σ is the mean charge exchange cross-section, N_S the number density of solar wind protons, $\mathbf{V}$ the solar wind velocity, and v_0 the rate of photoionization per hydrogen atom at $R_0 = 1$ AU.

The number density N_H and the bulk velocity $\mathbf{V}_H$ can be calculated from

$$N_H = \int f_H d\mathbf{v} \qquad \text{and} \qquad \mathbf{V}_H = \frac{1}{N_H}\int \mathbf{v} f_H d\mathbf{v}$$

Let m be the proton mass, $\rho_H = mN_H$ be the mass density, and $\mathbf{u} = \mathbf{v} - \mathbf{V}_H$ be the intrinsic velocity. The zeroth moment and the first moment of f_H with respect to $\mathbf{u}$ are

$$\int m f_H d\mathbf{v} = \rho_H \tag{3}$$

and

$$\int m\mathbf{u} f_H d\mathbf{v} = 0 \tag{4}$$

The second, third, and fourth moments of f_H with respect to $\mathbf{u}$ respectively define P_{ij}, Q_{ijk}, and R_{ijkl}

$$\int m u_i u_j f_H d\mathbf{u} = P_{ij} \tag{5}$$

$$\int m u_i u_j u_k f_H d\mathbf{u} = Q_{ijj} \tag{6}$$

$$\int m u_i u_j u_k u_l f_H d\mathbf{u} = R_{ijkl} \tag{7}$$

Because of the symmetry, the number of distinct components in P_{ij} is 6, in Q_{ijk} is 10, and in R_{ijkl} is 15. The contraction of P_{ii} and Q_{ijj} gives

$$P_H = \frac{1}{3}\sum_i P_{ii} = \frac{m}{3}\int \mathbf{u}^2 f_H d\mathbf{v} \tag{8}$$

$$q_i = \frac{1}{2}\sum_j Q_{ijj} = \int \frac{m}{2}\mathbf{u}^2 u_i f_H d\mathbf{v} \tag{9}$$

Here $P_H = N_H k T_H$ is the isotropic partial pressure of interstellar hydrogen, k is the Boltzmann constant, and q_i is the heat flux vector. We may separate P_{ij} into P_H and p_{ij} a deviation components of the pressure tensor

$$p_{ij} = P_{ij} - P_H\delta_{ij} \tag{10}$$

If the distribution function f_H is Maxwellian,

$$f_H^0 = N_H\left(\frac{m}{2\pi kT_H}\right)^{3/2}\exp\left(-\frac{m\mathbf{u}^2}{2kT_H}\right) \tag{11}$$

p_{ij} and q_i are zero. However, under the influence of the solar force field and the ionization process, the distribution function may be distorted from Maxwellian distribution. As a consequence, p_{ij} and q_i may deviate from zero.

Moment equations are obtained by multiplying the Boltzmann equation (1) through by $\Phi = mu_iu_j...u_n$ and integrating. It is convenient for the integration procedure if we use **x**, **u**, and t instead of **x**, **v**, and t as the independent variables of the Boltzmann equation. Since $\mathbf{V}_H$ is a function of **x** and t, **u** is a function of **v**, **x**, and t. In terms of **x**, **u**, and t, we can write the Boltzmann equation as

$$\frac{Df_H}{Dt} + \mathbf{u}\cdot\nabla f_H + \left\{(1-\mu)\nabla\left(\frac{GM}{R}\right) - \frac{D\mathbf{V}_H}{Dt} - \mathbf{u}\cdot\nabla\mathbf{V}_H\right\}\cdot\nabla_u f_H = -\beta f_H \tag{12}$$

Here $\frac{D}{Dt} = \frac{\partial}{\partial t} + \mathbf{V_H}\cdot\nabla$ is the material derivative and ∇_u is the vector operator in **u**-space. On the right-hand side, the ionization rate β is independent of **u**, but it can be a function of **x** and t. If we multiply equation (12) by $\Phi d\mathbf{u}$ and integrate over **u** we can obtain

$$\begin{gathered}\frac{D}{Dt}\int\Phi f_H d\mathbf{u} + \frac{\partial}{\partial x_i}\int\Phi u_i f_H d\mathbf{u} + \left\{\frac{DV_{Hi}}{Dt} - (1-\mu)\frac{\partial}{\partial x_i}\left(\frac{GM}{R}\right)\right\}\int\frac{\partial\Phi}{\partial u_i}f_H d\mathbf{u} \\ + \frac{\partial V_{Hi}}{\partial x_i}\int\Phi f_H d\mathbf{u} + \frac{\partial V_{Hi}}{\partial x_j}\int u_j\frac{\partial\Phi}{\partial u_i}f_H d\mathbf{u} = -\beta\int\Phi f_H d\mathbf{u}\end{gathered} \tag{13}$$

For $\Phi = m$, mu_k, and $m\mathbf{u}^2/2$, we can obtain from (13) the continuity equation, the equation of motion, and the energy equation

$$\frac{D\rho_H}{Dt} + \rho_H\varepsilon = -\beta\rho_H \tag{14}$$

$$\rho_H\frac{DV_{Hk}}{Dt} + \frac{\partial P_H}{\partial x_k} + \frac{\partial p_{ik}}{\partial x_i} - \rho_H(1-\mu)\frac{\partial}{\partial x_k}\left(\frac{GM}{R}\right) = 0 \tag{15}$$

$$\frac{3}{2}\frac{DP_H}{Dt} + \frac{5}{2}P_H\varepsilon + \frac{1}{2}p_{ij}\varepsilon_{ij} + \frac{\partial q_i}{\partial x_i} = -\frac{3}{2}\beta P_H \tag{16}$$

where $\varepsilon = \frac{\partial V_{Hi}}{\partial x_i}$ and $\varepsilon_{kl} = \frac{\partial V_{Hk}}{\partial x_l} + \frac{\partial V_{Hl}}{\partial x_k} - \frac{2}{3}\varepsilon\delta_{kl}$. If $p_{ik} = q_i = 0$, these equations form a closed system of fluid equations. Otherwise, the system has more unknown variables than the number of equations.

We can obtain higher moment equations for $\Phi = mu_ku_l$ and $\Phi = mu_k\mathbf{u}^2/2$

$$\begin{gathered}\frac{Dp_{kl}}{Dt} + p_{kl}\varepsilon + p_{jl}\frac{\partial V_{Hk}}{\partial x_j} + p_{jk}\frac{\partial V_{Hl}}{\partial x_j} - \frac{2}{3}p_{ij}\frac{\partial V_{Hj}}{\partial x_i}\delta_{kl} \\ + \frac{\partial}{\partial x_i}(Q_{ikl} - \frac{2}{3}q_i\delta_{kl}) + P_H\varepsilon_{kl} = -\beta p_{kl}\end{gathered} \tag{17}$$

and

$$\frac{Dq_k}{Dt}+q_k\varepsilon+q_j\frac{\partial V_{Hk}}{\partial x_j}+Q_{klj}\frac{\partial V_{Hl}}{\partial x_j}+\frac{5}{2}P_H\left\{\frac{DV_{Hk}}{Dt}-(1-\mu)\frac{\partial}{\partial x_k}\left(\frac{GM}{R}\right)\right\}$$
$$+p_{lk}\left\{\frac{DV_{Hl}}{Dt}-(1-\mu)\frac{\partial}{\partial x_l}\left(\frac{GM}{R}\right)\right\}+\frac{1}{2}\frac{\partial R_{ikll}}{\partial x_i}=-\beta q_k \tag{18}$$

The system of moment equations is still not closed due to the presence of higher moment terms Q_{klj} and R_{ikll}.

Using Grad's thirteen moment approximation [1949], the distribution function can be approximated by a three terms expansion

$$f_H=\left\{1+\frac{1}{5}\frac{mN_H}{P_H^{\,2}}q_k\left(\frac{m}{kT_H}\mathbf{u}^2-5\right)u_k+\frac{m}{2kT_H}\frac{p_{kl}}{P_H}u_ku_l\right\}f_H^0 \tag{19}$$

This distribution function satisfies all definitions for N_H, $\mathbf{V}_H$, T_H, p_{kl}, and q_k given in equations (3), (4), (9), and (10) and it offers flexibility to describe some special flow problems. When p_{kl} and q_k are not zero, the spherical symmetry of Maxwellian distribution is distorted into a skewed ellipsoid as shown in Figure 1.

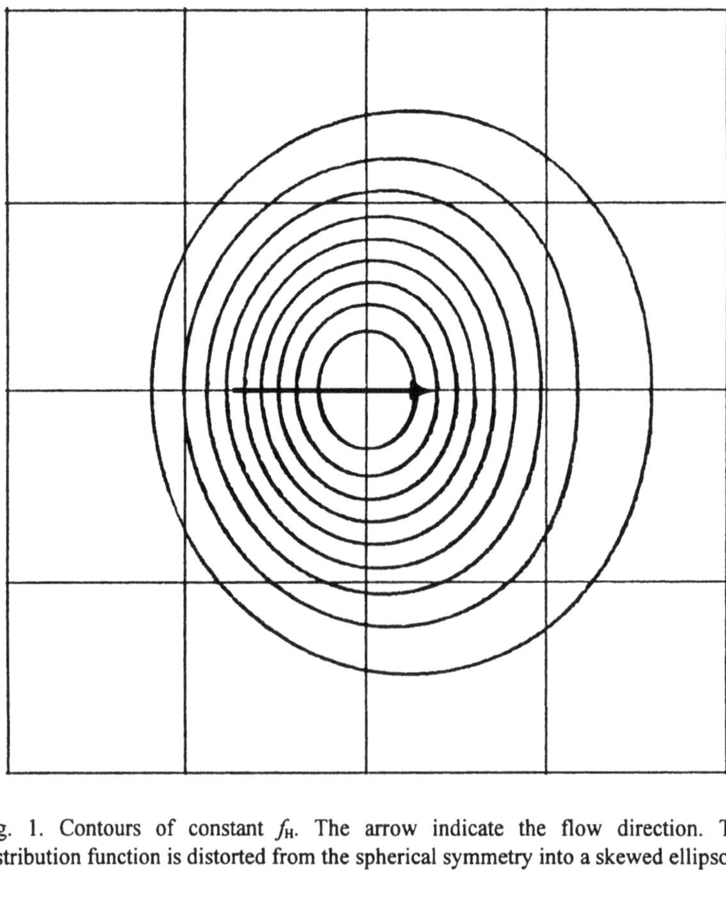

Fig. 1. Contours of constant f_H. The arrow indicate the flow direction. The distribution function is distorted from the spherical symmetry into a skewed ellipsoid.

Using the distribution function of thirteen moment approximation [equation (19)], the higher moments Q_{ikl} and R_{ikll} can be expressed as functions of low moments

$$Q_{ikl} = \frac{2}{5}(q_i\delta_{kl} + q_l\delta_{ik} + q_k\delta_{il}) \tag{20}$$

$$R_{ikll} = \frac{kT_H}{m}(5P_H\delta_{ik} + 7p_{ik}) \tag{21}$$

We can write $Q_{111} = \frac{6}{5}q_1$, $Q_{122} = \frac{2}{5}q_1$, $Q_{123} = 0$, $R_{11kk} = \frac{kT_H}{m}(5P_H + 7p_{11})$, $R_{12kk} = 7\frac{kT_H}{m}p_{12}$, etc. If we replace Q_{klj} and R_{ikll} by lower moments, moment equations (14)-(18) form a closed system of equations. The number of unknown variables equals the number of equations.

3. Recent progress of the fluid model

We can write the equation of continuity (14) in the form of conservation of mass

$$\frac{\partial \rho_H}{\partial t} + \nabla\cdot(\rho_H \mathbf{V}_H) = -\beta\rho_H \tag{22}$$

Multiplying (22) by $m\mathbf{V}_H$ and adding the result to equation (15), we obtain the equation for conservation of momentum

$$\frac{\partial}{\partial t}(\rho_H \mathbf{V}_H) + \nabla\cdot(\rho_H \mathbf{V}_H \mathbf{V}_H) = -\nabla P_H - \nabla\cdot\ddot{\mathbf{p}} + \rho_H(1-\mu)\nabla\left(\frac{GM}{R}\right) - \beta\rho_H \mathbf{V}_H \tag{23}$$

We can write the energy equation (16) as an equation of energy conservation or as an entropy equation. If we take the dot product of (15) with $\mathbf{V}_H$, multiply (22) with $V_H^2/2$, and add the results to equation (16), we obtain the equation for conservation of energy

$$\begin{aligned}\frac{\partial}{\partial t}\left(\rho_H\frac{V_H^2}{2} + \frac{3}{2}P_H\right) + \nabla\cdot\left[\left(\rho_H\frac{V_H^2}{2} + \frac{3}{2}P_H\right)\mathbf{V}_H\right] = \\ -\nabla\cdot\mathbf{q} - \nabla\cdot\left[P_H\mathbf{V}_H + \ddot{\mathbf{p}}\cdot\mathbf{V}_H\right] + \rho_H(1-\mu)\mathbf{V}_H\cdot\nabla\left(\frac{GM}{R}\right) - \beta\left(\rho_H\frac{V_H^2}{2} + \frac{3}{2}P_H\right)\end{aligned} \tag{24}$$

From the equation of continuity (22) we can write

$$\nabla\cdot\mathbf{V}_H = -\beta - \frac{1}{N_H}\left(\frac{\partial N_H}{\partial t} + \mathbf{V}_H\cdot\nabla N_H\right)$$

Substituting this into equation (16), we obtain the entropy equation

$$\frac{3}{2}P_H\frac{D}{Dt}\left(\ln\frac{P_H}{N_H^{5/3}}\right) = -\ddot{\mathbf{p}}:\nabla\mathbf{V}_H - \nabla\cdot\mathbf{q} + \beta P_H \tag{25}$$

This equation describes the relationship between P_H and N_H following the fluid motion. Due to pressure deviation, heat flux, and ionization process the flow of interstellar hydrogen in the heliosphere is not isentropic.

If the distribution function f_H is Maxwellian, the pressure is isotropic and the moment equations are a closed system of traditional fluid equations. The system has been used to study the steady flow of interstellar hydrogen [Whang, 1996] under the assumptions (a) $\mu = 1$, (b) β is inversely proportional to R^2, and (c) the flow is uniform at infinity on the upwind side. The fluid solution is much less time-consuming than the kinetic theory solutions.

As the flow of interstellar hydrogen approaches the Sun, hydrogen atoms are ionized to produce pickup protons [Axford, 1972; Holzer, 1977, 1989; Gloeckler, 1993; Isenberg, 1986; Whang et al., 1996]. In order to study the solar wind structures in the outer heliosphere, the global distribution of the number density and the partial pressure of pickup protons over a very large region with the radius of the order of 100 AU is critically needed. Since the production rate of pickup protons is directly proportional to the local density of neutral hydrogen, we need high-resolution solution of neutral hydrogen for the study the global pickup protons. The moment equations can be used to calculate high-resolution solutions of the hydrogen number density in the heliosphere. Using the closed system of moment equations derived in this paper, one can obtain solutions with non-Maxwellian distribution function and anisotropic pressure.

Acknowledgments

This work was supported by National Science Foundation under grant ATM-9310444.

References

Axford, W. I.: 1972, *Solar Wind, NASA Spec. Publ., SP-308*, 609.
Fahr, H. J.: 1979, *Astron. Astrophys.*, **77**, 101.
Fahr, H. J. and Ripken, H. W.: 1984, *Astron. Astrophys.*, **139**, 551.
Fahr, H. J. et al.: 1986, *Space Sci. Rev.*, **43**, 329.
Gloeckler, G. et al.: 1993, *Science*, **261**, 70.
Grad, H.: 1949, *Comm. Pure Appl. Math.*, **2**, 331.
Holzer, T. E.: 1977, *Rev. Geophys. Space Phys.*, **15**, 467.
Holzer, T. E.: 1989, *Annu. Rev. Astron. Astrophys.*, **27**, 199.
Isenberg, P. A.: 1986, *J. Geophys. Res.*, **91**, 9965,.
Lallement, R. et al.: 1985, *Astron. Astrophys.*, **150**, 21.
Osterbart, R. and Fahr, H. J.: 1992, *Astron. Astrophys.*, **264**, 260.
Ripken, H. W. and Fahr, H. J.: 1983, *Astron. Astrophys.*, **122**, 181.
Thomas, G. E.: 1978, *Ann. Rev. Earth Planet. Sci.*, **6**, 173.
Vasyliunas, V. M. and Siscoe, G. L.: 1976, *J. Geophys. Res.*, **81**, 1247.
Whang, Y. C.: 1996, in press, *Astrophys. J.*, **468**, September 10.
Whang, Y. C. et al.: 1996, *Space Sci. Rev.*, this issue
Wu, F. M. and Judge, D. L.: 1979, *Astrophys. J.*, **231**, 594.

PICKUP PROTONS IN THE HELIOSPHERE

Y. C. WHANG[1], L. F. BURLAGA[2] and N. F. NESS[3]

Abstract. We calculate the conditions of pickup protons inside the termination shock. Outside 50 AU the partial pressure of pickup protons is greater than the magnetic pressure by a factor of > 10, and greater than the partial pressure of solar wind protons by a factor of > 100. Thus, pickup protons have a significant dynamical influence on the structures of the solar wind in the outer heliosphere.

The interstellar hydrogen atoms can penetrate through the heliosphere and become ionized by interactions with solar wind protons and the solar radiation. The newly created protons are subsequently "picked up" by the solar wind electromagnetic fields and convected outward by the solar wind. Detection of pickup protons has been reported by Gloeckler et al. [1993].

The bulk speed of the interstellar neutral hydrogen and its thermal speed are very small compared with the solar wind speed V. In a frame of reference moving with the solar wind, a newly created proton initially moves toward the sun with a speed $u \approx V$. The Lorentz force changes this motion into a gyration about the local interplanetary magnetic field with a large pitch angle. These interstellar pickup protons are then convected outward by the solar wind. The distribution of pickup protons is unstable to the growth of electromagnetic waves. The effect of plasma waves produces a strong pitch angle diffusion, the density of the pickup protons in velocity space are rapidly isotropized in the solar wind frame. The newly created pickup protons have a thermal speed of $u \approx V$ which is much greater than the thermal speed of the solar wind protons.

Because interstellar pickup protons do not thermally assimilate into the population of the solar wind protons quickly, interstellar pickup protons and solar wind protons should be treated as two proton species [Isenberg, 1986]. Thus, the solar wind plasma consists of electrons, solar wind protons, and interstellar pickup protons. The three species have the same solar wind bulk velocity **V**, but they may have different temperatures. They are respectively denoted by subscripts i, S, and e. P_i, P_S, P_e are partial pressures, N_i, N_S, N_e are number densities and T_i, T_S, T_e are temperatures.

Available plasma and magnetic field data obtained from Voyager in 1978 through 1991 have been studied to generate the best-fit power law representation of the solar wind conditions inside the termination shock [Whang, et al., 1995]. The 14-year average solar wind speed $\langle V \rangle = 440$ km/s and the best-fit power law solutions of the solar wind parameters are $N_S = 7.63 \times (R_0/R)^{2.03}$ protons cm^{-3}, $P_S = 1.32 \times 10^{-10} \times (R_0/R)^{2.85}$ dyn cm^{-2}, $T_S = 1.25 \times 10^5 \times (R_0/R)^{0.81}$ K, $B^2/8\pi = 8.73 \times 10^{-11} \times (R_0/R)^{2.01}$ dyn cm^{-2}. Here B is the magnitude of the magnetic field, R is the heliocentric distance and

[1] Catholic University of America, Washington, DC 20064, U.S.A.
[2] NASA Goddard Space Flight Center, Greenbelt, MD 20771, U.S.A.
[3] Bartol Research Institute, University of Delaware, Newwark, DE 19716, U.S.A.

Space Science Reviews **78**: 393–398, 1996.

R_0 = 1 AU. Figure 1 shows the monthly averages of Voyager data and the best-fit power law representation.

The variations in N_i and P_i of the interstellar pickup protons are governed by

$$\frac{\partial N_i}{\partial t} + \nabla \cdot (N_i \mathbf{V}) = \beta N_H \tag{1}$$

and

$$\frac{d}{dt}\left(\frac{3}{2}P_i\right) + \frac{5}{2}P_i \nabla \cdot \mathbf{V} = \beta \frac{mV^2}{2} N_H \tag{2}$$

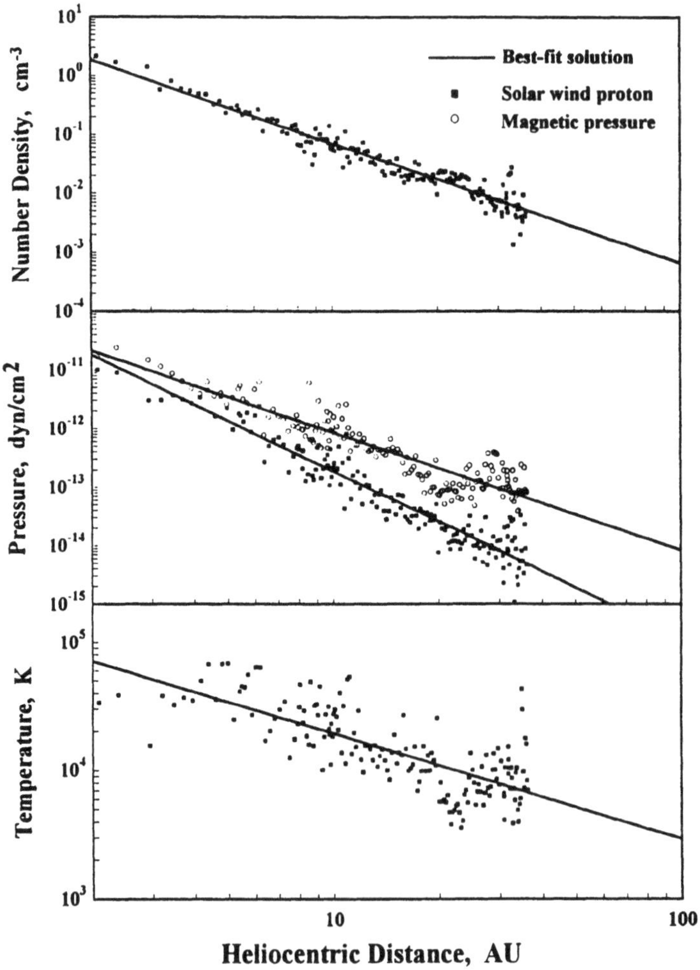

Figure 1. Monthly averages of Voyager data from 1978 through 1991 and the best-fit power law representations: (top) proton number density, (middle) magnetic pressure and proton pressure, and (bottom) proton temperature.

Here m is the proton mass, N_H is the number density of the interstellar neutral hydrogen, and β is the total ionization rate of a hydrogen atom by charge exchange ionization and by photoionization. Let ν_0 be the rate of photoionization per hydrogen atom at $R_0 = 1$ AU, and σ be the mean charge exchange cross-section, we can write

$$\beta = \nu_0\left(\frac{R_0^2}{R^2}\right) + \sigma N_S V \tag{3}$$

In the supersonic region of the outer heliosphere, $N_S V$ is approximately proportional to R^{-2}. Let $\beta_0 = \sigma N_{S0} V + \nu_0$, then we can write

$$\beta = \beta_0\left(\frac{R_0^2}{R^2}\right) \tag{4}$$

In a steady-state spherically symmetric solar wind, we can write (1) as

$$\frac{d}{dR}(R^2 N_i V) = \beta_0 R_0^2 N_H \tag{5}$$

The averaged radial solar wind speed V tends to decrease slightly with R. Making use of an approximation that $V \cong$ constant, we can write

$$\nabla \cdot \mathbf{V} = \frac{2V}{R}$$

Now we can express (2) as

$$\frac{d}{dR}(R^{10/3} P_i) = \frac{1}{3}\beta_0 R_0^2 m\, V R^{4/3} N_H \tag{6}$$

Let $N_{H\infty}$ and $V_{H\infty}$ be the number density and the bulk speed of interstellar hydrogen at infinity. Because the inflow of interstellar hydrogen is the source of the pickup protons flow and because the thermal energy carried by each newborn pickup proton is $mV^2/2$, the number flux of pickup protons $N_i V$ should be proportional to $N_{H\infty} V_{H\infty}$ and the energy flux of pickup protons $(3/2)P_i V$ should be proportional to the product of $N_{H\infty} V_{H\infty}$ and $mV^2/2$. We may write the solution of N_i and P_i in the following form

$$N_i = \frac{N_{H\infty} V_{H\infty}}{V} F_n \tag{7}$$

and

$$P_i = \frac{m N_{H\infty} V_{H\infty} V}{3} F_p \tag{8}$$

Here F_n and F_p are dimensionless functions to be determined. Making use of (7) and (8) we can write (5) and (6) as

$$\frac{d}{dR}\left(R^2 F_n\right) = \lambda \frac{N_H}{N_{H\infty}} \tag{9}$$

and

$$\frac{d}{dR}\left(R^{10/3} F_p\right) = \lambda R^{4/3} \frac{N_H}{N_{H\infty}} \tag{10}$$

Here $\lambda = \beta_0 R_0^2 / V_{H\infty}$ is the ionization scale length. Once the density of the interstellar hydrogen is given, we can study (9) and (10) for the solutions of F_n and F_p, and subsequently calculate the number density and the partial pressure of the interstellar pickup protons.

The flow of interstellar hydrogen is symmetric about a heliocentric axis in the direction of the Sun's motion relative to the interstellar medium. When the Mach number of the interstellar hydrogen flow relative to the Sun $M_{H\infty}$ approaches infinity, the number density of interstellar hydrogen in the heliosphere has been studied using a particle dynamics approach [Fahr, 1968; Axford, 1972; Vasyliunas and Siscoe, 1976]. The solution is a function of a parameter μ that is the ratio of the repulsion force on a hydrogen atom by solar radiation to the gravitational attraction force. When $\mu = 1$, the number density of interstellar hydrogen is

$$N_H = N_{H\infty}\, exp\left(-\frac{\lambda\phi}{R\,sin\,\phi}\right) \tag{11}$$

where ϕ is the heliocentric polar angle measured from the axis of symmetry pointing in the upwind direction. Equation (11) is valid in the limit of $M_{H\infty} \rightarrow \infty$ or zero interstellar temperature. This solution is known as a cold hydrogen model. The real $M_{H\infty}$ of the interstellar hydrogen flow is of the order of 2. Making use of (11), Equations (9) and (10) can be integrated to give

$$F_n = \left(\frac{\lambda}{R}\right)^2 \frac{\phi}{sin\,\phi}\Gamma\left(-1, \frac{\lambda\phi}{R\,sin\,\phi}\right) \tag{12}$$

and

$$F_p = \left(\frac{\lambda}{R}\right)^{10/3} \left(\frac{\phi}{sin\,\phi}\right)^{7/3} \Gamma\left(-\frac{7}{3}, \frac{\lambda\phi}{R\,sin\,\phi}\right) \tag{13}$$

where Γ is the gamma function. The two dimensionless function F_n and F_p are functions of R/λ and ϕ. These similarity solutions can be used to calculate the solutions of N_i and P_i at any ionization scale length. The derivation of the above solutions follows the ideas developed by several previous authors [Vasyliunas and Siscoe, 1976; Burlaga et al., 1994; Isenberg, private communication, 1995].

We calculate N_i, P_i and T_i for pickup protons along the axis of symmetry in the upwind direction assuming that interstellar hydrogen approaches the Sun with a bulk velocity of $V_{H\infty} = 25.3$ km/s [Witte et al. 1996], and $N_{H\infty} = 0.125$ cm^{-3} [Gloeckler, 1996]. Using $\sigma = 2 \times 10^{-15}$ cm^2 [Fite, 1962], $\nu_0 = 0.9 \times 10^{-7}$ s^{-1} [Möbius, 1993], $V = 440$ km/s, and $N_{S0} = 7.63$ protons/cm^3, we obtain $\beta_0 = 7.68 \times 10^{-7}$ s^{-1} and the ionization scale length $\lambda = 4.54$ AU.

Figure 2 shows that outside 50 AU the partial pressure of pickup protons P_i is greater than the magnetic pressure by a factor of > 10, and greater than the partial pressure of solar wind protons by a factor of > 100. Thus, pickup protons have a significant dynamical influence on the structures of the solar wind in the outer heliosphere. These theoretical results support the results determined by analyzing pressure balanced structures [Burlaga et al., 1994, 1996]. This conclusion holds over a wide range of the polar angle ϕ in the region where $R > 10\,\lambda$ because the variation of P_i with respect to ϕ is small in that region. The effects of pickup protons on the termination shock, and on the heliopause have been studied by several authors [Whang, et al., 1995; Pauls, et al., 1995; Lee, 1996].

Figure 3 shows that throughout the outer heliosphere, the number density of pickup protons N_i is less than the number density of the solar wind protons (top panel) and the temperature of pickup protons T_i is much higher than the temperature of the solar wind protons (bottom panel). We calculate the mean temperature for the mixture of three species of charged particles

$$T = \frac{N_S T_S + N_i T_i + N_e T_e}{N_S + N_i + N_e}$$

by assuming that $N_e = N_S$, and $T_e = T_S$. The open circles represent the mean temperature. It shows that heating by the pickup protons outside 10 AU causes a radial increase in the mean temperature of the solar wind.

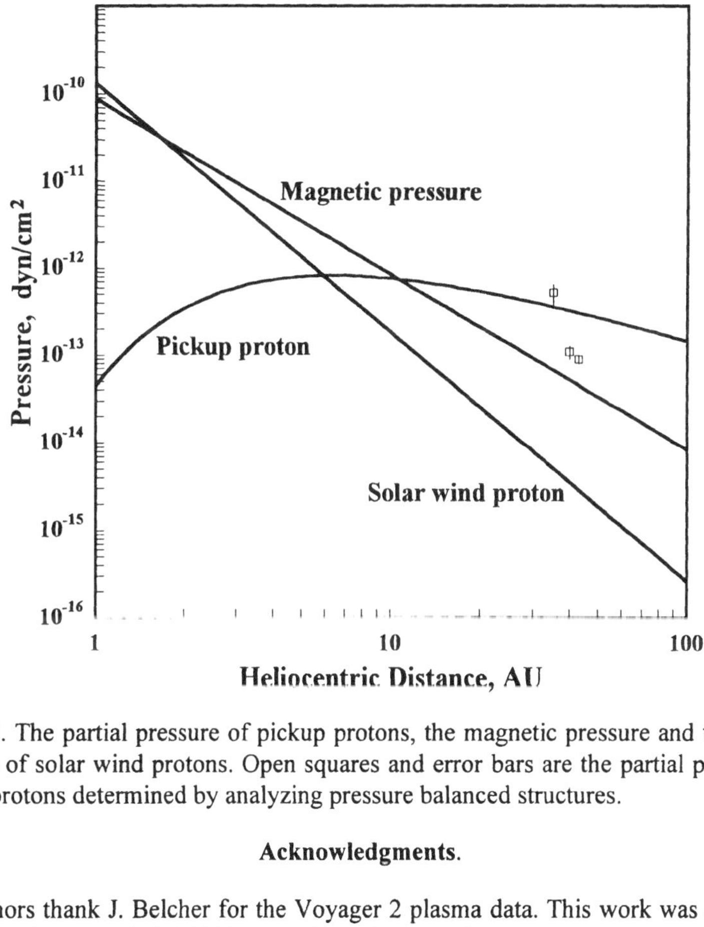

Figure 2. The partial pressure of pickup protons, the magnetic pressure and the partial pressure of solar wind protons. Open squares and error bars are the partial pressure of pickup protons determined by analyzing pressure balanced structures.

Acknowledgments.

The authors thank J. Belcher for the Voyager 2 plasma data. This work was supported by NSF under grant ATM-9310444 and by NASA under JPL contract 956167.

References

Axford, W. I.: 1972, in Solar Wind, *NASA Spec. Publ., SP-308*, 609.
Baranov, V. B.: 1990, *Space Sci. Rev.*, **52**, 89.
Baranov, V. B. and Malama Y. G.: 1993, *J. Geophys. Res.*, **98**, 15,157.

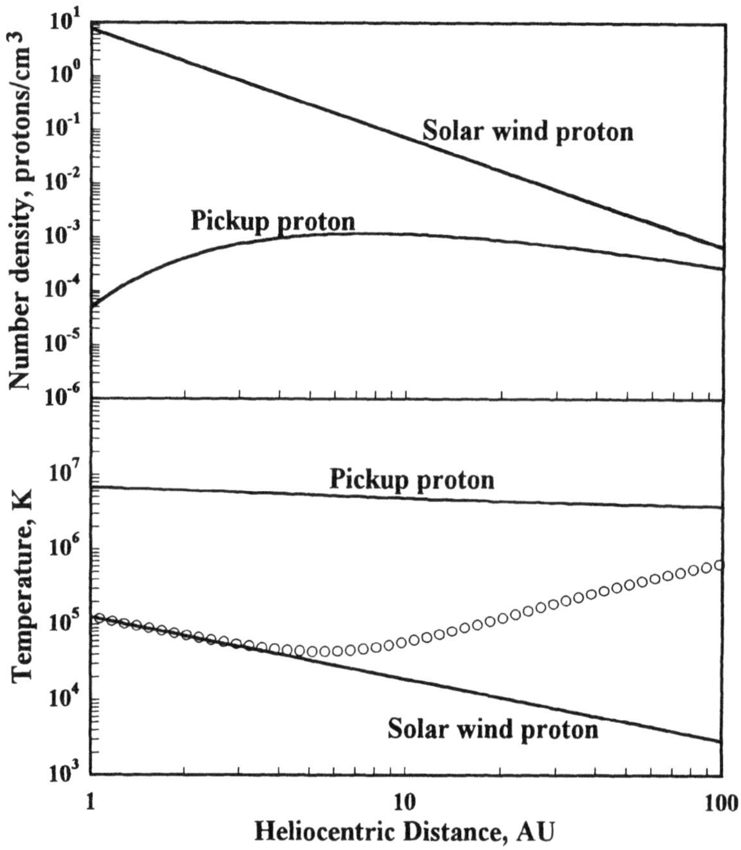

Figure 3. The number densities and the temperatures of pickup protons and the solar wind protons.

Burlaga, L. F. et al.: 1996, submitted to *J. Geophys. Res.*

Fahr, H. J.: 1968, *Astrophys. Space Sci.*, **2**, 474.

Fite, W.: 1962, *Proc. R. Soc. London, Ser. A.*, **268**, 527.

Gloeckler, G. et al.: 1993, *Science*, **261**, 70.

Gloeckler, G.: 1996, *Space Sci. Rev.*, this issue.

Holzer, T. E.: 1972, *J. Geophys. Res.*, **77**, 5407.

Isenberg, P. A.: 1986, *J. Geophys. Res.*, **91**, 9965.

Lallement, R. et al.: 1993, *Science*, **260**, 1095.

Lee, M. A.: 1996, in *Cosmic Winds and the Heliosphere*, edited by J. R. Jokipii et al., University of Arizona Press, Tucson, in press.

Möbius, E.: 1993, in *Landolt-Bornstein Tables of Physics*, edited by H. H. Voigt, pp. 186-191, Springer-Verlag, New York.

Pauls, H. L. et al.: 1995, *J. Geophys. Res.*, **100**, 21,595.

Vasyliunas, V. M. and Siscoe G. L.: 1976, *J. Geophys. Res.*, **81**, 1247.

Whang, Y. C. et al.: 1995, *J. Geophys. Res.*, **100**, 17,015.

Witte, M. et al.: 1996, *Space Sci. Rev.*, this issue.

Author Index

Space Science Reviews **78**: 399, 1996.

MIX
Papier aus verantwortungsvollen Quellen
Paper from responsible sources
FSC® C105338

FSC
www.fsc.org

If you have any concerns about our products,
you can contact us on
ProductSafety@springernature.com

In case Publisher is established outside the EU,
the EU authorized representative is:
Springer Nature Customer Service Center GmbH
Europaplatz 3, 69115 Heidelberg, Germany

Printed by Libri Plureos GmbH
in Hamburg, Germany